Kohei Arai
Editor

Proceedings of the Future Technologies Conference (FTC) 2023, Volume 4

Springer

Editor
Kohei Arai
Faculty of Science and Engineering
Saga University
Saga, Japan

ISSN 2367-3370 ISSN 2367-3389 (electronic)
Lecture Notes in Networks and Systems
ISBN 978-3-031-47447-7 ISBN 978-3-031-47448-4 (eBook)
https://doi.org/10.1007/978-3-031-47448-4

This Springer imprint is published by the registered company Springer Nature Switzerland AG
The registered company address is: Gewerbestrasse 11, 6330 Cham, Switzerland

Paper in this product is recyclable.

Editor's Preface

It is a matter of great privilege to present the proceedings of eighth Future Technologies Conference (FTC), 2023 held in San Francisco, USA. The conference witnessed significant international participation where researchers from far and wide assembled to present their unique and original studies across diverse domains including, though not limited to computing, data science, artificial and ambient intelligence, machine learning, e-learning and cybersecurity.

With rapid changes, the technology becomes obsolete very swiftly. Hence, continuous research and advancements in the scientific field are becoming a necessary norm. The scientific community is therefore delving deep and making several ground-breaking discoveries to awe the world. To help the scientists and academia to reach out to the world and showcase their studies, The Science and Information (SAI) Organization conducts FTC every year. This conference uncovers research challenging the status quo and offering new perspectives to analyze and solve difficult issues in future.

FTC received 490 papers from researchers all over the world, out of which 163 were carefully selected for their originality, applicability and presentation. The thundering success of the conference could not have been possible without the support of many people. The learned keynote speakers and presenters added the spark to the entire event. The cumbersome task of reviewing and selecting papers by the technical committee was accomplished seamlessly due to the immense expertise of its members. The Session Chair's role in smooth conduct of sessions was worth appreciating. This token of thanks would be incomplete without the mention of the organizing committee who looked into the minute aspects to make the event a grand success.

With this edition of proceedings too, we hope to kindle interest about current scientific developments and its future implications in our readers. With this pious aim in mind, we seek your continued support to help us grow exponentially.

Kind Regards,

Kohei Arai
Conference Program Chair

Contents

LaksNet: An End-to-End Deep Learning Model for Self-driving Cars in Udacity Simulator

Lakshmikar R. Polamreddy(✉) and Youshan Zhang

Yeshiva University, New York City, NY 10016, USA
lpolamre@mail.yu.edu, youshan.zhang@yu.edu

Abstract. The majority of road accidents occur because of human errors, including distraction, recklessness, and drunken driving. One of the effective ways to overcome this dangerous situation is by implementing self-driving technologies in vehicles. In this paper, we focus on building an efficient deep-learning model for self-driving cars. We propose a new and simple CNN model called 'LaksNet' consisting of four convolutional layers and two fully connected layers. We conducted extensive experiments using our LaksNet model with the training data generated from the Udacity simulator. Our model outperforms many existing pre-trained ImageNet and NVIDIA models in terms of the duration of the car for which it drives without going off the track on the simulator.

Keywords: Self Driving Cars · Steering Angles · Pre-trained ImageNet Models · Udacity Simulator

1 Introduction

WHO (World Health Organization) [25] reported that approximately 1.3 million people die each year as a result of road traffic crashes. To reduce the accident rate, We could exploit various machine learning algorithms to drive a vehicle without human intervention. End-to-end machine learning models can be built and trained on huge sets of available data in the form of images generated by sensors and cameras. Then, these trained models will drive the vehicles in such a way as to minimize accidents. As we have not yet collected the real-time data, we have made use of the simulation environment developed by Udacity for its nano degree program on self-driving cars. We utilized this platform to generate training data and to test the performance of our model.

Previous work proposed to develop deep learning models for end-to-end control of a self-driving car, and they mostly utilized NVIDIA model architecture [3] and also never compared its performance with the available pre-trained ImageNet models. Some scholars proposed a simple architecture than that of NVIDIA but could not establish that their model performance was better by comparing the results. In addition, they did not measure the performance of the models in

K. Arai (Ed.): FTC 2023, LNNS 816, pp. 1–13, 2023.
https://doi.org/10.1007/978-3-031-47448-4_1

terms of the car run time on the track. Our proposed work overcomes all these limitations.

The major focus of our research is to first check the performance of the NVIDIA model and to develop a model better than that of NVIDIA in terms of predicting the steering angles of a self-driving car. We have utilized the Udacity simulator for driving the car in autonomous mode on the track. We first trained the NVIDIA model with our dataset and then measured its performance. As mentioned in the results Sect. 4.2, it could drive the car for 120 s without going off the track. Then, we investigated if any of the existing pre-trained ImageNet models would show better performance than that of the NVIDIA model. Finally, we developed our own model to achieve these two objectives. Firstly, our model achieves state-of-the-art performance. Secondly, we build a model with a smaller number of convolutional and fully connected layers so that it requires a smaller number of parameters for training. After conducting extensive experiments, we achieved these two objectives with our LaksNet model.

The remainder of the paper is organized as follows. Section 2 covers the unique contributions made by various researchers in the field of autonomous driving. Section 3 explains about the proposed LaksNet model architecture and how it differs from that of NVIDIA. Section 4 covers the datasets considered for training and testing of all the models and their results. Section 5 discusses how hyper-parameter tuning was carried out to arrive at our model architecture. Section 6 captures concluding remarks of the proposed work and scope for future work.

2 Related Work

Khan et al. [12] looked at the pros and cons of the implementation of autonomous vehicles. They discussed the benefits such as safety, congestion, traffic management, and the adoption of autonomous vehicle technology by various sectors like mining, freight transportation, and the military industry. Shalev et al. [21] proposed a white-box, interpretable, mathematical model for safety assurance called Responsibility-Sensitive Safety (RSS) for self-driving cars and also the design of a system that adhered to safety assurance requirements and was scalable to millions of cars. Ko et al. [13] proposed a method for key points estimation and point instance segmentation approach for lane detection called Point Instance Network (PINet), which can localize the drivable area on the road. Pan et al. [18] proposed a novel realistic translation network to make the model trained in a virtual environment be workable in the real world. They concluded that by using synthetic real images as training data in reinforcement learning, the agent generalizes better in a real environment than pure training with virtual data or training with domain randomization. Wu et al. [26] YOLOP (You Only Look Once for Panoptic Driving Perception) for a high-precision and real-time perception system that can assist the vehicle in making reasonable decisions while driving. They presented a YOLOP network to perform traffic object detection, drivable area segmentation, and lane detection simultaneously. Li et al. [15] proposed a Stereo R-CNN-based 3D object detection for autonomous driving by fully exploiting the sparse and dense, semantic, and geometry information in stereo imagery.

Bojarski et al. [3], as part of the NVIDIA research team, proposed a new CNN architecture for end-to-end deep learning for self-driving cars. In a new automotive application, they have used convolutional neural networks (CNNs) to map the raw pixels from a front-facing camera to the steering commands for a self-driving car. Zhenye-Na [28] also implemented NVIDIA model architecture for training the model for self-driving vehicles using Udacity's simulation environment to generate training data and test the model. In addition to this, Smolyakov et al. [22] explored various architectures of convolutional neural networks in order to obtain good results with a minimum number of parameters for predicting the steering angles to drive the car in autonomous mode in the Udacity simulation environment.

Santana and Hotz [20] investigated video prediction models based on autoencoders and RNNs. Instead of learning everything in an end-to-end way, they first trained the autoencoder with generative adversarial network-based cost functions to generate realistic-looking images of the road and then trained the RNN transition model in the embedded space. Behley et al. [1] proposed three benchmark experiments based on the KITTI dataset for (i) semantic segmentation of point clouds using a single scan, (ii) semantic segmentation using multiple past scans, and (iii) semantic scene completion to understand LIDAR sequences useful for learning the environment around a self-driving car. Vora et al. [24] proposed PointPainting: a sequential fusion method for 3D object detection in the case of self-driving cars. They mentioned that Point Painting works by projecting lidar points into the output of an image-only semantic segmentation network and appending the class scores to each point. The appended (painted) point cloud can then be fed to any lidar-only method. Accurate depth estimation is a key prerequisite in many robotics tasks, including autonomous driving. Guizilini [7] proposed a novel self-supervised monocular depth estimation method combining geometry with a new deep network, PackNet, learned only from unlabeled monocular videos. They leveraged the novel symmetrical packing and unpacking blocks to jointly learn to compress and decompress detail-preserving representations using 3D convolutions. Bertoni et al. [2] proposed an approach to fundamentally tackle the ill-posed problem of 3D human localization from monocular RGB images for autonomous driving. They addressed the ambiguity in the task by predicting confidence intervals through a loss function based on the Laplace distribution. Gu et al. [6] proposed an approach to learning a long short-term memory (LSTM)-based model for imitating the behavior of Waymo's self-driving model. The proposed model was evaluated based on Mean Absolute Error (MAE) and the experimental results have shown that the LSTM model outperformed several baseline models in driving action prediction. Image Segmentation has been an active field of research as it has a wide range of applications, ranging from automated disease detection to self-driving cars. Jadon [11] summarized existing well-known loss functions widely used for Image Segmentation and listed out the cases where their usage can help in fast and better convergence of a model, and they also introduced a new log-cosh dice loss function. Liao et al. [16] developed KITTI-360 for autonomous driving, the successor of the popular KITTI dataset. KITTI-360 is a suburban driving dataset that comprises richer input modalities, comprehensive semantic instance annotations, and accurate localization to facilitate research at the intersection of vision, graphics, and robotics. They also created a tool to label 3D scenes with bounding primitives and developed a model that transfers this information into the 2D image domain, resulting in over 150k images. Yang et al. [27] proposed a multi-task learning framework to predict the steering angle and speed control simultaneously in an end-to-end manner by taking previous feedback speeds and visual recordings as inputs. Moghadam and Elkaim [17] proposed multi-modal architecture that includes the environmental modeling of ego surrounding and training a deep reinforcement learning (DRL) agent that yields consistent performance in stochastic highway driving scenarios and obtained the high-level sequential commands (i.e. lane change) to send them to lower-level controllers. Deruyttere et al. [5] considered a problem in an autonomous driving setting, where a passenger requests an action that can be associated with an object found in a street scene. Their work presented the Talk2Car dataset, which was the first object referral dataset that contains commands written in natural language for self-driving cars.

3 Methods

Firstly, the images and corresponding steering angles collected from the simulator in training mode are passed onto the CNN model as inputs. Secondly, the CNN model is trained for a sufficient number of epochs till the convergence in loss is achieved. Thirdly, the outputs from the CNN model, i.e., steering angles, are passed onto the simulator in autonomous mode. Immediately, the car drives on its own on the selected track in the simulator.

As the model is expected to predict the steering angle values, we have considered regression loss, and this is calculated using the equation below.

$$\text{Regression Loss} = \frac{1}{n}\sum_{i=1}^{n}(y_i - \hat{y_i})^2 \tag{1}$$

where y_i is the actual steering angle and $\hat{y_i}$ is the predicted steering angle

3.1 Model Architecture

Our LaksNet model consists of four convolutional layers, four max-pooling layers, two dropout layers, and two fully connected layers, with one being an output layer, as shown in Fig. 1. We use 3×3 kernels for the first three convolutional layers and 5×5 kernels for the last convolutional layer. The first fully connected layer takes 576 input parameters and gives 256 output parameters, and the second fully connected layer results in a single output because it is the final output layer. The total training parameters of this model are 274,017.

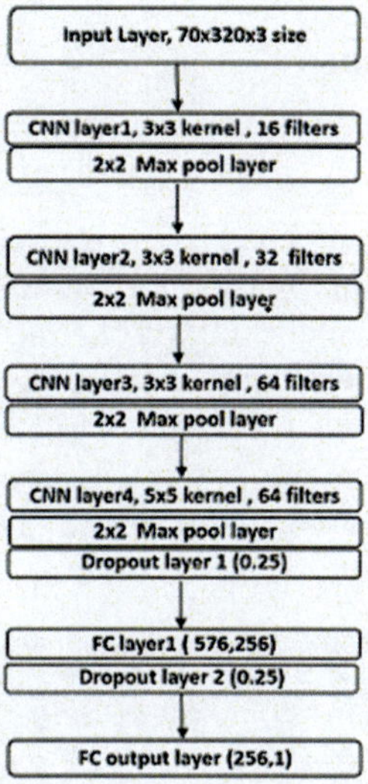

Fig. 1. LaksNet model architecture.

On the other hand, the model developed by NVIDIA architecture [3] consists of five convolutional layers, one dropout layer, and four fully connected layers, with one being an output layer. The total training parameters of this model are 559,419. In addition to these two model architectures, we utilized the below nine ImageNet pre-trained models to check the performance of the self-driving car in the simulation environment (AlexNet [14], GoogleNet [23], MobileNetv2 [19], ResNet50 [8], SqueezeNet [10], DenseNet201 [9], NasnetLarge [29], ResNet101 [8], and Xception [4]).

3.2 Implementation Details

Our model is implemented based on the following configurations:

1. Processor: 12th Gen Intel(R) Core(TM) i7-12700H 2.30 GHz
2. Installed RAM: 16.0 GB
3. System type: 64-bit operating system, x64-based processor
4. GPU: NVIDIA GeForce RTX 3060, 6.0 GB

Our model is trained based on the following hyperparameters:

1. Number of epochs: 50
2. Batch size: 32
3. Optimizer: Adam
4. Learning rate: 0.1

4 Results and Datasets

4.1 Datasets

The Udacity simulation environment has been used to generate training data. It has two different tracks that can be used for training and testing. After selecting either track1 or track2 and then clicking on the record button, a folder of images is created taken from the center, right and left cameras at each instant of time. These images contain details like a portion of the car on the track, a track with borders, and an environment outside the track. A sample of the images generated in training mode at one particular instant of time is shown in Fig. 2.

Fig. 2. Images from the left, front, and right cameras.

In addition to these images, the Udacity simulation generates a drivinglog.csv file that contains seven columns. The first three columns contain the path to the images of the front, right and left cameras. Column 4 shows the steering angle values - zero corresponds to the straight direction, a positive value indicates a right turn and a negative value indicates a left turn. Columns 5, 6, and 7 indicate acceleration, deceleration, and speed of the vehicle, respectively. Some rows of this CSV file are shown in Table 1.

Table 1. Drivinglog.csv file generated by the simulator.

Column 1	Column 2	Column 3	Column 4	Column 5	Column 6	Column 7
center_216.jpg	left_216.jpg	right_216.jpg	0	0.3325	0	0.2858
center_316.jpg	left_316.jpg	right_316.jpg	0	0.6327	0	0.8770
center_424.jpg	left_424.jpg	right_424.jpg	−0.1215	0.9266	0	1.8474
center_535.jpg	left_535.jpg	right_535.jpg	−0.4860	1	0	3.2320
center_637.jpg	left_637.jpg	right_637.jpg	−0.1827	1	0	4.4072
center_739.jpg	left_739.jpg	right_739.jpg	0	1	0	5.5639

We generated 33096 images from the simulator and combined this with the training data of 97330 images available on the GitHub page of Zhenye-Na [28]. In addition to this, we have generated a separate set of images for validation purposes. $70 \times 320 \times 3$ is the size of the images generated in the simulator.

4.2 Results

The block diagram shown in Fig. 3 highlights the testing approach of the model to drive the car in the autonomous mode of the simulation environment. The training images and the respective steering angles generated from the Udacity simulator are fed into the CNN model to train it. After training for a required number of epochs, the model will predict steering angle values that are passed onto the simulator. Then, we notice that the car drives on its own on the track in the simulation environment in autonomous mode.

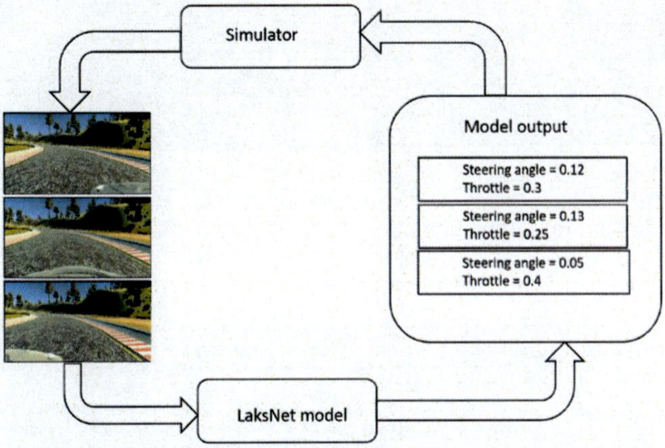

Fig. 3. Block diagram of testing approach in the simulator.

During the test stage, we notice the steering angles on the terminal and also the car moving in the simulator window during the same time. A snapshot of this is shown in Fig. 4. We can also save the images taken by the front camera during autonomous mode.

After training each of the models, we obtained the results shown in Table 2. Run time (in seconds) refers to the duration for which the car runs at an average speed of 10 miles per hour without going off the track. These results highlight that the LaksNet model has shown better performance than that of NVIDIA and the other pre-trained models. Table 3 shows the actual steering angles, predicted steering angles, and MSE (Mean Square Error) of all the models for 30 test images generated separately from the Udacity simulator. Though the MSE of the LaksNet model is more than that of the NVIDIA model, it could drive the car on track for more time. This might be due to the fact that it did not overfit the data.

5 Discussion

As we are inspired by NVIDIA model architecture for end-to-end learning of a self-driving car, we first trained this model with our data set and noticed that the

Fig. 4. A snapshot of steering angles predicted by the model in the terminal and simulator window in autonomous mode.

Table 2. Car run time (in secs) on the track in the simulator for each model.

Model	Run time (in secs)
AlexNet [14]	50
GoogleNet [23]	50
MobileNet [19]	8
ResNext50 [8]	3
SqueezeNet [10]	12
DenseNet201 [9]	7
NasnetLarge [29]	4
ResNet101 [8]	14
Xception [4]	5
NVIDIA [3]	120
LaksNet	**150**

car ran properly on the track in the simulator for 120 s in autonomous mode and then deviated from the track. Then, we decided to check the performance of the available pre-trained models for this purpose. Surprisingly, these complex models have shown poor performance when compared to the NVIDIA model. This might be due to the fact that all these pre-trained models are designed for classification tasks to identify a particular category instead of predicting numerical values, but our project is a regression problem as we are predicting continuous numerical values (steering angles). Secondly, each of the similar images can have different angles, which also confuses these complex models. Among these models, Alexnet and Googlenet have given better results when compared to the other pre-trained models because these two are simpler models in terms of the number of convolutional layers.

Table 3. Predicted steering angles and MSE by each model for 30 test images (ANet: Alexnet; GNet: Googlenet; MNet: Mobilenet; R50: Resnext50; SqNet: Squeezenet; D201: Densenet201; NasL: Nasnetalarge; R101: Resnet101).

Actual angles	**LaksNet**	NVIDIA	ANet	GNet	MNet	R50	SqNet	D201	NasL	R101	Xception
−0.341	−0.348	−0.079	−0.133	−0.288	−0.122	−0.040	0.000	0.053	3.315	−0.039	−0.005
0.000	−0.355	−0.082	−0.033	−0.123	0.078	−0.200	0.000	0.025	3.294	0.110	0.026
0.000	−0.363	−0.098	−0.289	−0.201	−0.100	−0.216	0.000	0.014	3.458	0.018	0.009
−0.072	−0.375	−0.186	−0.053	−0.144	−0.202	−0.121	0.000	0.002	3.507	0.001	0.084
−0.418	−0.355	−0.079	−0.035	−0.148	−0.073	−0.109	0.000	0.037	3.459	−0.073	−0.029
−0.116	−0.350	−0.094	−0.206	−0.105	−0.089	0.015	0.000	−0.006	3.563	−0.009	0.001
0.000	−0.325	−0.032	−0.225	−0.191	−0.108	−0.126	0.014	0.032	3.421	−0.190	−0.005
0.000	−0.251	−0.060	−0.169	−0.121	−0.064	−0.200	0.000	−0.032	3.511	−0.152	−0.018
0.000	−0.280	−0.030	0.001	−0.311	−0.290	−0.070	0.0229	0.032	3.576	−0.088	−0.037
0.000	−0.375	−0.036	−0.197	−0.014	−0.109	−0.112	0.025	0.014	3.652	−0.074	−0.034
0.000	−0.360	−0.032	−0.297	−0.210	−0.275	−0.138	0.000	−0.017	3.742	−0.108	−0.008
0.000	−0.383	−0.068	−0.023	−0.156	−0.091	−0.112	0.000	0.041	3.632	−0.030	0.681
0.000	−0.417	−0.024	−0.032	−0.094	−0.093	−0.318	0.000	−0.006	3.742	−0.182	0.004
−0.071	−0.464	−0.059	−0.130	−0.011	−0.101	−0.017	0.000	−0.075	3.745	0.001	0.029
−0.391	−0.395	−0.064	0.001	−0.073	−0.312	−0.124	0.000	−0.041	3.978	−0.103	−0.013
−0.279	−0.402	−0.055	−0.057	−0.123	−0.047	−0.072	0.000	0.029	3.722	−0.065	−0.049
0.000	−0.264	−0.045	−0.044	−0.038	−0.057	0.006	0.000	0.144	3.901	−0.105	−0.012
0.000	−0.219	−0.029	−0.020	−0.006	−0.071	−0.056	0.000	0.000	3.840	−0.151	0.085
0.000	−0.206	−0.049	−0.147	−0.105	−0.318	−0.071	0.000	0.066	3.818	0.015	0.003
0.000	−0.238	−0.048	−0.065	−0.123	−0.028	−0.023	0.000	0.076	3.703	−0.020	−0.001
0.000	−0.237	−0.056	−0.012	−0.119	−0.026	−0.040	0.000	0.053	3.794	0.101	−0.003
0.000	−0.280	−0.042	−0.228	−0.023	−0.039	−0.045	0.000	0.092	3.583	0.015	0.001
0.000	−0.278	−0.044	0.034	−0.069	−0.054	−0.050	0.000	0.011	3.625	0.022	−0.014
0.000	−0.251	−0.036	−0.135	−0.116	−0.038	−0.239	0.000	0.051	3.692	−0.057	0.011
0.000	−0.335	−0.053	−0.007	−0.075	−0.049	−0.231	0.000	0.083	3.771	0.015	−0.026
0.000	−0.305	−0.040	−0.184	−0.178	−0.081	−0.074	0.000	−0.026	3.762	−0.037	−0.013
0.000	−0.483	−0.051	−0.068	−0.151	−0.181	−0.096	0.000	0.088	3.651	−0.052	−0.014
0.000	−0.342	−0.051	−0.077	−0.167	−0.089	−0.092	0.000	0.116	3.557	0.026	−0.019
0.000	−0.381	−0.074	−0.244	−0.214	−0.131	−0.010	0.000	−0.027	3.518	−0.042	−0.006
0.000	−0.345	−0.083	−0.151	−0.108	−0.063	−0.038	0.000	−0.113	3.589	−0.093	−0.018
MSE	0.091	0.014	0.031	0.023	0.021	0.047	0.066	0.023	13.682	0.020	0.033

As the pre-trained models have not met the expectations set by the NVIDIA model, we have tried to build our own CNN models that are more competent. One of our CNN models, with just four convolutional layers and two fully connected layers, could drive the car on the track properly for 150 s, which is more than that of the NVIDIA model. To arrive at this model, we have built many models and checked their performances.

5.1 Hyper-parameter Tuning

We first explored a CNN model with seven convolutional layers using 3×3 filters. We initially thought that adding more convolutional layers would make the model learn better and would show better results, but this model could drive properly for just 20 s. We observe that adding more layers will extract features up to a certain extent and eventually overfit the data instead of extracting features. So, we built a few more models with five and six convolutional layers using bigger kernels of size 5×5 and 7×7. We observed that the model with five convolutional layers using 5×5 filters and another model using a combination of 7×7 and 5×5 filters had shown a major improvement in their performance by driving the car properly on the track for 90 s. Then, we built another model with just 3×3 convolutional layers using 7×7 filters, but it could drive the car properly for just 50 s. Later, when we reduced the filter size to 3×3 in this model, it could drive the car properly for 120 s, meeting the expectations of the NVIDIA model. Therefore, we noticed that models with a smaller number of layers and using small filter sizes are giving better results in this case, as shown below in Table 4.

Table 4. Run time of the car for various model configurations.

Model configuration	Run time (in secs)
Seven layers, all 3×3 filters	20
Five layers, all 5×5 filters	90
Five layers, three 7×7 filters and two 5×5 filters	90
Three layers, all 7×7 filters	50
Three layers, all 3×3 filters	120

In an attempt to achieve better results than that of the NVIDIA model, we have experimented with various models with 3 or 4 convolutional layers using 3×3 or 5×5 or a combination of 3×3 and 5×5 filters. Finally, our LaksNet model with four convolutional layers, using 3×3 filters for the first three layers and a 5×5 filter for the last convolutional layer, has achieved better results driving the car properly for 150 s. In this model, we only used two fully connected layers, with one being an output layer. When we tried this model with 3 or 4 fully connected layers, it was overfitting and did not yield better results.

In the LaksNet model, we added a max pooling layer after each convolutional layer to reduce the height and width of the images. Another alternative way is the usage of an average pooling layer. In the fully connected block, we added a dropout layer after the last convolutional layer and another dropout layer after the first fully connected layer to avoid over-fitting. We applied the ReLU activation function for all the convolutional and fully connected layers, except the output layer, to fire neurons. The other suitable alternative to ReLU is the ELU activation function.

Regarding loss function, we used the MSE loss function as the output is just one value, and the loss is measured by calculating the difference between the actual and the predicted steering angle values. Accordingly, weights will be adjusted during back propagation and this process continues based on the number of epochs. In addition to the data augmentation process, we normalized the data to avoid saturation so that gradients would work better for the model training. We also performed random rotations on the images and cropped the images as well. This is required for better training of the model. As part of generating the training data from the simulator, one has to take care, while driving manually, to ensure the car always moves in the center of the track. Otherwise, the training data will be erroneous and reduce the model performance during predictions. In addition to this, a large training data is required for better learning of the model in terms of turns on the track and surrounding environment.

6 Conclusion

In this paper, we first train the NVIDIA model and other pre-trained models with 134K images generated from the Udacity simulator and demonstrate that the NVIDIA model outperformed the pre-trained models in terms of car run time on the track. Secondly, we built the LaksNet model with only four convolutional and two fully connected layers using a small number of training parameters and achieved better results than that of the NVIDIA model. While the NVIDIA model was able to drive the car for 120 s before going off-track, LaksNet showed significant improvements by driving the car for 150 s. Our future work will focus on building a model that will run the car for a long time without going off the track under the Udacity simulator. In addition, we would like to predict acceleration values along with steering angles and ensure that the car will always run in the middle of the track.

References

1. Behley, J., et al.: A dataset for semantic scene understanding of LiDAR sequences. In: Proceedings of the IEEE/CVF International Conference on Computer Vision, pp. 9297–9307 (2019)
2. Bertoni, L., Kreiss, S., Alahi, A.: Monoloco: monocular 3D pedestrian localization and uncertainty estimation. In: Proceedings of the IEEE/CVF International Conference on Computer Vision, pp. 6861–6871 (2019)
3. Bojarski, M., et al.: End to end learning for self-driving cars. arXiv preprint arXiv:1604.07316 (2016)
4. Chollet, F.: Xception: deep learning with depthwise separable convolutions. In: Proceedings of the IEEE Conference on Computer Vision and Pattern Recognition, pp. 1251–1258 (2017)
5. Deruyttere, T., Vandenhende, S., Grujicic, D., Van Gool, L., Moens, M.-F.: Talk2Car: taking control of your self-driving car. arXiv preprint arXiv:1909.10838 (2019)
6. Zhicheng, G., Li, Z., Di, X., Shi, R.: An LSTM-based autonomous driving model using a Waymo open dataset. Appl. Sci. **10**(6), 2046 (2020)
7. Guizilini, V., Ambrus, R., Pillai, S., Raventos, A., Gaidon, A.: 3D packing for self-supervised monocular depth estimation. In: Proceedings of the IEEE/CVF Conference on Computer Vision and Pattern Recognition, pp. 2485–2494 (2020)

8. He, K., Zhang, X., Ren, S., Sun, J.: Deep residual learning for image recognition. In: Proceedings of the IEEE Conference on Computer Vision and Pattern Recognition (CVPR), pp. 770–778 (2016)
9. Huang, G., Liu, Z., Van Der Maaten, L., Weinberger, K.Q.: Densely connected convolutional networks. In: Proceedings of the IEEE Conference on Computer Vision and Pattern Recognition, pp. 4700–4708 (2017)
10. Iandola, F.N., Han, S., Moskewicz, M.W., Ashraf, K., Dally, W.J., Keutzer, K.: SqueezeNet: AlexNet-level accuracy with 50x fewer parameters and <0.5 MB model size. arXiv preprint arXiv:1602.07360 (2016)
11. Jadon, S.: A survey of loss functions for semantic segmentation. In: 2020 IEEE Conference on Computational Intelligence in Bioinformatics and Computational Biology (CIBCB), pp. 1–7. IEEE (2020)
12. Khan, F., Kumar, R.L., Kadry, S., Nam, Y., Meqdad, M.N.: Autonomous vehicles: a study of implementation and security. Int. J. Electr. Comput. Eng. (2088–8708) **11**(4), 3013–3021 (2021)
13. Ko, Y., Lee, Y., Azam, S., Munir, F., Jeon, M., Pedrycz, W.: Key points estimation and point instance segmentation approach for lane detection. IEEE Trans. Intell. Transp. Syst. **23**(7), 8949–8958 (2021)
14. Krizhevsky, A., Sutskever, I., Hinton, G.E.: ImageNet classification with deep convolutional neural networks. In: Advances in Neural Information Processing Systems, pp. 1097–1105 (2012)
15. Li, P., Chen, X., Shen, S.: Stereo R-CNN based 3D object detection for autonomous driving. In: Proceedings of the IEEE/CVF Conference on Computer Vision and Pattern Recognition, pp. 7644–7652 (2019)
16. Liao, Y., Xie, J., Geiger, A.: KITTI-360: a novel dataset and benchmarks for urban scene understanding in 2D and 3D. IEEE Trans. Pattern Anal. Mach. Intell. **45**(3), 3292–3310 (2022)
17. Moghadam, M., Elkaim, G.H.: A hierarchical architecture for sequential decision-making in autonomous driving using deep reinforcement learning. arXiv preprint arXiv:1906.08464 (2019)
18. Pan, X., You, Y., Wang, Z., Lu, C.: Virtual to real reinforcement learning for autonomous driving. arXiv preprint arXiv:1704.03952 (2017)
19. Sandler, M., Howard, A., Zhu, M., Zhmoginov, A., Chen, L.: MobileNetV 2: inverted residuals and linear bottlenecks. In: Proceedings of the IEEE Conference on Computer Vision and Pattern Recognition, pp. 4510–4520 (2018)
20. Santana, E., Hotz, G.: Learning a driving simulator. arXiv preprint arXiv:1608.01230 (2016)
21. Shalev-Shwartz, S., Shammah, S., Shashua, A.: On a formal model of safe and scalable self-driving cars. arXiv preprint arXiv:1708.06374 (2017)
22. Smolyakov, M.V., Frolov, A.I., Volkov, V.N., Stelmashchuk, I.V.: Self-driving car steering angle prediction based on deep neural network an example of CarND Udacity simulator. In: 2018 IEEE 12th International Conference on Application of Information and Communication Technologies (AICT), pp. 1–5. IEEE (2018)
23. Szegedy, C., et al.: Going deeper with convolutions. In: Proceedings of the IEEE Conference on Computer Vision and Pattern Recognition, pp. 1–9 (2015)
24. Vora, S., Lang, A.H., Helou, B., Beijbom, O.: PointPainting: sequential fusion for 3D object detection. In: Proceedings of the IEEE/CVF Conference on Computer Vision and Pattern Recognition, pp. 4604–4612 (2020)
25. **WHO. Road traffic injuries. WHO report (2022)**
26. Wu, D., et al.: YOLOP: you only look once for panoptic driving perception. Mach. Intell. Res. **19**, 550–562 (2022)

27. Yang, Z., Zhang, Y., Yu, J., Cai, J., Luo, J.: End-to-end multi-modal multi-task vehicle control for self-driving cars with visual perceptions. In: 2018 24th International Conference on Pattern Recognition (ICPR), pp. 2289–2294. IEEE (2018)
28. Zhenye-Na. End-to-end learning for self-driving cars—Udacity's simulation env. (2019)
29. Zoph, B., Vasudevan, V., Shlens, J., Le, Q.V.: Learning transferable architectures for scalable image recognition. In: Proceedings of the IEEE Conference on Computer Vision and Pattern Recognition, pp. 8697–8710 (2018)

Impact of Learning Rate on Noise Resistant Property of Deep Learning Models

Omobayode Fagbohungbe(✉) and Lijun Qian

CREDIT Center Prairie View A&M University, Prairie View, TX 77446, USA
ofagbohungbe@pvamu.edu

Abstract. The interest in analog computation has grown tremendously recently due to its fast computation speed and excellent energy efficiency. These characteristics of analog computation are very important for edge and IoT devices in the sub-watt power envelope for deep learning inferencing. However, significant performance degradation suffered by deep learning models due to the inherent noise present in analog computation can limit their use in mission-critical applications. Hence, there is a need to understand the impact of critical model hyperparameters' choice on the resulting model noise-resistant property. This need is critical as the insight obtained can be used to design deep learning models that are robust to analog noise. This paper investigates the impact of the learning rate, a critical model design choice, on the noise-resistant property. The study is achieved by training deep learning models using different learning rates. After that, the models are injected with analog noise, and the resulting models' noise-resistant property is examined by measuring the performance degradation due to the analog noise. The results showed there exists a sweet spot of learning rate values that achieves a good balance between model prediction performance and model noise-resistant property. Furthermore, a theoretical explanation of the observed phenomenon is provided.

Keywords: Deep Learning · Hardware Implemented Neural Network · Analog Computation · Additive White Gaussian Noise · Learning Rate · Analog Inference

1 Introduction

Deep neural network (DNN) models continue to enjoy wide application in very challenging and complex real-time cognitive applications such as machine translation, autonomous driving, object detection, recommender systems, etc., due to their superlative ability to detect and recognize complex features in text, images, and speech. The inherently scalable nature of this ability of DNN models has enabled the design and deployment of very large models in commercial and industrial application [1–4]. Despite their impressive success, DNN models

K. Arai (Ed.): FTC 2023, LNNS 816, pp. 14–30, 2023.
https://doi.org/10.1007/978-3-031-47448-4_2

Lecture Notes in Networks and Systems 816

The series "Lecture Notes in Networks and Systems" publishes the latest developments in Networks and Systems—quickly, informally and with high quality. Original research reported in proceedings and post-proceedings represents the core of LNNS.

Volumes published in LNNS embrace all aspects and subfields of, as well as new challenges in, Networks and Systems.

The series contains proceedings and edited volumes in systems and networks, spanning the areas of Cyber-Physical Systems, Autonomous Systems, Sensor Networks, Control Systems, Energy Systems, Automotive Systems, Biological Systems, Vehicular Networking and Connected Vehicles, Aerospace Systems, Automation, Manufacturing, Smart Grids, Nonlinear Systems, Power Systems, Robotics, Social Systems, Economic Systems and other. Of particular value to both the contributors and the readership are the short publication timeframe and the world-wide distribution and exposure which enable both a wide and rapid dissemination of research output.

The series covers the theory, applications, and perspectives on the state of the art and future developments relevant to systems and networks, decision making, control, complex processes and related areas, as embedded in the fields of interdisciplinary and applied sciences, engineering, computer science, physics, economics, social, and life sciences, as well as the paradigms and methodologies behind them.

Indexed by SCOPUS, INSPEC, WTI Frankfurt eG, zbMATH, SCImago.

All books published in the series are submitted for consideration in Web of Science.

For proposals from Asia please contact Aninda Bose (aninda.bose@springer.com).

require millions of multiply and accumulate operations that consume a lot of computing and memory resources. This requirement makes them unattractive for deployment in application areas with power envelopes of sub-Watt or very few Watts, generally referred to as the IoT or edge computing realm [5–7]. Although several methods, such as low precision quantization techniques and weight pruning [8,9] methods have been introduced to reduce the arithmetic complexity and memory requirement of DNN models and many digital accelerators proposed, the progress has been very limited. The limitation can be attributed to the von-Neumann nature of the proposed platforms [5–7]. The von-Neumann nature of this platform means there is a continuous need for data movement and communication between the memory and compute unit, leading to high latency and high energy cost [5,6]. Hence, there is a need for hardware accelerators that are fast, reliable, and energy-efficient as an alternative to these von Neumann digital hardware [7,10].

This requirement has led to significant interest in analog accelerators, a form of analog computation which uses in-memory computing [5,7]. The interest in these accelerators can be attributed to the fact that they can potentially help perform matrix-vector multiplication, a significant computation load in DNN, with unprecedented speed and energy efficiency [5,6,11]. Leveraging on the physical characteristics of these resistant-based or charge-based memory devices and arranging them in a crossbar array configuration, a matrix-vector multiplication can be carried out with $O(1)$ time complexity using Ohm's and Kirchhoff's laws, compared to the $O(N^2)$ time complexity in digital accelerators [5,6,11]. With the computation happening in the memory unit, data transfer is substantially reduced, data parallelism is increased, and significant improvement in energy efficiency is achieved [5,6,12].

However, computation in analog accelerators is very imprecise due to the inherent presence of noise or variability such as device-to-device variability, write noise, and conductance drift [5,6,12]. This noise can severely impact the performance of DNN models, as demonstrated in [7,13]. Therefore, there is a need to design DNN models with excellent noise-resistant property to minimize the impact of this noise on the model performance. To achieve this objective, there is a need to investigate and understand the impact of various model hyperparameters on the noise-resistant property of DNN models. This investigation is critical as this could help maintain the delicate balance between the model inference performance and its noise-resistant property. Although many methods have been proposed to improve the noise-resistant property of DNN models [5,7,10], most of these methods did not investigate the impact or leverage on model hyperparameters to achieve these goals. Hence, this research seeks to know if model hyperparameters affect a model's noise-resistant property and if they can be explored to design models more resistant to noise. Specifically, the impact of the learning rate on the resulting model noise-resistant property is investigated in this paper. The choice of learning rate is informed by its significant importance in the model training process as it determines if the model training process will converge and the convergence speed.

Noise, either those inherently present in the training process or externally injected during the training process, has played a critical role in designing DNN models by helping the training algorithm converge to or close to the global minima. One of the popular methods to improve the noise-resistant property of the DNN model is the noise injection method, which injects either the model input, model weight, or gradient of the model weight with noise during the forward propagation during training alone or in combination with other methods like knowledge distillation [5,7,10]. In fact, the default noise-resistant property of DNN models can be attributed to the inherent presence of noise in the training process in the form of an error term with zero mean [14,15]. The zero mean value of the noise is due to the constant batch size during the training process. The error term is mathematically defined below in Eq. (1) where α is the learning rate, ∇ is the gradient, $L_D(x)$ is the loss function over all the dataset, $L_i(x)$ is the loss function for a single data point i, $\nabla L_{SGD}(x)$ is the estimated true loss gradient, and B is the Batch size. This error term introduces some noise into the computed gradient of the weight, influencing the weights and properties of the resulting model. The upper bound of the noise of the gradient due to the error term is defined in Eq. (2) where C is given in Eq. (3) [14].

$$\alpha\nabla_{SGD}(x) = \alpha\nabla L_D(x) + \frac{\alpha}{B}\sum_{i\in B}(\nabla L_i(x) - \nabla L_D(x)) \tag{1}$$

$$E[\|\alpha\nabla L_B(x) - \alpha\nabla L_{SGD}(x)\|^2] < \frac{\alpha^2}{|B|}C \tag{2}$$

$$C = E[\|\nabla L_i(x) - \nabla L_D(x)\|^2] \tag{3}$$

It can be inferred that the learning rate affects the power of the inherent noise present in the training process, which influences the noise-resistant property of the resulting DNN model. Hence, there is a range of learning rate values that strikes the delicate balance between model inference performance and the model noise-resistant property. With other factors fixed, the learning rate value below the sweet spot reduces the inherent noise's power, potentially leading to a model with low noise-resistant value, long training time, and poor inference performance. Learning rate values greater than the sweet spot increases the inherent noise's power, leading to a model with poor noise-resistant property and excellent model inference performance. Therefore, selecting the appropriate learning rate value influences the gradient noise, resulting in DNN models of excellent noise-resistant property and outstanding inference performance.

To validate the postulations above, DNN models are trained with different learning rates, and the noise-resistant property of the resulting model is measured. The noise-resistant property is measured by injecting additive noise into all the model layers, and new inference accuracy due to the noise is measured and normalized using the model inference accuracy in the absence of noise. The normalized accuracy represents the degradation due to the injected noise. This paper makes the following contributions:

1. Provided theoretical explanation and insights on how learning rate affects the noise-resistant property of DNN models;
2. Established that there is a range of learning rate values that maintain the delicate balance between model inference performance and model noise-resistant property;
3. Discovered that for a fixed learning rate, the choice of model optimizer would impact the noise-resistant property of the resulting DNN model.

The remainder of this paper is organized as follows: The detailed proposed methodology for this work is discussed in Sect. 2. Results and analysis are presented in Sect. 3. Further discussions and related works are reviewed in Sect. 4 and Sect. 5 concludes the paper.

Table 1. The details of the DNN models and dataset used in the experiments.

Model Name	Dataset	# of Classes	Input Dimension	# Images per class
ResNet20	[CIFAR10, CIFAR100]	[10,100]	32*32*3	[6000,600]
ResNet32	[CIFAR10, CIFAR100]	[10,100]	32*32*3	[6000,600]
ResNet44	[CIFAR10, CIFAR100]	[10,100]	32*32*3	[6000,600]
ResNet56	[CIFAR10, CIFAR100]	[10,100]	32*32*3	[6000,600]

2 Proposed Methodology

This section details the procedures followed in performing the various experiments to study the impact of the learning rate on the noise-resistant property of the DNN model. The discussion covers various headings such as dataset used, model design and training, and software and hardware used.

2.1 Dataset

The CIFAR10 and CIFAR100 dataset [16], designed, compiled, and labeled by the Canadian Institute for Advanced Research out of the 80 million tiny images dataset, are used in this work. The datasets are very similar in that they contain colored images of dimension $32 \times 32 \times 3$ and are used for image classification tasks. Furthermore, they contain 60,000 colored images, which can be subdivided into 50,000 training images and 10,000 testing/inference images. The difference is that the CIFAR10 dataset contains images that can be grouped into 10 mutually exclusive classes without semantic overlap. However, the CIFAR100 contains images that can be grouped into 100 non-mutually exclusive classes with some form of semantic overlap, as the dataset contain just 20 superclasses.

2.2 Model Design and Training

ResNet [17] model of various depths trained on CIFAR10 and CIFAR100 datasets as detailed in Table 1 is used in this work. ResNet models are a form of convolutional neural network based on the residual learning framework, which eases the training and optimization of deeper models. It achieves this by reformulating the constituent layers of a model as a learning residual function with reference to the layers inputs instead of learning unreferenced function [17]. The choice of ResNet models is influenced by their popularity and excellent performance on various classification tasks of varying complexity. Furthermore, the models used in this work are trained from scratch until convergence is achieved by minimizing the cost function, which is the categorical cross-entropy. The cost function minimizes the error between the ground truth and the predicted label. This is achieved using the glorot-uniform method as an initializer, Adams optimization algorithm as the optimizer, and performing Data augmentation during training to prevent overfitting.

2.3 Model Training Stage

The four DL models, ResNet20, ResNet34, ResNet44, and ResNet56, are trained from scratch using six different learning rates (0.0005, 0.000625, 0.00075, 0.001, 0.00125, and 0.00150) until convergence. This results in 24 trained models for each dataset.

2.4 Model Inference Stage

The performance of the various models trained in Sect. 2.3 in the absence and presence of noise is evaluated in this section. The evaluation uses the appropriate metrics to understand the learning rate's impact on the model's noise-resistant property. The performance metric of the models in this work is the model inference accuracy (%), as all the models under consideration are classification models. The inference accuracy of the models in the absence of noise is the baseline inference value of the models, and it is essential to calculate the performance degradation due to the presence of noise measured using the normalized inference accuracy defined in Eq. (8).

The performance of the models in the presence of noise is measured by injecting the model's weight with noise. The noise in this work is modeled as white additive Gaussian noise of zero mean and a standard deviation of σ_{noise}, which is also defined as the power/energy of the noise as done in [7,13]. The value of σ_{noise} of a noise added to a particular layer is selected using the signal-to-noise ratio (SNR) or noise form factor (η) and the standard deviation of the weights in that layer as defined in Eqs. (4)–(6). An additive Gaussian noise of zero mean and σ_{noise} value equivalent to the noise form factor of 1%, 5%, 10%, 20%, 30%, and 40% is injected into all the model weights to evaluate the model noise-resistant property in this paper. It should be noted that the model baseline value, which is the model performance in the absence of noise, is equivalent to additive Gaussian noise of zero mean and η value of zero, and it is denoted by a_o.

$$\sigma_{noise} = \frac{\sigma_w}{SNR} \tag{4}$$

$$\eta = \frac{1}{SNR} \tag{5}$$

$$\sigma_{noise} = \eta \times \sigma_w \tag{6}$$

After that, the impact of the noise is evaluated by measuring the new inference accuracy value of the DNN model on the test dataset. The evaluation is performed using six noise form factor values of 1%, 5%, 10%, 20%, 30%, and 40%, for all the variants of each of the DNN models in Table 1. The variants are obtained by using different learning rate values of 0.0005, 0.000625, 0.00075, 0.001, 0.00125, and 0.00150, respectively, to train each of the DNN models. It should be noted that the steps to calculate the inference accuracy due to the presence of white noise of σ_{noise} value of X is repeated multiple times due to the stochastic nature of the noise and average calculated as stated in Eq. (7) where a_i is the inference accuracy at instant i and N is the number of times the experiment is performed. Furthermore, the inference accuracy due to the presence of noise is then normalized using the baseline inference accuracy value to get the normalized inference accuracy as stated in Eq. (8) where A^x, a^x, and a_o are the normalized classification accuracy, classification accuracy due to the presence of noise of η value of x%, and the baseline classification accuracy respectively. The normalized inference accuracy is essential as it is a fairer way of comparing the performance of the various DNN models due to noise as it captures the percentage change in model inference accuracy.

$$a^x = \frac{\sum_{i=1}^{N} a_i}{N} \tag{7}$$

$$A^x = \frac{a^x}{a_o} \tag{8}$$

2.5 Software and Hardware

All the models used in this work are trained and tested using Keras deep learning framework installed on the NVIDIA V100-DGXS-32GB GPU.

3 Results and Analysis

The noise-resistant property of various models trained with different learning rate values in Sect. 2 is compared in this section. The noise-resistant property is evaluated by injecting additive analog noise into all the model's weights and evaluating the model's inference and the normalized inference accuracy.

Table 2. Comparison of the baseline inference accuracy of DL models trained with learning rates of 0.0005, 0.000625, 0.00075, 0.001, 0.00125, and 0.00150 using CIFAR10 and CIFAR100 dataset without noise injection to the weights of the trained model during testing. The performance metric is the model classification accuracy.

Dataset	CIFAR10						CIFAR100					
	Learning Rate						Learning Rate					
Model Name	0.0005	0.000625	0.000725	0.001	0.00125	0.0015	0.0005	0.000625	0.000725	0.001	0.00125	0.0015
Resnet_20	91.11%	91.39%	91.87%	91.27%	91.53%	91.49%	65.22%	66.05%	66.79%	66.89%	67.37%	68.37%
Resnet_32	91.51%	92.13%	91.18%	91.72%	89.471%	91.54%	67.78%	68.27%	68.60%	69.39%	68.70%	67.68%
Resnet_44	92.11%	92.63%	92.26%	92.24%	91.62%	91.23%	68.69%	69.36%	70.07%	69.38%	69.57%	68.77%
Resnet_56	92.24%	92.62%	92.39%	91.94%	91.75%	91.72%	69.10%	69.93%	70.01%	69.79%	69.82%	68.79%

Table 3. Comparison of the average normalized inference accuracy of ResNet20, ResNet32, ResNet44, and ResNet56 models trained with learning rate values of 0.0005, 0.000625, 0.00075, 0.001, 0.00125 and 0.00150 with CIFAR10 and CIFAR100 dataset. The average normalized accuracy is defined in Eq. (9).

Dataset	CIFAR10						CIFAR100					
	Learning Rate						Learning Rate					
Model Name	0.0005	0.000625	0.000725	0.001	0.00125	0.0015	0.0005	0.000625	0.000725	0.001	0.00125	0.0015
ResNet20	72.28%	71.93%	71.57%	68.15%	60.87%	55.26%	51.77%	53.81%	52.50%	53.30%	45.50%	41.36%
ResNet32	74.36%	75.53%	68.90%	60.67%	53.72%	38.48%	54.02%	55.60%	53.87%	39.11%	32.14%	17.84%
ResNet44	71.80%	72.61%	65.09%	35.23%	43.93%	30.48%	57.52%	56.97%	55.20%	24.79%	17.67%	14.17%
ResNet56	72.67%	62.81%	59.81%	42.24%	26.47%	25.93%	58.83%	55.38%	39.45%	25.83%	3.67%	15.67%

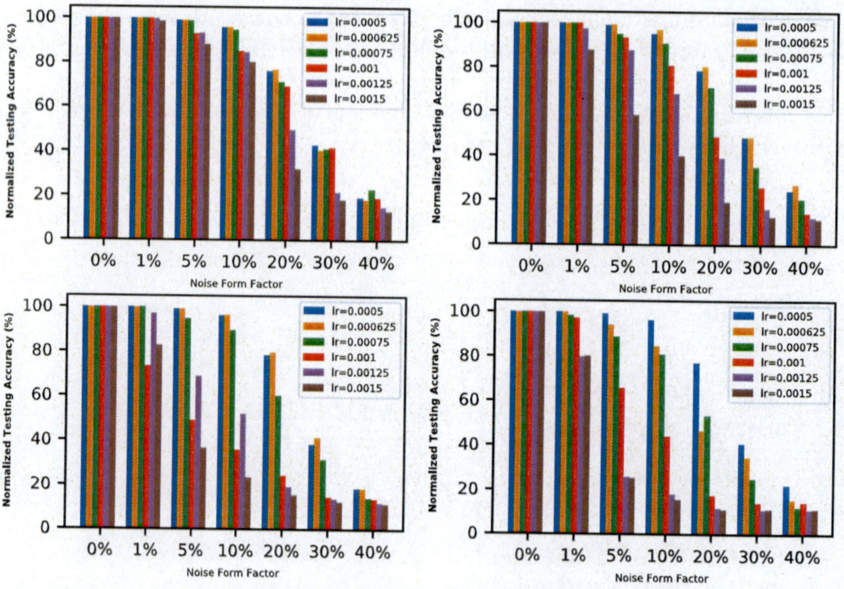

Fig. 1. The comparative study of noise resistant property of various models trained with CIFAR10 dataset with learning rates of 0.0005, 0.000625, 0.00075, 0.001, 0.00125, and 0.00150, respectively. The noise-resistant property is measured with Normalized Test Accuracy. The figure shows the plot of the noise form factor against the Normalized Test Accuracy for ResNet20 (top left), ResNet32 (top right), ResNet44 (bottom left), and ResNet56 (bottom right), respectively. The noise form factor of 0 is equivalent to when no noise is injected into the models.

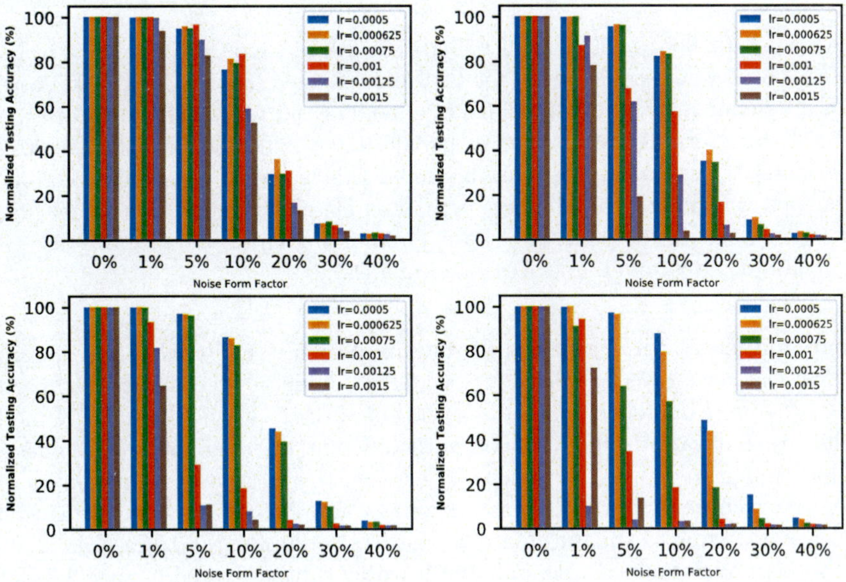

Fig. 2. The comparative study of noise resistant property of various models trained with CIFAR100 dataset with learning rates of 0.0005, 0.000625, 0.00075, 0.001, 0.00125, and 0.00150, respectively. The noise-resistant property is measured with Normalized Test Accuracy. The figure shows the plot of the noise form factor against the Normalized Test Accuracy for ResNet20 (top left), ResNet32 (top right), ResNet44 (bottom left), and ResNet56 (bottom right), respectively. The noise form factor of 0 is equivalent to when no noise is injected into the models.

Baseline Performance. Table 2 presents the baseline inference accuracy of the DNN models given in Table 1. The baseline inference accuracy is obtained when evaluating trained DNN models on the testing dataset without noise, equivalent to injecting with Gaussian noise of zero mean and zero standard deviation. The baseline inference accuracy of a DNN model trained on CIFAR10 is higher than that of the corresponding DNN model trained on CIFAR100. The higher number of classes and the classes' overlapping nature mean that the classification task on the CIFAR100 dataset is more complex than on the CIFAR10 dataset. Furthermore, the reduction in the number of data per class for the CIFAR100 dataset and the small dimensions of the image means that the performance of the DNN models is limited.

Also, for both CIFAR10 and CIFAR100 datasets, using the lower or higher value of the learning rate did not substantially give any performance advantage to the trained DNN models. For example, the ResNet20 model has a baseline inference value of 91.11%, 91.39%, 91.87%, 91.27%, 91.53%, and 91.49% for learning rate value of 0.0005, 0.000625, 0.00075, 0.001, 0.00125 and 0.00150 respectively. A similar observation is also made for ResNet20 trained on the CIFAR100 dataset with a baseline inference value of 65.22%, 66.05%, 66.79%, 66.89%, 67.37%, and 68. 37% for learning rate values of 0.0005, 0.000625, 0.00075, 0.001, 0.00125, and 0.00150, respectively. It must be noted that a higher learning rate confers an advantage in that the training process is faster, although the model might converge at the saddle point or the local minimum. This observation is also true for ResNet33, ResNet44, and ResNet56 models.

Performance in the Presence of Analog Noise. The impact of the learning rate on the noise-resistant property of a DNN model is analyzed by comparing the normalized inference accuracy of the six variants of the same DNN model architecture obtained when trained with different learning rates. The normalized learning rate is used to reasonably capture the percentage change in model inference accuracy due to the noise. Figures 1 and 2 show the normalized inference accuracy due to the injected noise for ResNet20, ResNet32, ResNet34, and ResNet56 models trained with different learning rates as stated in Sect. 2.3 using CIFAR10 and CIFAR100 datasets, respectively. It can be observed that the normalized inference accuracy, which is a measure of the noise-resistant property of the DNN model, decreases with the noise form factor, which is a measure of the power of the injected noise. Furthermore, the degradation in the model's performance also occurred gracefully, speaking to the noise-resistant property of DNN models. This observation is the same irrespective of the model's type, the classification task's complexity, and the learning rate used in training the model, although the degradation rate varies. The observation can be attributed to the stronger tendency of the noise to corrupt the model weight due to its higher magnitude, which increases with an increase in noise power/noise form factor. For example, ResNet56 trained with a learning rate of 0.00075 on the CIFAR10 dataset has normalized accuracy values of 98.34%, 88.81%, 80.91%, 53.32%, 25.25%, and 12.27% when the model's weights are injected with the noise of noise form factors of 1%, 5%, 10%, 20%, 30%, and 40%, respectively. The same model, trained and tested with the CIFAR100 dataset, has normalized accuracy values of 91.12%, 63.87%, 56.98%, 18.33%, 4.29%, and 2.09% when injected with the noise of noise form factors of 1%, 5%, 10%, 20%, 30%, and 40%, respectively.

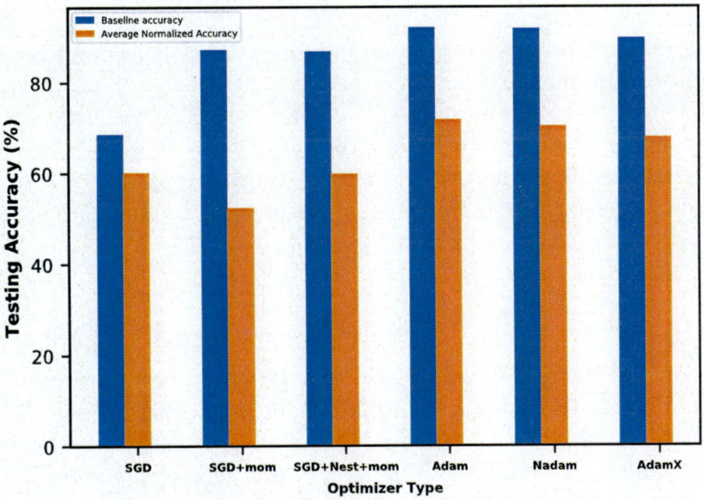

Fig. 3. Comparative study of the effect of different optimizers (with or without adaptive learning rate) on model inference performance and model noise-resistant property using ResNet20 trained on CIFAR10 dataset. The optimizers used are stochastic gradient descent (SGD), SGD with momentum, SGD with Nesterov momentum, Adaptive Moment Estimation (Adam), Nesterov-accelerated Adaptive Moment Estimation (Nadam), and AdamX algorithms. The model inference performance is measured using the baseline inference accuracy, and the model noise-resistant property is measured using the average normalized accuracy as defined in Eq. (9).

Impact of Varying Learning Rates. The performance of the variants of each of the DNN models in Table 1, measured using the noise-resistant property, is compared in this section. The performance of the variants is obtained by training the same model architecture with different learning rates using Adaptive Moment Estimation (Adam) optimizer[18]. After that, the obtained models are injected with additive Gaussian noise, and performance is measured using the normalized inference accuracy. The results are shown in Figs. 1 and 2. It can be observed that the noise-resistant property of a DNN model of specific architecture is affected by the learning rate. The observation holds for all the DNN models, irrespective of the learning rate values and datasets. For example, a normalized inference accuracy value of 95.93%, 96.17%, 94.78%, 85.35%, 84.93%, and 80.52% is obtained for the ResNet20 models trained on the CIFAR10 dataset with learning rates of 0.0005, 0.000625, 0.00075, 0.001, 0.00125 and 0.00150 respectively when injected with Gaussian noise of noise form factor of 10%. This trend is also observed for variants of the ResNet20 DNN model trained on the CIFAR100 dataset trained with learning rates of 0.0005, 0.000625, 0.00075, 0.001, 0.00125, and 0.00150, which have normalized inference accuracy values of 76.55%, 81.26%, 79.43%, 83.33%, 58.78%, and 52.48% when injected with additive Gaussian noise of noise form factor of 10%. The results show that the noise-resistant property of a DNN model can be tuned by varying the learning rate without compromising too much on the model performance.

A new metric, average normalized inference accuracy, defined in Eq. (9), is used to summarize the performances of the variant of each model in the presence of noise and stated in Table 3 where A_i is the normalized inference accuracy at a noise form factor, and M is the number of the non-baseline noise factor, which is 6.

$$A_{avr} = \frac{\sum_{i=1}^{N} A_i}{M} \tag{9}$$

Table 3 shows that the learning rate that gives the best performance in the presence of analog noise varies from one model to another. While the learning rate value of 0.0005 gives the best noise-resistant property for ResNet20 and ResNet56 trained on the CIFAR10 dataset, the learning rate value of 0.000625 gives the best noise-resistant property for ResNet32 and ResNet44 trained on the CIFAR10 dataset. The observation also applies to these models trained on the CIFAR100 dataset, where the learning rate value of 0.000625 gives the best performance for ResNet20 and ResNet32, and the learning rate value of 0.0005 gives the best noise-resistant property for ResNet44 and ResNet56.

Impact of Different Optimizers (With or Without Adaptive Learning Rate). A comparison study of the noise-resistant property of DNN models trained with adaptive learning rate and the fixed global learning rate is done in this section. The comparison is made by training the ResNet20 model with a global learning rate of 0.00075 using stochastic gradient descent (SGD), SGD with momentum [19], SGD with Nesterov momentum, Adaptive Moment Estimation (Adam), Nesterov-accelerated Adaptive Moment Estimation (Nadam) [20], and AdamX optimizers. The Adam, Nadam, and AdamX optimizers are types of the adaptive learning rate, as they use localized learning rates for each parameter during training compared with SGD, SGD with momentum, and SGD with Nesterov momentum that uses fixed and equal learning rates for all parameters. The result of our comparative study is shown in Fig. 3, which shows a plot of baseline inference accuracy and average normalized accuracy for the ResNet20 model trained using a global learning rate value of 0.00075 using the selected optimizers. It can be observed that the different optimizers do affect both the model inference performance and its noise-resistant property. The result showed that the ResNet20 model has a baseline inference accuracy value of 68.51%, 87.09%, 86.65%, 91.87%, 91.57%, 89.47% using SGD, SGD with momentum, SGD with Nesterov momentum, Adam, Nadam and AdamX optimizers, respectively. Average normalized accuracy of 52%, 60%, 60%, 72%, 70%, and 68% are also obtained for the resulting ResNet20 model trained using SGD, SGD with momentum, SGD with Nesterov momentum, Adam, Nadam, and AdamX optimizers, respectively. The result shows that the models with adaptive learning rates have better model inference performance and noise-resistant property than models with fixed learning. The improved performance might be

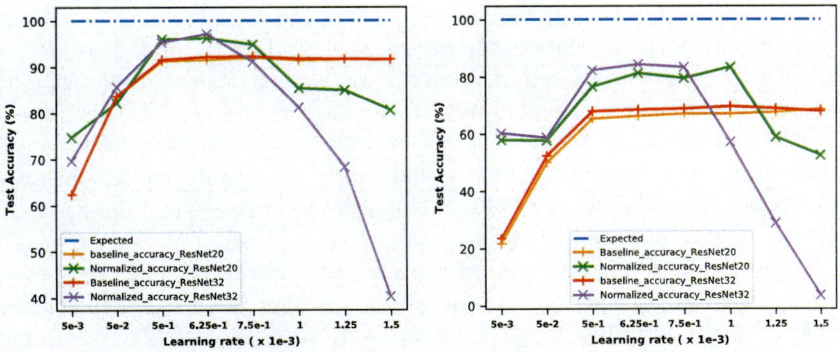

Fig. 4. The plot shows the tradeoff between model inference performance and its noise-resistant property as the learning rate is increased for CIFAR10 (left) and CIFAR100 (right) for ResNet20 and ResNet32. The model inference accuracy is measured using the baseline inference accuracy, and the model noise-resistant property is measured using the average normalized accuracy defined in Eq. (9). The x-axis contains the learning rate value (not drawn to scale). Adam optimizer is used for all the experiments.

due to the dependence of the effective local learning rate in the adaptive learning rate algorithm on the past learning rate, which increases the learning rate used for training. This means that the power of the gradient noise present during training is higher for adaptive learning rate scenarios as the resulting learning rate influences the power of the noise.

Tradeoff Between Prediction Accuracy and Model Noise Resistant Property. The plots of the baseline inference accuracy for ResNet20 and Resnet32 in the absence of noise for various learning rate values when trained CIFAR10 and CIFAR100 datasets using the Adam optimizer are shown in Fig. 4. A similar plots of the normalized inference accuracy for ResNet20 and Resnet34 due to the injection of the noise of noise form factor of 10% on the same datasets are also shown in Fig. 4. The learning rate values used here are 0.000005, 0.00005, 0.0005, 0.000625, 0.00075, 0.001, 0.00125, and 0.00150. Extra learning rate values (0.000005 and 0.00005) were investigated here to capture better the trend at lower learning rate values. For the CIFAR10 dataset (leftmost image in Fig. 4), it is observed that at a learning rate value of 5e-5 and lower, the model inference performance measured using the baseline inference accuracy and the noise-resistant property when injected with the noise of noise factor of 10% is adversely affected for ResNet20 and ResNet34. This learning rate region is not desirable even if the noise-resistant property is perfect, as the baseline inference performance is poor. At a learning rate value of 1e-3 and above, improved inference performance is obtained for both models as measured by the baseline inference accuracy. However, the noise-resistant property of the model, as measured by the normalized inference accuracy, trends downward with an increase in the learning rate. This region will have been desirable if an excellent noise-resistant property is not desired as the baseline inference accuracy value is good. However, the need for excellent noise-resistant property for analog devices means that this region is not desirable as this property is the most desirable as training models in this region might be faster. The region with great noise-resistant property and excellent inference performance is the most desirable for models for analog hardware as this region finds a balance between performance and noise-resistant property. This desired region is the region with a learning rate value between 0.0005 and 0.001 in Fig. 4 for ResNet20 and ResNet34: the baseline inference accuracy and normalized inference accuracy are high in this region, finding a balance between performance and noise-resistant property is achieved. A similar trend is also observed for ResNet20 and ResNet34 trained on the CIFAR100 dataset, as shown in the rightmost image in Fig. 4. The three regions establish that changing the learning rate also impacts the noise-resistant property and inference performance of the DNN model. Hence, there is a need to select the learning rate value in the region where there is a balance between model performance and model noise-resistant property.

Impact of Task Complexity on Model Noise Resistant Property. Table 3 showed that the average normalized inference accuracy of DNN models due to the injection of noise into the model weight is affected by the complexity of the task. The table showed that ResNet20, ResNet32, ResNet44, and ResNet56 trained on the CIFAR10 dataset with a learning rate of 0.000625 has an average normalized inference accuracy of 71.93%, 75.53%, 72.61%, and 62.81%, respectively. These values are higher than 53.81%, 55.06%, 56.97%, and 55.38% normalized inference accuracy values obtained for the same models trained on the CIFAR100 dataset. This performance gap is noticed irrespective of the DNN model type and the learning rates used in training the models. These show that the complexity of the model's task strongly influences its performance in the presence of Gaussian noise, as the CIFAR100 dataset is a more complex classification task than CIFAR10. The complexity of the CIFAR100 dataset is because it contains 100 classes, with most of the classes overlapping compared to the ten non-overlapping classes in the CIFAR10 dataset.

4 Related Works and Discussions

The learning rate is the most critical hyperparameter for stochastic gradient-based optimization problems as it influences if convergence is possible and the rate of convergence. It works by influencing the size of the steps to take in the direction of the negative gradient of the parameters to be estimated. Therefore, there is a need to find an effective method to select and design an appropriate value for the learning rate at each iteration. This need has given rise to many heuristics for estimating a reasonable learning rate at each iteration by either increasing the learning rate when suitable or decreasing the learning rate when near a local minimal [21]. For example, the authors in [22] introduced the learning rate annealing method, which automatically reduces the learning rate size based on the number of epochs. It is also possible to decrease the learning rate when the validation accuracy appears to have plateaus [21]. These approaches are critical to prevent the parameter values from oscillating back and forth around the minimal. Also, there is a recent paradigm that involves allowing the global learning rate to vary cyclically between reasonable boundary values during training [23]. This method automatically eliminates the need to tune the learning rate and needs no extra calculation. In recent times, there has been an introduction of adaptive learning rate [24,25] such as Adam [26], Adadelta [21], RMSprop, etc. instead of a global learning rate, and this new method can be used in combination with the Cyclical Learning Rates. The need for an adaptive learning rate is because different dimensions in the model parameter vectors interact with the cost function in entirely different ways, and using a localized learning rate for this dimension can significantly improve the training process. Also, the authors in [27] demonstrated the use of a large learning rate to achieve improved speed and performance. The authors argued that a large learning rate provides some form of regularization for model training when other forms of regularization are reduced to maintain an optimal balance in regularization. The work in [28] also showed that learning rate in conjunction with the batch size and gradient covariance influences the minima found by SGD during model training.

The need to design models with excellent noise-resistant properties for deployment on analog accelerators has led to studies on noise-resistant properties of popular DNN models and novel methods to design and train models with excellent noise-resistant properties [10,13]. The authors in [10] and [5] proposed using additive and multiplication noise injection methods to increase the noise-resistant property of DNN models. These noise injection methods are combined with knowledge distillation to achieve even better noise-resistant property in [7]. Also, the impact of Batch Normalization on the noise-resistant property of DNN models is investigated, and a novel DNN design and training method that finds a balance between model inference performance and model noise-resistant property is proposed in [29]. A comparative performance of various BatchNorm types on the model noise-resistant property is also done in [30]. A method that conditions DNN models by exposing them to a noisy computation environment is proposed in [31,32]. A chip-in-the-loop method that adapts pre-trained model weights for the inference-only chip is introduced in [33]. The use of a generalized fault-aware pruning technique to improve model resilience to hardware fault is also discussed in [34]. The use of error correction code [35] only or in combination with other methods such as reinforcement learning [36] can also be used to ensure models are resistant to fault during computation.

This work is different from existing work on learning rate as it does not propose new ways of selecting learning rate values [24] or design learning rate for model training [23,26]. To the best of our knowledge, it is the first paper to investigate the impact of the learning rate on the noise-resistant property of DNN models. This work is significant as the noise-resistant property of DNN models can be tuned by increasing/decreasing the learning rate. This paper shares some similarities with [7,10] in that computational noise is modeled as analog noise.

It is still very different as this research did not explore the use of external analog noise to improve the DNN model noise-resistant property as done in those works. Furthermore, the insight from this work can help select appropriate learning rate values that maintain the delicate balance between model inference performance and its noise-resistant property. It is also very different from the work in [34–36] where model noise is modeled as digital noise.

5 Conclusion

The impact of the learning rate on the analog noise-resistant property of DNN models is studied in this work. The finding is significant because the learning rate is a critical model training hyperparameter. The study is achieved by training DNN models with different learning rates and comparing the resulting models' inference performances when analog noise is injected (additive white Gaussian noise) into the model parameters. The experimental results demonstrate that the choice of learning rate value and the optimizer type significantly impact the noise-resistant property of DNN models. Specifically, this work established a range of learning rate values that maintain the delicate balance between model prediction accuracy and model noise-resistant property. The model inference performance and noise-resistant property are adversely affected at a learning rate below this range of values. At a learning rate value above this range, the resulting model has excellent model inference performance but poor noise-resistant property. In addition, through theoretical analysis of the model training process, it is clear that the impact of learning rate on the model noise-resistant property is due to its influence on the inherent noise present in the model training process, which is responsible for the noise-resistant property of DNN models.

Acknowledgment. This work was supported by the Laboratory Directed Research and Development program at Sandia National Laboratories, a multimission laboratory managed and operated by National Technology and Engineering Solutions of Sandia LLC, a wholly owned subsidiary of Honeywell International Inc. for the U.S. Department of Energy's National Nuclear Security Administration under contract DE-NA0003525. This paper describes objective technical results and analysis. Any subjective views or opinions that might be expressed in the paper do not necessarily represent the views of the U.S. Department of Energy or the United States Government.

References

1. Burr, G.W., Sebastian, A., Ando, T., Haensch, W.: Ohm's law + kirchhoff's current law = better ai: neural-network processing done in memory with analog circuits will save energy. IEEE Spectr. **58**(12), 44–49 (2021)
2. Onasami, O., Adesina, D., Qian, L.: Underwater acoustic communication channel modeling using deep learning (2022)
3. Yang, B., et al.: A joint energy and latency framework for transfer learning over 5g industrial edge networks. IEEE Trans. Indust. Inf. **18**, 531–541 (2022)

4. Fagbohungbe, O., Reza, S.R., Dong, X., Qian, L.: Efficient privacy preserving edge intelligent computing framework for image classification in IoT. IEEE Trans. Emerg. Topics Comput. Intell. (2021)
5. Kariyappa, S., et al.: Noise-resilient DNN: tolerating noise in PCM-based AI accelerators via noise-aware training. IEEE Trans. Electron Devices **68**(9), 4356–4362 (2021)
6. Dazzi, M., Sebastian, A., Benini, L., Eleftheriou, E.: Accelerating inference of convolutional neural networks using in-memory computing. Front. Comput. Neurosci. 63 (2021)
7. Zhou, C., Kadambi, P., Mattina, M., Whatmough, P.N.: Noisy machines: understanding noisy neural networks and enhancing robustness to analog hardware errors using distillation. arXiv preprint arXiv:2001.04974 (2020)
8. Han, S., Mao, H., Dally, W.J.: Deep compression: compressing deep neural networks with pruning, trained quantization and Huffman coding. arXiv preprint arXiv:1510.00149 (2015)
9. Luo, J.-H., Wu, J., Lin, W.: Thinet: a filter level pruning method for deep neural network compression. In: Proceedings of the IEEE International Conference on Computer Vision, pp. 5058–5066 (2017)
10. Joshi, V., et al.: Accurate deep neural network inference using computational phase-change memory. arXiv preprint arXiv:1906.03138 (2019)
11. Fasoli, A., et al.: Optimization of analog accelerators for deep neural networks inference. In: 2020 IEEE International Symposium on Circuits and Systems (ISCAS), pp. 1–5 (2020)
12. Chen, A., et al.: Enabling high-performance DNN inference accelerators using non-volatile analog memory (invited). In: 2020 4th IEEE Electron Devices Technology Manufacturing Conference (EDTM), pp. 1–4 (2020)
13. Fagbohungbe, O.I., Qian, L.: Benchmarking inference performance of deep learning models on analog devices. arXiv preprint arXiv:2011.11840 (2020)
14. Bjorck, J., Gomes, C.P., Selman, B.: Understanding batch normalization. In: NeurIPS (2018)
15. Merolla, P., Appuswamy, R., Arthur, J., Esser, S.K., Modha, D.: Deep neural networks are robust to weight binarization and other non-linear distortions. arXiv preprint arXiv:1606.01981 (2016)
16. Krizhevsky, A., et al.: Learning multiple layers of features from tiny images (2009)
17. He, K., Zhang, X., Ren, S., Sun, J.: Deep residual learning for image recognition. arXiv preprint arXiv:1512.03385 (2015)
18. Kingma, D.P., Ba, J.: Adam: a method for stochastic optimization. arXiv preprint arXiv:1412.6980 (2014)
19. Qian, N.: On the momentum term in gradient descent learning algorithms. Neural Netw. **12**(1), 145–151 (1999). https://www.sciencedirect.com/science/article/pii/S0893608098001166
20. Dozat, T.: Incorporating Nesterov momentum into Adam. In: Proceedings of the 4th International Conference on Learning Representations, pp. 1–4 (2016)
21. Zeiler, M.D.: Adadelta: an adaptive learning rate method (2012)
22. Robbins, H.E.: A stochastic approximation method. Ann. Math. Stat. **22**, 400–407 (2007)
23. Smith, L.N.: Cyclical learning rates for training neural networks (2017)
24. Schaul, T., Zhang, S., LeCun, Y.: No more pesky learning rates (2013)
25. Gulcehre, C., Sotelo, J., Moczulski, M., Bengio, Y.: A robust adaptive stochastic gradient method for deep learning (2017)

26. Kingma, D.P., Ba, J.: Adam: a method for stochastic optimization (2017)
27. Smith, L.N., Topin, N.: Super-convergence: very fast training of neural networks using large learning rates (2018)
28. Jastrzebski, S., et al.: Three factors influencing minima in SGD (2018)
29. **Fagbohungbe, O., Qian, L.: Effect of batch normalization on noise resistant property of deep learning models (2021)**
30. Fagbohungbe, O.I., Qian, L.: L1 batch normalization and noise resistant property of deep learning models (2021)
31. Bo, G.M., Caviglia, D.D., Valle, M.: An on-chip learning neural network. In: Proceedings of the IEEE-INNS-ENNS International Joint Conference on Neural Networks (IJCNN 2000). Neural Computing: New Challenges and Perspectives for the New Millennium, vol. 4, pp. 66–71 (2000)
32. Schmid, A., Leblebici, Y., Mlynek, D.: Mixed analogue-digital artificial-neural-network architecture with on-chip learning. In: Proceedings of the IEE Circuits, Devices and Systems, vol. 146, pp. 345–349 (2000)
33. Schmitt, S., et al.: Neuromorphic hardware in the loop: training a deep spiking network on the brainscales wafer-scale system. arXiv preprint arXiv:1703.01909 (2017)
34. Zhang, J.J., Basu, K., Garg, S.: Fault-tolerant systolic array based accelerators for deep neural network execution. IEEE Design Test **36**(5), 44–53 (2019)
35. Upadhyaya, P., Yu, X., Mink, J., Cordero, J., Parmar, P., Jiang, A.: Error correction for hardware-implemented deep neural networks (2019)
36. Huang, K., Siegel, P., Jiang, A.: Functional error correction for robust neural networks (2020)

Third Generation Neural Nets and Their Applications in Multi-modal Deep Learning: A Survey

Ismail El Sayad(✉), Mandeep Pannu, Josh Gourde, and Mohammed Al Nakshabandi

School of Computing, University of the Fraser Valley, Abbotsford, BC, Canada
{ismail.elsayad,mandeep.pannu,mohammed.alnakshabandi}@ufv.ca,
josue.gourde@student.ufv.ca

Abstract. Deep learning has demonstrated remarkable success in a variety of fields in the past decade and experienced remarkable developments. Despite improved performance, the models are more complex and take longer to train. One of the most promising paradigms for efficient deep learning algorithms is spiking neural networks (SNNs). Spiking neural networks (SNNs) are one of the most promising paradigms for efficient deep learning algorithms. They bridge the gap between neuroscience and machine learning by combining biologically realistic neuron models with machine learning techniques. Multimodal deep learning approaches offer the potential to train models capable of learning complex regulatory dynamics across industries, especially as the availability of diverse datasets grows. However, the lack of comprehensive literature detailing the progress made in this area, as well as the challenges and prospects of SNN models, creates a barrier to their further exploration and application. In this paper, we aim to review biological neurons comprehensively and present various spike-based neuron models. Additionally, the article covers several recent applications of SNNs in different industries using different multi-modal data sets, including biomedical data analysis, human authentication, speech and emotion recognition, intrusion detection in IoT system and automotive event detection. In conclusion, the paper discusses future perspectives.

Keywords: Spiking Neural Networks · Spatio-Temporal Pattern Recognition · Biomedical Data Analysis · Human Authentication · Emotion Recognition · Automotive Event Detection · Biological Neural Network

1 Introduction

As artificial neural networks (ANNs) [1] have evolved over the past decade from first generation multi-layer perceptrons (MLPs) to advanced techniques in second generation deep neural networks (DNNs) [23,30], many state-of-the-art techniques have been added to second generation deep neural networks (DNNs). Data

K. Arai (Ed.): FTC 2023, LNNS 816, pp. 31–45, 2023.
https://doi.org/10.1007/978-3-031-47448-4_3

annotated with a high level of precision and GPUs with general-purpose capabilities are critical to this achievement. However, concerning energy efficiency and online learning capabilities, artificial neural networks (ANNs) still lag behind biological neural networks. Various attempts have been made to reduce power consumption in deep learning models. Several techniques have been developed to discover more compact networks capable of achieving similar performance with significantly reduced complexity and fewer parameters, including quantization [9,46] and pruning [21,22]. The structure, computations, and learning rules of ANNs and DNNs are fundamentally different from those of biological neural networks. It is possible that SNNs, the third generation of neural networks, will be able to overcome the bottlenecks that are associated with ANNs. Biological neural networks use discrete electrical signals to communicate between neurons, such as spikes, and work continuously. SNNs are highly compatible with time conditions because of their functional similarity to biological neural networks. Though SNNs still lag behind DNNs in terms of their performance, the gap is closing on some tasks. SNNs generally exhibit lower energy consumption during operation. However, due to the non-differentiable nature of spike operations and the intricate dynamics of neurons, training SNNs is often challenging. As a variety of data sets become increasingly available, multi-modal deep learning approaches offer the potential for training deep learning models that can learn complex regulatory dynamics across industries [15,17,38]. A brief overview of SNN models and recent trends in learning rules and network architectures is provided in this survey. The article also discusses a few recent applications of SNNs in a variety of industries using different multi-modal data sets.

2 SNNs: Fundamentals

2.1 Biological Neuron

The central nervous system of a living organism comprises specialized cells known as neurons, which transmit and receive signals through electrical impulses. It is estimated that the human brain contains around 200 billion neurons. Moreover, neurons exist in various types. Typically, neurons are composed of a cell body or soma, a nucleus, dendrites, and an axon [2].

Information is received from the dendrites, causing a voltage buildup in the cell body. As soon as this membrane potential reaches a certain threshold, an electrical impulse is produced and transmitted to other neurons by the axon. In response to a spike, a neuron returns to a lower potential known as the resting potential. There is a short period of time, known as the refractory period, following the generation of a spike, when a neuron cannot generate another spike. Synapses are connections between the axon terminals and dendrites of neurons as shown in Fig. 1.

Presynaptic neurons release neurotransmitters into the synaptic cleft when an electrical impulse reaches the synapse. As the brain learns, synaptic connections are strengthened or weakened, or new synaptic connections are formed or dissolve synaptic connections that are no longer needed.

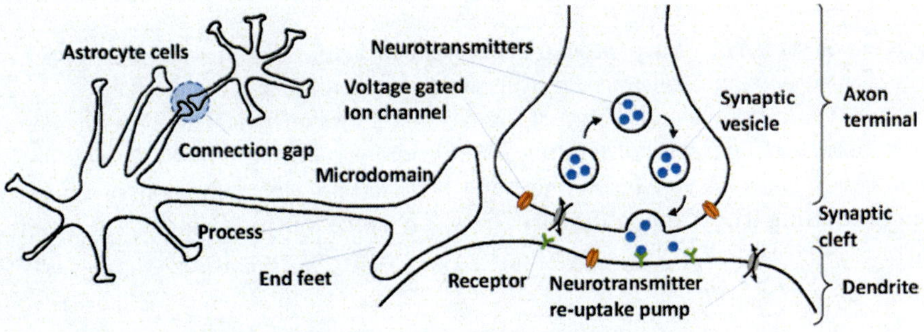

Fig. 1. Neuronal synapse along with astrocyte cells [2].

2.2 SNNs Structure

Inspired by the biological neuron, synapses that are connected by presynaptic (source) and postsynaptic (target) neurons establish the SNN architecture. Presynaptic neurons control postsynaptic neurons by modulating their activity conductance [33]. Unlike conventional ANNs, SNNs encode and transmit information using spikes. SNNs typically have multiple forward passes instead of a single forward propagation, because each input is presented for a prespecified length of time (T). When presynaptic neurons fire, they send signals to their postsynaptic counterparts via synaptic currents, which are proportional to the synapse's weight, or conductance.

3 Training Strategies

The SNN is a fault-tolerant, energy-efficient, and data-efficient solution based on biological principles. There are still a few barriers that need to be overcome in order to achieve promising results, despite their promising results. A biological neural network is able to acquire knowledge through a variety of mechanisms, including spikes that process and transmit information at a rapid pace. Nevertheless, it remains unclear how they acquire that knowledge and the methods by which they acquire it. As a result, learning strategies are often coupled with various elements of neural networks, including how information is encoded, how neurons work, and how the general architecture is constructed. The learning process in SNNs is, therefore, challenging, and an optimal solution needs to be found.

3.1 Supervised Learning

Backpropagation (BP). Many backpropagation (BP) algorithms have been used with classical ANNs. However, there is a need to develop appropriate strategies to apply BP to spike-neuron networks due to inherent incompatibilities. A number of strategies have proven successful in achieving this goal. Approximations for activation functions are implemented in almost all solutions.

Less than two decades ago [7] introduced SpikeProp, the first successful supervised learning rule for spiking neural networks. The SpikeProp uses a classical Error Back Propagation method. It relies on the definition of an error-function using the time difference between spike output time t_j^d and actual output spike time t_j^a of an neuron j. The error-function is defined as follows $E = \frac{1}{2}\sum_{j\in J}\left(t_j^a - t_j^d\right)^2$.

Several variants of SpikeProp have been developed following its successful implementation, including; SuperSpike [45] proposed an improved method of SpikeProp by utilizing the derivative of the membrane potential to enhance its performance. Additionally, modifications such as Back Propagation (BP) with momentum [42], QuickProp [29], Resilient Propagation [29] were proposed to improve classification accuracy and convergence rate in comparison to the original SpikeProp algorithm. However, a significant drawback of using these rules is that output neurons can only emit one spike at most, and another disadvantage is the lack of bio-plausibility when training SNNs with BP.

Prescribed Error Sensitivity (PES). The Neural Engineering Framework (NEF) is utilized to apply biological constraints (NEF) [4], when designing networks of single neurons. This framework often employs biologically plausible supervised learning rules, like the Prescribed Error Sensitivity (PES). The PES learning rule on decoders is represented in the below equation where $\mathbf{e_j}$ are the encoders of the post-population, κ is a learning rate, n is the number of neurons in the pre-population, and a_j is the gain of the post-population [5,10].

$$\Delta\omega_{ij} = -\frac{\kappa}{n}\alpha_j \mathbf{e_j} \cdot \mathbf{E} a_i$$

3.2 Unsupervised Learning

Spike-Time-Dependent Plasticity (STDP). A Hebbian unsupervised learning mechanism known as spike-time-dependent plasticity (STDP) [37] adjusts synaptic weight based on pre- and post-synaptic spikes' temporal order. Despite the absence of post-synaptic spikes, the synaptic weight is increased when pre-synaptic spikes occur before post-synaptic spikes. This phenomenon is known as long-term potentials (LTPs). During long-term depression (LTD), the synaptic weight is decreased if the spike's arrival timing is reversed.

$$\Delta w = \begin{cases} A_+ \exp\left(\frac{t_{\text{pre}} - t_{\text{post}}}{\tau_+}\right) & \text{if } t_{\text{pre}} \leq t_{\text{post}} \\ -A_- \exp\left(-\frac{t_{\text{pre}} - t_{\text{post}}}{\tau_-}\right) & \text{if } t_{\text{pre}} > t_{\text{post}} \end{cases}$$

It is biologically unrealistic and unstable to have synaptic strength increase and decrease infinitely, as shown in the above equation. A biological neuron can regulate its own excitability by decreasing the strength of each synapse in relation to the activity of the network by decreasing the strength of each synapse.

$A_{\pm}$ weight dependence can be implemented to implement this phenomenon of homeostatic scaling. A weight modification is standardized by applying an exponential formula to the synaptic weight based on the current synaptic weight [43].

$$\begin{cases} A_{+}(w) = \eta_{+} \exp\left(w_{init} - w\right) \\ A_{-}(w) = \eta_{-} \exp\left(w - w_{init}\right) \end{cases}$$

Here, $\eta_{\pm}$are learning rates, and w_{init} describes the initial weight.

Probabilistic STDP. [28] introduces a probabilistic variant of simplified STDP. This approach incorporates an exponential function of the current synaptic weight magnitude to adjust the synaptic weight dynamically. This study demonstrated that probability STDP performs well regardless of the amount of complexity in the model of spiking neurons.

$$\Delta w = \begin{cases} \eta_{+} \exp(-w) & \text{if } t_{\text{pre}} \leq t_{\text{post}} \\ -\eta_{-} & \text{if } t_{\text{pre}} > t_{\text{post}} \end{cases}$$

Reward Modulated STDP (R-STDP). The corresponding synapse undergoes potentiation in response to a positive reward, whereas it experiences depression when the reward is negative. According to [18] dopaminergic neurons demonstrate two distinct firing patterns. In the absence of stimuli, their firing rate is characterized as slow, referred to as background firing. However, when these neurons are stimulated, they exhibit a transient burst of firing, followed by a period of inactivity. This modulation is achieved by introducing the eligibility trace z for pre- and post-synaptic spikes as follows:

$$\frac{dw}{dt} = \eta r(t) z_{i,j}(t)$$

$$\frac{dz_{i,j}}{dt} = -\frac{z_{i,j}(t)}{\tau} + \text{STD}\, P(t)$$

where $r(t)$ is the reward per time t, z is the eligibility trace.

4 Applications

4.1 Biomedical Data Analysis

A multimodal dataset is used by Tan et al. [39] to solve the emotion recognition problem using SNNs. To classify emotional valence, the authors use NeuCube framework [19], which utilizes an evolving SNN architecture [24]. A general scheme of the approach based on NeuCube is presented in Fig. 2.

This study utilizes multimodal data for emotion recognition, including facial expressions and physiological signals. Facial expressions are crucial in conveying information during face-to-face interactions. [13]. Emotions like fear, anger, sadness, pleasure, and disgust can be detected through facial expressions, each

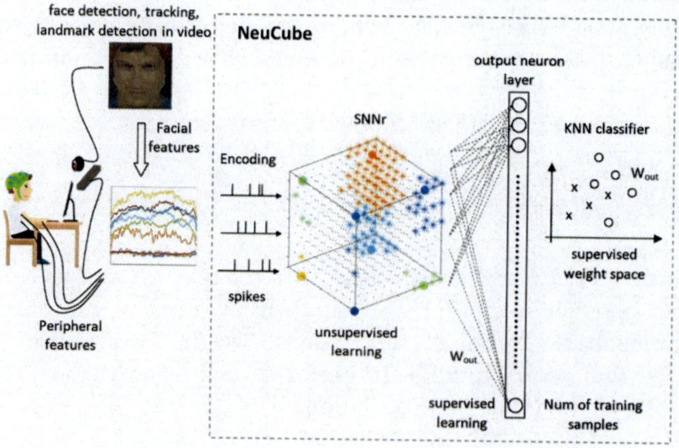

Fig. 2. Emotion valence classification [39].

having its characteristic expression [14]. Apart from the non-physiological signals mentioned earlier, various physiological signals, such as electroencephalography (EEG), electrocardiogram (ECG), electromyogram (EMG), Galvanic skin response (GSR), skin temperature, and pupillary diameter, are employed for affect detection. These noninvasive physiological signals are typically collected through wearable sensors. ECG signals are analyzed to reveal heart rate (HR) and heart rate variability (HRV). Barnea et al. [3] Skin temperature, primarily determined by autonomic nervous system activity, can be utilized as an indication of emotions. GSR, another measure of autonomic activity, captures electrical potential changes in the skin through a negligible amount of current, making it an accurate indicator of arousal as it measures sweat gland activity on the skin [10,31].

The cross-validation is conducted following the Leave-One-Subject-Out (LOSO) principle. The proposed approach achieves a 73.15% accuracy in classifying binary valence. This study demonstrates the feasibility of employing SNN (Spiking Neural Networks) for emotion recognition by utilizing multimodal data.

4.2 Human Authentication Using Audiovisual Data

A system is presented by Wysoski et al. [41] that integrates data from different sources to perform pattern recognition by taking into account some biological aspects connected to sensory integration. Adaptive SNNs are used to process each modality separately. Fast spiking neurons are used to integrate modalities. Specialized adaptive SNNs are used for each modality. Specialized adaptive SNNs are used for each modality.

Visual System Model. Based on four layers of spiking neurons, the visual system is modeled. Figure 3 shows the network architecture used in this work that combines opinions of being/not being a desired face over several frames. L1 and L2 neurons respond to image contrast and orientation, respectively. In L3, complex cells respond to specific patterns. Depending on the excitations it receives, L4 accumulates opinions.

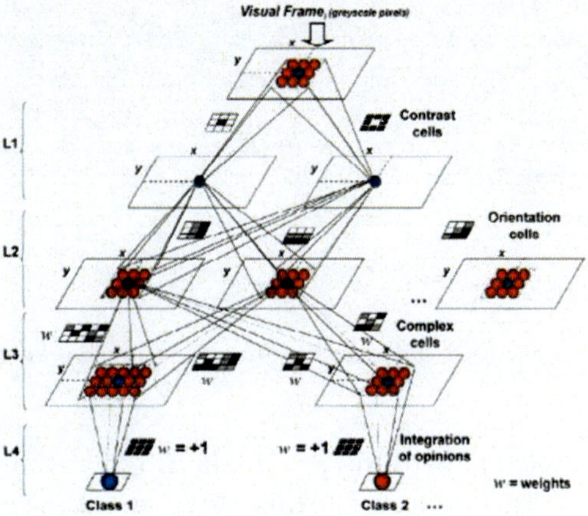

Fig. 3. Visual System Model Architecture [39].

Auditory System Model. The auditory system model architecture is illustrated in Fig. 4. This setup has proven to be very effective for solving short-sentence text-independent problems (typically composed of input utterances ranging from 3 s to 1 min), primarily because long-term dependencies among frames are difficult to extract and train [6,8,34].

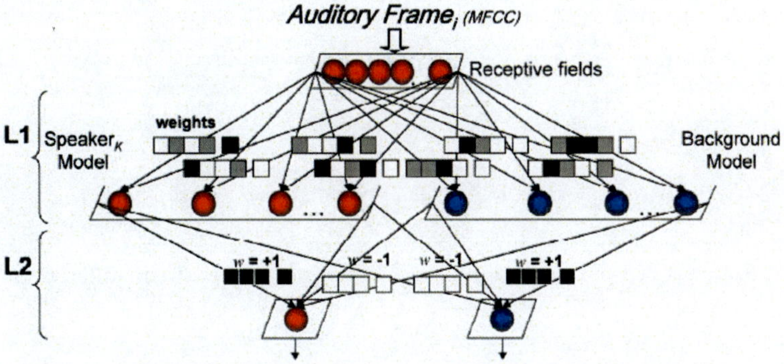

Fig. 4. Auditory System Model Architecture [39].

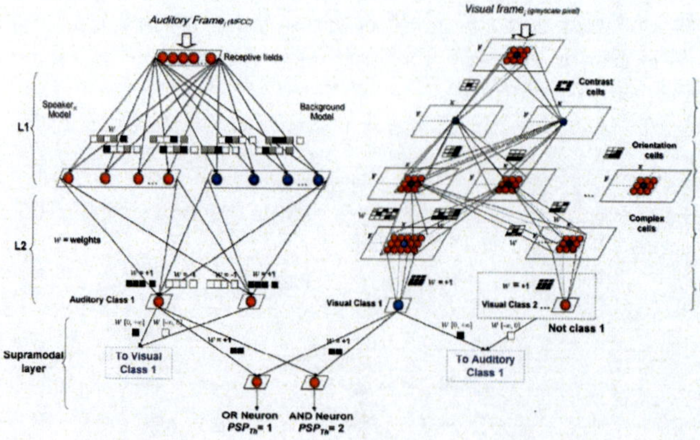

Fig. 5. Crossmodal integration of audiovisual modalities[39].

Integrating Audiovisual Modalities. Figure 5 shows the crossmodal audio-visual integration architecture. By having supramodal neurons with modulation factors mod = 1 and setting all connection weights W to 1, the OR gate $PSP[Th]$ can be implemented for two modalities. In this analysis, VidTimit dataset [35] was used. This allows the system to authenticate 35 users automatically by matching their facial expressions to their utterances. The integration performs better than the individual modes as shown in Fig. 6.

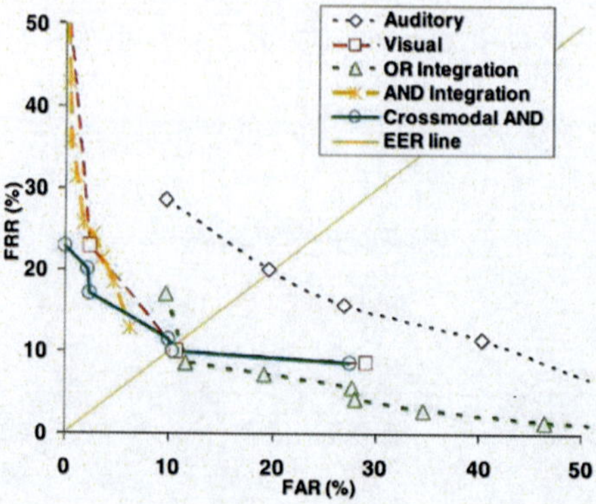

Fig. 6. Analysis of the integration of individual modes and their differences [39].

4.3 Speech Emotion Recognition

Mansouri-Benssassi et al. [26] propose a new method of approaching Speech Emotion Recognition (SER). In their study, SNNs are used for a cross-modal enhancement approach. In multisensory audio-visual processing, visual information precedes, enhances, and predicts auditory information, based on how auditory information is processed in the brain. The methodology consists of the following steps:

- Feature Extraction: Mel-scale spectrograms and MFCCs are two commonly used parameters in SER [40].
- SNN Architecture for SER: The experiments were done using an SNN simulator named BRIAN [16]. The input layer has two groups of neurons representing audio and visual modalities. The size of the input neurons is proportional to the input features. The input layer is connected to a convolutional layer with a stride window. Features and cross-modal connections are learned using unsupervised STDP.
- Multimodal Integration Baseline: To compare their early cross-modal enhancement approach, the authors created a multimodal integration approach with an early fusion approach [26].

The proposed methodology aims to improve the SER performance by leveraging cross-modal information from visual data using SNNs. The Ryerson Audio-Visual Database of Emotional Speech and Song (RAVDESS) [25] is a collection of emotional speech and song recordings that cover six basic emotions. The data set offers audio, video, and audio-video recordings for each emotional state. The emotions in question are happy, sad, angry, fearful, surprised, and disgust. The eNTERFace dataset [27] is a acted dataset with 42 English speakers from 14 different countries, of which 81% are males and 19% are females. The results show that the proposed approach outperforms traditional approaches that only use audio data, as well as other methods that incorporate visual information through neural networks. The authors conclude that the early integration of visual information through SNNs can significantly improve speech emotion recognition accuracy.

4.4 Intrusion Detection in IoT System

Zarzoor et al. [44] use a combination of an SNN and a decision tree (DT) to provide a light-weight alternative to intrusion detection systems (IDS) on internet of things (IoT) devices (IDS-SNNDT). First, the authors pre-processed data from the IoT Botnet Dataset 2020 bynormalizing the data to a float value in the range [0,1]. Next, the entropy method and information gain is used with a decision tree to identify relevant features from the IoT Botnet Dataset 2020 (narrowing from 49 features down to 19) [44]. Finally, a 4 layer SNN using the non-leaky integrate neurons fire (NLIF) model is trained. The SNN is comprised of 4 layers: the input layer, two hidden layers and an output layer [44]. The Gaussian Receptive Field (GRF) is then used to encode information into firing times for the input layer. The hidden layers use a back-propagation algorithm to update learning weights and alleviate some errors. Finally, an ROC algorithm is applied to the output layer in order to determine whether there is an attack or not. Since the values are normalized to a range from 0 to 1, an attack is found when the value exceeds 0.5.

The BoT-IoT dataset [20] was generated by constructing a practical network setup in the Cyber Range Lab of UNSW Canberra, which involved a mixture of legitimate and botnet traffic. The dataset is available in various formats, such as the initial pcap files, argus files, and csv files, and the data files were categorized according to attack type and subtype for easier labeling. The method proposed by [44] (IDS-SNNDT) was tested against two other methods; namely, the IDS-DNN and the IDS-SNNTLF methods. The metrics used in this test were the latency, the power usage and the accuracy of detection. While the accuracy of the IDS-SNNDT was slightly behind that of the IDS-DNN, it surpassed the accuracy of the IDS-SNNTLF (as shown in Fig. 7). The IDS-SNNDT also managed to be faster and less resource intensive than both of the other methods tested (as shown in Fig. 8 and Fig. 9).

Method	Accuracy of Detection									
	DoS		DDoS		TDF		TK		OS	
	Train	Test	Train	Test	Train	Test	Train	Test	Train	Test
IDS-DNN	95.59	95.59	92.18	92.11	94.26	94.15	99.55	99.66	99.90	99.98
IDS-SNNDT	94	94.5	91.99	90.04	99.66	98.03	99.88	99.80	99.88	99.70
IDS-SNNTLF	90	92.54	90.90	89.90	88.22	90.10	94.66	92.34	95.88	96.20

Fig. 7. Comparison of the accuracy between the IDS-SNNDT, IDS-DNN and IDS-SNNTLF methods.

4.5 Automotive Event Detection

Cordone et al. [11] propose SNNs for fast and efficient automotive embedded applications using different modality of data. The authors utilise recent advancements in spike back-propagation, such as surrogate gradient learning [32] and the Spiking Jelly framework, as well as their new voxel cube event encoding as follows. Event sensors output data with microsecond temporal resolution, data is then converted into a dense representation. Each time window is divided into smaller micro time bins and events belonging to a micro time bin are stored in the channels dimension. This provides finer temporal information and reduces the number of steps, maintaining temporal precision compared to voxel grids. The authors designed four types spiking neural networks : Spiking VGG, Spiking SqueezeNet, Spiking MobileNet and Spiking DenseNet. The SSD object detection framework generates feature maps for object detection by employing a backbone and multiple predictor heads. To train the network, focal loss was used, and three extra blocks of convolutions were used to generate smaller feature maps. Figure 10 illustrates the architecture of spiking DenseNet + SSD.

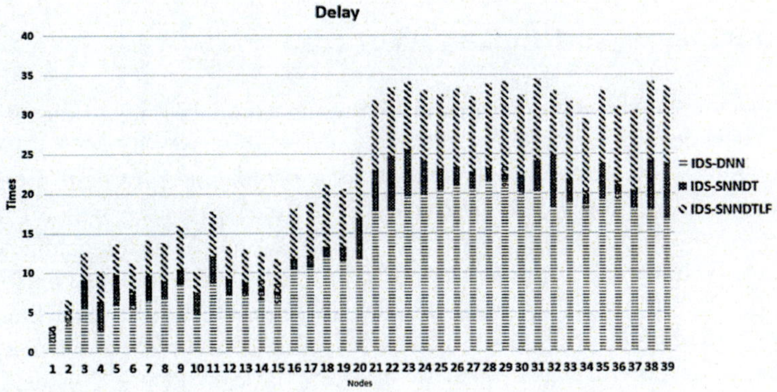

Fig. 8. Comparison of the latency between the IDS-SNNDT, IDS-DNN and IDS-SNNTLF methods [44].

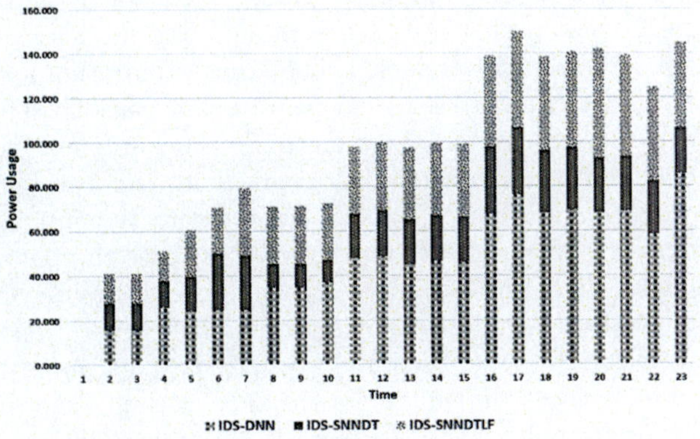

Fig. 9. Comparison of the power usage across the three methods: IDS-SNNDT, IDS-DNN and IDS-SNNTLF [44].

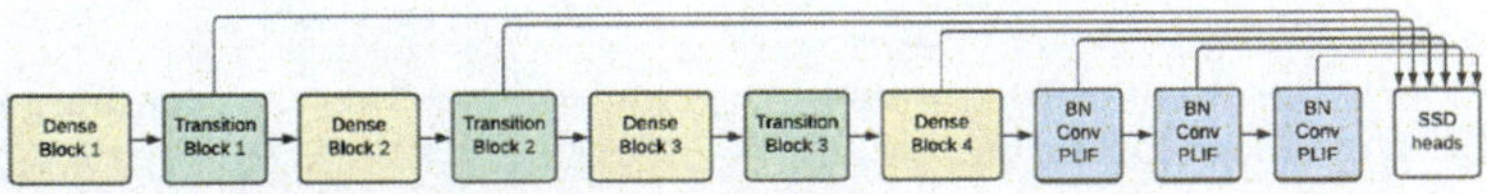

Fig. 10. Spiking DenseNet + SSD model [11].

Two automotive classification datasets are used by the authors: Prophesee NCARS dataset [36], Prophesee GEN1 Detection [12]. The experimental results demonstrate how spiking neural networks have progressed rapidly in recent years. After being restricted to small datasets for a long time, spiking neural networks now show strong performance on temporal data.

5 Conclusion and Future Work

In conclusion, deep learning has achieved remarkable success in various fields over the past decade, but models' complexity and training time have also increased. Spiking neural networks (SNNs) emerge as a promising paradigm for efficient deep learning algorithms, bridging the gap between neuroscience and machine learning by combining biologically realistic neuron models with machine learning techniques. While deep learning models have undergone significant performance improvements, they must catch up to biological neural networks regarding energy efficiency and online learning capabilities. SNNs, as the third generation of neural networks, offer potential solutions to address these challenges, using discrete electrical signals or spikes for communication between neurons, which makes them highly compatible with temporal conditions and more energy-efficient.

Multimodal deep learning approaches present a compelling avenue to leverage various data sets and enable complex regulatory dynamics across industries. However, the lack of comprehensive literature detailing the progress, challenges, and prospects of SNN models hinders their further exploration and application.

In this paper, we reviewed biological neurons and various spike-based neuron models. We explored recent applications of SNNs in different industries using diverse multimodal data sets, including biomedical data analysis, human authentication, speech and emotion recognition, intrusion detection in IoT systems, and automotive event detection. These applications showcase the potential of SNNs in handling complex and real-world data. Learning strategies for SNNs still need to be solved due to the non-differentiable nature of spike operations and complex neuron dynamics. Various supervised and unsupervised learning rules have been proposed to address these difficulties, with some variants showing promising results. The reviewed applications demonstrated the feasibility and advantages of using SNNs for emotion recognition, human authentication, speech emotion recognition, intrusion detection in IoT systems, and automotive event detection. SNNs have shown improved accuracy, energy efficiency, and temporal precision in handling time-sensitive data.

Further research and development are required to refine SNN training strategies, improve model performance, and enhance their applicability in diverse domains as we move forward. The progress in spiking neural networks holds great potential to advance the field of deep learning and contribute to more efficient and biologically inspired machine learning solutions.

References

1. Abiodun, O.I., Jantan, A., Omolara, A.E., Dada, K.V., Mohamed, N.A., Arshad, H.: State-of-the-art in artificial neural network applications: a survey. Heliyon 4(11), e00938 (2018)
2. Ahmed, K.: Brain-inspired spiking neural networks. In: Biomimetics. IntechOpen (2020)

3. Barnea, O., Shusterman, V.: Analysis of skin-temperature variability compared to variability of blood pressure and heart rate. In: Proceedings of the 17th International Conference of the Engineering in Medicine and Biology Society, vol. 2, pp. 1027–1028. IEEE (1995)
4. Bekolay, T.: Learning in large-scale spiking neural networks. Master's thesis, University of Waterloo (2011)
5. Bekolay, T., Kolbeck, C., Eliasmith, C.: Simultaneous unsupervised and supervised learning of cognitive functions in biologically plausible spiking neural networks. In: Proceedings of the Annual Meeting of the Cognitive Science Society (2013)
6. Bimbot, F., et al.: A tutorial on text-independent speaker verification. EURASIP J. Adv. Signal Process. **2004**, 1–22 (2004)
7. Bohte, S.M., Kok, J.N., La Poutré, J.A.: Spikeprop: backpropagation for networks of spiking neurons. In: ESANN, vol. 48, pp. 419–424. Bruges (2000)
8. Burileanu, C., Moraru, D., Bojan, L., Puchiu, M., Stan, A.: On performance improvement of a speaker verification system using vector quantization, cohorts and hybrid cohort-world models. Int. J. Speech Technol. **5**, 247–257 (2002)
9. Cai, Z., He, X., Sun, J., Vasconcelos, N.: Deep learning with low precision by half-wave gaussian quantization. In: Proceedings of the IEEE Conference on Computer Vision and Pattern Recognition, pp. 5918–5926 (2017)
10. Carter, J., Rego, J., Schwartz, D., Bhandawat, V., Kim, E.: Learning spiking neural network models of drosophila olfaction. Int. Conf. Neuromorph. Syst. **2020**, 1–5 (2020)
11. Cordone, L., Miramond, B., Thierion, P.: Object detection with spiking neural networks on automotive event data. In: 2022 International Joint Conference on Neural Networks (IJCNN), pp. 1–8. IEEE (2022)
12. De Tournemire, P., Nitti, D., Perot, E., Migliore, D., Sironi, A.: A large scale event-based detection dataset for automotive. arXiv preprint arXiv:2001.08499 (2020)
13. Denham, M.A., Onwuegbuzie, A.J.: Beyond words: using nonverbal communication data in research to enhance thick description and interpretation. Int. J. Qualitat. Methods **12**(1), 670–696 (2013)
14. Ekman, P.: An argument for basic emotions. Cognit. Emotion **6**(3–4), 169–200 (1992)
15. Gao, J., Li, P., Chen, Z., Zhang, J.: A survey on deep learning for multimodal data fusion. Neural Comput. **32**(5), 829–864 (2020)
16. Goodman, D.F.M., Brette, R.: Brian: a simulator for spiking neural networks in python. Front. Neuroinform. **5** (2008)
17. Guo, W., Wang, J., Wang, S.: Deep multimodal representation learning: a survey. IEEE Access **7**, 63373–63394 (2019)
18. Izhikevich, E.M.: Solving the distal reward problem through linkage of STDP and dopamine signaling. Cerebral Cortex **17**(10), 2443–2452 (2007)
19. Kasabov, N.K.: Neucube: a spiking neural network architecture for mapping, learning and understanding of spatio-temporal brain data. Neural Netw. **52**, 62–76 (2014)
20. Koroniotis, N., Moustafa, N., Sitnikova, E., Turnbull, B.: Towards the development of realistic botnet dataset in the internet of things for network forensic analytics: bot-iot dataset. Futur. Gener. Comput. Syst. **100**, 779–796 (2019)
21. Le, Q.V., Ngiam, J., Coates, A., Lahiri, A., Prochnow, B., Ng, A.Y.: On optimization methods for deep learning. In: Proceedings of the 28th International Conference on International Conference on Machine Learning, pp. 265–272 (2011)
22. Li, G., Qian, C., Jiang, C., Lu, X., Tang, K.: Optimization based layer-wise magnitude-based pruning for DNN compression. In: IJCAI, pp. 2383–2389 (2018)

23. Li, Y., Hao, Z., Lei, H.: Survey of convolutional neural network. J. Comput. Appl. **36**(9), 2508 (2016)
24. Lichtenauer, J., Soleymani, M.: Mahnob-hci-tagging database (2011)
25. Livingstone, S.R., Russo, F.A.: The Ryerson audio-visual database of emotional speech and song (ravdess): a dynamic, multimodal set of facial and vocal expressions in North American English. PLoS One **13**(5), e0196391 (2018)
26. Mansouri-Benssassi, E., Ye, J.: Speech emotion recognition with early visual cross-modal enhancement using spiking neural networks. In: 2019 International Joint Conference on Neural Networks (IJCNN)
27. Martin, O., Kotsia, I., Macq, B., Pitas, I.: The Enterface'05 audio-visual emotion database. In: 22nd International Conference on Data Engineering Workshops (ICDEW'06), p. 8. IEEE (2006)
28. Masquelier, T., Thorpe, S.J.: Unsupervised learning of visual features through spike timing dependent plasticity. PLoS Comput. Biol. **3**(2), e31 (2007)
29. McKennoch, S., Liu, D., Bushnell, L.G.: Fast modifications of the spikeprop algorithm. In: The 2006 IEEE International Joint Conference on Neural Network Proceedings, pp. 3970–3977. IEEE, (2006)
30. Mittal, S.: A survey on modeling and improving reliability of DNN algorithms and accelerators. J. Syst. Architect. **104**, 101689 (2020)
31. Nakasone, A., Prendinger, H., Ishizuka, M.: Emotion recognition from electromyography and skin conductance. In: Proceedings of the 5th International Workshop on Biosignal Interpretation, pp. 219–222 (2005)
32. Neftci, E.O., Mostafa, H., Zenke, F.: Surrogate gradient learning in spiking neural networks: bringing the power of gradient-based optimization to spiking neural networks. IEEE Signal Process. Magazine **36**(6), 51–63 (2019)
33. Nunes, J.D., Carvalho, M., Carneiro, D., Cardoso, J.S.: Spiking neural networks: a survey. IEEE Access **10**, 60738–60764 (2022)
34. Reynolds, D.A., Quatieri, T.F., Dunn, R.B.: Speaker verification using adapted gaussian mixture models. Digit. Signal Process. **10**(1–3), 19–41 (2000)
35. Sanderson, C., Paliwal, K.K.: Identity verification using speech and face information. Digit. Signal Process. **14**(5), 449–480 (2004)
36. Sironi, A., Brambilla, M., Bourdis, N., Lagorce, X., Benosman, R.: Hats: histograms of averaged time surfaces for robust event-based object classification. In: Proceedings of the IEEE Conference on Computer Vision and Pattern Recognition, pp. 1731–1740 (2018)
37. Song, S., Miller, K.D., Abbott, L.F.: Competitive Hebbian learning through spike-timing-dependent synaptic plasticity. Nat. Neurosci. **3**(9), 919–926 (2000)
38. Stahlschmidt, S.R., Ulfenborg, B., Synnergren, J.: Multimodal deep learning for biomedical data fusion: a review. Brief. Bioinform. **23**(2), bbab569 (2022)
39. Tan, C., Ceballos, G., Kasabov, N., Subramaniyam, N.P.: Fusionsense: emotion classification using feature fusion of multimodal data and deep learning in a brain-inspired spiking neural network. Sensors **20**(18), 5328 (2020)
40. Tarunika, K., Pradeeba, R.B., Aruna, P.: Applying machine learning techniques for speech emotion recognition. In: 2018 9th International Conference on Computing, Communication and Networking Technologies (ICCCNT), pp. 1–5. IEEE (2018)
41. Wysoski, S.G., Benuskova, L., Kasabov, N.: Adaptive spiking neural networks for audiovisual pattern recognition. In: Ishikawa, M., Doya, K., Miyamoto, H., Yamakawa, T. (eds.) ICONIP 2007. LNCS, vol. 4985, pp. 406–415. Springer, Heidelberg (2008). https://doi.org/10.1007/978-3-540-69162-4_42

42. Xin, J., Embrechts, M.J.: Supervised learning with spiking neural networks. In: Proceedings of the International Joint Conference on Neural Networks (Cat. No. 01CH37222) (IJCNN'01), vol. 3, pp. 1772–1777. IEEE (2001)
43. Yamazaki, K., Vo-Ho, V.-K., Bulsara, D., Le, N.: Spiking neural networks and their applications: a review. Brain Sci. **12**(7), 863 (2022)
44. Zarzoor, A.R., Shiltagh Al-Jamali, N.A., Abdul Qader, D.A.: Intrusion detection method for internet of things based on the spiking neural network and decision tree method. Int. J. Electric. Comput. Eng. **13**(2), 2278 (2023)
45. Zenke, F., Ganguli, S.: Superspike: supervised learning in multi-layer spiking neural networks. arXiv preprint arXiv:1705.11146 (2017)
46. Zhang, D., Yang, J., Ye, D., Hua, G.: Lq-nets: learned quantization for highly accurate and compact deep neural networks. In: Proceedings of the European Conference on Computer Vision (ECCV), pp. 365–382 (2018)

Neuroevolution-Based Fine Tuning of the Quantised Deep Learning Models

Marcin Pietron(✉)

AGH-UST ACK Cyfronet AGH Cracow, Cracow, Poland
pietron@agh.edu.pl

Abstract. Deep learning models are the most efficient models in many machine learning tasks. The main disadvantage when using them in IoT, mobile devices, independent autonomous or real-time systems is their complexity and memory size. Therefore, much research has concentrated on compression techniques of deep learning architectures. One of the most popular technique is quantisation. In most of the works, the quantisation is done based on the nearest neighbour quantisation technique. This work focuses on improving the quantisation efficiency in pretrained and quantised models. This approach has the potential to improve the final accuracy of quantised models. The main postulate of the work is that final quantisation states of the network based on nearest neighbour rounding does not guarantee optimal accuracy. In the presented work, the neuroevolution strategy is used as an optimisation approach. The neuroevolution in each iteration changes the values of the small percentage of weights. It shifts theirs values to different quantisation states. The work shows that proposed neuroevolution with an appropriate set of operators and parameters can fast improve the accuracy of the quantised models. The results are presented for popular architectures such as VGG and Resnet for image classification and detection. Additionally, simulations were carried out for the autoencoder architecture.

Keywords: Deep Learning · Quantization · Neuroevolution · Fine Tuning

1 Introduction

Low bit quantisation is a process which enables the adaption of deep neural architectures in embedded systems and helps to decrease theirs computational complexity. There are many approaches to decreasing bit precision while minimising the drop in accuracy [1,2,5,7,11,12,15,16,21]. In [1,17] authors show how quantisation can decrease execution time by reducing the bit-width format of deep learning models. Many of these techniques are based on complex optimisation algorithms which are time consuming and very often difficult for parallelisation. Creating a fast and efficient quantisation method which can improve the model accuracy presents a significant challenge. The quantisation methods

K. Arai (Ed.): FTC 2023, LNNS 816, pp. 46–55, 2023.
https://doi.org/10.1007/978-3-031-47448-4_4

can be divided according to several factors. The first factor is the stage in which it is performed. It can be done during the training process or just after it. The second factor is the type of quantisation. It can be linear or nonlinear quantisation. In this work, the presented method is run on linearly quantised model. The input of the presented algorithm is the pretrained and quantised model. The main assumption of the solution is that the quantization of the input model is done by rounding to the nearest neighbour. This is the approach used in most quantization algorithms [1,5–7,15].

In presented approach, the novel fine tuning approach is performed. It mutates the weights by value which is multiplication of least significant bit. The small percentage of weights are mutated in each iteration. The neuroevolution process is run on each layer separately. In addition, the algorithm is equipped with a sensitivity analysis of the tested model layers. It iterates through the layers from the least to the most sensitive. This feature allows to stop the process if some user defined drop in accuracy is achieved. Another approach with a similar input assumption is the algorithm presented in the paper [13]. It also elaborates that an approximation to the nearest neighbor does not guarantee optimal performance of the model. The main goal of the work was to create a methodology that is scalable and able to quickly improve accuracy of quantised models.

2 Related Work

Quantisation is one of the most efficient techniques for the compression of deep learning models [1,7]. Common strategies include the quantisation of all coefficients in a single layer with a specified number of bits to represent the integer and fractional components [2,5] based on the range of values of the coefficients set. Another strategy is to represent coefficients and data by integer numbers with an appropriate scaling factor. Many quantisation approaches in the literature adopt linear [1,5,7] or non-linear approaches including clustering [12]. Quantisation can be performed during model training [7] or can be run on a pretrained model [1,12]. Recently, several methods for low-bit representation have been designed [8,9,11,15,20]. Many of these can not be run without some significant degradation in accuracy. One known advantage of quantisation is the fact that it facilitates the adoption of deep neural networks in specialised hardware accelerators with limited arithmetic bit-width and memory space [17,19].

The authors in [10] show that a two-level neuroevolution strategy scheme can outperform human-designed models in some specific tasks, for example, language modelling and image classification. In [3], a novel neuroevolutionary method for the optimisation of the architecture and hyperparameters of convolutional autoencoders is presented. In [14], it is shown that a genetic algorithm could evolve autoencoders that can reproduce the data better than manually created autoencoders with more hidden units. The first approach of co-evolutionary neuroevolution-based multivariate anomaly detection system is presented in [4]. The authors show that proposed neuroevolutionary-based solution can outperform other human designed models on a well known benchmarks. These works show that neuroevolution can be very efficient method in deep learning architecture exploration.

Fine-tuning is quite a popular technique for improving the accuracy of the pretrained models. The most popular technique is gradient-based fine-tuning [18, 22]. The non-gradient approach is quite rare but can yield significant improvements as shown in [13]. In [13] the authors give the same input assumption is shown in this work. The authors in [13] try to find the near optimal state of the model which allows to have quantisation weight values which are not rounded to the nearest neighbour.

In the presented work, a novel approach is proposed. The fine tuning based on neuroevolution is used for improving the accuracy of pretrained models.

3 Quantisation

After network distillation by the pruning process, quantisation can be performed as the next step of reducing the complexity of the model. Quantisation is the procedure of constraining values from a continuous set or a more dense domain to a relatively discrete set. It is possible to define general mapping from a floating-point data $x \in \mathcal{S}$ to a fixed-point $q \in \mathcal{Q}$ using a function $f_{\mathcal{Q}} : \mathcal{S} \rightarrow \mathcal{Q}$ as follows (assuming signed representation):

$$q = f_{\mathcal{Q}}(x) = \mu + \sigma \cdot \mathrm{round}(\sigma^{-1} \cdot (x - \mu)). \tag{1}$$

In our case, $\mu = 0$ and $\sigma = 2^{-\mathbf{frac\_bits}}$ where:

$$\mathbf{int\_bits} = \mathrm{ceil}(\log_2(\max_{x \in \mathcal{S}} |x|)) \tag{2}$$

and

$$\mathbf{frac\_bits} = \mathbf{total\_bits} - \mathbf{int\_bits} - 1. \tag{3}$$

The number of MAC operations in the layer is equal to $P_i{\cdot}C_i{\cdot}D_i{\cdot}H_i{\cdot}W_i$, where P_i is the number of neurons in an output feature map, C_i and D_i are numbers of input and output channels, and W_i and H_i are the width and the height of the filter. The memory footprint of the single layer is $C_i \cdot D_i \cdot H_i \cdot W_i$. The quantised model has reduced the number of MAC operations. The same situation applies in the case of the memory footprint. The quantisation reduces the complexity further. The bit width of data decreases the number of cycles needed to run multiplication operations. In the case of the 8-bit quantisation (both weights and activations in 8-bit format), the number of MAC operations of the baseline floating point configuration is reduced by 1/9. If weights are further reduced to 4-bit, the number of cycles in full 8-bit configuration are reduced by more than 1/2. For 8-bit weights and 16-bit half precision activations, the reduction ratio is 2/9.

4 Neuroevolution Fine Tuning

Neuroevolution is an optimisation algorithm based on a genetic approach which helps to find more optimal neural architectures. In this work, the neuroevolution strategy is adapted for fine tuning the quantised weights in order to improve the accuracy of the model. The optimisation process is run on a pretrained quantised model. The pretrained model is quantised using linear quantisation with nearest neighbour rounding. The whole process in divided into two phases. The first phase is sensitivity analysis and the second is fine tuning the quantised layers by through the use of the neuroevolution approach.

4.1 Sensitivity Analysis

The goal of the sensitivity analysis is to set up the ranking list of the most sensitive layers. The whole process is shown in Algorithm 1. At the beginning of the analysis, the empty list is created (line 1). The list will be used for storing the accuracy measured by quantising the specific layer. A single layer in each iteration is then quantised into a low bit format (loop in lines 2–8). First, the copy of the original model is created (line 3). Then, the layer l is quantised (line 4). The θ_l weight tensor is transformed to a low bit format θ_l^q. The quantised layer is inserted to the whole model weight tensor (line 5). Finally, the inference is run in order to check the accuracy. The accuracy is added to the list (line 7). When all layers are run in the loop, the ranking is returned (line 9).

Algorithm 1. Sensitivity analysis

Require: ψ – desired bit-width
Require: Θ – weights of a model
Require: F_Θ – model
1: $\Lambda \leftarrow \emptyset$ {List for top1 metric}
2: **for** θ_i **in** Θ **do**
3: $\quad F'_\Theta \leftarrow copy(F_\Theta)$
4: $\quad \theta_i^q \leftarrow \mathrm{q}(\theta_i, \psi{=}4)$ {quantize layer to four bit}
5: $\quad \Theta' \leftarrow \{\theta_0, \theta_1, ..., \theta_i^q, ..., \theta_N\}$
6: $\quad a_i \leftarrow \mathrm{eval}(F'_{\Theta'})$
7: $\quad \Lambda \leftarrow \Lambda \cup a_i$
8: **end for**
9: **return** $\mathrm{argsort}(\Lambda)$

The algorithm goes through all the layers and quantises them to a specified bit width - ϕ. In the presented approach, the bit-width is set to 4. In line 1, the algorithm initialises the list in which it stores the accuracy values.

4.2 Fine Tuning Pretrained Model

The deep learning network is defined as a sequence of layers :

$$F_\Theta(X) = f_{\theta_L}(f_{\theta_{L-1}}...(f_{\theta_0}(X))) \tag{4}$$

The trainable parameters (weights) are defined as a following list:

$$\Theta = \{\theta_0, \theta_1, ..., \theta_L\} \tag{5}$$

The quantized version of the weight tensor is as follows:

$$\Theta^q = \{\theta_0^q, \theta_1^q, ..., \theta_L^q\} \tag{6}$$

Each layer weight tensor can be quantized (using q function) with specified bit width - ϕ:

$$\theta_i^q = q(\theta_i, \phi) \tag{7}$$

The pretrained quantised model is represented as F_{Θ^q}. It is an input to the neuroevolution strategy which is described in Algorithm 2. The algorithm iterates through all the layers in the model (line 1–20). It begins from the less sensitive layers. It takes the weight tensor of the layer (line 2) and adds the copies of it to the initial population P (line 4). The initial population is just a list of copies of the quantised l layer:

$$P = \{\Theta_l^q, \Theta_l^q, ..., \Theta_l^q\} \tag{8}$$

Then, mask is generated using binomial distribution (line 9). The mask is a binary tensor:

$$M_i \in \{0, 1\}^{SH_{\theta_i}} \tag{9}$$

The shape SH_{θ_i} of the mask is the same as the shape of weight tensor θ_i. The mask has p percentage of 1 values. The weights position for which the mask is set to 1 will be change during the mutation step (line 12):

$$\theta_i = \theta_i + M_i \odot \vec{q} \tag{10}$$

The vector $\vec{q}$ is generated by following equation (line 11):

$$\vec{q} = \vec{r} \cdot \sigma_l \tag{11}$$

where $sigma_l$ is least significant bit value for the specific l layer and $\vec{r}$ is defined as:

$$\vec{r} = \{r_0, r_1, ..., r_N\} \tag{12}$$

Each r_i value is a randomly generated number from the predefined set (line 10):

$$r_i \in \{-2, -1, 1, 2\} \tag{13}$$

The -1 and 1 values are generated with probability equal to 40% each. The -2 and 2 are generated with 10% each. Finally the mutated layer is added to the model weight tensor (line 15) and is evaluated. In line 18, the accuracy values in the population are sorted. The best candidates are taken to the next iteration (line 19).

Algorithm 2. Fine tuning

Require: r – ranking list
Require: P_S – population size
Require: Θ – weights of a model
for l **in** r **do**
    $P \leftarrow \emptyset$
    for i **in** P_S **do**
        $P \leftarrow P \cup copy(\theta_l^q)$
    end for
    $A \leftarrow \emptyset$
    for i **in** I **do**
        for j **in** P_S **do**
            $M_l \leftarrow \{B(n,p)\}, B(n,p)\}, ..., B(n,p)\}$ {Bernoulli distribution}
            $\vec{r} \leftarrow$ sample vector from $\{-2, -1, 1, 2\}$
            $q = \vec{r} \cdot \sigma_l$
            $P[j] = P[j] + M_l \odot q$
            $P \leftarrow P \cup P[j]$ {add mutated layer to the population}
            $\theta_l = P[j]$
            $\Theta = \{\theta_0, ..., \theta_l, ..., \theta_N\}$
            $A \leftarrow A \cup eval(F_\Theta)$
        end for
        $sorted \leftarrow argsort(A)$
        $P \leftarrow P[sorted[: P_S]]$ {take P_S best candidates}
    end for
end for
return P

5 Results

The presented results show that the proposed solution can significantly improve the results achieved by quantisation based on rounding to the nearest neighbour. The baseline floating point results are shown in Table 1. There are four models and four datasets. The FasterRCNN is a object detection model. It is tested on a VOC Pascal dataset. The baseline mAP (mean average precision) is 71.2%. The remaining three models: Resnet18, VGG16 are the models for image classification and CNN autoencoder - 12 layer autoencoder from [4] for time series anomaly detection. The Resnet18 and VGG16 are simulated on CIFAR100 and

Table 1. Baseline results.

	CIFAR100	VOC Pascal	ImageNet	SWAT
FasterRCNN	-	71.18	-	
Resnet18	75.0	-	69.75	
VGG16	70.4	-	71.59	-
CNN AE	-	-	-	82.0

ImageNet. The CNN autoencoder is tested on SWAT benchmark. For CIFAR100 and ImageNet the metric is top1. The metric for SWAT is F1.

Table 2. 8 bit linear quantisation results.

	CIFAR100	VOC Pascal	ImageNet	SWAT
FasterRCNN	-	71.04	-	-
Resnet18	74.9	-	69.5	-
VGG16	70.4	-	71.25	-
CNN AE	-	-	-	81.98

Table 3. Linear quantisation results: 4 bit Resnet18 and VGG16 for CIFAR100, 6 bit Resnet18 and VGG16 for ImageNet, 5 bit CNN AE, 6 bit FasterRCNN.

	CIFAR100	VOC Pascal	ImageNet	SWAT
FasterRCNN	-	64.0	-	-
Resnet18	73.5	-	62.04	-
VGG16	68.9	-	54.73	-
CNN AE	-	-	-	80.70

In Table 2 8-bit quantisation results for all models are presented, in Table 3 4-bit Resnet18 and VGG16 on CIFAR100, 6-bit Resnet18 and 5-bit VGG16 on ImageNet, 6-bit FasterRCNN and 5-bit CNN AE results are described. In Table 2 and 3 linear quantisation with rounding to nearest neighbour was used (Sect. 3). It can be observed that in the case of 8 bit quantisation, a less than 1% drop in accuracy was achieved in all models and datasets. In the case of lower bit quantisation, the drop is significantly higher for 6-bit Resnet18 (7.71%) and 5-bit VGG16 (16.86%) on ImageNet. The same can be observed for 6-bit Faster-RCNN where the drop is 7.18%. The small drop is for 5-bit CNN AE (1.3%), and for Resnet18 and VGG16 on CIFAR100 (less than 1.5%). In Table 4 and 5 the results are presented after applying fine tuning neuroevolution approach. In simulations for Resnet18 and VGG16, the population size was set to 32, the number of iterations was 64, and 2% of weights were mutated in each candidate. For CNN AE the population was 24, iterations number was set to 30. In the case of FasterRCNN the simulations were run with 16 iterations and 16 solutions in population. The FasterRCNN achieves a drop in accuracy of less than 1% in the cases of both 6 and 8-bit quantisation. The fine tuned 8 bit Resnet18 and VGG16 have better results on CIFAR100 than floating point version (+0.2%). The 4 bit fine tuned versions are just 0.2% worse than baseline floating point counterparts. For CNN autoencoder the accuracy is very close the floating point version (82.09% for 8 bit, which is better by 0.09%, and 81.93% for 5 bit version, which is better by more than 1% from the baseline 5 bit quantisation). The highest drops in accuracy are observed for 6-bit and 5-bit quantisation for ImageNet, but the proposed fine tuning significantly improves the baseline quantisation results (by more than 6% in case of Resnet18 and more than 14.5% in case of VGG16) and achieves accuracy about 1.5% and 2% worse than floating point baseline versions for Resnet18 and VGG16, respectively. For 8-bit quantisation the results are also improved and very close to the floating point results (less than 0.1%).

The obtained results show that the proposed method allows to achieve results very close to the original non-quantized model. Only for the ImageNet dataset it gives worse results than the method presented in [13], which is able in some cases to regain the accuracy in 4 bit format. The main advantage of proposed method is definitely better scalability and better time efficiency than [13].

Table 4. 8 bit fine tuned results.

	CIFAR100	VOC Pascal	ImageNet	SWAT
FasterRCNN	-	71.34	-	-
Resnet18	75.2	-	69.7	-
VGG16	70.6	-	71.5	-
CNN AE	-	-	-	82.09

Table 5. Fine tuned linear quantisation results: 4 bit Resnet18 and VGG16 for CIFAR100, 6 bit Resnet18 and VGG16 for ImageNet, 5 bit CNN AE, 6 bit FasterRCNN.

	CIFAR100	VOC Pascal	ImageNet	SWAT
FasterRCNN	-	70.5	-	-
Resnet18	74.8	-	68.2	-
VGG16	70.2	-	69.6	-
CNN AE	-	-	-	81.93

6 Conclusions and Future Work

The results described in this work show that the proposed solution is quite efficient. The simulations run on different datasets with various models showing that evolutionary fine tuning can boost accuracy and minimize the drop after nearest neighbour quantisation. This is a good alternative for gradient fine tuning. The next advantage is its scalability. It can be fully parallelised. In the next step the algorithm will be incorporated with the new features like finding different bit formats for each channel in a single layer. Future works will concentrate on improving the method by combining gradient with non-gradient fine tuning and adapting similar methodology for nonlinear quantisation. The more architectures like MobileNet and Transformer-based will be explored.

References

1. Al-Hami, M., Pietron, M., Casas, R., Wielgosz, M.: Methodologies of compressing a stable performance convolutional neural networks in image classification. Neural Process. Lett. **51**, 105–127 (2020). https://doi.org/10.1007/s11063-019-10076-y
2. Anwar, S., Hwang, K., Sung, W.: Fixed point optimization of deep convolutional neural networks for object recognition. In: IEEE International Conference on Acoustics, Speech and Signal Processing (ICASSP), pp. 1131–1135 (2015)
3. Dimanov, D., Balaguer-Ballester, E., Singleton, C., Rostami, S.: MONCAE: multi-objective neuroevolution of convolutional autoencoders. In: Neural Architecture Search Workshop, ICLR (2021)
4. Faber, K., Pietron, M., Zurek, D.: Ensemble neuroevolution-based approach for multivariate time series anomaly detection. Entropy **23**(11), 1466 (2021)
5. Gysel, P., Motamedi, M., Ghiasi, S.: Hardware-oriented approximation of convolutional neural networks. arXiv preprint arXiv:1604.03168 (2016)
6. Gysel, P.: Ristretto: hardware-oriented approximation of convolutional neural networks (2016)
7. Han, S., Pool, J., Tran, J., Dally, W.: Learning both weights and connections for efficient neural network (2015)
8. Jung, S., et al.: Joint training of low-precision neural network with quantization interval parameters. arXiv arXiv:1808.05779 (2018)
9. McDonnell, M.D.: Training wide residual networks for deployment using a single bit for each weight. In: ICLR (2018)
10. Miikkulainen, R., et al.: Evolving deep neural networks. CoRR abs/1703.00548, March 2017
11. Mishra, A., Nurvitadhi, E.: WRPN: wide reduced-precision networks. In: ICLR (2018)
12. Pietron, M., Karwatowski, M., Wielgosz, M., Duda, J.: Fast Compression and Optimization of Deep Learning Models for Natural Language Processing (2019)
13. Nagel, M., Amjad, R.A., van Baalen, M., Louizos, C., Blankevoort, T.: Up or down? Adaptive rounding for post-training quantization. In: Proceedings of ICML (2020)
14. Okada, H.: Neuroevolution of autoencoders by genetic algorithm. Int. J. Sci. Eng. Invest. **6**, 127–131 (2017)
15. Park, E., Ahn, J., Yoo, S.: Weighted-entropy-based quantization for deep neural networks. In: IEEE Conference on Computer Vision and Pattern Recognition (CVPR), July 2017
16. Park, J., et al.: Faster CNNs with Direct Sparse Convolutions and Guided Pruning (2016)

17. Pietron, M., Zurek, D., Sniezynski, B.: Speedup deep learning models on GPU by taking advantage of efficient unstructured pruning and bit-width reduction. J. Comput. Sci. **67**, 101971 (2023)
18. Ro, Y., Choi, J., Heo, B., Choi, J.Y.: Rollback ensemble with multiple local minima in fine-tuning deep learning networks. IEEE Trans. Neural Netw. Learn. Syst. **33**(9), 4648–4660 (2022)
19. Xu, K., An, J., Zhang, D., Liu, L., Liu, L., Wang, L.: GenExp: multi-objective pruning for deep neural network based on genetic algorithm. Neurocomputing **451**, 81–94 (2021)
20. Zhang, D., Yang, J., Ye, J., Hua, G.: Learned quantization for highly accurate and compact deep neural networks. arXiv arXiv:1807.10029 (2018)
21. Zhang, Y., Dong, Z.Y., Kong, W., Meng, K.: A composite anomaly detection system for data-driven power plant condition monitoring. IEEE Trans. Industr. Inf. **16**, 4390–4402 (2019)
22. Zhou, Y., Liao, L., Gao, Y., Wang, R., Huang, H.: TopicBERT: a topic-enhanced neural language model fine-tuned for sentiment classification. IEEE Tran. Neural Netw. Learn. Syst. **34**, 380–393 (2021)

Performance Evaluation of Swin Vision Transformer Model Using Gradient Accumulation Optimization Technique

Sanad Aburass[1(✉)] and Osama Dorgham[2,3]

[1] Department of Computer Science, Maharishi International University, Fairfield, IA, USA
Saburass@miu.edu

[2] Prince Abdullah bin Ghazi Faculty of Information and Communication Technology, Al-Balqa Applied University, 19117, Al-Salt, Jordan
o.dorgham@bau.edu.jo

[3] School of Information Technology, Skyline University College, University City of Sharjah, P.O. Box 1797, Sharjah, United Arab Emirates
o.dorgham@skylineuniversity.ac.ae

Abstract. Vision Transformers (ViTs) have emerged as a promising approach for visual recognition tasks, revolutionizing the field by leveraging the power of transformer-based architectures. Among the various ViT models, Swin Transformers have gained considerable attention due to their hierarchical design and ability to capture both local and global visual features effectively. This paper evaluates the performance of Swin ViT model using gradient accumulation optimization (GAO) technique. We investigate the impact of gradient accumulation optimization technique on the model's accuracy and training time. Our experiments show that applying the GAO technique leads to a significant decrease in the accuracy of the Swin ViT model, compared to the standard Swin Transformer model. Moreover, we detect a significant increase in the training time of the Swin ViT model when GAO model is applied. These findings suggest that applying the GAO technique may not be suitable for the Swin ViT model, and concern should be undertaken when using GAO technique for other transformer-based models.

Keywords: Image Classification · Optimization · Swin ViT · Transformers · Vision Transformers

1 Introduction

Image classification is a fundamental task in computer vision, which involves assigning a label to an image based on its content. This task has many practical applications, such as object recognition, facial recognition, and medical imaging [1, 2]. In recent years, deep learning methods, especially CNNs, have achieved remarkable success in image classification [3–5]. CNNs are neural networks specifically designed for processing and analyzing images, and they can capture complex patterns and features from images. CNNs consist of multiple layers, including convolutional, pooling, and fully

K. Arai (Ed.): FTC 2023, LNNS 816, pp. 56–64, 2023.
https://doi.org/10.1007/978-3-031-47448-4_5

connected layers. The convolutional layer is the core component of the CNN, which extracts features from the input image by applying a set of filters to the image. The pooling layer is used to down-sample the feature maps, reducing the spatial size of the output. The fully connected layer is used to classify the image based on the extracted features [6, 7]. Despite the success of CNNs in image classification, they have some limitations. CNNs are imperfect for modeling long-range dependencies in images [8, 9], which are crucial for understanding the context and relationships between different objects in an image. Transformers, on the other hand, are attention-based models that excel at capturing long-range dependencies in sequences, such as natural language processing [10]. Transformers have also shown promising results in image classification, especially for large datasets such as ImageNet [11]. The recent ViT model applies the Transformer architecture to image classification. ViT replaces the CNN's convolutional layers with a set of self-attention layers, which allow the model to attend to all the image pixels simultaneously, capturing the global context of the image. The Swin ViT is a recent improvement to the ViT model that addresses the limitation of long-range dependencies by using a hierarchical architecture. Swin divides the image into non-overlapping patches, which are processed by a series of self-attention layers. The resulting features are then aggregated using a Swin block, which captures both local and global dependencies [12–14]. Gradient Accumulation Optimization (GAO) is a technique that can be used to improve training efficiency in deep learning models. It involves accumulating the gradients over multiple mini-batches before updating the weights. This technique helps to reduce memory usage and allows for larger batch sizes, leading to faster convergence. However, its effectiveness depends on several factors like the number of the mini-batches and the learning rate, and it may not always lead to better results [15]. We have implemented the gradient accumulation optimization on the Swin ViT model and conducted experiments to measure its performance on image classification tasks using the CIFAR10 [16] and MNIST [17] datasets. Experiments involve training the Swin ViT model with and without gradient accumulation and comparing their accuracy and training time performance.

Our contributions summarized as follows: we present the possibility of applying GAO technique for classification model as in Swin ViT. We evaluate the performance (i.e. accuracy and time) of Swin Vision Transformer (ViT) model using gradient accumulation optimization (GAO) technique. To the best of our knowledge, this paper is the first to provide realistic performance evaluation of swin ViT model using such an optimization technique. The content of the paper can be summarized as follows. Section 2 presents the methodology and describes the implementation of our work. Section 3 presents the results and model evaluation. Finally, Sect. 4 presents the conclusions of this paper.

2 Methodology

2.1 Data Acquisition References

The CIFAR10 and MNIST datasets are two popular datasets used in the field of machine learning and computer vision for classification tasks. The CIFAR10 dataset consists of 60,000 32×32 color images, with 10 different classes, each containing 6,000 samples. These classes include objects such as airplanes, automobiles, birds, cats, dogs, and more.

On the other hand, the MNIST dataset consists of 70,000 28 × 28 grayscale images of handwritten digits, with 10 different classes, each containing 7,000 samples. The classes in this dataset represent digits ranging from 0 to 9. Both datasets have been widely used in research for classification tasks, with many models achieving high levels of accuracy on these datasets.

2.2 Vision Transformers

ViT and Swin Transformer are two popular models for image classification tasks. Both models consist of two main components: the Transformer encoder and the MLP head [18]. ViT Model: The ViT model takes an image as input and transforms it into a sequence of fixed-length vectors. The Transformer encoder is composed of L layers and consists of four main steps. Firstly, the image is split into a sequence of non-overlapping patches:

$$x_i = w_{patch} \times patch_i \tag{1}$$

where x_i is the ith patch, $patch_i$ is the representation of the ith patch, w_{patch} is a learnable weight matrix.

Secondly, learnable position embeddings are added to each patch to encode the spatial information of the image:

$$x_i = x_i \times pos_i \tag{2}$$

where pos_i is the learnable position embedding for patch i.

Thirdly, multi-head self-attention mechanism and feedforward neural networks are applied to the input embeddings:

$$x_{i'} = Multi\ Head\ Att(x_i) \times x_{i''} = FFN(x_{i'}) \times x_{i'''} = Layer\ Norm(x_{i'} + x_{i''}) \tag{3}$$

where *MultiHeadAtt* is the multi-head self-attention mechanism, *FFN* is the feedforward neural network, and *LayerNorm* is the layer normalization function.

Lastly, the output embeddings are aggregated by taking the mean or max pooling over the sequence dimension:

$$Z = Pooling(x_{1'''}, x_{2'''}, x_{3'''}, \ldots\ldots x_{N'''}) \tag{4}$$

where N is the number of patches and *Pooling* is the mean or max pooling operation. The MLP head takes the output of the Transformer encoder as input and performs linear projection, activation, dropout, and linear projection to obtain the final classification result. The Swin Transformer model takes an image as input and transforms it into a sequence of fixed-length vectors. The Swin Transformer encoder is composed of K groups, and each group contains a set of non-overlapping patches. The patches in each group are processed by multi-layer Shifted Windows to generate a set of Swin Transformer blocks. Each Swin Transformer block consists of a Shifted Window Attention (SWA) layer, a local window-based feedforward network (LWFFN) layer, and a residual connection [19]. The output of each block is passed as input to the next block within the same group, and the output of the last block in each group is passed as input to the first block in the next group. The output embeddings are aggregated by taking the mean or max pooling over the sequence dimension:

$$Z = Pooling(x_{1L}, x_{2L}, x_{3L}, \ldots\ldots .x_{NL}) \tag{5}$$

where L is the number of Swin Transformer blocks in each group, and x_{il} is the output of the lth block in group K. Then, the Swin Transformer head performs linear projection, activation, dropout, and linear projection to obtain the final classification result.

In summary, both ViT and Swin ViT models use a Transformer encoder to transform images into fixed-length vectors and an MLP head for classification. The ViT model uses a multi-head self-attention mechanism and feedforward neural networks, while the Swin ViT model uses multi-layer Shifted Windows to generate a set of Swin Transformer blocks. Both models can be trained using backpropagation with stochastic gradient descent (SGD) or other optimization methods. Figures 1 and 2 show the architectures of ViT and Swin-ViT respectively.

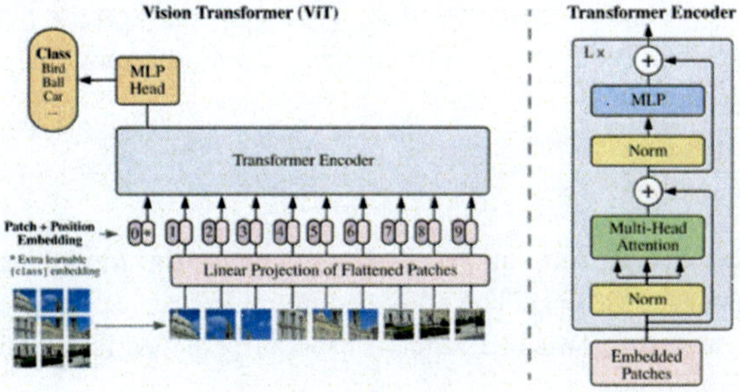

Fig. 1. The architecture of Vision Transformers [18].

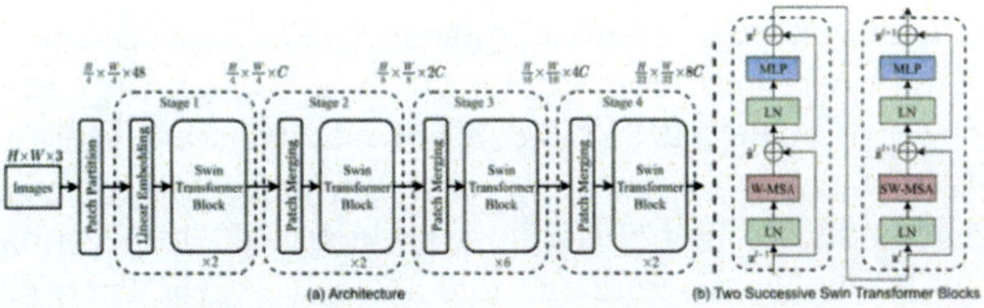

Fig. 2. The architecture of Swin Transformers [19].

2.3 Gradient Accumulation Optimization

Gradient accumulation optimization, also known as gradient accumulation over multiple small batches, is a technique used in deep learning to overcome the limitations of GPU memory while training deep neural networks. This technique allows the model to accumulate gradients over multiple small batches before updating the model's parameters. In this way, memory usage during training is reduced, while the model's accuracy is improved [15].

The basic idea behind gradient accumulation optimization is to perform multiple forward and backward passes on small batches of data before updating the model's parameters. Suppose we have a batch size of B, and we want to accumulate gradients over N batches. In that case, we split the original batch into N smaller batches of size B/N and perform forward and backward passes on each of these smaller batches. The gradients obtained from each backward pass are then accumulated over the N batches before updating the model's parameters.

Mathematically, the gradient accumulation optimization can be expressed as follows:

1. For each training step t, split the batch into N smaller batches of size B/N, and perform forward and backward passes on each of these smaller batches.
2. Accumulate the gradients obtained from each of the N backward passes:

$$\Delta\theta^{(t)} = \sum\nolimits_{i=1}^{N} \Delta\theta_i^{(t)} \tag{6}$$

where $\Delta\theta^{(t)}$ is the gradient obtained from the backward pass on the ith smallest batch. After accumulating the gradients over N batches, update the model's parameters using the accumulated gradient:

$$\theta^{(t+1)} = \theta^{(t)} - \eta\Delta\theta^{(t)} \tag{7}$$

where η is the learning rate, and $\theta^{(t)}$ and $\theta^{(t+1)}$ are the model parameters before and after the update, respectively. The above equations illustrate the process of gradient accumulation optimization. This technique is especially useful for training large models with limited GPU memory.

2.4 Experiment Setup

In our experimental setup, we utilized the resources provided by Google Colab, which offers a cloud-based environment for machine learning development. We leveraged the computational power of a GPU and 25 GB of RAM to train and evaluate our models efficiently. To build our models, we utilized Python, a popular programming language in the machine learning community, and TensorFlow, a widely used deep learning framework that provides high-level APIs for building and training deep neural networks. We chose TensorFlow because of its ease of use, its extensive documentation, and its ability to run on both CPUs and GPUs. Our experimental setup allowed us to run our experiments smoothly and efficiently, enabling us to focus on model development and analysis.

3 Results and Discussion

In this study, we applied gradient accumulation optimization on Swin ViT and compared its performance with the standard Swin ViT on the CIFAR10 and MNIST datasets. The results showed that using the optimization led to a decrease in accuracy and a significant increase in training time, as shown in Figs. 3 and 4, unlike the uplifting performance for applying GAO on [15]. We believe that the reason behind these results is overfitting, as the training accuracies were much higher than the testing accuracies, as shown in Figs. 5 and 6. Overfitting occurs when a model learns to fit the training data too closely, leading to poor generalization performance on new, unseen data. It can be caused by a variety of factors, such as model complexity, insufficient data, or inappropriate optimization strategies. In our case, we are dubious that the gradient accumulation optimization led to overfitting because it allowed the model to learn from the same data multiple times before updating the weights, which may have caused the model to become too specialized to the training set. One possible solution to this problem is to use regularization techniques to prevent overfitting. Regularization refers to a set of techniques that aim to reduce the model's variance by adding constraints or penalties to the optimization objective. For example, we could use L2 regularization to penalize large weights, or dropout to randomly remove units during training to prevent co-adaptation. Another approach is to use early stopping, where we stop training when the validation performance starts to deteriorate, to avoid overfitting.

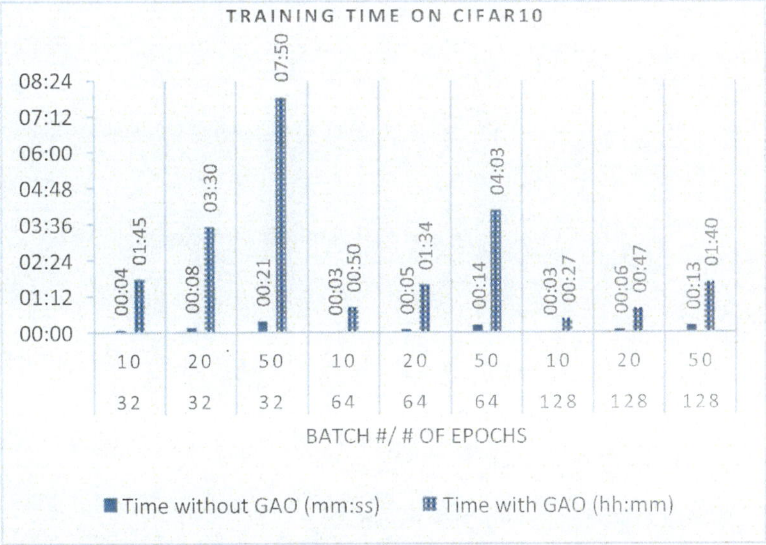

Fig. 3. Training Time of Swin ViT before and after applying GAO on CIFAR10.

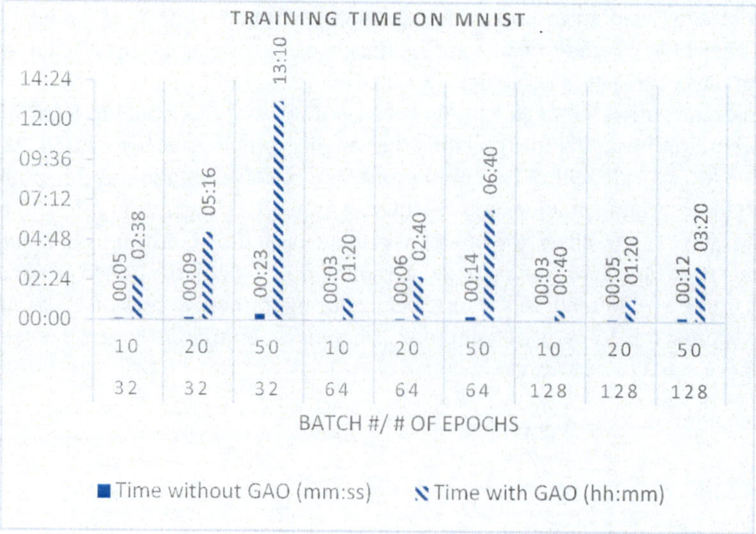

Fig. 4. Training Time of Swin ViT before and after applying GAO on MNIST.

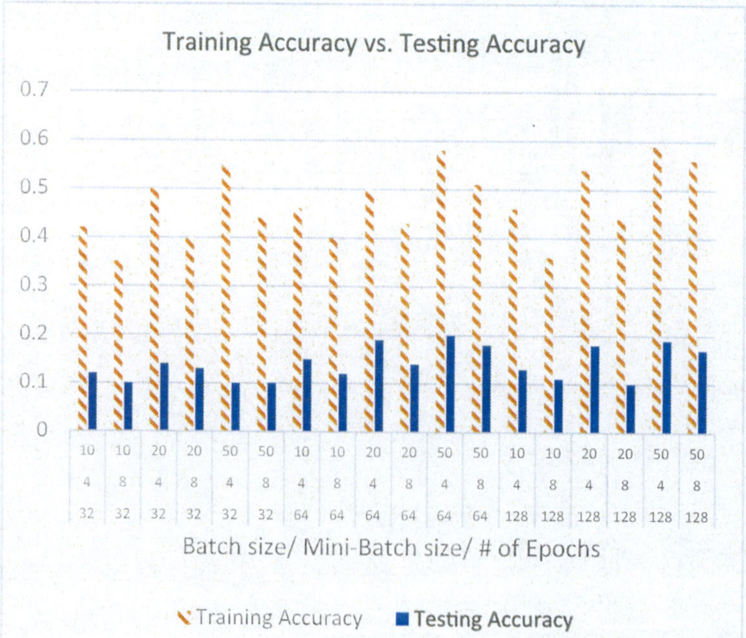

Fig. 5. Training accuracy and Testing accuracy of Swin ViT after applying GAO on CIFAR10.

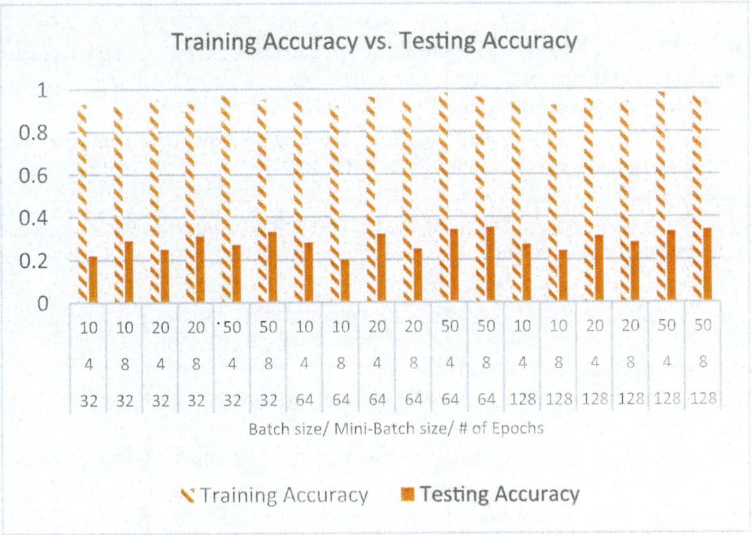

Fig. 6. Training accuracy and testing accuracy of Swin ViT after applying GAO on MNIST.

4 Conclusion

Our study evaluated the effectiveness of gradient accumulation optimization on the Swin ViT model. Our results indicate that the application of this optimization technique resulted in a considerable reduction in accuracy and significantly increased the training time compared to the standard Swin Transformers. Thus, caution should be exercised when using gradient accumulation optimization for the Swin ViT model, and other transformer-based models. Overall, our findings provide insights into the performance of gradient accumulation optimization and its potential impact on transformer-based model. Also, our study suggests that gradient accumulation optimization may not be an effective strategy for improving the performance of Swin ViT on the CIFAR10 and MNIST datasets. The observed decrease in accuracy and increase in training time may be due to overfitting caused by the optimization. Future research could explore alternative optimization strategies or regularization techniques to improve the performance of Swin ViT on these datasets.

References

1. Ryalat, M.H., Dorgham, O., Tedmori, S., Al-Rahamneh, Z., Al-Najdawi, N., Mirjalili, S.: Harris hawks optimization for COVID-19 diagnosis based on multi-threshold image segmentation. Neural Comput. Appl. **35**(9), 6855–6873 (2023). https://doi.org/10.1007/s00521-022-08078-4
2. Fisher, M., Dorgham, O., Laycock, S.D.: Fast reconstructed radiographs from octree-compressed volumetric data. Int. J. Comput. Assist. Radiol. Surg. **8**(2), 313–322 (2013). https://doi.org/10.1007/s11548-012-0783-5

3. Khan, A., Sohail, A., Zahoora, U., Qureshi, A.S.: A survey of the recent architectures of deep convolutional neural networks. Artif. Intell. Rev. **53**(8), 5455–5516 (2020). https://doi.org/10.1007/s10462-020-09825-6
4. AbuRass, S., Huneiti, A., Al-Zoubi, M.B.: Enhancing convolutional neural network using Hu's moments. Int. J. Adv. Comput. Sci. Appl. **11**(12), 130–137 (2020). https://doi.org/10.14569/IJACSA.2020.0111216
5. Aburass, S., Huneiti, A., Al-Zoubi, M.B.: Classification of transformed and geometrically distorted images using convolutional neural network. J. Comput. Sci. **18**(8), 757–769 (2022). https://doi.org/10.3844/jcssp.2022.757.769
6. Yao, X., Wang, X., Wang, S.H., Zhang, Y.D.: A comprehensive survey on convolutional neural network in medical image analysis. Multim. Tools Appl. **81**(29), 41361–41405 (2022). https://doi.org/10.1007/s11042-020-09634-7
7. Li, Z., Yang, W., Peng, S., Liu, F.: A Survey of Convolutional Neural Networks: Analysis, Applications, Prospects
8. Azad, R., Al-Antary, M.T., Heidari, M., Merhof, D.: TransNorm: transformer provides a strong spatial normalization mechanism for a deep segmentation model. IEEE Access **10**, 108205–108215 (2022). https://doi.org/10.1109/ACCESS.2022.3211501
9. Dorgham, O., Naser, M.A., Ryalat, M.H., Hyari, A., Al-Najdawi, N., Mirjalili, S.: U-NetCTS: U-Net deep neural network for fully automatic segmentation of 3D CT DICOM volume. Smart Health **26**, 100304 (2022). https://doi.org/10.1016/j.smhl.2022.100304
10. Nandi, A., De, A., Mallick, A., Middya, A.I., Roy, S.: Attention based long-term air temperature forecasting network: ALTF Net. Knowl. Based Syst. **252**, 109442 (2022). https://doi.org/10.1016/j.knosys.2022.109442
11. Zhang, Q., Xu, Y., Zhang, J., Tao, D.: ViTAEv2: vision transformer advanced by exploring inductive bias for image recognition and beyond. Int. J. Comput. Vis. **131**(5), 1141–1162 (2023). https://doi.org/10.1007/s11263-022-01739-w
12. Lin, T., Wang, Y., Liu, X., Qiu, X.: A survey of transformers. AI Open **3**, 111–132 (2022). https://doi.org/10.1016/j.aiopen.2022.10.001
13. Khan, S., Naseer, M., Hayat, M., Zamir, S.W., Khan, F.S., Shah, M.: Transformers in vision: a survey (2021). https://doi.org/10.1145/3505244
14. Liu, Y., et al.: A Survey of Visual Transformers. arXiv preprint http://arxiv.org/abs/2111.06091 (2021)
15. Hermans, J.R., Spanakis, G., Möckel, R., Noh, Y.-K., Zhang, M.-L.: Accumulated Gradient Normalization (2017)
16. N.V., A., Krizhevsky, H.G.: The CIFAR10 Dataset. https://www.cs.toronto.edu/~kriz/cifar.html. Accessed 11 Mar 2023
17. Deng, L.: The MNIST database of handwritten digit images for machine learning research [best of the web]. IEEE Signal Process. Mag. **29**(6), 141–142 (2012). https://doi.org/10.1109/MSP.2012.2211477
18. Dosovitskiy, A., et al.: An Image is Worth 16x16 Words: Transformers for Image Recognition at Scale (2020). http://arxiv.org/abs/2010.11929
19. Liu, Z., et al.: Swin Transformer: Hierarchical Vision Transformer using Shifted Windows (2021). http://arxiv.org/abs/2103.14030

A Deep Learning Approach to Predict Batting Strikes Played by a Batsman for Different Bowling Deliveries

Akna L. Delgodaarachchi(✉), K. D. Sandaruwan, Shiromi M. K. D. Arunathilake, and N. D. Kodikara

University of Colombo, Colombo, Sri Lanka
{ald,dsr,sda,ndk}@ucsc.cmb.ac.lk

Abstract. In Cricket, players and coaches employ various techniques to win matches. They anticipate the opposing team's batsmans' performance in various scenarios and prepare their bowlers appropriately. Similarly, they anticipate the strengths and weaknesses of their teams' batters against the opposing team's bowlers and advise them to try various techniques in certain situations. This study proposes a novel CNN model combined with a GRU network to predict batter's strikes for different bowling deliveries in Cricket, using videos of previously played matches. The proposed model was tested using a new dataset of 428 bowling delivery videos bowled to an English batsman, Joeseph Root, during test Cricket matches. On the prepared dataset, the proposed model attained an overall accuracy of 91.67, classifying four different batting strikes: the Cut shot, the Drive shot, the Flick shot, and the Sweep shot.

Keywords: Cricket · Batting · Bowling Deliveries · Deep Learning · LSTM · GRU · CNN · Predictive Classification

1 Introduction

The game of Cricket is one of the most popular watched games worldwide. According to recent research [14], the Indian subcontinent is home to 90% of cricket enthusiasts. Being one of the most crowd-pleasing sports, Cricket involves every Asian country in some way or another. Many tournament series are played across various countries throughout the year, and the World Cup is held every four years throughout all cricket-playing nations. Cricket is played in various formats [1], including One-Day Internationals (ODIs), Twenty20 (T20) matches, and Test matches. A cricket team winning a match is important for its country and its people. As a sport, it is one of the few things that draw people from various ethnicities together. Moreover, its tremendous popularity and fan base bring a business element too.

Before going into a match, every cricket team and its coaches usually do bowling and batting analysis and predict opponent players' bowling deliveries

K. Arai (Ed.): FTC 2023, LNNS 816, pp. 65–78, 2023.
https://doi.org/10.1007/978-3-031-47448-4_6

and batting styles. Then a game plan is drafted, including batter vs bowler match-ups and batting partnerships. The most generic approach used by the Cricket teams is benchmarking videos of previously played games. When making predictions by benchmarking videos, constantly monitoring and analyzing data from those video frames is complex, and it consumes time and needs experience. The number of videos an actual human can analyze is limited. Predictions done by coaches and experienced players will depend on the number of videos they have watched and analyzed to date. Besides, they can not be tested until the team goes to play.

There can be instances where a particular batter has never played with some of the bowlers from the opponent team, especially in the cases of new players. In such instances, only the experienced players and the coaches can predict the batter's performance for deliveries from the new bowlers that the batter has never played with and vice versa. However, the same issues discussed with human predictions will be there.

This research proposes a model to predict batting strikes of a specific batter for different bowling deliveries using a deep learning approach and video data of previously played matches to address the issues with human predictions. Cricket is a sport that is frequently determined by the smallest of margins: a single run can be enough to win a match. Our individualized predictions of a batter's strikes for different bowling styles from different bowlers can provide crucial information to coaches and captains regarding pre-match and in-game tactical decisions.

Once the model is trained for a specific player, it will also be capable of predicting batting strikes for deliveries from bowlers whom the batter has never played previously. It is a massive advantage in the instances of new players.

This article is organized as follows: In Sect. 2, a detailed summary of similar research work are included. Section 3 explains the methodology followed when conducting the research and Sect. 4 describes the experiments conducted to study the validity of the proposed model. Section 5 contains the results obtained through the conducted experiments followed by the Sect. 6 where we discuss the conclusion of this research.[1].

2 Related Work

When the overall sports industry is considered, researchers have conducted several pieces of research on predictions and classification concerning different elements played in different games, including previous match outcomes, player performance indicators, and opponent information as visual or non-visual data. Following is a detailed review of such classification and prediction studies in Cricket and other sports.

The [7] is a study done by developing a framework to forecast future events in sports of Water polo and Basketball using direct visual inputs. The study compares the performance of two systems. A hand-engineered system that extracts

[1] 979-8-3503-1321-5/23/$31.00 ©2023 IEEE.

features from raw visual inputs and an end to end neural network system is the two systems. They have discovered that the hand-engineered system performs similarly to humans in water polo and surpasses humans in Basketball. In [19] researchers have predicted NBA game results using previously played games. Several machine learning methods, such as Naive Bayes, artificial neural networks, and Decision Trees, were chosen to derive models using different learning strategies, and the accuracies were compared. The research [18] predicts the future direction of a Tennis ball according to the pose and position of the player, using YouTube Tennis match videos. The proposed model achieves an accuracy of 64.2% using particle filters for tracking players and detecting court lines.

Cricket team prediction [12] and Predicting the outcome of ODI cricket matches [10] were two pieces of research where the overall team's success was predicted. In [12], it was found that Random Forest is more reliable than using Decision Tree classifiers. They predict the team's success based on players' past records. They were concerned with predicting the best playing 11 instead of only predicting the performance of a single player. In addition to player's career statistics, the study [10] estimates the outcome of a cricket match by analyzing the strengths of the two sides, which is derived from basic match details and the outcomes of the toss. Only the top nine ODI-playing teams have been included in this research. Both the researches use data from the 'ESPNcricinfo' website [3].

The likelihood of a specific batter hitting a specific bowler and bowl type in a particular game situation is forecasted in this study [8]. Even though it is similar to this research work, the input data used was not visual data, and they predicted the shot location, whether it was a boundary shot. 'CrickShotClassify' [16] was a convolutional neural network(CNN) based approach to classifying different cricket batting shots from offline videos. Dataset used consists of videos of whole deliveries, including the batting strikes. They achieved an overall accuracy of 93% using the proposed model.

Cricket shot detection from videos [15] also recognizes and classifies various types of bat shots played in Cricket. For three classes of right-handed shots, this model achieves an accuracy of 83.098%, while achieving 65.186% accuracy for three left-handed shots. Research [11] proposed a technique to classify eight different batting shots based on a convolution neural network with both 2D and 3D convolutions. It has recognized shots with the highest accuracy of 90%. However, the classification accuracy for the Hook shot was 85%. Similar to [15] training of this study was also solely based on batting frames of videos of different teams and matches.

Research conducted using video footage sourced from the University of West Indies [13] is a study that introduces a novel method to recognize cricket strokes from pre-recorded videos utilizing bat detection and tracking. In this work, 'Otu's Thresholding method is used to separate the foreground, the bat and the batter from the background. Kalman Filter was used in isolating the bat.

A study [6] was conducted using different computer vision-based algorithms by dividing the bowling crease in Cricket into different regions to detect whether a particular delivery is a No ball. Their proposed system takes video data from four cameras placed at both ends of the pitch and analyses it to detect the overstep No ball. A different study [17] uses wearable sensors to identify cricket shot s and to generate live commentary. In particular, sensors like accelerometer and gyroscope were used to determine the shots. When collecting data for shot classification, they have used a microphone embedded into a smartphone and a sound sensor in order to know whether the batter hit the ball.

In the light of the literature mentioned above, it is evident that previous research has mainly concentrated on classification tasks rather than predictions and has ignored chiefly the bowling trajectories when forecasting a batters' batting strikes. As a result, this study uses videos of bowlers' bowling trajectories to forecast a batters' performance.

3 Methodology

This section provides a detailed explanation on the preparation of the data set and construction of the deep learning model.

In this study, we propose a deep learning model to predict the batting shots for different bowling deliveries from different bowlers. A new dataset of videos of bowling instances delivered to English batsman Joeseph Root was created from scratch to evaluate the proposed model. A video consists of a sequence of frames. Therefore this research uses a sequence of image frames to train the prediction model instead of a single image.

3.1 Data Set Generation

Selecting Benchmark Data Set. The batting strike of a batsman depends on several factors apart from the bowling delivery. The match type, batter's skill level, pitch conditions, placement of fielders, number of overs played and overs yet to play can affect the selection of batting strike. Bowling style is the primary consideration in this research, and it is captured in videos. To minimize the effect of the other factors to the utmost only videos from tournaments in Test cricket format were used. Considering the availability of data, top-order English batter Joseph Root's (Joe Root) videos were acquired and processed.

Creation of Data Set. Videos of bowling deliveries bowled to Joe Root were cropped from the original videos using a professional video editing software Adobe Premiere Pro [9]. The time when the bowler releases the ball was considered the start time, and the time the ball bounces off the ground after pitching and before the batter hits the ball was considered the end time. The frames generated from those trimmed videos included excessive irrelevant data such as the presence of the non-striker, fielders, wicket-keepers, and the advertisements painted on the field, as shown in Fig. 1. Therefore the video frames were manually cropped, aiming to reduce the noisiness. Only the bowler, ball, and pitch area were extracted. The number of frames from each video fragment was taken as necessary to train the model.

Fig. 1. Examples data frames before excluding noise.

Dataset of bowling delivery videos was classified according to the batting strike played by Joe Root for each delivery and saved with the label of the shot played for each delivery. Figure 2 demonstrates some of the examples from the dataset, which comprises 428 videos that belong to four different shot classes used in this study.

Since the dataset is really small, for cross-validation purposes, 10% of data was classified for validation purposes, and 10% of the total dataset was allocated for testing. Figure 3 demonstrates the data distribution according to the shot class.

Frame Selection Process. The input to the proposed model is a sequence of image frames from the created videos from the dataset. Since using the whole sequence of frames for predictions is challenging with resource availability, only a small set of frames was selected to represent the whole videos of bowling deliveries. Randomly selecting frames from the videos (image sequences) resulted in an irregular sequence of frames. Irregular sequences of images can affect the final accuracy of the model. Each frame has got at least a minor contribution to the predictions.

Therefore distributively sampled different numbers of frames from each video from the dataset were considered in all evaluations of experiments. If the specified number of frames is eight, the videos were divided into eight windows. One frame from each window was sampled and checked for a specified percentage of the difference between adjacent frames. Frames are then prepossessed, resized and created in a batch using the python package Keras-video-generator [2]. It captures videos from files using OpenCV [5], and it also generates validation and testing data by separating the total data set.

3.2 Developing the Machine Learning Model

The proposed model needs to process time-series data chronologically ordered to detect motions, actions, and directions. Convolutions are generally used when working with images in machine learning, and LSTM is ideal for filtering relevant values from inputs ordered in time. The model also needs to repeat several action

(a) Cover Drive

(b) Cut Shot

(c) Sweep Shot

(d) Flick Shot

Fig. 2. Examples of collected data.

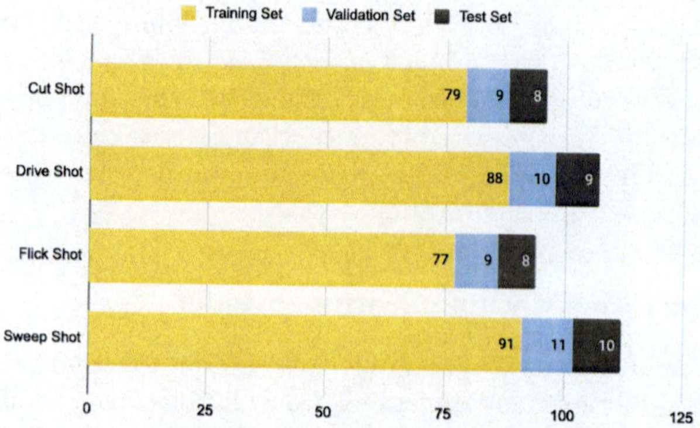

Fig. 3. Data splitting statistics.

detections to predict a batting strike being played. The basic structure of the developed model is as follows:

1. To identify features, a CNN is formed and time distributed.
2. To address 'time series', the time distributed output will introduce into GRU or LSTM.
3. The decision will then be made using a DenseNet to 'classify' it.

In summary, the DenseNet will decide, and the CNN will detect the features in time series data. Hence a CNN with several convolutions is created with batch normalization. The generalized model architecture is depicted in Fig. 4

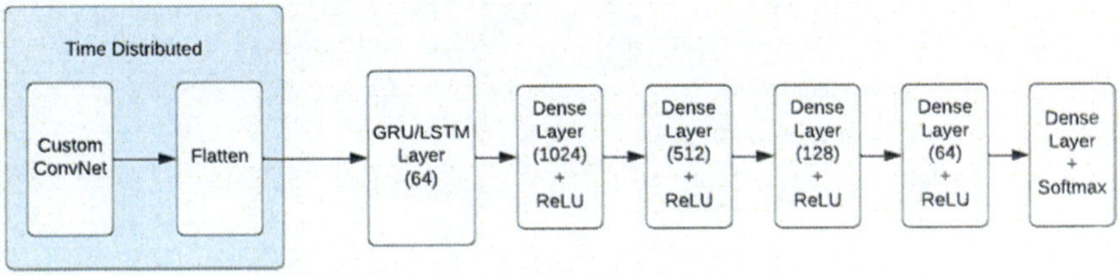

Fig. 4. Overview of the model architecture.

Implementation. This study was conducted using a Keras time distributed layer to apply a time distributed wrapper to every temporal slice of input. The architecture takes 112×112 input images. Moreover, the subsequent four convolutional blocks were connected with four fully connected layers. The first and third fully connected layers contained 1024 hidden units, while the second and fourth had 512 layers, signifying the number of output classes as shown in Fig. 5

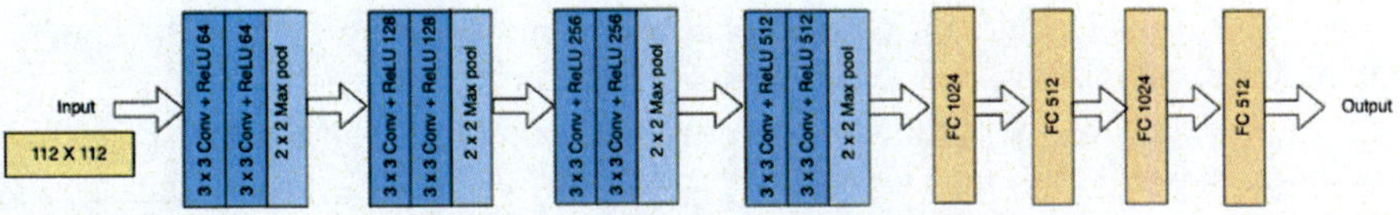

Fig. 5. Keras time distributed layer implementation.

Several experiments were conducted using a GRU and an LSTM layer separately to track temporal changes across video frames. The LSTM was highly efficient, but the GRU, which was conceptually the same, used less memory. However, it had the disadvantage of being less efficient. In addition, four sequential dense layers with hidden units of 1024, 512, 128, and 64 were added to the model, and ReLU was employed as the activation function, which accelerated the model capacity by introducing non-linearity. Finally, a softmax classifier assigned a probability value to each bowling delivery.

After implementing the model, experiments were carried out with various combinations and fine-tuning of parameters through trial and error to achieve adequate accuracy. All the experiments were conducted using Google Colaboratory. Table 1 provides a basic description of the model.

Table 1. Description of the model parameters

Properties	Description
Optimizer	Adam
Activation Functions	ReLU and Softmax
Input Shape	112×112
Number of Output Classes	4

4 Experiments

This section gives an overview on the experiments conducted to study the validity of the proposed model.

Various experimental scenarios were carried out in order to evaluate the performance of the developed model on the specified dataset by tuning several hyperparameters expecting to achieve higher accuracy.

The experiments found that the Adam optimizer worked best with a learning rate of 0.0001 for a batch size of 3 and 12 image frames to predict the batting strikes of Joe Root.

The model was initially developed with an SGD optimizer with a learning rate of 0.001 and a batch size of one. When training the model with five input frames, the model could only approach an accuracy of 22.7% for 50 epochs. When the number of epochs was increased up to 250, there was nearly a 6% increase in accuracy. Keeping other parameters constant, the number of frames was increased to nine and trained, expecting better results. From that, it was found that there is a positive impact of the number of frames on the model's accuracy. Figure 6 illustrates different number of frames used when conducting the experiments.

Furthermore, experiments were conducted, increasing the batch to three and eight. Further increasing the batch size resulted in lesser accuracy and the model took longer to train.

Due to slight accuracy increases when using the SGD optimizer, an Adam optimizer was introduced to the model to investigate any growth in the accuracy. Even with the Adam optimizer, the increase in accuracy was insufficient when trained with a learning rate of 0.001. Therefore experiments were conducted, increasing the learning rate for the optimizer. Then a significant change

Fig. 6. Different number of frames used.

in accuracy was visible. Furthermore, the validation loss was still decreasing for 250 epochs. Therefore by training the model for 350 epochs, we could attain an accuracy of 89.3%. Additionally, another experiment was performed, increasing the number of frames to 12. Then the model achieved a good accuracy of 91.67%. Table 2 lists the experiments conducted, along with the accuracies obtained for each experiment conducted. An input shape of 112×112 was used for all the experiments.

Table 2. Summary of conducted experiments.

Number of Epochs	Number of Frames	Optimizer	Batch Size	Learning Rate	Accuracy Obtained (%)
50	5	SGD	1	0.001	22.7
250	5	SGD	1	0.001	28.2
250	9	SGD	1	0.001	34.2
250	9	SGD	3	0.001	38.4
250	9	SGD	8	0.001	30.5
250	9	Adam	3	0.001	41.6
250	9	Adam	3	0.0001	87.9
350	9	Adam	3	0.0001	89.3
350	12	Adam	3	0.0001	91.7

4.1 Hardware Configurations

The overall experiment were conducted on Google Colaboratory environment. NVIDIA Tesla K80 was the GPU memory of Google colaboratory. Keras with a TensorFlow [4] back-end was used for the model implementations.

5 Results and Discussion

In this section, the results obtained by carrying out the aforementioned experiments are discussed.

The highest accuracy attained using the developed model was 91.67% from 350 epochs for the created dataset of 428 videos. The training and the validation accuracy increases steadily with the number of epochs, as demonstrated in Fig. 7 and in Fig. 8. A variety of callback functions were used to monitor the accuracies and to identify and stop training the model after no improvement was visible in the model predictions.

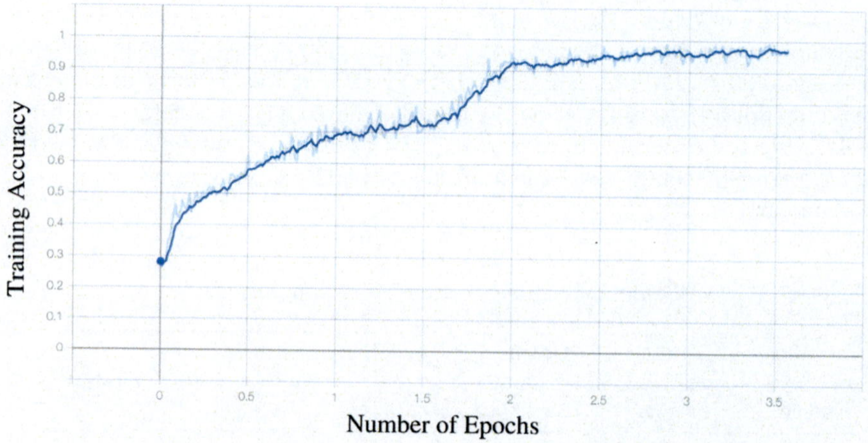

Fig. 7. Model performance: training accuracy against epochs.

The Fig. 9 demonstrates how the training loss decreased to a very low value depicting a good learning rate.

The Table 3 provides a summary of the evaluation matrices for each shot class predicted.

According to the F1 score summary given in Table 3 the model works best for the drive shot class because providing the highest F1 score. As precision measures stated in the summary, all the shots predicted as cut shot and flick shot were correct predictions.

The confusion matrix depicted in Fig. 10 the model falsely predicts three shots, two flick shots and one cut shot. Flick shots were predicted as Sweep shots

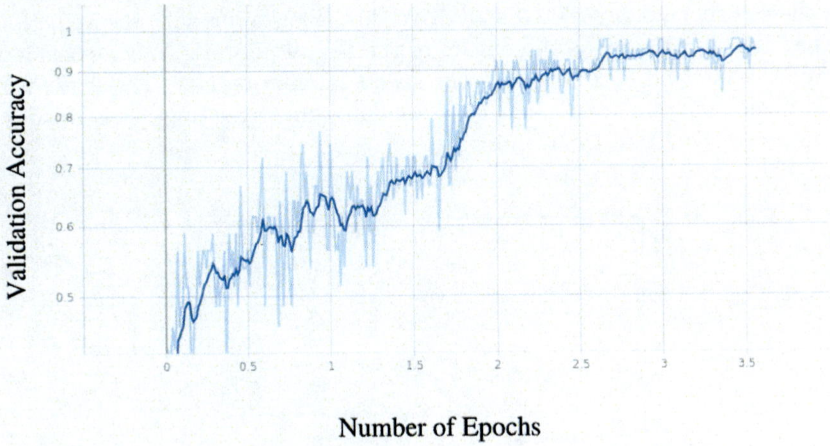

Fig. 8. Model performance: validation accuracy against epochs.

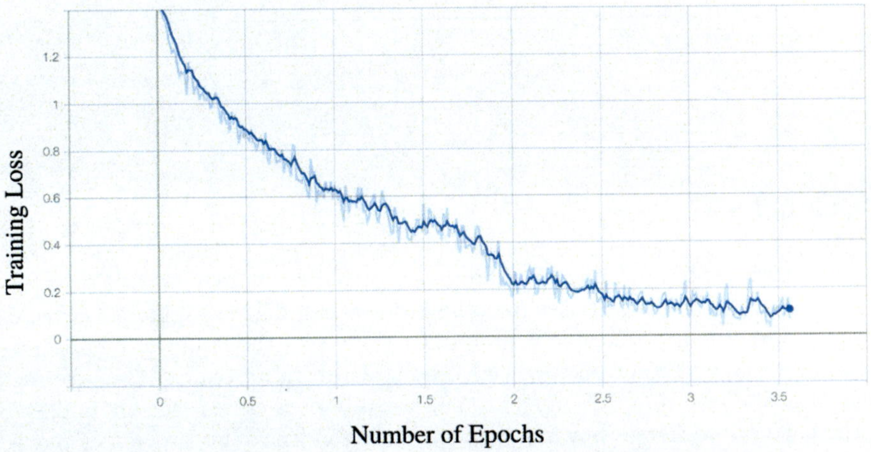

Fig. 9. Model performance: training loss against epochs.

Table 3. Summary of the model evaluation on the test set.

Shot Class	F1 Score	Precision %	Recall %	Accuracy
Cut Shot	0.93	100	88	91.67%
Flick Shot	0.86	100	75	
Sweep Shot	0.92	85	100	
Drive Shot	0.95	90	100	

and the cut shot was predicted as a drive shot. Reason for this misprediction is unclear and it may not be the nature of the neural network. When considering the dataset cut and flick shots has the lowest number of data. Therefore, it can be assumed that the error could be lessened by incorporating more video data of those two shots into the training set.

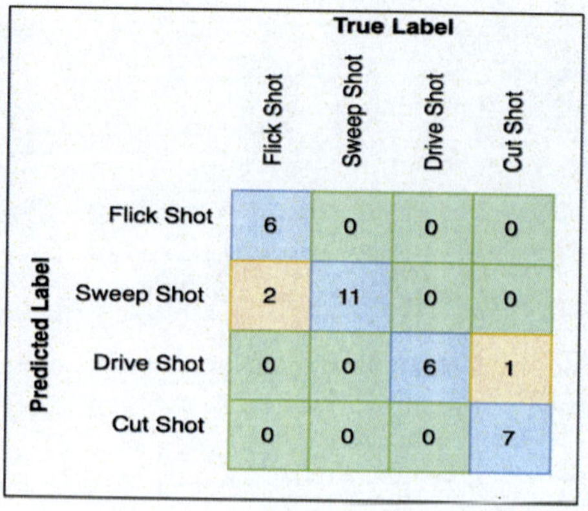

Fig. 10. Confusion matrix.

6 Conclusion

This research proposes a novel method with considerably high accuracy to predict a batsman's batting strikes for different bowling deliveries, given the video of the bowling delivery. The novel approach addresses the issues of human predictions, which is the current practice used by the Cricket teams. The proposed method uses a custom CNN model to successfully predicts batting shots played by the batter for different bowling deliveries using videos of his previously played matches. Machine learning techniques of deep CNN with GRU and LSTMs were used to handle time distributed data from the videos.

Due to the unavailability of data required to evaluate the proposed model, a new dataset of 428 videos of uneven lengths was prepared from scratch. It consists of bowling deliveries bowled to English batter Joe Root, labeled with the name of the batting strike Joe Root played, to train and evaluate the model. As a proof of concept, only four batting strikes, namely the cut shot, drive shot, flick and sweep shot, were considered in this study. The model attained an accuracy of 91.67% after 350 epochs.

This model can be implemented for Cricket players of any country upon the availability of data. Moreover, the proposed method can be implemented to make predictions in various other sports as well.

7 Future Work

The model developed in this research is trained only to predict four different batting strikes of a single batter. It can be extended to detect other types of shots. Moreover, this research only focuses on Test cricket tournaments. This model can be tested with other types of cricket matches as well. Additionally, with the popularity of shorter cricket formats such as T20, various unconventional shots are now being used. That is also an opportunity for improvement to incorporate unorthodox shots.

A significant improvement suggested is integrating the parameters considered constant when conducting this research. The placement of fielders, pitch conditions, air resistance and humidity, match situation can be appended as labels to the data set of bowling deliveries apart from the shot's name and could be taken into consideration.

Regarding our prediction model, it would be fascinating to investigate and adequately validate the model with a larger data set that requires more processing power than was available for this study. It was tough to train and tune the model on local machines due to the quantity of data in the data set. We were able to train these models on our data set in particular; nevertheless, iterating over the hyper-parameters of these models proved challenging and could be optimized with additional computational power.

In future work, experiments can also be carried out by combining attention neural networks to increase the accuracy of the model by identifying critical points or the regions in the image frames, which as a result, will allow us to work with fewer frames and parameters.

References

1. International cricket council
2. Keras-video-generators
3. Live cricket scores, match schedules, latest cricket news, cricket videos
4. Tensorflow
5. Bradski. The OpenCV library. Dr. Dobb's J. Softw. Tools **25**(11), 120–123 (2000)
6. Chowdhury, A.Z.M.E., Rahim, Md.S., Rahman, Md.A.U.: Application of computer vision in cricket: foot overstep no-ball detection. In: 2016 3rd International Conference on Electrical Engineering and Information Communication Technology (ICEEICT), pp. 1–5. IEEE (2016)
7. Felsen, P., Agrawal, P., Malik, J.: What will happen next? Forecasting player moves in sports videos. In: Proceedings of the IEEE International Conference on Computer Vision, pp. 3342–3351 (2017)
8. Gürpınar-Morgan, W., Dinsdale, D., Gallagher, J., Cherukumudi, A., Lucey, P.: You cannot do that ben stokes: dynamically predicting shot type in cricket using a personalized deep neural network. arXiv preprint arXiv:2102.01952 (2021)
9. James, G.: Adobe premier pro (1996)
10. Jhanwar, M.G., Pudi, V.: Predicting the outcome of ODI cricket matches: a team composition based approach. In: MLSA@ PKDD/ECML (2016)

11. Khan, M.Z., Hassan, M.A., Farooq, A., Khan, M.U.G.: Deep CNN based data-driven recognition of cricket batting shots. In: 2018 International Conference on Applied and Engineering Mathematics (ICAEM), pp. 67–71. IEEE (2018)
12. Patil, N., Sequeira, B.H., Gonsalves, N.N., Singh, A.A.: Cricket team prediction using machine learning techniques. Available at SSRN 3572740 (2020)
13. Roopchand, R., Pooransingh, A., Singh, A.: Bat detection and tracking toward batsman stroke recognition. In: 2016 8th International Conference on Computational Intelligence and Communication Networks (CICN), pp. 256–260. IEEE (2016)
14. Samiuddin, O.: ICC survey reveals over a billion fans - 90% in subcontinent, June 2018
15. Semwal, A., Mishra, D., Raj, V., Sharma, J., Mittal, A.: Cricket shot detection from videos. In: 2018 9th International Conference on Computing, Communication and Networking Technologies (ICCCNT), pp. 1–6. IEEE (2018)
16. Sen, A., Deb, K., Dhar, P.K., Koshiba, T.: CricShotClassify: an approach to classifying batting shots from cricket videos using a convolutional neural network and gated recurrent unit. Sensors **21**(8), 2846 (2021)
17. Sharma, R.A., Sankar, K.P., Jawahar, C.V.: Fine-grain annotation of cricket videos. In: 2015 3rd IAPR Asian Conference on Pattern Recognition (ACPR), pp. 421–425. IEEE (2015)
18. Shimizu, T., Hachiuma, R., Saito, H., Yoshikawa, T., Lee, C.: Prediction of future shot direction using pose and position of tennis player. In: Proceedings Proceedings of the 2nd International Workshop on Multimedia Content Analysis in Sports, pp. 59–66 (2019)
19. Thabtah, F., Zhang, L., Abdelhamid, N.: NBA game result prediction using feature analysis and machine learning. Ann. Data Sci. **6**(1), 103–116 (2019)

Computational Empathy Using Facial Emotion Recognition: An Update

Khalid Alharbi and Sudhanshu Semwal(✉)

Department of Computer Science, University of Colorado at Colorado Springs, 1420 Austin Bluffs Pkwy, Colorado Springs, CO 80918, USA
{kalharbi,ssemwal}@uccs.edu

Abstract. Facial emotion recognition plays a vital role in computational empathy enabling the nature and effectiveness of human-machine interaction. Although facial emotion has received much attention over the last years as the number of applications has increased, it remains a complicated issue due to facial features and ethnic and cultural differences in facial expressions. This short survey reviews the updated deep learning techniques used for facial emotion recognition. It explores and classifies deep learning models and datasets used for facial emotion recognition. The outcome of the reviews indicates that facial emotion recognition limits complexity and accuracy. This paper also points out the research directions for future work.

Keywords: Computational Empathy · Facial Emotion Recognition · Deep Learning · Human-Machine Interaction

1 Introduction

Computational Empathy uses facial emotion recognition to enhance human-machine interaction. It advances the nature and effectiveness of such interaction by determining a human's thoughts, behavior, and feeling via his/her facial expressions. Facial emotion recognition increases rapidly in various application domains, such as human-Machine interaction, Virtual Reality, Advanced Driver Assistant Systems, Medical Therapy, and entertainment. However, facial emotion recognition remains a complicated issue due to variations in facial features and ethnic and cultural differences in facial expressions. Nevertheless, facial emotion recognition remains a complicated issue due to variations in facial characteristics and ethical and artistic differences in facial expressions. This short survey paper explores the up-to-date research on Deep Convolutional Neural Networks (D-CNN) used for facial emotion recognition. It covers a set of 14 articles [1–14] published in a time from 2019–2023. This paper answers the following research questions that are extracted from other surveys:

- **RQ1**: What D-CNN models are used for facial emotion recognition?
- **RQ2**: What feature extraction techniques are used?

K. Arai (Ed.): FTC 2023, LNNS 816, pp. 79–88, 2023.
https://doi.org/10.1007/978-3-031-47448-4_7

- **RQ3**: What dataset structures are used for facial emotion recognition?

We build a literature classification scheme based on these research questions to categorize the 14 articles collected into sub-category. These 14 articles were selected carefully after a detailed scan regarding several criteria, including the most citations, providing details of the approach used, analyzing the results, and presenting the findings. This short survey explains each sub-category and discusses the limitations of facial emotion recognition along with points to future research directions.

The remains of this paper are arranged as follows. In Sect. 2, we briefly describe the recent facial emotion recognition techniques found in the literature. Section 3 responds to the research question, discusses the limitations of facial emotion recognition, and points to the research directions to advance facial emotion recognition over the limitations. Section 4 concludes this paper, clarifies the findings, and presents a research plan to be conducted on facial emotion recognition.

2 Literature of Facial Emotion Recognition

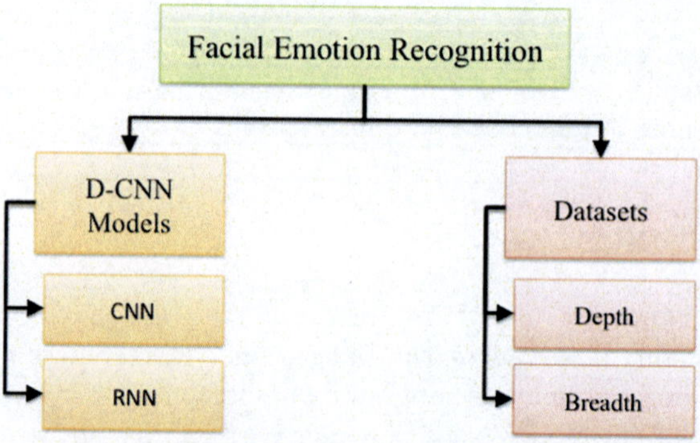

Fig. 1. Facial emotion recognition literature classification scheme.

We built the classification scheme to filter the 14 articles on facial emotion recognition (FER) and categorize them into sub-categories. The classification scheme is designed based on the research questions mentioned above. As illustrated in Fig. 1, the FER literature classification scheme consists of two main categories: D-CNN models and datasets. The D-CNN models include two subcategories, Convolutional Neural Networks (CNN), Recurrent Neural Networks (RNN) while the dataset includes depth and breadth datasets.

2.1 Deep Convolutional Neural Networks (DCNN)

A CNN model filters an input image through convolution layers to produce a feature map combined into fully connected networks representing a facial expression. A loss function such as *softmax* algorithm then recognizes the facial expression based on a particular output. The DCNN models are grouped into two subcategories: Convolutional Neural Networks (CNN) and Recurrent Neural Networks (RNN). The CNN is a feed-forward network that handles fixed-size inputs and generates fixed-size outputs, while the RNN manages arbitrary lengths of inputs and output. Unlike a feed-forward network, RNN utilizes internal memory to process inputs and produces time-series information. Table 1 illustrates the model architecture based and accuracy of the DCNN models covered in this short survey.

Table 1. Deep CNN models.

DCNN Archit.	Ref.	Architecture Based	Accuracy
CNN	[1]	Backbone	69.10%
	[2]	Assembled	73.5 %
	[3]	Assembled	72.48%
RNN	[5]	LSTM	82.12%

Convolutional Neural Networks (CNN). Pasquale Foggia *et al.* [1] consider three CNN architectures, MobileNet, ResNet, and SENet, to evaluate user profiling. The approach is established on a multi-task CNN to deal with gender, age, ethnicity, and emotion recognition from face images. The multi-task CNN, as shown in Fig. 2, shares convolutional layers among the tasks to minimize the processing time and memory space. The classification layers are then determined for the various tasks. The results show that multi-task CNN is influential compared to single-task CNNs. It achieves a processing speed-up between 2.5 and 4 times and memory space reduction between 2 and 4 times while preserving accuracy. Dorota Kaminska *et al.* [2] deal with the weakness of discrimination between dominant and complementary emotions. The authors thus present a two-stage recognition model comprising two classification processes: coarse recognition and fine recognition. The coarse recognition extracts appearance features using CNN. Fine recognition divides symmetrical emotions into clusters. It then trains a binary classifier for each cluster. The method limits noise handling. The model hardly differentiates between the dominant and complementary emotions. They create a facial expression dataset (iCV Multi-Emotion Facial Expression Dataset (iCV-MEFED)) that includes 31,250 images of 115 persons. These images are distributed based on gender, age of 18 to 37, ethnicity, and hairstyle. They use compound emotions, shown in Table 2, to present the dominant-complementary emotion combinations.

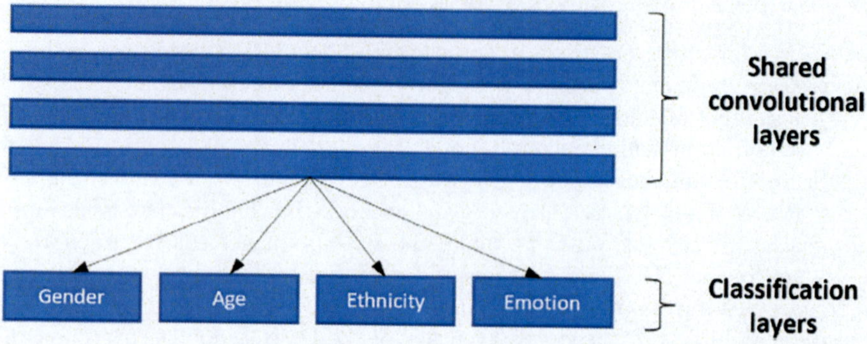

Fig. 2. Multi-task CNN model. [1]

Table 2. 49 Dominant-complementary emotions, the 50th is natural.[2]

	Angry	Contempt	Disgust	Fear	Happy	Sadness	Surprise
Angry	**angry**	contemptly angry	disgustingly angry	fearfully angry	happily angry	sadly angry	surprisingly angry
Contempt	angrily contempt	**contempt**	disgustingly contempt	fearfully contempt	happily contempt	sadly contempt	surprisingly contempt
Disgust	angrily disgusted	contemptly disgusted	**disgust**	fearfully disgusted	happily disgusted	sadly disgusted	surprisingly disgusted
Fear	angrily fearful	contemptly fearful	disgustingly fearful	**fearful**	happily fearful	sadly fearful	surprisingly fearful
Happy	angrily happy	contemptly happy	disgustingly happy	fearfully happy	**happy**	sadly happy	surprisingly happy
Sadness	angrily sad	contemptly sad	disgustingly sad	fearfully sad	happily sad	**sad**	surprisingly sad
Surprise	angrily surprised	contemptly surprised	disgustingly surprised	fearfully surprised	happily surprised	sadly surprised	**surprised**

Yousif Khaireddin and Zhuofa Chen [3] introduced a single-network (CNN) classification and applied it to the FER2013 dataset, which contains 35888 images of seven various emotions such as anger, fear, disgust, happiness, sadness, surprise, and neutral. The authors use VGGNet architecture and fine-tune its hyperparameters. They run experiments with 7 optimizers, including Stochastic gradient descent (SGD), SGD with Nesterov Momentum, Average SGD, Adam, Adam with AMSGrad, Adadelta, and Adagrad. They also run experiments with five different learning rate schedulers (Reduce Learning Rate on Plateau (RLRP), Cosine Annealing (Cosine), Cosine Annealing with Warm Restarts (CosineWR), One Cycle Learning Rate (OneCycleLR), and Step Learning Rate (StepLR)). The results show that the model performs best utilizing the SGD with Nesterov momentum, achieving an accuracy of 73.5 %. For the learning rate scheduler, the Reducing Learning Rate on Plateau (RLRP) performs best, achieving a validation accuracy of 73.59 % and a testing accuracy of 73.06 %. The confusion matrix applied on the FER2013 for testing illustrates the most reasonable classification of the "happiness" and "surprise" emotions, while the low classification accuracy is in "disgust" and "anger". Nyle Siddiqui *et al.* [11] present a framework for developing a facial emotion recognition model. The framework consists of lightweight CNN and a large dataset (i.e., AffectNet) for training. It is also a web application to achieve maximum accessibility to users for testing. Experiments and comparisons to other approaches are provided. The framework shows that the test accuracy increased by 8–10%. However, the approach limits the difference between frequent emotions (i.e., happy, neutral, angry) and infrequent emotions (i.e., contempt, disgust). Wang Xiaohua *et al.* [12] improve Generative Adversarial Networks (GANs), a D-CNN architecture, to generate one-to-many faces with different poses and expressions. The proposed technique is so-called StarGAN. They also add the Attention U-Net to the net to mitigate flaws in the face generated by StarGAN. An experiment was conducted on two emotional faces datasets (Karolinska Directed Emotional Faces and MMI Facial Expression). The result shows an enhancement for datasets.

Hengshun Zhou *et al.* [14] introduce an approach that extracts emotion features and feature fusion based on audio and video. The approach follows a pipeline, as shown in Fig. 3, including Audio and Video Processing, feature extraction, feature fusion, and classifier. The approach exploits various fusion methods (self-attention, relation-attention, and transformer). The authors applied three CNN models to extract facial emotion features from the video, VGGFace, ResNet18, and IR50. They use four emotion datasets for training, AffectNet, RAF-DB, FER+, and AFEW. The model has been submitted to the EmotiW 2019 Challenge. It ranked second with 62.48% accuracy.

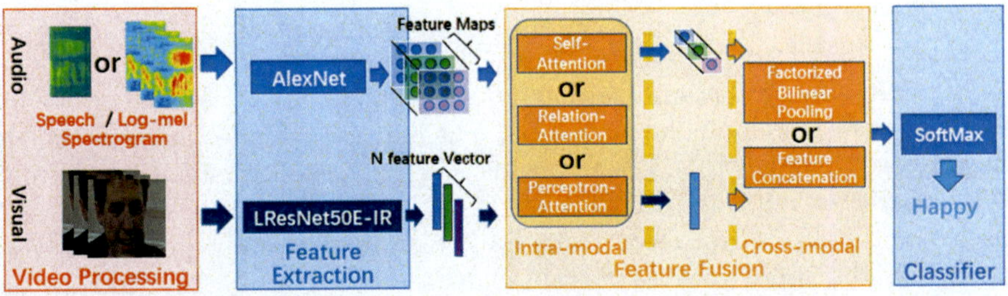

Fig. 3. Audio-video emotion recognition model. [14]

Recurrent Neural Networks (RNN). Shuai Liu *et al.* [5] illustrates an approach to solve the Shortcomings of existing fusion methods that limit the affection of changing emotions modalities. The authors use Long-Term Short-term Memory (LSTM) neural network, a composite RNN with a linear self-loop, to capture time-dependent attributes. They improve the discrimination between modalities by constructing an attention network with complementary modalities. They experiment with their approach on public datasets, MELD and IEMOCAP datasets, and compare the results to the traditional methods. The results show that the proposed approach is accurate and performs better in both the rate and confusion matrix metrics.

2.2 Datasets

The datasets are categorized into two types (depth and breadth). A depth dataset contains a limited number of subjects, each including many images. It executes a model to handle a wide range of intraclass variations, such as lighting, age, and pose. A breadth dataset comprises many subjects; each has few images. It assures a model to handle the variable appearance of various people. Table 3 describes some public datasets used for facial emotion recognition.

Table 3. Facial emotions datasets.

Dataset Type	Ref.	Dataset Name	#Identities	#Images	Diversity
Depth	[6]	JAFFE	10	213	very low
	[6]	Cohn-Kanade CK	100	500	low
	[6]	FER2013	123	30,000	low
	[4]	UTKFace	2284	20,000	medium
	[6]	IMDb-Wiki	20,284	523,051	above medium
Breadth	[11]	AffectNet	450,000	1,000,000	above medium

Depth Datasets. Shraddha Mane and Gauri Shah [6] use six datasets to evaluate different techniques in facial analysis. The datasets include Japanese Female Facial Expression (JAFFE), Cohn-Kanade (CK), Persistent Systems Ltd. (PSL), FER2013, IMDb-Wiki, and the Real-World dataset. They describe these datasets based number of images, color scale, and resolution. Many experiments were conducted on these datasets for evaluation. The experiments have face recognition utilizing Amazon Rekognition, transfer learning, facial landmarks, facial expression recognition, Histogram of Oriented Gradient (HOG) descriptors, and Gender Identification. The results show that the CNN model performs well on large datasets. For gender identification, the dataset consists of 500K images, and the accuracy was 96%. The dataset has 1200 images for facial expression recognition, and the accuracy was significantly lower at 66%. Asad Khattak *et al.* [4] propose a DCNN model to classify facial images for detecting age and gender from facial expressions. For training and testing, they use three facial expression datasets (JAFFE, CK+, and UTKFace). The JAFFE is a Japanese female facial expression dataset comprising 213 images of 10 Japanese models. The CK+ dataset consists of 327 video sequences by 118 participants labeled with seven facial expressions. The UTKFace dataset comprises 20k+ facial images of age and gender, with a range of ages from 0 to 116. They use *Python* library (Keras employing TensorFlow). The model achieves an accuracy of 95.65% for emotion recognition, 98.5% for age recognition, and 99.14% for gender recognition. However, the approach limits the classification of human ages into four categories, experiments on age and gender recognition with only one dataset in a single domain, and uses small-size datasets. Pranav E. *et al.* [9] create a Deep Convolutional Neural Network (DCNN) model that classifies five different human facial emotions. The dataset contains 2550 images manually collected. They use a *Python* library (Keras) to add convolution layers. They also use the Scikitlearn library to compute the confusion matrix, including precision, sensitivity, specificity, F1-score, and accuracy. The authors use an Adam optimizer to reduce the loss function. The model classifies 5 facial emotions from the image dataset and achieves an accuracy of 78.04%, primarily for happy, angry, and neutral emotions. The model limits the recognition of emotions from videos with a sequence of images that cannot use for real-time applications. Hsuan-Chia Yang *et al.* [13] present an open dataset (a small dataset) of doctor-patient interactions. It allows doctors to analyze patients' emotions. They can reach out to the patient through empathic care. To avoid violating patient privacy, they utilize the Deepfake technique to de-identify video recordings of patients' clinical meetings. The technique substitutes a patient's face and manipulates it into an unrecognizable person's image or video that has identical facial expressions.

Breadth Datasets. Fezile Ozdamli *et al.* [8] provide an approach that employs computer vision algorithms and deep learning algorithms for distance learning (online learning). The approach focuses on understanding students' emotions and abnormal behaviors during educational sessions and e-exams. They use OpenCV Library for image acquisition, pre-processing the images (e.g., noise cutback), feature extraction, and detection and segmentation. The authors use four datasets for different purposes. They use the Mini Xception CNN to predict facial emotions. The dataset used to train face verification consists of 13,500 labeled face images. They use a training emotion recognition dataset that includes 32,300 images of faces with different facial expressions at 48×48 pixels in grayscale, training the gaze tracking dataset. It contains 1380 images with a resolution of 600×800 pixels, training the head movement dataset that consists of 2800 images of 200 people at 640×480 pixels, and the dataset (consists of 80 object categories) for detecting objects depicted in the camera. The authors experimented on students' class attendance. The result of the performance and accuracy for the deep metric learning algorithm and the ResNet-34 algorithm used in the system is about 99.38%. However, the approach has several inabilities, including automatically detecting the head and face movements of students who may express the same emotion to another and the poor internet connection in rural areas. Nyle Siddiqui *et al.* [11] propose a framework for developing facial emotion recognition models that can be compared and contrasted against one another in a constant, standardized fashion. They use CNN, trained on the AffectNet dataset of 300,000 images at 224×224 pixels. These images have been manually annotated and verified to represent eight facial emotions (neutral, happy, sad, surprise, fear, disgust, anger, and contempt). The framework limits that it has not exercised to other datasets. It also limits the real-time operation over the cost of accuracy. Ninad Mehendale [7] introduces an approach for facial emotion recognition consisting two-level CNN framework. The first level clears the background from the picture. The second level extracts the facial feature expressions by tracking down relevant points of importance. The author operates on five different types of facial expressions. The model was trained and tested on a facial expression dataset of 10,000 images (154 persons). The author experiments with the model on more than 750K images using extended Cohn-Kanade expression, Caltech faces CMU and NIST datasets. The model correctly highlights the emotion with 96% accuracy. The model aims to predict the learning of students and lie detectors.

3 Discussion

According to this short survey covering 14 articles related to facial emotion recognition, the three research questions can be answered as:

- **RQ1**: What D-CNN models are used for facial emotion recognition?
  Most techniques of facial emotion recognition use CNN more than RNN. The techniques use CNN, either a single-network CNN such as VGGNet and Xception or a generative network (GANs). The techniques that RNN is rare. One article [5] uses Long-Term Short-term Memory (LSTM), a type of RNN. Although RNN introduces promising results, it requires a high computational capability to execute.
- **RQ2**: What feature extraction techniques are used?
  Three extraction techniques are used for facial emotion recognition (geometric features, appearance features, or a hybrid). The geometric features extraction technique uses facial components (i.e., landmark points) to construct a feature vector. Some techniques utilize position and angle to specify the facial landmark points. The appearance features extraction technique utilizes different face regions to extract different types of information. Such techniques deliver low accuracy because they limit the local variations of the facial components.
- **RQ3**: What dataset structures are used for facial emotion recognition?
  Most of the datasets used for facial emotion recognition are public datasets such as AffectNet, JAFFE, CK, PSL, FER2013, and IMDb-Wiki. These datasets are depth structured, containing few subjects, which usually need more diversity. This leads to model overfitting when it applies to real-world scenarios. The breadth datasets also need more diversity, although they contain many subjects. It is because these subjects are related to the topic of interest; for instance, in [8] collect data from students who attended a class to understand students' emotions using facial emotion recognition.

Limitations: The recent DCNN models of facial emotion recognition are promising for computational empathy, emerging for various application domains. Facial emotion recognition still limits noise handling and differentiates between the dominant and complementary emotions. Furthermore, the DCNN models were trained and tested on datasets that either lack diversity of people's faces or struggle with small sample sizes of facials. This limitation impacts the application of the models and leads to overfitting and less accuracy, especially when they run in real-life applications. Thus, creating a large and diverse dataset is vital for computational empathy. It also limits automatically detecting head and face movements as well as the real-time scenarios over the cost of accuracy. This short survey finds that facial emotion recognition lacks a diverse dataset for computational empathy. This usually results in model overfitting and less accuracy when a facial emotion recognition model runs in real-life scenarios.

Research Directions: An effective facial emotion recognition model that uses a Deep Convolutional Neural Network (D-CNN) must handle people's emotions as variations as possible with facial features, race, gender, age, ethnicity, and culture. Thus, the model should recognize the combination of facial emotions (see Table 2). The model should also detect head and face movements for real-time applications such as video games and online exams. This requires building a dataset consisting of a large number of images, labeled as emotion type and demographical information such as the region where a person lives, race, culture, gender, age range, etc. The facial emotion recognition model can be used in various applications, including video-game, robotics, and healthcare. Video game techniques and tools, such as Unity-3D, with users engaged across the world can be used to collect diverse data for evaluating emotion recognition model.

4 Conclusion

We briefly investigated the research area of computational empathy using facial emotion recognition. We scanned the publications of time 2019–2023 that cover 14 papers. This set of literature explores the facial emotion recognition techniques using deep learning, the popular datasets used for training and testing facial emotion recognition models, and the limitations these techniques have to overcome. We concluded from this research that a large, diverse dataset for facial emotion recognition is required. Building an effective and accurate model for real-time scenarios is also needed for computational empathy application. Thus, we narrow the future research direction into these two objectives: create diverse datasets and build real-time facial emotion recognition models.

References

1. Foggia, P., Greco, A., Saggese, A., Vento, M.: Multi-task learning on the edge for effective gender, age, ethnicity and emotion recognition. Eng. Appl. Artif. Intell. **118**, 105651 (2023)
2. Kaminska, D., et al.: Two-stage recognition and beyond for compound facial emotion recognition. Electronics **10**(22), 2847 (2021)
3. Khaireddin, Y., Chen, Z.: Facial emotion recognition: state of the art performance on FER2013. arXiv preprint arXiv:2105.03588 (2021)
4. Khattak, A., Asghar, M.Z., Ali, M., Batool, U.: An efficient deep learning technique for facial emotion recognition. Multim. Tools Appl. **81**(2), 1649–1683 (2022)
5. Liu, S., Gao, P., Li, Y., Weina, F., Ding, W.: Multi-modal fusion network with complementarity and importance for emotion recognition. Inf. Sci. **619**, 679–694 (2023)
6. Mane, S., Shah, G.: Facial recognition, expression recognition, and gender identification. In: Balas, V.E., Sharma, N., Chakrabarti, A. (eds.) Data Management, Analytics and Innovation. AISC, vol. 808, pp. 275–290. Springer, Singapore (2019). https://doi.org/10.1007/978-981-13-1402-5_21

7. Mehendale, N.: Facial emotion recognition using convolutional neural networks (FERC). SN Appl. Sci. **2**(3), 1–8 (2020). https://doi.org/10.1007/s42452-020-2234-1
8. Ozdamli, F., Aljarrah, A., Karagozlu, D., Ababneh, M.: Facial recognition system to detect student emotions and cheating in distance learning. Sustainability **14**(20), 13230 (2022)
9. Pranav, E., Kamal, S., Satheesh Chandran, C., Supriya, M.H.: Facial emotion recognition using deep convolutional neural network. In: 2020 6th International Conference on Advanced Computing and Communication Systems (ICACCS), pp. 317–320. IEEE (2020)
10. Sharma, P., et al.: Student engagement detection using emotion analysis, eye tracking and head movement with machine learning. In: Reis, A., et al. (eds.) Technology and Innovation in Learning, Teaching and Education: Third International Conference, TECH-EDU 2022, Lisbon, Portugal, 31 August–2 September 2022, Revised Selected Papers, pp. 52–68. Springer, Cham (2022). https://doi.org/10.1007/978-3-031-22918-3_5
11. Siddiqui, N., Reither, T., Dave, R., Black, D., Bauer, T., Hanson, M.: A robust framework for deep learning approaches to facial emotion recognition and evaluation. In: 2022 Asia Conference on Algorithms, Computing and Machine Learning (CACML), pp. 68–73. IEEE (2022)
12. Wang, X., Gong, J., Min, H., Yu, G., Ren, F.: LAUN improved StarGAN for facial emotion recognition. IEEE Access **8**, 161509–161518 (2020)
13. Yang, H.-C., et al.: How can research on artificial empathy be enhanced by applying deepfakes? J. Med. Internet Res. **24**(3), e29506 (2022)
14. Zhou, H., et al.: Exploring emotion features and fusion strategies for audio-video emotion recognition. In: 2019 International Conference on Multimodal Interaction, pp. 562–566 (2019)

Classification and Characterization of EEG Biomarkers to Predict Parkinson's Disease Through Quantitative Feature Analysis Using Deep Learning

Vatsal Sivaratri(✉)

Thomas Jefferson High School for Science and Technology, Alexandria, VA 22312, USA
vatsal.sivaratri@gmail.com

Abstract. Parkinson's Disease (PD) is a prevalent and challenging neurodegenerative disease, affecting nearly 10 million people globally. Current diagnostic methods rely on observations of symptoms and medical history but have proven to be fallible, with misdiagnosis rates ranging from nearly 10%–20%. Furthermore, a lack of understanding in the field of PD poses a great hurdle in determining a well-fitted treatment for each patient. As of now, PD has no cure, making early detection critical for appropriate intervention to slow its progression. Electroencephalograms (EEGs) are a novel approach in the field of neurodegenerative diseases that provide crucial information on a patient's brain's electrical activity, which can be crystalized into new observable symptoms of PD, as well as serve in diagnosing the disease earlier and more accurately. In this project, we verify the applicability of EEG biomarkers to characterize and predict PD through quantitative feature analysis. We draw from a multitude of feature extraction techniques, namely Spectral Fast Fourier Transform (FFT) and Cycle-by-Cycle Analysis. We evaluate each feature for its ability to differentiate the two populations. We then implement models to classify this optimized feature set, which attained a 98.92% training accuracy and 81.20% validation accuracy, outperforming current medical diagnostics. These results indicate that the proposed technique has strong potential in differentiating PD from electro-neurophysiological data, ultimately improving treatment outcomes for millions worldwide.

Keywords: Parkinson's Disease · Electroencephalography · Deep Learning

1 Introduction

Parkinson's Disease (PD) is the second-most common neurodegenerative disease globally, after Alzheimer's Disease. It is a progressive idiopathic disease caused by a degeneration of dopaminergic nerve cells in the substantia nigra pars compacta (SN), which plays a critical role in the modulation and coordination of motor functions in the brain, and the development of neuronal Lewy Bodies.

This impaired neural activity has been characterized by various manifestations in motor control, namely, extreme rigidity and stiffness, as well as tremors, slowness of

K. Arai (Ed.): FTC 2023, LNNS 816, pp. 89–103, 2023.
https://doi.org/10.1007/978-3-031-47448-4_8

movement, and postural instability. Risk factors for the disease include age, family history, and various environmental exposure [5]. The exact etiology of PD remains elusive, but it is hypothesized to be multifactorial, involving both genetic and environmental factors [16].

Recent studies on PD conducted in recent years have substantially advanced our understanding of the disease and identified methods to investigate its pathology further. For instance, studies examining the immune responses of PD have suggested that the activation of the peripheral immune system synergistically drives degeneration, although the mechanisms that link these two are not yet fully understood, and the sequences of responses with respect to PD remain unclear [5, 14]. Furthermore, genomics research has demonstrated promise, with correlations between PD and SNCA, LRRK2, VPS35, EIF4G1, DNAJC13, and CHCHD2, among other genes. In addition, miscellaneous studies on the metagenomics of PD have established associations with other imbalances, such as gut microbiomes [11]. These imbalances are postulated to be risk factors for the disease, underscoring the multifactorial nature of PD's pathogenesis.

The current gold standard for testing for this disease is entirely clinical, performed through an examination by a medical professional. Doctors weigh in symptoms, family history, and other factors to form an accurate diagnosis [17]. Unfortunately, usage of this method has proven to be fallible, with nearly 10–20% of cases being misdiagnosed, depending on clinician experience [3]. Common misdiagnoses include general movement-impairing conditions that have a similar physical effect in the elderly, such as arthritis, or other neurodegenerative diseases. Furthermore, this method is invasive, and time and cost-inefficient, demanding overqualified professionals to conduct lengthy examinations that are uncomfortable and unnecessary for the patient. Above all, it fails to facilitate the biggest relief for PD patients. Due to the ongoing loss of brain cells being irreversible, PD currently has no cure, and early intervention is the next best step towards slowing the progression of the disease and prolonging a patient's life span. Unfortunately, unstandardized tests lack this timely manner, with many patients having to wait nearly 1 to 2 years before receiving a diagnosis.

Electroencephalography (EEG) is a neuroimaging technique that is used to roughly measure the electrical activity of groups of neurons in the brain. In a conventional test, a specialized cap equipped with small metal discs, referred to as electrodes, is attached to the scalp of the brain. Thereafter, the patient is requested to assume a resting position with their eyes closed for a duration of approximately 20 to 40 min to reduce the likelihood of interference that may undermine the accuracy of the results. Videos are routinely recorded to document movements that may complement the test, leading to a more precise diagnosis.

Due to their ability to effectively measure general neurological activity, EEG biomarkers have become quite useful in identifying certain conditions. Specifically, the characterization of brain injury, seizures, sleep disorders, inflammation, and general encephalopathy all have benefited from using EEGs. The primary objective of this study is to make a significant contribution to the ongoing research on PD by quantitatively evaluating current feature extraction algorithms to analyze EEG readings in both PD patients and healthy controls (HC). These features include the fast fourier transform (FFT) and temporal cycle-by-cycle analysis. We will implement Deep Learning to evaluate how well these features differentiate the two groups. We will filter out low-correlation features and subsequently train multiple deep learning models to assess the practicality and accuracy of EEG-based diagnosis for PD. Our aim is to acquire valuable insights into the disease and ultimately make substantial contributions towards improving the lives of millions of individuals worldwide affected by PD.

2 Existing Research

Past research done in the field of PD within the last two decades has revealed a considerable amount of information about what we know about the disease, including some information regarding its effect on EEG readings. Soikkeli et al. (1991) found a significant difference in the absolute amplitude (V) values of EEG power spectra between control and PD [15]. Specifically, demented patients showed the highest amplitudes in delta and theta, which are lower frequencies usually occurring during deep relaxation or sleep. Consequently, they also had the lowest alpha and beta amplitudes, which indicate alertness. These findings suggest that EEGs are often slowed in PD patients.

Additionally, several clinical and demographic variables have been linked to the quality of life in patients with PD. A study surveyed 233 patients and computed a Nottingham Health Profile (NHP), a measurement of the quality of life, for each patient, where a higher score indicated greater distress. Higher distress scores were found in PD patients, and the variables that predicted the greatest score were depressive symptoms, self-reported insomnia, and a low degree of independence [6]. Based on these results, it is possible that there are additional factors that contribute to the quality of life of individuals with PD, and further research is needed to understand better the full range of clinical and demographic variables that may be associated with this outcome.

Another study by Eeden et al. (2003) found that the incidence of PD varies significantly by both sex and race/ethnicity [4, 18]. 91% more men suffered from PD than women, and the rate per 100,000 people was highest in Hispanics, followed by non-Hispanic Whites, Asians, and Blacks. These results provide insight into the sociodemographic makeup of PD patients, which is helpful when understanding what populations are at greater risk.

In recent years, machine learning (ML) techniques have been increasingly utilized in the field of PD research. A literature review conducted by Mei et al. (2021) identified 209 studies published up until February 14, 2020, that explored ML techniques for diagnosing PD [10]. These studies utilized various data types, such as speech recordings, movement, and Brain MRI data, to identify patterns that distinguish PD patients from HC or patients with other movement disorders. A majority of these studies attained accuracies of at least 90%, surpassing that of the current survey and examination-based diagnostics. The promising results of numerous studies emphasize the potential of ML in tracking multiple parkinsonism-inflicted differences in patients and invite more ML-based methodologies to further our understanding.

3 Related Studies

In the field of PD, machine learning (ML) has emerged as a commonly used technique to evaluate the effectiveness of biomarkers for disease identification. EEGs have been established as a reliable measurement method for PD diagnosis [9]. Deep learning, particularly neural networks such as convolutional neural networks (CNNs), recurrent neural networks (RNNs), and probabilistic neural networks (PNNs), have demonstrated high accuracy rates of over 97% in classifying PD [7, 8, 20]. Other ML models, including support vector machines (SVMs), K-nearest neighbors (KNNs), decision trees (DTs), random forests (RF), naive bayes (NB), and logistic regression (LR), have also shown promising accuracy rates [1, 20].

The performance of ML models in differentiating populations is influenced by specific parameters and architecture choices. For example, Ruffini et al. (2019) explored DL for rapid eye movement, an early sign of α-synucleinopathies like PD [13]. Their CNN model with 4 hidden layers, dropout regularization, max-pooling layers, and a cross-entropy loss function achieved an average accuracy of 79 with a standard deviation of 1, and an AUC of 0.87 with a standard deviation of 0.1. Although these results were relatively lower compared to previous research, it can be inferred that adjusting parameters such as the number of hidden layers or choosing a different loss function could impact the model's performance. In another study by Oh et al. (2018), a 13-layer CNN with 4 1-D convolutional layers, 4 max-pooling layers, and an Adam optimizer achieved an accuracy of 88.25% [12]. Therefore, considering successful settings from previous research becomes crucial when designing experiments.

In addition to model selection and tuning, the extraction, selection, and formatting of features significantly impact ML performance. Preprocessing techniques that have proven successful in previous studies have shed light on effective methodologies. For instance, Yuvuraj et al. (2016) explored PD diagnosis using Higher Order Spectra (HOS) features in EEG signals, a technique for studying nonlinear features of signals [20]. The inclusion of this additional preprocessing step noticeably improved the accuracy of their models, with the highest average accuracy reaching 99.62% using an SVM model. These findings, along with other studies, emphasize the crucial role of preprocessing techniques and methods in influencing ML results.

4 Materials and Methods

4.1 Dataset

In this study, our dataset consisted of 99 patients diagnosed with PD and 48 HC participants [14]. This dataset, approved by the University of Iowa, included EEG recordings captured during the execution of the Simon Task—a widely utilized measure of interference and conflict resolution. During the task, participants were instructed to respond to visual stimuli by pressing a rightward or leftward key, irrespective of the stimuli's location on the screen (left or right). By examining reaction times and accuracy separately for congruent (matching stimulus location and response hand) and incongruent (mismatching stimulus location and response hand) trials, we aimed to investigate potential differences in cognitive conflict processing between individuals with PD and HC. The congruency information for each patient was reported in the patient metadata file, which was later considered as a feature.

Regarding the gender distribution, it was noted that the PD group exhibited an imbalance, consisting of 67 male patients and 32 female patients. We recognized the established differences in EEG readings between males and females, which raised slight concerns. To address this potential confounding factor, we included gender as a feature in our analysis. By accounting for the differences in average neuroelectric activity between genders, we aimed to mitigate any potential impact on model performance. In the control group, the distribution of males and females was relatively balanced, with 26 males and 22 females included in the dataset.

Recordings were roughly anywhere from 7–14 min, measured at a sampling frequency of 500 Hz. A total of 63 electrodes were used to measure the electrical activity of neurons in each patient, with one stimulus. Raw arrays from the.fdt files were truncated to 197180 points to account for the shortest recording length. This was to account for the lack of dynamic array lengths in certain models. The shortened pre-extracted data represents roughly 6.572 min of each reading.

Additional information, such as behavior was documented for each patient. These extra observations or information didn't fit under other categories and were mostly omitted. However, Test scores, such as the Montreal cognitive assessment (MoCA) and the Unified Parkinson's disease rating scale (UPDRS) were also recorded in the dataset to demonstrate general cognitive ability, though it is unclear whether these were conducted by the researchers or surveyed for extra information. For the purpose of this study, all aforementioned extra features are unnecessary, and will be omitted, with the exception of the MoCA score. It is notable that the average PD MoCA score was 24.465, which is a

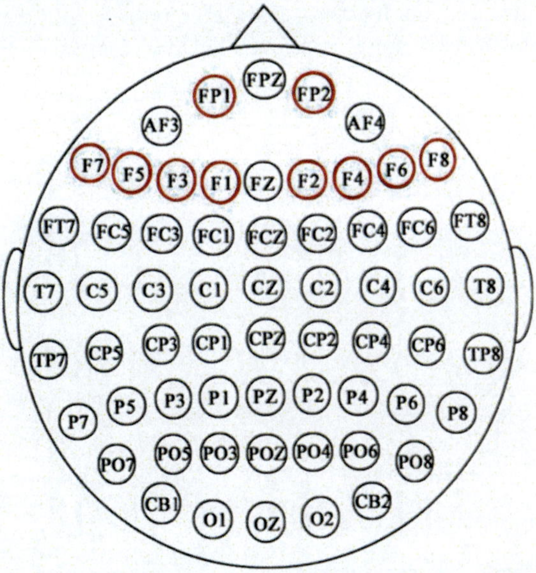

Fig. 1. Diagram of 62-EEG System, with Annotated Electrodes of Interest.

little lower than normal cognition. Including this additional information likely benefited our results.

4.2 Region of Interest

The dataset used in this study consisted of 62 electrodes placed on the scalp according to the internationally recognized 10–20 system. To ensure optimal electrical contact with the skin, conductive gel was used during electrode attachment. This placement allowed for the recording of electrical activity from specific regions of the brain, including the frontal, temporal, parietal, and occipital lobes, as well as the midline structures.

Interpreting the electrical activity of the basal ganglia and substantia nigra located deep within the brain, can be challenging due to their size and location. The proximity of surrounding regions can lead to potential misinterpretations or an overrepresentation of activity in these areas.

To address this, our analysis in this study involved averaging the signals from all channels that started with 'FP' or 'F', with the exception of those that contained 'Z' as a suffix' (Fig. 1), as those also include the central lobe. The prefrontal cortex plays a role in dopamine signaling, which indicates that it also decreases in overall neural activity in the presence of PD. By comparing this specific brain section individually between different populations, we aim to present a more comprehensive understanding of the potential of electrical activity as an indicator of motor degeneration.

4.3 Cycle-by-Cycle Analysis

The cycle-by-cycle algorithm (i.e. bycycle) is a novel approach for analyzing neural oscillations in the time domain. It aims to extract the properties of the oscillatory component, even in the presence of noise, by segmenting the signal into individual cycles and

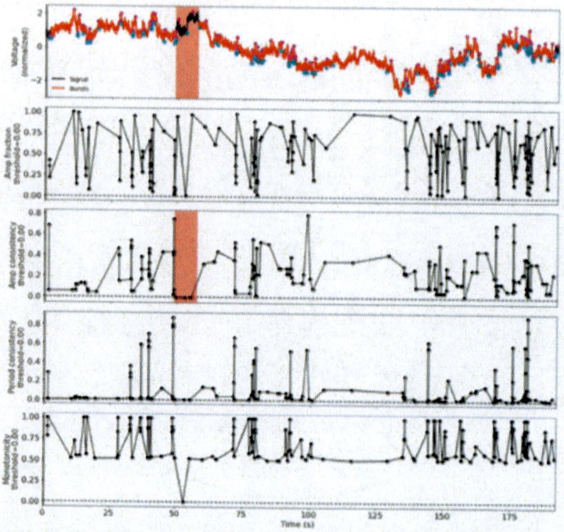

Fig. 2. Annotated feature computation and burst detection on F3 node in PD patient from bycycle library.

analyzing them separately [2]. A cycle is measured by the distance between one peak to another or one full wave. The algorithm's advantage over the conventional methods is that it allows for better analysis of waveform shape properties of neural oscillations. It is also helpful in studying amplitude and frequency effects.

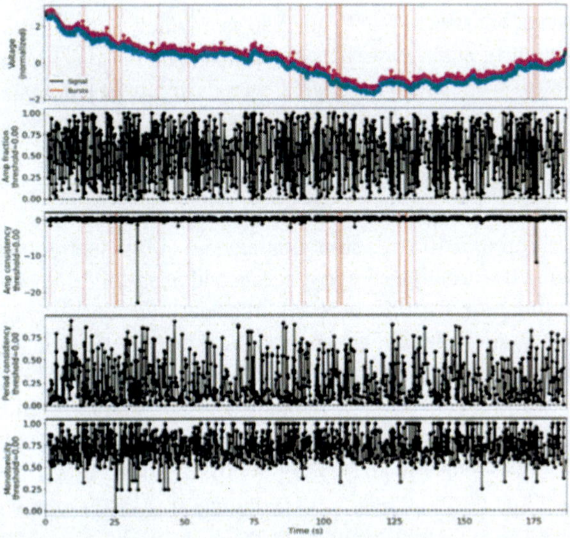

Fig. 3. Annotated feature computation and burst detection on F3 node in healthy control from bycycle library.

Figure 2 and Fig. 3 show 5 of 24 different feature computations of two patients; one control and one PD, generated by the plot burst detect summary function. The features ordered from top to bottom are voltage, amp fraction, amp consistency, period consistency, and monotonicity. The method takes one node reading as input and identifies aperiodic or non-cyclic activity in the wave, indicated by a red rectangle.

When observing the two summaries, we see the algorithm's ability to detect aperiodic or non-cyclic activity in the neural oscillations of the two experimental groups. The red rectangles in the figures indicate where bursts of activity were detected. It is evident from the figures that the voltage over time in PD patients is much more erratic, with a much larger window of irregularity and more bursts indicated by a solid red line. In contrast, the healthy control's voltage over time is more consistent and shows fewer bursts of activity. This is a clear indication of the disrupted neural activity in patients with PD.

In this project, we leveraged the power of bycycle to compute various features from the EEG data, including amplitude, period, symmetry, and more. These features provide crucial measurements that can be tested to identify differences in the neural activity between Parkinson's disease (PD) patients and healthy controls. We computed bycycle data frames for each patient in the experimental groups and averaged them with respect to time, or by rows. We then ran 24 two-sample t-tests for differences in population means of each feature and observed the p-values to make conclusions. By analyzing these features, we can uncover distinctive patterns and irregularities in the neural activity of PD patients.

4.4 FFT Spectral Analysis

Spectral Fast Fourier Transform (FFT) analysis is a widely used method for examining the frequency content of signals, such as electroencephalogram (EEG) data. It enables the transformation of a time-domain signal into its frequency-domain representation, providing insights into the distribution of power across different frequency bands. By applying the FFT algorithm to EEG signals, we can uncover the magnitude and phase of frequency components, revealing valuable information about the spectral characteristics of neural oscillations.

The FFT algorithm transforms a time-domain signal into its frequency-domain representation, allowing us to analyze the magnitude and phase of different frequency components. The resulting power spectrum reveals the amplitude or power of each frequency component present in the signal. This analysis enables the identification of frequency bands associated with specific neural activities, such as delta (0.5–4 Hz), theta (4–8 Hz), alpha (8–13 Hz), beta (13–35 Hz), and gamma (above 35 Hz) oscillations.

Previous research has demonstrated alterations in the spectral characteristics of EEG signals in PD patients compared to healthy controls. These alterations provide valuable insights into the neurobiological underpinnings of PD and can serve as potential biomarkers for diagnosis and monitoring. For instance, studies have reported increased beta band power and reduced alpha band power in PD patients, reflecting abnormal oscillatory activity and disrupted neural networks. These spectral changes are believed to be associated with motor and cognitive impairments observed in PD.

In our study, we aim to incorporate Spectral FFT analysis as part of our analytical framework, alongside the bycycle algorithm. By examining the frequency-specific characteristics of EEG signals, we can further explore the relationship between neural oscillations and PD. This comprehensive analysis allows us to identify distinctive spectral patterns, detect frequency-specific abnormalities, and unravel the underlying mechanisms of PD.

To perform Spectral FFT analysis, we apply the FFT algorithm to the pre-processed EEG data. The resulting power spectrum provides information about the magnitude and distribution of different frequency components. We add this transformed data to our feature set to investigate the spectral alterations associated with PD, where each datapoint is a column. By quantifying the power in these frequency bands, we can extract relevant features for classification and gain insights into the neural dynamics of PD.

4.5 Feature Reduction

Due to the multi-dimensional and multi-factorial nature of our compiled feature set, there likely exist features that are unnecessary, and reduce the effectiveness of our models. Thus, to achieve the best model performances, we implemented feature reduction as an extra step in preprocessing. Feature reduction aims to select or transform the most informative and relevant features while discarding redundant or noisy ones. In our study, we employed two commonly used techniques: t-SNE (t-Distributed Stochastic Neighbor

Embedding), SVD (Singular Value Decomposition) and PCA (Principal Component Analysis).

To scale our data, we implemented standard scaling, a technique used to normalize/standardize the data to a normal distribution with a mean (μ) of 0 and a standard deviation (σ) of 1. This step in data preprocessing is important as it brings all features to a common scale, typically contributing to improved model performance. In the case of our data, the standard scaler processes the data feature-wise, meaning that each column is separately transformed.

t-SNE, also known as t-distributed stochastic neighbor embedding, is a statistical technique widely employed to visualize data with high dimensionality. Its primary objective is to assign each data point a position in a two or three-dimensional map. By extending the principles of Stochastic Neighbor Embedding, t-SNE represents high-dimensional objects as points in a lower-dimensional space. It ensures that objects with similarities are positioned close to each other, while dissimilar objects are placed farther apart with a higher probability.

The algorithm is composed of two key stages: constructing a probability distribution for pairs of high-dimensional objects based on their similarity and defining a similar probability distribution for points in the lower-dimensional map. Through the minimization of the Kullback-Leibler divergence between these distributions, t-SNE successfully captures the underlying structure and intricate relationships present within the data.

SVD, or Singular Value Decomposition, is a mathematical technique employed for feature reduction and dimensionality reduction. It decomposes a matrix into three separate matrices: U, Σ, and V*. The singular values in Σ represent the importance or significance of each component, and the corresponding left and right singular vectors provide information about the direction or pattern associated with each component. SVD has various applications, including computing the pseudoinverse, matrix approximation, determining the rank, range, and null space of a matrix. It finds utility in fields such as least squares fitting of data, process control, and particularly, signal processing.

PCA, or Principal Component Analysis, is another commonly used technique for feature reduction and dimensionality reduction. It transforms the original feature space into a new set of orthogonal variables called principal components.

Each principal component is a linear combination of the original features and captures the maximum variance in the data. By selecting a subset of the most informative principal components, we can effectively reduce the dimensionality of the data while preserving the most significant patterns or structures.

In our study, we employed all three techniques: t-SNE, SVD, and PCA; as part of our feature reduction process. After creating reduced feature sets of the original features, we were left with 5 inputs for our models: Our raw feature set (2170624 features/dimensions), 3 separately reduced sets (3 features/dimensions each), and the raw EEG data (11 $\times$ 197180 data points for each patient).

4.6 Model Selection

For the classification of the extracted features, we explored a range of traditional machine learning algorithms, including Support Vector Machines (SVMs), K-Nearest Neighbors (KNNs), Decision Trees (DTs), Random Forests (RF), Naive Bayes (NB), and Logistic

Regression (LR). These models have demonstrated promising accuracy rates in PD classification tasks in previous studies [1, 9, 20]. We employed appropriate parameter tuning techniques and evaluated the performance of each model based on accuracy, precision, recall, and F1-score.

In addition to traditional machine learning algorithms, we also considered deep learning models to exploit the capabilities of neural networks in processing the raw EEG data. Specifically, we explored Dense Neural Networks (DNNs). Usage of this specific network is advantageous, as their connected access to features in previous layers enables them to capture complex and nonlinear patterns in EEG features.

To facilitate a comprehensive comparison, we also introduced experimentation with the raw EEG data points in addition to the preprocessed feature set. For this purpose, we utilized the Keras sequential class, which allowed us to flexibly modify the layers of the deep models we implemented. We tested a DNN model to explore its performance and determine its suitability for the task at hand.

5 Results

Table 1. Model validation accuracy.

Algorithm	Without Feature Reduction	t-SNE	SVD	PCA	Raw Data
SVM	70.00	70.00	70.00	70.00	NaN
KNN	66.67	56.67	66.67	36.67	NaN
DT	56.67	70.00	70.00	70.00	NaN
RF	70.00	60.00	56.67	50.00	NaN
Gaussian NB	46.67	70.00	43.33	76.67	NaN
LR	56.67	70.00	63.33	66.67	NaN
DNN	NaN	81.20	67.52	66.66	67.50

Table 1 summarizes the accuracy results of the different models used in our study with various types of inputs. The models included Support Vector Machines (SVM), K-Nearest Neighbors (KNN), Decision Trees (DT), Random Forests (RF), Naive Bayes (NB), Logistic Regression (LR), and a Dense Neural Network (DNN). The non-deep learning models were trained on the non-reduced feature set as well as the feature sets obtained through t-SNE, SVD, and PCA dimensionality reduction techniques. The highest average accuracy achieved across all input types was obtained by the SVM model, with an accuracy of 70.00%.

The DNN model's performance was evaluated both the optimized feature sets and the raw data. However, the model performed best when the features were extracted and optimized. Figures 4 and 5 present the accuracy and loss metrics of the DNN model over approximately 400 epochs using the t-SNE feature reduced data. The model showed rapid improvement in the number of valid classifications during the initial 50 epochs.

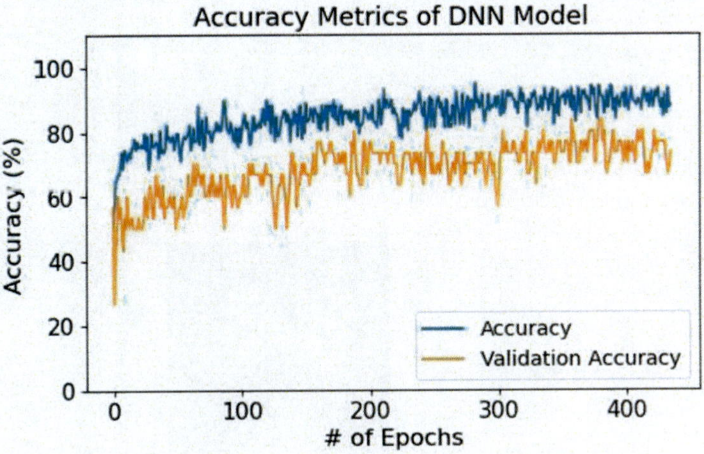

Fig. 4. Accuracy metrics of DNN model over ~400 Epochs.

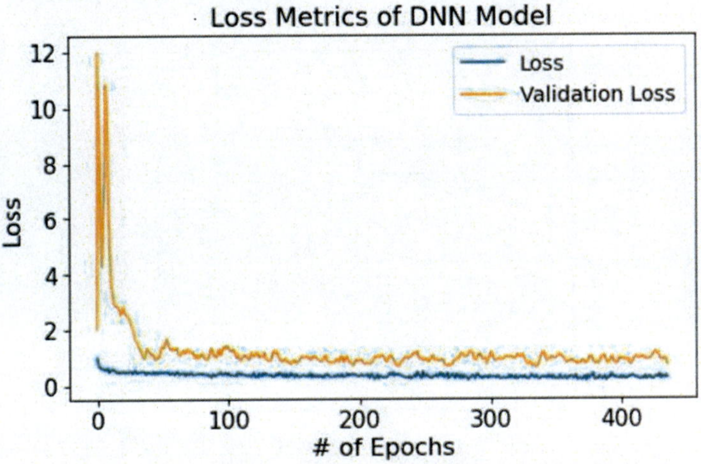

Fig. 5. Loss metrics of DNN model over ~400 Epochs.

Both the training and validation accuracy continued to increase for the next several hundred epochs, albeit with occasional fluctuations. These fluctuations suggest that the model's accuracy may not be entirely stable, indicating the need for further evaluation and refinement.

The loss pattern observed in the DNN model provides additional insights into its performance for PD classification. The steep decrease in training loss during the initial 50 epochs indicates that the model effectively learned from the available PD data. However, the pattern of the validation loss suggests that the model encountered challenges in generalizing well to unseen data.

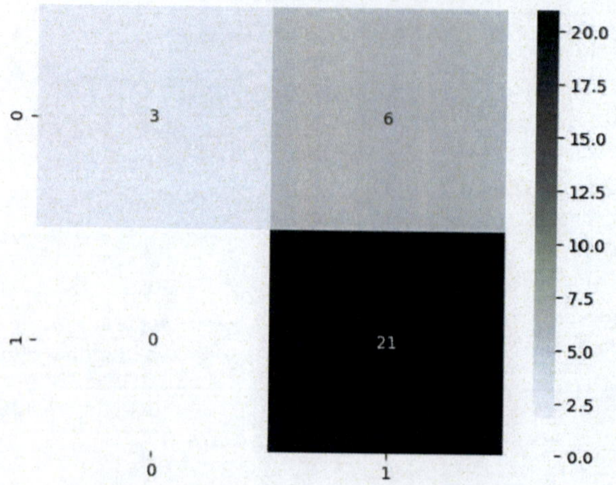

Fig. 6. Confusion matrix of DNN model.

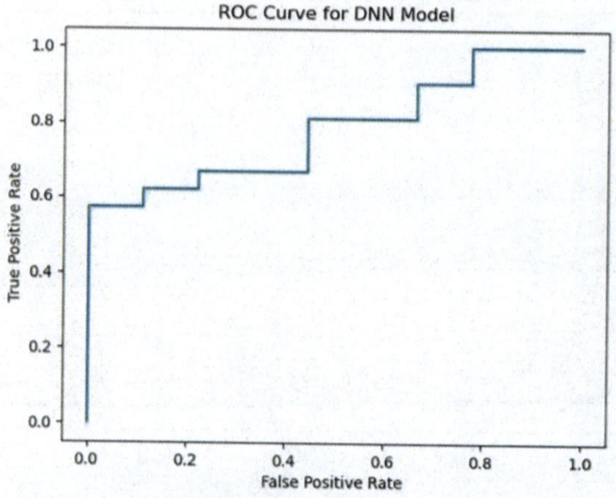

Fig. 7. ROC curve of DNN model.

To gain further insights into the performance of the models, we analyzed the confusion matrix and the receiver operating characteristic (ROC) curve for the DNN model. Figure 6 displays the confusion matrix, which shows the distribution of predicted labels compared to the true labels. The matrix provides information about the model's performance in terms of true positive (TP), true negative (TN), false positive (FP), and false negative (FN) predictions. A well-performing model would exhibit a high number of TPs and TNs and a low number of FPs and FNs.

The ROC curve, depicted in Fig. 7, illustrates the trade-off between the true positive rate (sensitivity) and the false positive rate (1-specificity) across different classification thresholds. Ideally, a desirable ROC curve would be situated closer to the top-left corner, indicating a high true positive rate and a low false positive rate. However, in our analysis, the ROC curve appeared to be positioned more towards the center for a significant portion of its trajectory, which is considered highly undesirable. This indicates that the model's performance in distinguishing between Parkinson's disease and healthy controls may not be optimal, as it exhibits relatively high false positive rates and/or low true positive rates across a range of classification thresholds.

Overall, the results highlight the performance of the SVM model as the most accurate among the traditional machine learning models tested. The DNN model demonstrated great potential, but further refinement and evaluation are necessary to enhance its stability and generalizability.

6 Discussion

The use of the bycycle library in this study proved valuable in capturing differences in the electrical activity of PD patients compared to healthy controls. The library's ability to compute various features and generate informative graphs and figures allows for visual observation of aperiodic activity or bursts, aiding in the identification of patterns and irregularities. This highlights the potential of the bycycle algorithm as a cost-effective and efficient tool for diagnosing PD and minimizing inaccuracies or guesswork.

The volt rise and volt decay features showed noticeably lower values compared to other features. This supports the notion that there exists a slowing of readings and activity in PD patients and an increase in delta and theta frequencies, which is crucial when trying to look for more specific biomarkers in EEG.

Upon performing a feature selection, we were able to remove features that were not contributing to the overall performance of the model, thereby making the model very specific to respond only to correlated and important features. By eliminating irrelevant or redundant features, the model became more focused and responsive to the correlated and important features. This step was crucial in building a precise and efficient model for PD classification.

One notable benefit of incorporating Spectral Fast Fourier Transform (FFT) analysis into our analytical framework is the ability to examine the frequency specific characteristics of EEG signals. By applying the FFT algorithm to the preprocessed EEG data, we transformed the time-domain signals into their frequency-domain representations, which enabled us to analyze the magnitude and phase of different frequency components and obtain a power spectrum that reveals the amplitude or power of each frequency component present in the signal.

7 Conclusion

The results of this study yield valuable information that benefits the growing field of neurodegenerative disease research. Firstly, a statistically significant difference in over half of the features computed by the bycycle library underscores its ability to highlight differences in the electrical activity of PD and Controls. This finding is significant, as it shows promise for a solution that is more cost-effective and efficient and minimizes guesswork or inaccuracies in a diagnosis. Additionally, its graphs and figures can be particularly useful in visually observing aperiodic activity or bursts and studying patterns that may benefit future research.

Future studies that may be conducted to build upon this research include continuing to implement the bycycle library to other neurodegenerative diseases or mental conditions that can be modeled by changes in electrical activities, namely Alzheimer's disease, epilepsy, traumatic brain injury (TBI), and various sleep disorders. More importantly, we can increase the region of interest and model electrical activity in different parts of the brain or study the brain as a whole.

In addition, it may be fruitful to expand the demographic of the study. For instance, past research has suggested that ethnic variation is a determinant of heterogeneity in PD. Thus, it may be helpful to study differences in the neuro-physiological data in PD patients with respect to race to test for key differences. We could also observe other socioeconomic factors and see their relations with PD, which can improve precision medicine. Due to the low patient count, it wasn't feasible to stratify patients based on demographical factors. However, as researchers continue to explore EEGs for PD, data will likely become more diverse, allowing for a more thorough analysis.

Finally, Future studies with similar methodologies as presented in this paper could further explore the application of Higher Order Spectra (HOS) analysis in PD research. HOS analysis extends the traditional spectral analysis by capturing higher-order statistical dependencies, and by incorporating HOS techniques, researchers can potentially uncover more intricate patterns and nonlinear interactions in the data, thereby enhancing our understanding of the underlying neurophysiological mechanisms.

References

1. Chaturvedi, M., et al.: Quantitative EEG (QEEG) measures differentiate Parkinson's disease (PD) patients from healthy controls (HC). Front. Aging Neurosci. (2017). https://doi.org/10.3389/fnagi.2017.00003
2. Cole, S., Voytek, B.: Cycle-by-cycle analysis of neural oscillations. J. Neurophysiol. **122**(2), 849–861 (2019). https://doi.org/10.1152/jn.00273.2019
3. Hess, C.W., Okun, M.S.: Diagnosing Parkinson disease. CONTINUUM: Lifelong Learn. Neurol. **22**(4), 1047–1063 (2016). https://doi.org/10.1212/CON.0000000000000345
4. Juarez, J., Corsi-Cabrera, M.: Sex differences in interhemispheric correlation and spectral power of EEG activity. Brain Res. Bullet. **38**(2), 149–151 (1995). https://doi.org/10.1016/0361-9230(95)00081-O
5. Kalia, L.V., Lang, A.E.: Parkinson's disease. The Lancet **386**(9996), 896–912 (2015). https://doi.org/10.1016/S0140-6736(14)61393-3
6. Karlsen, K.H., Larsen, J.P., Tandberg, E., Maland, J.G.: Influence of clinical and demographic variables on quality of life in patients with Parkinson's disease. J. Neurol. Neurosurg. Psychiatry **66**(4), 431–435 (1999). https://doi.org/10.1136/jnnp.66.4.431
7. Lee, S., Hussein, R., McKeown, M.J.: A deep convolutional-recurrent neural network architecture for Parkinson's disease EEG classification. IEEE (n.d.). https://doi.org/10.1109/GlobalSIP45357.2019.8969309
8. Lee, S., Hussein, R., Ward, R., Jane Wang, Z., McKeown, M.J.: A convolutional-recurrent neural network approach to resting-state EEG classification in Parkinson's disease. J. Neurosci. Methods **361**, 109282 (2021). https://doi.org/10.1016/j.jneumeth.2021.109282
9. Maitin, A.M., Romero Munoz, J.P., Garcıa-Tejedor, A.J.: Survey of machine learning techniques in the analysis of EEG signals for Parkinson's disease: a systematic review. Appl. Sci. **12**(14), 6967 (2022). 10. 3390/app12146967

10. Mei, J., Desrosiers, C., Frasnelli, J.: Machine learning for the diagnosis of Parkinson's disease: a review of literature. Front. Aging Neurosci. (2021). https://doi.org/10.3389/fnagi.2021.633752
11. Nuytemans, K., Theuns, J., Cruts, M., Van Broeckhoven, C.: Genetic etiology of Parkinson disease associated with mutations in the SNCA, PARK2, PINK1, PARK7, and LRRK2 genes: a mutation update. Hum. Mutat. **31**(7), 763–780 (2010). https://doi.org/10.1002/humu.21277
12. Oh, S.L., et al.: A deep learning approach for Parkinson's disease diagnosis from EEG signals. Neural Comput. Appl. **32**(15), 10927–10933 (2018). https://doi.org/10.1007/s00521-018-3689-5
13. Ruffini, G., et al.: Deep learning with EEG spectrograms in rapid eye movement behavior disorder. Front. Neurol. **10** (2019). https://doi.org/10.3389/fneur.2019.00806
14. Singh, A., Cole, R.C., Espinoza, A.I., Wessel, J.R., Cavanagh, J.F., Narayanan, N.S.: Evoked midfrontal activity predicts cognitive dysfunction in Parkinson's disease. MedRXiv (2022). https://doi.org/10.1101/2022.07.26.22278079
15. Soikkeli, R., Partanen, J., Soininen, H., Paakkonen, A., Riekkinen, P.: Slowing of EEG in Parkinson's disease. Electroencephalogr. Clin. Neurophysiol. **79**(3), 159–165 (1991). https://doi.org/10.1016/0013-4694(91)90134-P
16. Su, X., Federoff, H.J.: Immune responses in Parkinson's disease: interplay between central and peripheral immune systems. Biomed. Res. Int. **2014**, 1–9 (2014). https://doi.org/10.1155/2014/275178
17. Tolosa, E., Wenning, G., Poewe, W.: The diagnosis of Parkinson's disease. Lancet Neurol. **5**(1), 75–86 (2006). https://doi.org/10.1016/S1474-4422(05)70285-4
18. Van den Eeden, S.K.: Incidence of Parkinson's disease: variation by age, gender, and race/ethnicity. Am. J. Epidemiol. **157**(11), 1015–1022 (2003). https://doi.org/10.1093/aje/kwg068
19. Zheng, W.-L., Lu, B.-L.: Investigating critical frequency bands and channels for EEG-based emotion recognition with deep neural networks. IEEE Trans. Auton. Ment. Dev. **7**(3), 162–175 (2015). https://doi.org/10.1109/TAMD.2015.2431497
20. Yuvaraj, R., Rajendra Acharya, U., Hagiwara, Y.: A novel Parkinson's disease diagnosis index using higher-order spectra features in EEG signals. Neural Comput. Appl. **30**(4), 1225–1235 (2016). https://doi.org/10.1007/s00521-016-2756-z

Meta-heuristic Algorithms in IoT-Based Application: A Systematic Review

Alaa Hassan(✉)

University of Calgary, Calgary, Canada
eng.alaa.hassan@ieee.org

Abstract. The Internet of Things (IoT) has become increasingly popular as a result of the development of technologies like big data analytics, blockchain, artificial intelligence, machine learning, and deep knowledge. The Internet of Things (IoT)-based systems offer a logical and automated framework for sound decision-making and the automation of numerous chores to make life easier for people. Meta-heuristic algorithms use collective intelligence to tackle challenging issues and are self-organized, decentralized algorithms. Meta-heuristic algorithms have recently gained notoriety for handling several IoT-related problems. The meta-heuristic techniques employed to create IoT-based apps are carefully examined in this paper. Broad categories and descriptions of existing meta-heuristic-based algorithms have been provided. Also demonstrated are the numerous applications of IoT-based systems that employ meta-heuristic methods. In order to demonstrate the fresh potential of the scholars, the current research subjects are also discussed. The current IoT trends and projected future advancements are summarized in the last section. Researchers looking into meta-heuristic algorithms and Internet of Things-based systems will discover new paths after reading this paper.

Keywords: Meta-heuristic Algorithms · Swarm Intelligence · Internet-of-Things

1 Introduction

One of the popular methods for resolving various challenging real-world issues is the use of meta-heuristics. The intriguing behaviors seen in nature are the true inspiration for these algorithms. Researchers have been exploring practical problem-solving algorithms that rely on self-organized systems in recent years due to the increased complexity of optimization-based challenges [1, 2]. The behavior of physical phenomena, and living things like fish, birds and hummingbirds inspire meta-heuristics. They exhibit intelligent behaviors at different group levels and are defined by interactions among local people [3]. Many meta-heuristics have been developed and effectively used to address a variety of applications. Also, the newly proposed algorithms are still being researched and must demonstrate their effectiveness. Many reviews have been suggested for the different meta-heuristic algorithm uses. In contrast to previous research published in the literature, this study collects the application breadth and offers the correct conclusions for IoT-based applications. The current work aids researchers in focusing their investigation on a decisive path.

K. Arai (Ed.): FTC 2023, LNNS 816, pp. 104–116, 2023.
https://doi.org/10.1007/978-3-031-47448-4_9

Moreover, researchers may leverage the significant discoveries in the literature to develop several heuristics for resolving applications. IoT has received much attention recently and it can be used in various productions, including wearable technology, traffic monitoring systems, and medicine [4]. "Internet-of-Things" (IoT) refers to related technologies, including sensors, actuators, and data processing for creating applications and services. IoT aims to give little devices the intelligence they need to make decisions, sense their environment, and work with other machines, items, and even people. It is a field of study where a network connects tangible objects like machinery and people. The speedy information sharing makes it possible for new services and applications. There is a rise in the volume, diversity, and velocity of data as the IoT network grows. IoT systems face several problems, including dynamic characteristics, device mobility, and wireless communications.

On the other hand, meta-heuristics have the adaptability and durability qualities that make them an effective design paradigm for handling complicated IoT-based challenges. As a result, sophisticated IoT systems implemented in real-time activities and modeled using an IoT-based method to accomplish specific objectives might draw inspiration from meta-heuristic algorithms. The combination of meta-heuristics with IoT for addressing real-time challenges is shown in Fig. 1. Using these straightforward guidelines for personal conduct and interpersonal interactions ultimately achieves a global optimum. The capacity to self-organize must be altered to scale effectively in such a diverse setting. According to recent research, many algorithms are developed in the main application domains. This work includes a comprehensive literature assessment that identifies the algorithms used in the systems up to this point [5]. The present study's goal has three components. Initially, a thorough explanation of algorithms and their use in diverse applications have been given. The many benefits of meta-heuristics in IoT-based systems are demonstrated in the next stage. Lastly, key IoT meta-heuristics characteristics, current obstacles, research concerns, and enhancements for future study are presented.

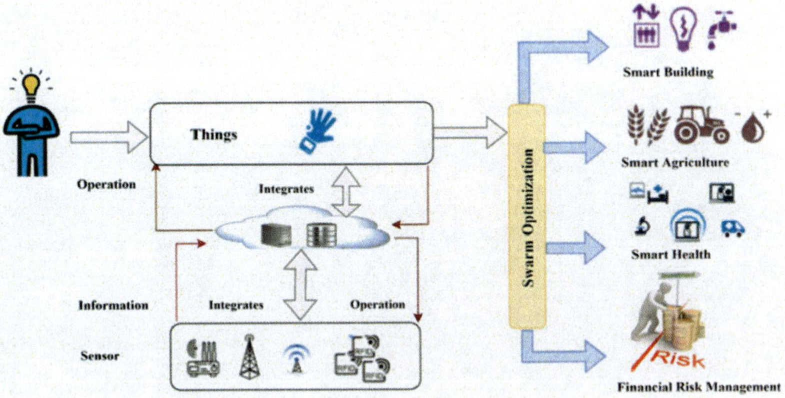

Fig. 1. The algorithm integration in IoT systems.

2 Related Work

This section describes how algorithms are broadly considered in the literature. Depending on what they are, the algorithms are divided into five groups. Several meta-heuristic-based methods are discussed together with their IoT-based applications.

2.1 Meta-heuristic Algorithms

Many optimization issues are arising in various domains, including bioinformatics, computer vision, and the Internet of Things [6]. Thus, only small-scale instances can be treated by precise mathematical techniques. Instead of losing up, the researchers reasoned that they could discover a workable solution using available approximation approaches within the allotted time. Depending on how they use randomization, these algorithms may be divided into meta-heuristics and heuristics. Heuristic algorithms, instead of meta-heuristics, are more problem-dependent, which is a significant distinction between the two algorithms. Just a few specialized issues can be solved using these techniques.

On the other hand, because meta-heuristic algorithms employ the so-called "black box" optimizer, they may be related to almost any optimization problem. According to recent research, many nature-inspired metaheuristics algorithms are now in use, including swarm intelligence, and physical-based algorithms. A systematic review of the literature on swarm intelligence in the Internet of Things technology is performed in this work. Eight well-known databases were used to gather the literature: Google Scholar, IEEE Explorer, and Taylor Francis. There is an investigation into the most relevant reviews, referenced articles, and hot places. Figure 2's information depicts a survey report summary of numerous papers.

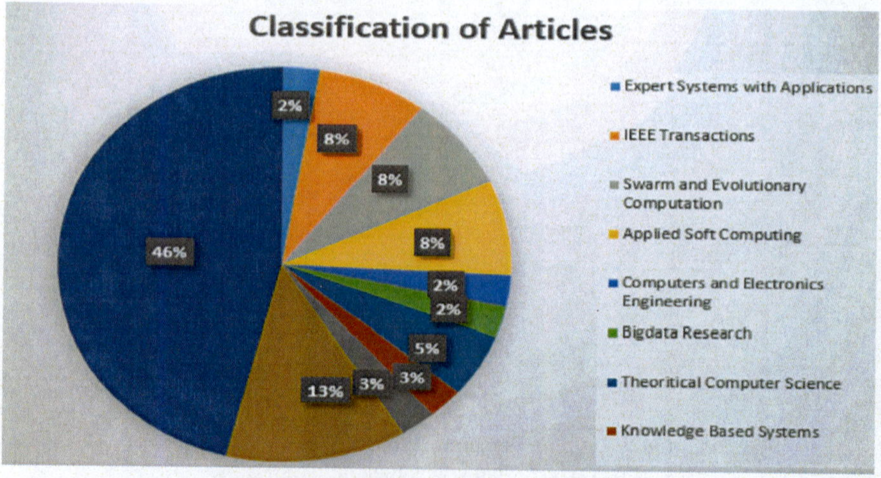

Fig. 2. The papers in different databases.

2.2 Classification of Meta-heuristic Algorithms

More than a thousand of algorithms have been created to address real-world optimization issues in various fields. These existing algorithms have been categorized and summarised in this section. The known meta-heuristics are categorized in Fig. 3 depending on how they behave in real-world situations.

Classification of Meta-heuristic Algorithms. These algorithms begin with a population set; then, a search procedure is carried out over a series of generations until the best workable answer is found [7, 8]. The optimizer [9], evolutionary algorithm [10], acceptable agent evolutionary algorithm [11], and genetic programming [12, 13] are only a few of the algorithms that have been developed for use in evolutionary computing. In addition to providing answers to problems like the traveling salesperson problem, and training artificial neural networks, evolutionary-based methods can be used to solve problems that are more commonly encountered. These algorithms often address many real-world issues in engineering, agriculture, basic science, and several sectors.

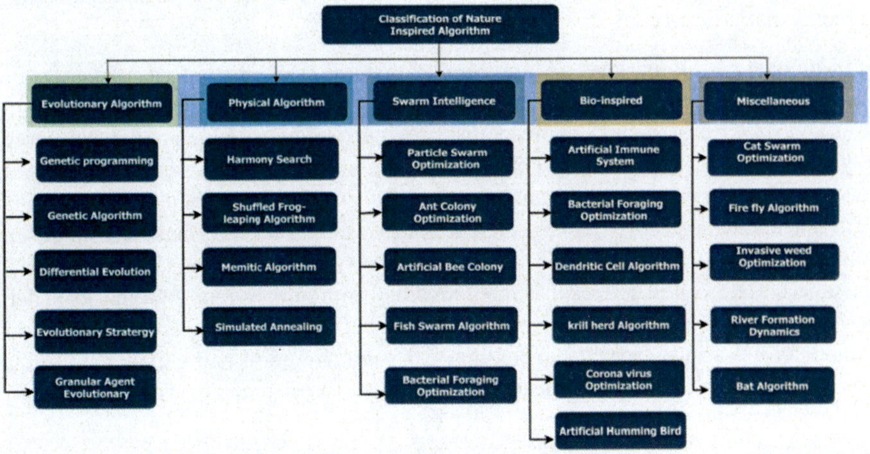

Fig. 3. Meta-heuristics Algorithms.

Physics Heuristics Algorithms. The physical laws of the cosmos, such as the laws of gravity and newtons law, inspire the physics-based approaches [14, 15]. Different behavior is shown at various temperatures via algorithms-based on simulated annealing. The material is initially warmed during the physical process, and then the temperature is progressively lowered to remove flaws, which reduces the system's energy consumption. Finding the ideal solution is more effective using simulated annealing. Moreover, these algorithms need a beginning point technique [16, 17]. Ray optimization, the black hole method, local gravitational search, electron search, and the shuffling frog jumping algorithm are well-known techniques in simulated annealing. These algorithms have shown encouraging outcomes in addressing issues from the actual world [18, 19]. Further testing has been done to see whether basic physics-based behavior may be incorporated into current applications, including structural alignment, information retrieval from the web, temporal frequency analysis, vehicle routing issues, and software testing [20, 21]. However, the benefits offered by distributed physics systems may only sometimes be effectively tapped into by these applications.

Swarm Intelligence-Based Meta-heuristics Algorithms. The structured collective activity of social animals or insects is used by swarm intelligence, an artificial intelligence method, to solve problems, according to Collins Dictionary [22]. Moreover, biological research is where swarm intelligence's roots may be found. The systems that get their inspiration from collective intelligence are the focus of swarm intelligence. In addition, interactions amongst homogenous agents are what generate this collective intelligence. Many intelligent behaviors have inspired the development of various algorithms. The biological swarm intelligence may be seen in herds of birds, ant colonies, fish schools, mosquitoes searching for hosts, social spiders, and honey bee mating [23, 24]. Artificial bee colony and particle swarm optimization [25, 26] are a few of the most well-known swarm intelligence techniques; Ageist spider monkey optimization and artificial hummingbird algorithm [27–38]. These many applications aim to address various problems that arise in everyday life.

Meta-heuristics Algorithms. Algorithms with biological inspiration are becoming increasingly common in the contemporary environment. Most meta-heuristic algorithms are drawn from biologically evolved live creature behavior, which is what biologically inspired algorithms mean [8]. These algorithms are dispersed, decentralized, self-organized flexible in nature. The amount of data is growing dramatically as we go into the digital age. One problem is figuring out how to get helpful information from decentralized data. Also, it takes much work to discover the best option. There is a need for some clever strategies to solve this. Artificial immune system and and krill herd algorithm [39–46] are some examples of intelligent bio-inspired algorithms. Various researchers in various contexts, including text summarization and network optimization, have shown the use of bio-inspired algorithms. Experimentally and conceptually, the algorithms above redefine the parameters utilized to update and assess the various applications.

Miscellaneous Meta-heuristics Algorithms. Algorithms with various meta-heuristics are inspired by nature. Several nature-inspired metaheuristics have been used in multiple real-world applications during the last three decades. These techniques may replace several conventional real-time-based applications, which have been realized as well-known tools. The search space has been intensified and diversified in several recent meta-heuristics algorithms that academics have created. River formation dynamics, cuckoo search algorithm, and gravitational optimization [47–59] are some additional examples of nature-inspired algorithms. The algorithms competence seems to impact many real-world applications, as is further evident from the examination and discussion of the authors' suggested algorithms that were done above [60, 61]. These algorithms are created using observing, designing, converting mathematical modules, creating pseudocode, and testing as the final phases [62, 63]. These methods have addressed pattern recognition, and decision-making problems. Several algorithms used over the last four decades are categorized along with the application domains and the sources from which they originated. Examining a particular optimization approach was not prioritized in these investigations, either. Nonetheless, a wide range of them is taken advantage of.

3 An Overview of Meta-heuristic IoT Application

This part has provided a thorough analysis of algorithm used in IoT systems. Four main categories are used to group the apps.

3.1 Enhancing Health Care Services

Swarm intelligence (SI) in the IoT has become a research topic, offering keys to challenging problems, including the delivery of healthcare services [64]. Because it can quickly and cheaply extract patient data, collect sensor data, diagnose diseases, and provide other healthcare services, SI plays a significant role in supervising the healthcare sector in the smart cities. However, the main issue in smart cities is handling each stakeholder's request [65]. Additionally, several task scheduling methods are employed to improve speed and shorten the execution time of requested activities. The methods above will assist the various stakeholders in minimizing turnaround time and maximizing resources.

Moreover, intelligent optimization techniques like particle swarm have been introduced to help with the job scheduling challenge and to provide stakeholders with increased support. However, various frameworks and algorithms have been presented to enhance the functionality of the current healthcare system. There are some restrictions, the first of which is the issue with the data set. Large-scale data are needed for models to function well. Second, the model's performance is impacted by choosing the best strategy in various circumstances in this industry. In addition, there are a few outstanding problems and likely trends for future development, including information fusion and security.

3.2 Financial Risk Management

Several Swarm intelligence applications are created in recent years with the increasing use of the internet of things. Data is gathered in an unstructured manner since the growth of economic data is so rapid [66]. Commercial banks should be concerned about correctly preventing and managing different risks from various sources, including the internet. Thus, it is necessary to create some models enough to forecast behavior accurately. Credit risk assessment is required investor consent in financial risk management [67]. Most researchers have created models for assessing the risk of various bank loans [68]. As a result, various financial data from multiple companies are tested using these algorithms to predict bankruptcy. This body of literature gives insight into recently emerging issues in financial risk management, such as financial option pricing and feature selection attracting the interest of practitioners and scientists from various fields. These problems are complicated, but recent techniques focused on optimization attempted to address these concerns to some range.

3.3 Smart Agriculture

A possible collaborative swarm intelligence key is the intelligent agriculture system. It is subject to several limitations and topological complexity. The main restriction is finding the best route to work together [69]. There needs to be more than the current agricultural practices to meet the population's food needs because of how quickly the population is growing. Production needs to be improved in various ways to support the current farming methods. Organic farming has made the continued increase in agricultural output possible, which has produced encouraging results. Researchers have introduced robotics and artificial systems to help farmers overcome recent obstacles. The development of crop health on farms has revolutionized thanks to technological assistance for precision agriculture [70, 71].

Additionally, it necessitates data gathering, interpretation, and action—all of which can be automated in IoT-based systems. Various methods have been proposed for optimal path planning in agricultural settings. In addition, a variety of agricultural and cultural robots have been introduced. Rural areas can use agricultural robots to perform various activities like picking fruit, weeding, spraying pesticides, and growing. They include autonomous path planning. The enormous farming circumstances, meanwhile, necessitate several alternative options that can save time and money.

3.4 Infrastructural Development

Latest technologies have significantly impacted emerging nations, particularly in metropolitan areas. Additionally, the rapid development has destroyed the notions of time and space with which people are currently familiar [72]. Swarm intelligence-based IoT systems, such as smart cities, have recently been included in building infrastructure. Yet, it can be challenging to guarantee scalability, efficiency, security, and dependability in smart cities. Swarm intelligence is a viable technique in these circumstances. Modern software and embedded intelligence can all be used to simulate intelligent infrastructures. Developing modern infrastructures includes protection against natural disasters like tornadoes, earthquakes, and floods that cause catastrophic damage and can potentially destroy communication, bridge, and electrical infrastructure. Other methods, such as a parallel military based algorithm [73, 74], have been proposed to address these issues and ensure optimal infrastructure development. These algorithms are also used to detect damage early on and offer quick fixes to prevent more serious harm.

4 Research Challenges and Future Tracks

To chronicle, the existing research trends, difficulties, and future study orientations are the driving force for this review. A literature review was conducted to respond to this, and a research question was generated. Many of these queries include:

4.1 Questions Investigated

RQ-1: What are the core ideas behind meta-heuristics?

Meta-heuristic techniques have been identified as a critical asset in technology development today. The complete value of these resources is crucial because time, money, and resources are all finite. Choosing an ideal model is problematic because nonlinear optimization problems in the real world are expected. Meta-heuristics, therefore, aid in resolving these computationally challenging issues.

RQ-2: What benefit does recent meta-heuristics research have for the Internet of Things?

Swarm intelligence has a history with the Internet of Things systems. SI algorithms are preferred to address these problems because these systems are sophisticated and incorporate bright items. As a result, several intelligent SI algorithms can address these critical problems in these systems due to the mobility of the devices, the availability of information, the dynamic nature of the plans, and wireless communication. Artificial bee colonies, is SI-based algorithms that normalize IoT workflow and address several practical issues.

RQ-3: What technologies are employed in developing a meta-heuristic algorithm design pattern?

Study the tools and frameworks used from many fields to design meta-heuristics algorithms. Design patterns abstractly overcome the significant elements of critical practice adoption. For the construction of optimization algorithms, various computer scientists and academics have employed pattern-based viewpoints encompassing methodological, structural, behavioral, and component-based patterns.

RQ-4: What analytical methods are employed to evaluate the efficacy of meta-heuristic algorithms?

Assess the applicability of the validation techniques employed in light of the collected data type. The effectiveness of meta-heuristic algorithms has been validated using various assessment criteria, including precision, recall and multimodel benchmark functions.

RQ-5: In which journals have the research been published?

Multiple publications, conferences, books, and proceedings are considered for this. The study findings are displayed in Fig. 2 to identify where the research publications have been published.

4.2 Research Challenges and Future Directions

Throughout time, it will be necessary to use meta-heuristic methods to support real-time applications in IoT contexts. Before scaling to the next level, several challenges and future directions must be addressed, even when the meta-above heuristic strategies are effective. The literature is searched for the following research issues and future trends.

- Hardware and software Meta heuristic difficulties in IoT: With the use of smart devices in various industries during the past two decades, IoT has attracted significant interest. Lately, meta-heuristic optimization approaches have become necessary to develop smart systems for various industries, including agriculture and related fields. Yet, two significant obstacles exist to real-time meta-heuristic implementation using an IoT-based framework. First, the biggest obstacle to becoming smart, particularly in agriculture, is the lack of skilled sensors. Second, the system suffers from unreliable data transfer particularly in rural areas. Consequently, it is essential to provide a workable framework for evaluating how computationally challenging tasks perform in various contexts.

- Functional and interoperability challenges: No single organizational structure has recently been devised to manage dynamic and high-dimensional data. In addition, several security risks might arise during data transfer, including tampering, wiretapping, and replay attacks. Authenticity must all be guaranteed during data collection and transfer to achieve the most significant outcomes. Nonetheless, several optimization methods have been created to deal with some issues considerably. Capital investments may hamper the expansion of this industry in the various applications stated in the literature.
- Managing large-dimensional data: Many IoT devices generate much data, so handling it is essential. The "curse of dimensionality" also manifests as data dimensions grow. Smart infrastructure is required to automate various processes. These crucial steps will aid meta-heuristic approaches in providing satisfactory solutions to various dynamic situations. Also, due to low-cost infrastructure, platform service models are quickly gaining favor.

5 Conclusion

IoT has advanced quickly in several technologies, including big data analytics and machine learning, thanks to the development of new technology. These technologies can also be used in various industries, including agriculture, healthcare, risk management in the financial sector, and infrastructure development. The work demonstrates how meta-heuristic algorithms interact with novel concepts and emergent applications. Furthermore, it is undeniably true that meta-heuristic algorithms are among the effective paradigms that help provide trustworthy and reliable solutions to IoT-based system problems. This essay is laid out in three folds. Initially, a review and discussion of several algorithms based IoT applications have been held. The second fold contains information on several existing IoT systems that use meta-heuristic-based methods. Also, some descriptions of the study topics have been given from the studies that have been examined. The primary requirements of IoT systems are reviewed and summarised at the end of this study. Other swarm-based applications like big data and blockchain may be covered.

References

1. Parpinelli, R.S., Lopes, H.S.: New inspirations in swarm intelligence: a survey. Int. J. Bio-Inspired Comput. **3**(1), 1–16 (2011)
2. Pandey, A.C., Tripathi, A.K., Pal, R., Mittal, H., Saraswat, M.: Spiral salp swarm optimization algorithm. In: 2019 4th International Conference on Information Systems and Computer Networks (ISCON), pp. 722–727. IEEE (2019)
3. Garnier, S., Gautrais, J., Theraulaz, G.: The biological principles of swarm intelligence. Swarm Intell. **1**(1), 3–31 (2007)
4. Fortino, G., Trunfio, P.: Internet of Things Based on Smart Objects: Technology, Middleware, and Applications. Springer, Cham (2014). https://doi.org/10.1007/978-3-319-00491-4
5. Zedadra, O., Savaglio, C., Jouandeau, N., Guerrieri, A., Seridi, H., Fortino, G.: Towards a reference architecture for swarm intelligence-based internet of things. In: Fortino, G., Ali, A., Pathan, M., Guerrieri, A., Di Fatta, G. (eds.) Internet and Distributed Computing Systems. IDCS 2017. LNCS, vol. 10794, pp. 75–86. Springer, Cham (2018). https://doi.org/10.1007/978-3-319-97795-9_7

6. Johri, A., Tripathi, A., et al.: Parkinson's disease detection using deep neural networks. In: 2019 Twelfth International Conference on Contemporary Computing (IC3), pp. 1–4. IEEE (2019)
7. Holland, J.H.: Genetic algorithms. Sci. Am. **267**(1), 66–73 (1992)
8. Kar, A.K.: Bio-inspired computing–a review of algorithms and scope of applications. Expert Syst. Appl. **59**, 20–32 (2016)
9. Ma, H., Simon, D., Siarry, P., Yang, Z., Fei, M.: Biogeography-based optimization: a 10-year review. IEEE Trans. Emerg. Top Comput. Intell. **1**(5), 391–407 (2017)
10. Goldberg, D.E.: Genetic Algorithms. Pearson Education India, Noida (2006)
11. Beyer, H.-G., Schwefel, H.-P.: Evolution strategies–a comprehensive introduction. Nat. Comput. **1**(1), 3–52 (2002)
12. Pan, X., Jiao, L.: A granular agent evolutionary algorithm for classification. Appl. Soft Comput. **11**(3), 3093–3105 (2011)
13. Jaderyan, M., Khotanlou, H.: Virulence optimization algorithm. Appl. Soft Comput. **43**, 596–618 (2016)
14. Kirkpatrick, S., Gelatt, C.D., Vecchi, M.P.: Optimization by simulated annealing. Science **220**(4598), 671–680 (1983)
15. Mittal, H., Tripathi, A., Pandey, A.C., Pal, R.: Gravitational search algorithm: a comprehensive analysis of recent variants. Multimedia Tools Appl. **80**(5), 7581–7608 (2021)
16. Hatamlou, A.: Black hole: a new heuristic optimization approach for data clustering. Inf. Sci. **222**, 175–184 (2013)
17. Rashedi, E., Nezamabadi-Pour, H., Saryazdi, S.: GSA: a gravitational search algorithm. Inf. Sci. **179**(13), 2232–2248 (2009)
18. Kashan, A.H., Tavakkoli-Moghaddam, R., Gen, M.: Find-fix-finish-exploit-analyze (F3EA) meta-heuristic algorithm: an effective algorithm with new evolutionary operators for global optimization. Comput. Ind. Eng. **128**, 192–218 (2019)
19. Tabari, A., Ahmad, A.: A new optimization method: electro-search algorithm. Comput. Chem. Eng. **103**, 1–11 (2017)
20. Geem, Z.W., Kim, J.H., Loganathan, G.V.: A new heuristic optimization algorithm: harmony search. SIMULATION **76**(2), 60–68 (2001)
21. Eusuff, M., Lansey, K., Pasha, F.: Shuffled frog-leaping algorithm: a memetic meta-heuristic for discrete optimization. Eng. Optim. **38**(2), 129–154 (2006)
22. Blum, C., Merkle, D.: Swarm Intelligence: Introduction and Applications. Springer, Berlin (2008). https://doi.org/10.1007/978-3-540-74089-6
23. Dorigo, M., Blum, C.: Ant colony optimization theory: a survey. Theoret. Comput. Sci. **344**(2–3), 243–278 (2005)
24. Hu, X., Shi, Y., Eberhart, R.: Recent advances in particle swarm. In: Proceedings of the 2004 Congress on Evolutionary Computation (IEEE Cat. No. 04TH8753), vol. 1, pp. 90–97. IEEE (2004)
25. Karaboga, D., Akay, B.: A comparative study of artificial bee colony algorithm. Appl. Math. Comput. **214**(1), 108–132 (2009)
26. Li, X.: An optimizing method based on autonomous animats: fishwarm algorithm. Syst. Eng. Theor. Pract. **22**(11), 32–38 (2002)
27. Sharma, A., Sharma, A., Panigrahi, B.K., Kiran, D., Kumar, R.: Ageist spider monkey optimization algorithm. Swarm Evol. Comput. **28**, 58–77 (2016)
28. Rajakumar, B.: The lion's algorithm: a new nature-inspired search algorithm. Proc. Technol. **6**, 126–135 (2012)
29. Tripathi, A.K., Mittal, H., Saxena, P., Gupta, S.: A new recommendation system using map-reduce-based tournament empowered Whale optimization algorithm. Complex Intell. Syst. **7**(1), 297–309 (2021)

30. Sharma, N., Kaur, A., Sharma, H., Sharma, A., Bansal, J.C.: Chaotic spider monkey optimization algorithm with enhanced learning. In: Bansal, J., Das, K., Nagar, A., Deep, K., Ojha, A. (eds.) Soft Computing for Problem Solving. Advances in Intelligent Systems and Computing, vol. 816, pp 149–161. Springer, Singapore (2019). https://doi.org/10.1007/978-981-13-1592-3_11
31. Bhardwaj, S., Kim, D.-S.: Dragonfly-based swarm system model for node identification in ultra-reliable low-latency communication. Neural Comput. Appl. **33**(6), 1837–1880 (2021)
32. Bitam, S., Zeadally, S., Mellouk, A.: Fog computing job scheduling optimization based on bees swarm. Enterprise Inf. Syst. **12**(4), 373–397 (2018)
33. Chu, X., Wu, T., Weir, J.D., Shi, Y., Niu, B., Li, L.: Learning–interaction–diversification framework for swarm intelligence optimizers: a unified perspective. Neural Comput. Appl. **32**(6), 1789–1809 (2020)
34. Ahmed, W.A.E.M., Mageed, H.M.A., Mohamed, S.A., Saleh, A.A.: Fractional order darwinian particle swarm optimization for parameters identification of solar PV cells and modules. Alexand. Eng. J. **61**(2), 1249–1263 (2022)
35. Martínez-Álvarez, F., et al.: Coronavirus optimization algorithm: a bioinspired metaheuristic based on the COVID-19 propagation model. Big Data **8**(4), 308–322 (2020)
36. Liu, Y., Chai, Y., Liu, B., Wang, Y.: Bearing fault diagnosis based on energy spectrum statistics and modified mayfly optimization algorithm. Sensors **21**(6), 2245 (2021)
37. MiarNaeimi, F., Azizyan, G., Rashki, M.: Horse herd optimization algorithm: a nature-inspired algorithm for high-dimensional optimization problems. Knowl.-Based Syst. **213**, 106711 (2021)
38. Zhao, W., Wang, L., Mirjalili, S.: Artificial hummingbird algorithm: a new bio-inspired optimizer with its engineering applications. Comput. Methods Appl. Mech. Eng. **388**, 114194 (2022)
39. Dasgupta, D.: Artificial Immune Systems and their Applications. Springer, Cham (2012)
40. Del Ser, J., et al.: Bio-inspired computation: where we stand and what's next. Swarm Evol. Comput. **48**, 220–250 (2019)
41. Castelli, M., Manzoni, L., Mariot, L., Nobile, M.S., Tangherloni, A.: Salp swarm optimization: a critical review. Expert Syst. Appl. **189**, 116029 (2022)
42. Del Ser, J., Osaba, E., Sanchez-Medina, J.J., Fister, I.: Bioinspired computational intelligence and transportation systems: a long road ahead. IEEE Trans. Intell. Transp. Syst. **21**(2), 466–495 (2019)
43. Kallioras, N.A., Lagaros, N.D., Avtzis, D.N.: Pity beetle algorithm–a new metaheuristic inspired by the behavior of bark beetles. Adv. Eng. Softw. **121**, 147–166 (2018)
44. Passino, K.M.: Biomimicry of bacterial foraging for distributed optimization and control. IEEE Control. Syst. Mag. **22**(3), 52–67 (2002)
45. Greensmith. J., Aickelin, U.: The dendritic cell algorithm (Ph. D. thesis), University of Nottingham (2007)
46. Gandomi, A.H., Alavi, A.H.: Krill herd: a new bio-inspired optimization algorithm. Commun. Nonlinear Sci. Numer. Simul. **17**(12), 4831–4845 (2012)
47. Rabanal, P., Rodríguez, I., Rubio, F.: Using river formation dynamics to design heuristic algorithms. In: Akl, S.G., Calude, C.S., Dinneen, M.J., Rozenberg, G., Wareham, H.T. (eds.) Unconventional Computation. UC 2007. LNCS, vol. 4618. Springer, Berlin (2007). https://doi.org/10.1007/978-3-540-73554-0_16
48. Lam, A.Y., Li, V.O.: Chemical reaction optimization for cognitive radio spectrum allocation. In: 2010 IEEE Global Telecommunications Conference GLOBECOM 2010, pp. 1–5. IEEE (2010)
49. Cheng, S., Zhang, M., Ma, L., Lu, H., Wang, R., Shi, Y.: Brainstorm optimization algorithm for solving-knowledge spillover problems. Neural Comput. Appl. **35**, 12247–12260 (2021). https://doi.org/10.1007/s00521-020-05674-0

50. Bogar, E., Beyhan, S.: Adolescent identity search algorithm (AISA): a novel metaheuristic approach for solving optimization problems. Appl. Soft Comput. **95**, 106503 (2020)
51. Yang X-S., Deb, S.: Cuckoo search via Lévy flights. In: 2009 World Congress on Nature & Biologically Inspired Computing (NaBIC), pp. 210–214. IEEE (2009)
52. Nanda, S.J., Panda, G.S., and evolutionary computation (2013)
53. Jain, M., Singh, V., Rani, A.: A novel nature-inspired algorithm for optimization: squirrel search algorithm. Swarm Evol. Comput. **44**, 148 175 (2019)
54. Mafarja, M., Qasem, A., Heidari, A.A., Aljarah, I., Faris, H., Mirjalili, S.: Efficient hybrid nature-inspired binary optimizers for feature selection. Cogn. Comput. **12**(1), 150–175 (2020)
55. Mehrabian, A.R., Lucas, C.: A novel numerical optimization algorithm inspired by weed colonization. Ecol. Inf. **1**(4), 355–366 (2006)
56. Zhang, J., Xiao, M., Gao, L., Pan, Q.: Queuing search algorithm: a novel metaheuristic algorithm for solving engineering optimization problems. Appl. Math. Model. **63**, 464–490 (2018)
57. Yang, X.S.: Firefly Algorithms for multimodal optimization. In: Watanabe, O., Zeugmann, T. (eds.) Stochastic Algorithms: Foundations and Applications. SAGA 2009. LNCS, vol. 5792, pp. 169–178. Springer, Berlin (2009). https://doi.org/10.1007/978-3-642-04944-6_14
58. Yang, X.S.: A new metaheuristic bat-inspired algorithm. In: González, J.R., Pelta, D.A., Cruz, C., Terrazas, G., Krasnogor, N. (eds.) Nature Inspired Cooperative Strategies for Optimization (NICSO 2010). Studies in Computational Intelligence, vol. 284, pp. 65–74. Springer, Berlin (2010). https://doi.org/10.1007/978-3-642-12538-6_6
59. Moscato, P., et al.: On evolution, search, optimization, genetic algorithms, and martial arts: towards memetic algorithms. In: Caltech Concurrent Computation Program, C3P Report, vol. 826, p. 1989 Citeseer (1989)
60. Rahnamayan, S., Tizhoosh, H.R., Salama, M.M.: Opposition-based differential evolution. IEEE Trans. Evol. Comput. **12**(1), 64–79 (2008)
61. Liu, C., Yan, X., Liu, C., Wu, H.: The wolf colony algorithm and its application. Chin. J. Electron. **20**(2), 212–216 (2011)
62. Sharma, H., Bansal, J.C., Arya, K.: Fitness-based differential evolution. Memetic Comput. **4**(4), 303–316 (2012)
63. Sharma, N., Sharma, H., Sharma, A.: Beer froth artificial bee colony algorithm for job-shop scheduling problem. Appl. Soft Comput. **68**, 507–524 (2018)
64. Tsai, C.-W., Chiang, M.-C., Ksentini, A., Chen, M.: Metaheuristic algorithms for healthcare: open issues and challenges. Comput. Electr. Eng. **53**, 421–434 (2016)
65. Eichelberg, M., Aden, T., Riesmeier, J., Dogac, A., Laleci, G.B.: A survey and analysis of electronic healthcare record standards. Acm. Comput. Surveys. (Csur) **37**(4), 277–315 (2005)
66. Zhou, H., Sun, G., Fu, S., Liu, J., Zhou, X., Zhou, J.: A big data mining approach of PSO-based BP neural network for financial risk management with IoT. IEEE Access **7**, 154035–154043 (2019)
67. Wang, R., Yu, C., Wang, J.: Construction of supply chain financial risk management mode based on internet of things. IEEE Access **7**, 110323–110332 (2019)
68. Shepherd, C., Petitcolas, F.A.P., Akram, R.N., Markantonakis, K.: An exploratory analysis of the security risks of the internet of things in finance. In: Lopez, J., Fischer-Hübner, S., Lambrinoudakis, C. (eds.) Trust, Privacy and Security in Digital Business. TrustBus 2017. LNCS, vol. 10442, pp. 164–79. Springer, Cham (2017). https://doi.org/10.1007/978-3-319-64483-7_11
69. Botteghi, N., Kamilaris, A., Sinai, L., Sirmaçek, B.: Multi-agent path planning of robotic swarms in agricultural fields. In: ISPRS Annals of the Photogrammetry, Remote Sensing and Spatial Information Sciences, vol. 1, pp. 361–368. Copernicus GmbH (2020)

70. Albani, D., IJsselmuiden, J., Haken, R., Trianni, V.: Monitoring and mapping with robot swarms for agricultural applications. In: 2017 14th IEEE International Conference on Advanced Video and Signal Based Surveillance (AVSS), pp. 1–6. IEEE (2017)
71. Sinai, L.: Efficient path planning for multiple agents in agriculture fields (Master's thesis), University of Twente; (2020)
72. Zedadra, O., Guerrieri, A., Jouandeau, N., Spezzano, G., Seridi, H., Fortino, G.: Swarm intelligence and IoT-based smart cities: a review. In: Cicirelli, F., Guerrieri, A., Mastroianni, C., Spezzano, G., Vinci, A. (eds.) The Internet of Things for Smart Urban Ecosystems. Internet of Things. Springer, Cham (2019). https://doi.org/10.1007/978-3-319-96550-5_8
73. Tripathi, A.K., Sharma, K., Bala, M., Kumar, A., Menon, V.G., Bashir, A.K.: A parallel military-dog-based algorithm for clustering big data in the cognitive industrial internet of things. IEEE Trans. Ind. Inf. **17**(3), 2134–2142 (2020)
74. Tripathi, A.K., Sharma, K., Bala, M.: A novel clustering method using enhanced grey wolf optimizer and mapreduce. Big Data Res. **14**, 93–100 (2018)

Metaverse with the Internet of Things: Convergence of Physical and Cyber Worlds

Alaa Hassan(✉)

University of Calgary, Calgary, Canada
eng.alaa.hassan@ieee.org

Abstract. IoT has recently been researched in the Metaverse and the growth of the this will enable the realization of the Metaverse, including vast virtual worlds. This study highlights the IoT applications in the Metaverse, such as sociability, smart cities, collaborative healthcare, and education. We also thoroughly examine the pillar technologies: responsible artificial intelligence (AI) and digital twins that allow augmented reality (AR) and virtual reality (VR) in the IoT inspired Metaverse. We describe the industrial initiatives and the seven essential criteria for creating the Metaverse by the needs of the physical world: immersion, diversity, economy, civility, interaction, authenticity, and independence. Also, to finally realize the fusion of the digital worlds, the significant problems in the IoT inspired Metaverse are described in this survey.

Keywords: Meta-heuristic Algorithms · Internet-of-things · Swarm Intelligence

1 Introduction

Metaverse aims to create physical worlds where users may have fully experiences in the real world. In particular, two prominent apps are created to deliver social connections to the Metaverse's users. The user's status, such as the health problems that can trigger a reaction in the virtual world, can be instrumented by the medical IoT devices to count the user's body or a sensor-filled body suit [1–5]. In the virtual fitting room, IoT sensors monitor user movement; the IoT may also enhance the e-commerce experience. For example, the data from smartphone images or smart scales might be used to update the user's bodily information. This goes beyond typical online purchasing restrictions by letting users completely immerse themselves in a virtual store version. The latest Tactile Internet, which creates a network or network of networks allowing people or machines to access or manipulate real or virtual things in real-time remotely, may also use IoT data [6, 7]. In addition to facilitating data transmission between the physical and digital worlds, IoT data may provide AR/VR applications environment and situational understanding of physical objects [8]. For example, an AR gadget may respond to the customer's finger movements or activate a cyber-physical function in response to a real-world occurrence. A digital matching, or a digital statement of a particular physical entity's physical state and condition, is produced thanks to the IoT-enabled contact between the valid and virtual worlds [9]. To build a meaningful digital twin, the Metaverse strives to guarantee

K. Arai (Ed.): FTC 2023, LNNS 816, pp. 117–134, 2023.
https://doi.org/10.1007/978-3-031-47448-4_10

that the reflection is as close to the actual physical situation as possible. Because of this distinguishing feature, digital twins have become one of the Metaverse's most important applications. Digital twins may be constructed in professional contexts utilizing the Haptic Codecs (IEEE P1918.1.1) [10] to make group meetings more productive since users can converse with one another while using or displaying a replica of the hardware or software prototype. During technical training programs, digital twins assist engineers in doing 3D simulations of complex systems [11]. The digital twins may even construct a comprehensive maintenance workshop with a virtual reproduction of the maintained gadgets and specific mechanical equipment for mending the item. The digital twins provide a simulated setting linked to a real physical workplace where maintenance may be done remotely. Digital twins have the ability to virtualize a city for urban planning and construction, allowing inhabitants or corporate actors to use visual representation to carry out a development plan and identify prospective future urban projects [12]. The metaverse may be used by the digital economy application to create a range of transaction scenarios while achieving high trade efficiency. Figure 1 depicts the Metaverse's standard six application scenarios, which include collaborative healthcare [13], education [14], smart cities [15], entertainment [16], real estate [17], and sociability [18]. The metaverse age is soon arriving in healthcare. According to the World Economic Forum, the introduction of digital services will be one of the most crucial parts of healthcare reform during the next 10 years [13]. Traditional healthcare is a face-to-face, hands-on experience in which patients and doctors interact directly for treatment. Because of the Covid-19 outbreak, people are looking for methods that monitor patients remotely outside of the hospital [19, 20]. The Metaverse promises to speed up the development of telemedicine since patients may interact with their physicians while immersed in the Metaverse using their avatars. Those with impairments or restricted mobility may live better thanks to the metaverse. Those with paraplegia may live healthy lives in the Metaverse. Those with hearing or vision impairments, on the other hand, can benefit from Metaverse technology (such as AR or VR) to better communication. Furthermore, augmented reality (AR) appears as a feasible tool for strengthening knowledge- and skill-based medical students. For instance, several technology firms are creating surgical assisting technologies for on-demand surgical procedures, such as Microsoft Hololens [21]. The Metaverse may also support psychotherapy [18]. A peaceful and virtual environment where people with mental problems may connect and communicate with avatars can be developed using the Metaverse. In education, recently, there has been a substantial surge in the employment of Metaverse technology in this subject [14]. The increased use of AR/VR, which stresses visualization-based learning approaches, is transforming traditional teaching techniques [22]. The Metaverse is expected to extend the number of educational institutions capable of visualizing teaching materials and fostering comprehension of learning materials in order to create engaging and immersive learning environments for students. Furthermore, the metaverse can create virtual, secure, but realistic lab environments. Furthermore, certain learning materials and activities of an organ system in the human body or the universe, are difficult to experience personally or to understand [23]. AR/VR development may be done to make the learning materials easier for students to grasp and to help them get the necessary practice and experience regularly. Curiscope, an AR firm based in Brno, Czech Republic, developed the Virtuali-Tee [24]. Students may explore the inside of the human body with Virtuali-Tee, much like in an anatomy lab. Smart city: As a Metaverse-supporting technology, IoT-enabled digital twins may digitize real-world goods such as buildings, autos, houses, and municipal infrastructures. This allows for the construction of virtual cities. This technique may be useful in the construction of smart cities [15]. Designers and builders may easily develop a simulation environment that closely matches the actual world using digital twins. People, automobiles, traffic lights, and buildings might all be represented as digital twins in the smart city. IoT data is used to map the physical environment in order to construct a controllable and digital twin city. The Metaverse in the smart city is mainly used to optimize the allocation of facilities and resources. For example, Intel in Germany developed a smart parking system in which the meta-boundary platform is used to regulate local traffic using digital twins [25]. Entertainment: Technological advancements linked to the Metaverse have considerably improved gaming immersion, which may increase user experience and enjoyment. A Metaverse game is Roblox game with many active users [26]. [27] proposes a blockchain-based game, as well as the widespread use of personal data in the Metaverse. Large events such as exhibits, concerts, and book signings may be conducted in the Metaverse since it provides an immersive environment for this purpose. For example, in the Metaverse's virtual world, Korean pop sensations Asepa and Black Pink unveiled new songs [16]. Property has been highlighted as the next possible use for the Metaverse [17]. Customers may have a more realistic and immersive experience owing to AR/VR, which is one of the major technologies in Metaverse [28]. Real estate brokers, for example, may utilize VR to provide customers with a virtual house tour. The consumer benefits from being able to perform inspections in any situation, and there is no time spent leaving home. The Metaverse may also offer an immersive virtual environment and allow real-time interaction, making it a great candidate for deployment in interior design and architectural areas with AR/VR and AI technologies [29]. We can not only get a 360° panoramic picture of the inside and exterior of the house, but we can also change the interior architecture and furniture to see our dream home in real-time. Beike, for example, provides a virtual remodelling platform [30] where consumers may explore what they imagine their restored house might look like to help them grasp the potential of a property. Urbanbase [31] has developed the "AR Scale" 3D augmented reality presentation solution. Architects may show others practical concepts without producing actual models. Instead, they only utilize an augmented reality to display 3D models on the actual building site in order to assess compatibility with the surrounding environment and structures in advance. Socialization: The Metaverse presents a new application field in terms of human social forms [18]. The Metaverse transcends time and location, allowing for various social interactions that may bind individuals together [32]. Above the level of the actual world, people may pursue certain higher-level demands such as virtual meetings, dating, and workplaces. The value of telecommuting and distant social networking has emerged due to the COVID-19 epidemic [33]. The Metaverse may be used to build a more open social environment to overcome the drawbacks of conventional models and enhance the usefulness of telecommuting and distant social platforms.

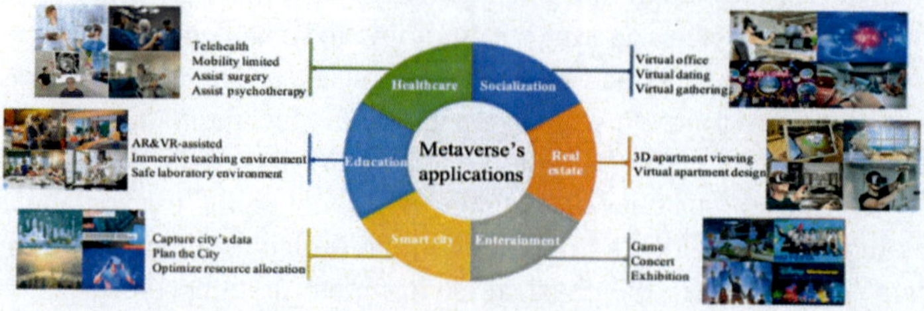

Fig. 1. The six typical applications of the Metaverse.

2 Existing Surveys and Tutorials

The Metaverse combines social activities and AI techniques by combining service demands ranging from mobile access to interacting with reality via virtual currency. A lesson in [34] presents a Metaverse taxonomy, which groups technologies based on implementations, and applications. Three studies focusing on Facebook Research are used to investigate the social impact and technological restrictions of the Metaverse. The authors examined network architecture and convergence [18]. We define three Metaverse properties: sociality and multi-technology dominance (which includes augmented reality and blockchain-based economic systems) (which blurs the line between the virtual and real worlds). The research [35] offers a digital twin that separates native material production, merging of the actual and virtual worlds, and Metaverse development into these three components. The ecosystems in the Metaverse are being built using various technologies, including AR/VR, HMI, AI, robots, blockchain, and MEC, which may be used for applications related to content generation and security. The poll [36] focuses on privacy and security concerns with the Metaverse. Similar to [37], the production of native content, the merger of physical and virtual reality, and digital twins are the three stages of the Metaverse's growth. AR/VR systems with IoT sensors may gather ambient data, facial expressions, eye or hand movements, and brainwave signals in an AI-enabled virtual world [38]. Threatening human safety and vital IoT infrastructures, adversaries may penetrate and manipulate AR/VR gadgets to monitor users' positions, biometric traits, and identities. The existing security protection options for the Metaverse, such as identity and user generated content control, are investigated to meet the application demand. The authors of [39] examine AI-based Metaverse frameworks such as reinforcement learning. Several well-known AI techniques are also being researched to improve the user's virtual experience in real-world applications such as health monitoring, intelligent transportation, industrial production, and online shopping. Future AI development is discussed nationally in [40], and it can redefine a nation's economic activities, output, and commercial demand. The important national AI frameworks, such as National Open Innovation Platforms, are listed by the authors using China as an example. The majority of China's IT firms are working to build the infrastructure necessary to provide AI solutions for clinical workflow, autonomous driving, online consumer credit services, and virtual assistants that can be customized. The privacy and security of users are at risk because of the massive amounts of data that the Metaverse apps. Monitoring tools should be created with humanism, equality, and variety to deal with users' inappropriate conduct in the Metaverse while simultaneously promoting appropriate behavior. Therefore, the tutorial in [41] proposes that three criteria—privacy, governance, and ethical consideration—be used to drive application creation in the Metaverse [42]; the development of virtual worlds across several decades, from the environment to the Metaverse. The survey highlights four features based on physical and virtual interaction: realism, ubiquity, and preserving the user's virtual identity (allowing the user to have a seamless, uninterrupted virtual experience), and scalability (allowing a large number of users to run the Metaverse applications concurrently). The authors demonstrate how blockchain and AI may be used to establish an economic system that includes digital assets and money [43]. Because of better data traceability and secrecy, Metaverse users may access immutable and verifiable services like as cryptocurrency exchanges [44]. The Metaverse uses AI to process personal data and stores the users' medical information on several servers across the globe. The services' capacity to scale may be increased, and harmful assaults can be avoided, thanks to decentralized processing and storage. The survey [25] explores AI approaches supported by 6G for in-the-moment user engagement. The physical layer, virtual layer, and technical layer are utilized to investigate the Metaverse's underlying architecture. To support large-scale AI models and boost resource allocation and computational power efficiency, three 6G-enabled architectures are provided: edge cloud-Metaverse, mobile edge cloud-Metaverse, and decentralized Metaverse.

3 Four Pillar Technologies of Metaverse

3.1 Responsible AI

Motivation. Even though AR/VR is at the forefront of the Metaverse, AI is a critical technology to make miracles happen. AI is very useful for generating predictions and computations based on data. It can help advance algorithms for avatar creation, natural language translation, and the creation of new worlds. As AI watches sensors that monitor our bioelectricity, it may improve our virtual reality communication. AI also increases inclusivity by providing services like picture recognition for those with visual impairments. Understanding and tracing the operation of the used AI methods are more critical as additional AI techniques and applications proliferate in the Metaverse. This can guarantee that the Metaverse operates as anticipated and upholds legal requirements. The Metaverse's construction is so complicated to understand. Deep neural networks, for example, are among the most difficult to comprehend. These models are created using data. As a result, biases in the Metaverse may be caused by demographic data, measurement inaccuracy, the data quality and data augmentation. These biases jeopardize the Metaverse's data quality, fairness, accountability, transparency, and explainability. Developing such a policy necessitates a deep understanding of AI techniques in the Metaverse, where AI choices must be used to encourage confidence and transparent AI for combatting negative social consequences [45].

Importance. Responsible AI [46], as represented in Fig. 2, is being created to fully comprehend AI decision-making processes while taking into account the Metaverse's ethical and socioeconomic implications. This is done to increase the Metaverse's trust and usage. Furthermore, the responsible AI assists engineers or data scientists that design

AI algorithms in describing what exactly happens within IoT applications and delivers a certain result [47]. Many responsible AI applications can offer trustworthy guidance on the boundaries of a learning system and specify the training data utilized [48]. The AI system is used in [49] to study a framework for responsible urbanization, which balances urban development's advantages, dangers, and effects. Responsible AI may consider how judgments based on machine learning can affect the environment and future urban development. Sustainability may thus be used to define responsible urban innovation, which is crucial for urbanization policies.

Important Technologies. In recent years, many explainable AI-based techniques [50] have been developed to strengthen control and supervision in the event of incorrect or undesirable results. Deep learning is used in [51] to identify drainage and gully photographs for flood monitoring in intelligent cities, where the image may be categorized into one of three obstacle categories. Explainable AI makes deep algorithms more understandable by using semantic representation to illustrate relationships between the coverage of different items within the image. Explainable AI may assess rivals' work and find possibilities to exceed them to aid product creation and marketing [52]. Explainable AI can accurately portray consumer thoughts and derive the product's competitive characteristics from user-generated reviews.

Uses in the Metaverse. The development of intelligent health systems in the Metaverse has extensively used responsible AI. Responsible AI's capabilities may influence the healthcare business, and patients may encounter a dynamically altered service environment [53]. To create responsible AI with the Metaverse's novel health systems and integrity must all be considered. The authors of [54] employ AI to develop a patient-centered, personalised, home-based rehabilitation aid system. Explainable AI-based techniques integrate ensemble learning and hybrid stacking with neighbours to fulfill individualization, interpretability, and system design needs. Synthetic datasets augment experimental observations while also reducing bias. Socially responsible AI [55] is being researched in order to prioritize Metaverse data and users while removing the disadvantaged individuals. Socially accountable [56] extends AI to detect early indicators of cognitive deterioration, explaining why a specific prognosis is made. Clinical indicators identifying hidden and overt behavioural aberrations and wandering behaviours are included in the explainable AI-based approach. To detect anomalies, the explainable AI-based system adapts to different persons and conditions. To allow the Metaverse, Industry 4.0 often entails hyperparameter optimization, finetuning, computing capabilities, and continuous training of massive data sets. Millions of gadgets generate and aggregate data on a daily basis. As a result, it is possible that hackers would modify crucial user data. As a result, an increasing number of industry 4.0 applications are experimenting with explainable AI leveraging Metaverse to increase risk assessment and automation [57]. Creating a risk assessment technique [58] for human-sensing devices, Human attention states are classified and mapped into four categories in IoT systems. Explainable artificial intelligence (AI) is being developed to detect abnormal brainwave patterns. It assesses the importance of each feature used in the learning model. IoT nodes are installed in remote farms to monitor metrics like humidity and temperature. Sensory data is used to adjust irrigation control systems. Explainable AI has the potential to deliver comprehensible data and automated irrigated [59], and this allows the users to test the model logic in irrigation management systems. COVID-19 can be recognized via 5G networks based on of blood or medical photographs of patients.

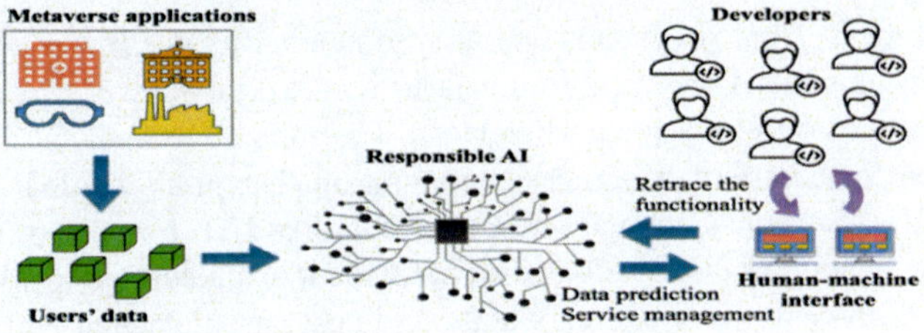

Fig. 2. Understanding the AI processes in the Metaverse.

3.2 High-Speed Data Communications

Motivation. The Metaverse provides real-time services to many mobile users. Because of the signals nature and human perception, networking AR/VR services rely on massive data interchange with very low latency—at most, 1 ms. Nevertheless, the majority of 4G technologies are unable to provide numerous mobile users with real-time data connections. It is difficult, for example, to reduce the latency of LTE-based wireless networks below 25 ms [34]. In recent years, 5G has been rapidly developed to allow real-time data exchange to connected devices [60, 61]. 5G provides more data capacity than 4G-based devices while maintaining rapid communication suitable for the Metaverse. 5G networks, in particular, enable a number of innovative technologies [64]. Following the widespread commercial deployment of 5G networks, 6G is expected to expand personalized communications in order to realize the model by connecting not only people and Internet of Things (IoT) devices and mobile robots [65].

Significance. The Metaverse's bandwidth and density may all be increased by high-speed transmission of data and service reliability. 5G, in particular, has the potential to boost data rates for mobile users in a uniform geographic area, [62, 66]. High data rates enable large-scale machine-to-machine connections as well as enhanced mobile broadband. Furthermore, 5G ensures the reliability and timeliness of critical services, with final latency as low as one nanosecond and reliability of 99.99%. 5G networks give ten times greater network speed, ten times less communications delay, 100 times greater data transfer capacity, and 100 percent more network efficiency than present 4G technologies. To satisfy the Metaverse's criterion for ubiquitous applications, 6G may be developed based on the present 5G architecture. Large IoT devices, for example, may enhance VR/AR capabilities by providing the necessary assets (such as power for computation, space for storage, visual processing capabilities, and connection resources) [63, 67]. For 6G, researchers are investigating the use of more frequency bands beyond 5G and 4G, such as sub-THz and THz, as well as VLC, to accomplish ultra-high data speeds and drastically minimize communication latency. Despite the well-known fact that a higher radio frequency may result in increased data loss, 6G benefits from the shorter data transmission distance that allows for more low-latency services in congested areas. To address the issue of localization, 6G employs ultra-massive MIMO with high-speed, low-latency machine-to-machine communication technologies.

Key Technologies. The IoT systems operating frequency enables several gigabits with high speed data amounts. Large-scale antenna arrays are used in several researched studies for the mmWave's high-frequency band [69]. The antenna components may be set up in a full-dimensional variety or a linear configuration. The mmWave IoT systems have propagation loss than low-frequency message tools like LTE because of the high carrier frequency. In addition, the mmWave signal finds it challenging to pass through and diffract around urban barriers like cars and buildings, and molecular and atmospheric absorption may further obstruct mmWave transmission. NOMA is another 5G option that shares time and frequency capacity by using power-domain or code-domain multiplexing. To improve network advantages, power-domain multiplexing distributes Internet of Things (IoT) devices with different transmit outputs that adapt to the time-varying channel quality. Another characteristic is the ability of Internet of Things gadgets to do code-domain multiplication, which generates specific codes that can be multiplexed over the same time-frequency resources. Even though code domain multiplexing uses more signal the bandwidths than energy domain multiplexing, it can boost spreading and shaping gain. NOMA may facilitate large connections in the Metaverse, where the amount of usable orthogonal resource is not restricted by the number of connected devices. To meet the demand for 100 Gbps data throughput and one-millisecond latency, THz of 6G may be deployed in IoT networks, allowing huge AR/VR coverage and scaling in the Metaverse. Because much of the band over 100 GHz remains unutilized, the THz range is 300 to 3 THz. THz communication still has challenges as new technology advances. For instance, meteorological factors like rain, snow, or atmospheric water vapor may readily attenuate THz signals.

Applications in Metaverse. Data transmissions in the Metaverse are shown in Fig. 3, which enables IoT applications, including virtual design and manufacturing, education, and agriculture monitoring. High-speed data transmission, in particular, allows students to participate virtually in classrooms where instructors from anywhere in the world eloquently display their subject-matter expertise through the video/audio stream. Virtual courses may control student involvement, enhance the educational process, and remove the most frequent obstacles to synchronous learning. Also, experts from various industrial production organizations may be able to collaborate online while designing items thanks to high-speed data connectivity. Engineers may develop a shared virtual 3D model using IoT data, AR, and VR. The Metaverse, which leverages high-speed data transmission, combines complementary technology and commercial design ideas to bring the real and digital worlds together. The industrial business specialized in designing hardware may also collaborate with hundreds of certified partners who offer management of supply chains services or software support. This will eventually result in a full platform for virtual three-dimensional creation and cooperation using physics-based digital models.

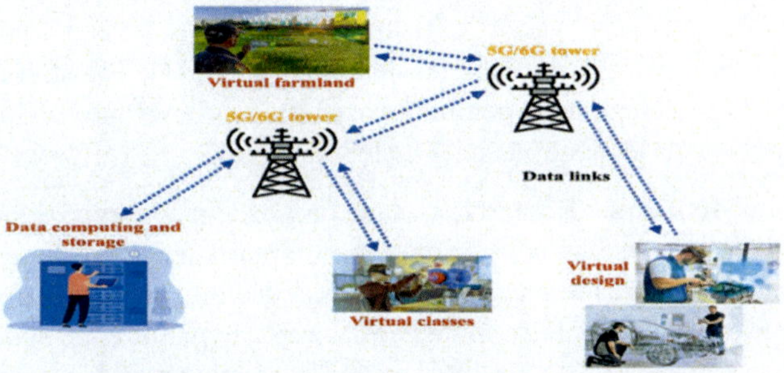

Fig. 3. The high-speed communication technology.

3.3 Cost-Effective Mobile Edge Computing

Motivation. Maintaining user participation and engagement in the Metaverse is critical for providing the same level of encounter as in reality. In this strategy, the time taken to respond to user input ought to be significantly reduced to levels below the human perception threshold. Mobile Edge computer (MEC), a new technology, hosts a combination of communication networks and computer resources near users to offer rapid response times. Figure 4 depicts how MEC extends a network and software-based environment that may fast approach or exceed what users consider to be a fair response time. It is feasible to achieve the Metaverse's criterion for a dynamic and fully immersive experience by using MEC. MEC may provide the Metaverse with ultra-low latency connections for undetectable reaction times and vast server capacity to power the virtual 3D world.
Significance. MEC is a possible means of connecting the Metaverse as 5G (which is expected to reduce last-mile delay to 1ms) and the impending 6G expand. It may enable single-hop border offloading via cellular-connected devices used by users, such as AR glasses, to improve the user interface in the Metaverse. MEC is notably important for outdoor The Metaverse services in order to grasp the distinctive local environment and establish close collaboration between neighbouring users or devices. 5G MEC servers, for example, may control AR content for nearby users and enabling real-time interactions between users for sociable AR apps [68, 70].
Key Technologies. MEC is implemented:

- Offloading and computing at the edge: Expensive computation activities (such as matrix multiplication) required for processing information and AI training of models can also be broken into smaller tasks and transferred to edge servers. In addition, rendering calculations are offloaded to the edge node. By joining the completed subtasks at the central node, the computing results can be continued.
- Edge caching: Reducing computation and transmission redundancy through restricting repeated use of popular content or calculations. The Metaverse can teach probabilistic models of document popularity distributions (similar to a field of vision). The popular field of vision can therefore be preserved on edge nodes close to users who seek a more significant reduction in graphics latency and processing expenditure.
- Protecting your privacy, Local model training: The danger of data breaches rises as more people join the Metaverse since the attack surface does too. One method for

safeguarding user privacy is federated learning (FL) [71]. FL offers a distributed machine-learning technique by combining user models developed on limited policies without explicitly disclosing personal material.

Applications in Metaverse. Researchers have proposed many approaches to improve the efficiency of Metaverse apps by using the minimal latency advantages associated with edge computing. For example, the MEC was added into the realm of the Metaverse to enhance the user experience [72]. In the MEC design, dynamic edge nodes are as close near users as feasible. Multiple nodes on the edge can be joined to assist the same user in completing their instructions, thereby overcoming the latency difficulties caused by user movement in the Metaverse. Nonetheless, user data needs to be transferred in order for several edge nodes to collaborate, which may have negative consequences such as user privacy breaches and problems with identity in the Metaverse. To protect users' privacy, a FL-based MEC design for the universe of Metaverse was released [73], in which locally trained models are transferred between edge nodes rather than user data. Furthermore, the edge nodes built a directed and acyclic network to improve the system's computational efficiency. The scientific community has focused on employing MEC to enhance the Metaverse experience. Dai et al. [74] created a 360-degree VR cache system utilizing view synthesis using MECCache servers in Clouds Radio Radio Network (C-RAN) to improve the quality experience (QoE) for wireless VR applications.

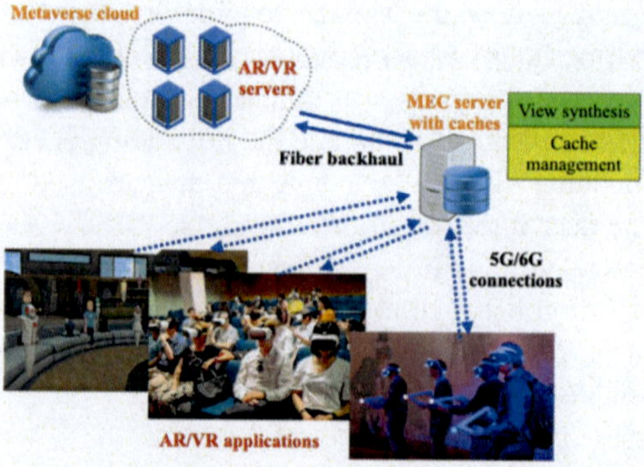

Fig. 4. The requested data of the user can be synthesized by the MEC server.

3.4 IoT-Based Digital Twins

Motivation. Building facilities, operational procedures and social services are just a few examples of things that may be created as virtual twins of real-world items or services. Digital twins are often employed to provide an entirely realistic purchasing practise in the real world. Digital duplicates of actual products and services may enhance data analytics, enabling businesspeople to simulate real-world situations before making expensive choices. Digital twins, digital representations of real-world objects, may synchronize resources, operations, and operating systems with the outside world and routine tasks like visualization, analysis, and prediction [75]. Via IoT connections, digital twins play a vital role in the interaction [76]. Changes to the real world are thus rejected in the digital world. Creating duplicates of the natural world to access the services that is important building extensions of the Metaverse. For instance, a 3D model of a complex system might be made by an engineer at various degrees of complexity (descriptive, informative, predictive, holistic, autonomous) with multiple purposes. By artificial intelligence (AI), engineers and service providers may remotely execute physical investigations thanks to digital twins [77]. Since David Gelernter introduced the idea of digital twins in his book [78], companies have had access to a never-before-seen perspective on how products are created, used, and performed. This makes it simpler to provide superior goods or services. The advancement of cutting-edge technology like artificial intelligence (AI) and the growing economic value of autonomous cars are the fundamental causes. In addition, a significant portion—over 20%—comprises industrial production and aviation manufacturing. Over 10% of the market is occupied by digital twins and innovative home sectors, all developing and promising futures. Many companies, like NVIDIA, are researching the relationship between the Metaverse and experiences for people heavily reliant on technology. According to Leeway Hertz [79], as more businesses and organizations employ digital twins. The production of sophisticated, multi-level, large-scale industrial goods or systems, including those for manufacturing [44, 80], and smart cities [45, 81], is the primary usage of the current digital twins due to their high development costs.

Significance. Digital twins are a digital depiction of an item or services that is present in the real world. Using a closed software entity or model, digital twins blend multiple IoT data to build composite representations of numerous real-world aspects. By taking a thing from the real world and producing an exact clone of the same item in the digital realm, users and engineers may imitate the behaviour of natural objects and predict their motions and reactions. As a result, digital twins enable strategic creation of products while avoiding costly blunders. Digital twins are an essential component of the Metaverse due of their individual traits. The IoT-enabled Metaverse allows users to build exact replicas of reality, bringing realism to the virtual realm and creating simulated worlds that are beyond their wildest dreams. Digital twins are useless with current information for IoT networks; this is a problem that artificial intelligence and machine learning have long grappled with. Customers may expect to use some amazing AR/VR apps and experience immersive services in 2022 and beyond, due to the Metaverse, wherein Internet and digital twins mingle and rejoice. First, we must debunk a myth: virtual scanning items are Metaverse digital twins. Digital twins may appear to be scanned things. But only if correct metadata and real-world stream of data of IoT devices are added in parallel. In intricate simulations, linking digitally represented physical components with overlay information is vital.

Key Technologies. According to their size and inclusivity, digital twins may be categorized as component twins and process twins [82].

- Component twins are the minor units of the total system and are made up of essential components that influence performance or other details with unstable versions [83].
- Asset twin: Compared to component twins, asset twins are on a higher level [84]. The information makes up an asset twin. Asset twins examine performance data, which

emerges from the interplay of various elements, to find areas for improvement and transform them into actionable insights.

- System twin: It works at a higher level, combining various assets to create a fully operational system [85]. System twins inform users of how different assets interact, improving performance [86].
- Process twin: By combining multi-dimensional process knowledge models, Process Twin uses techniques to improve machinery and production [40].

Applications in Metaverse. A digital twins system using VR and AI has been developed [87] to track and evaluate welder behavior. A data-driven digital twins architecture was looked at in [88] to improve the performance and functionality of health diagnostics in intelligent medical systems. To duplicate a virtual depiction of agricultural productivity, digital twins were developed in [89]. A wildfire area visualization was created in California. Fire behavior analysts may utilize visuals and data to make predictions that help firefighters more effectively put out real-world flames by using information from injury-free simulations. Businesses build digital twins of their manufacturing facilities and factories to spot flaws and mistakes during production. Optimizing the design, manufacturing, and labor processes involves visualizing every step of the production process. Customers of retail and e-commerce businesses may make purchases from home at virtual storefronts. Customers may interact with the things they want to buy thanks to immediate information on product availability. Holding client meetings in the Metaverse may be as simple as inviting the client to your virtual workplace and seating them in your digital conference room, saving both time and cash on travel while facilitating face-to-face interactions. By supplementing digital objects with links to data sources within the Metaverse, teams may gain the understanding they need about the organizations or people they will interact with in order to win meetings and support them proactively. Metaverse employs digital twins to improve productivity and profitability across the product creation cycle, from concept through after-production monitoring and service [90].

4 Open Issues in the Future IoT-Empowered Metaverse

4.1 Data Processing

Metaverse programs like AR and VR need the collection of real-time IoT data in order to investigate the activities or statuses of actual world items or procedures [83]. It is difficult to monitor equipment and gather data from all sides, including initial monitoring information, operating information, company information, and program data [83]. After collecting real-time IoT data, analysis and mining of data must be performed. In contrast to traditional simulation [84], the Metaverse as a whole may execute numerous simulation processes utilizing real-time data and deliver real-time feedback to the source object for improvement. To provide consumers and developers with a complete picture of how the virtual world is functioning, these simulation procedures depend on data modeling methods, including engineering simulations [85], and spatial computing [86, 87] may be used. This allows IoT devices to store and alter references to virtual environments and things. The Metaverse must provide the data variety and scalability needed for spatial computing. Also, most commercial AR/VR devices feature a variety of sensors that provide a significant quantity of data with diverse modalities and kinds.

In contrast to many independent viewpoints, the Meta-verse may use several data sources to produce a complete picture [88]. Critical data management concerns include dense information transmission, storage, security, and privacy. It is a good idea to minimize transmission data, reduce transfer delay, and secure the original content while developing a unified semantic representation of multiple data kinds [47], which is utilized to communicate with intelligent objects inside the Metaverse.

4.2 Security and Privacy

IoT systems are combined to enable remote virtualization. Integrating services, apps, or IoT systems from various businesses or organizations is likely to occur in the metaverse [89]. Since integrating many IoT devices necessitates coordination and data interaction between all parties, this poses severe problems. The reliable AI is anticipated to assist Metaverse designers and developers in understanding the black box of AI, enhancing the effectiveness of IoT system integration. However, these answers provide further details regarding the process that uses vulnerable consumers' private data. An attacker may leverage a privacy breach like this to, for instance, identify specific individuals and their medical issues based on characteristics like gender, race, and birthdate. Also, the attacker can introduce a tiny quantity of contaminated data into the AI model. As "big data" is often processed using an AI model, it is challenging to identify harmful data that has been tainted in the Metaverse application. Such undetectable contaminated data might influence more and more AI model functions as training time increases, ultimately leading to breakdowns in the Metaverse service.

4.3 Real-Time 3D Modeling

Powerful gaming machines are often needed for AR/VR-based computer games to analyze the data as quickly as feasible. The communication networks are also constrained in their ability to provide access to several games concurrently. For instance, the typical data transfer time supported by current 4G technology is 50ms (or 20 frames per second), which might make for an unpleasant immersive experience. With 5G, the communication lag may be as little as 10ms. The IoT network should guarantee that numerous users' AR/VR apps are closely coordinated and have minimal latencies [89]. This criterion provides that consumers will always be able to have seamless, immersive experiences. This suggests that network requirements go above 5G (B5G) or perhaps 6G [90]; on the opposite hand, huge data surveillance and real-time processing may raise the demand for bandwidth in the cloud.

5 Conclusion

This paper discusses IoT-inspired uses for healthcare and socializing. The Metaverse's four pillar technologies for AI, fast data connection, affordable MEC, and digital twins, as well as their usual IoT applications, are thoroughly discussed. The four pillar technology may link IoT applications to the Metaverse, which attempts to integrate both the digital and physical worlds. For each pillar technology, we focus on the motivations, significance, important developments, and typical Metaverse applications. We also expand on the fundamental remaining issues for further creation and execution in the IoT-inspired Metaverse, such as data processing, privacy and security, real-time 3D modeling, scalable cyber realms, connected mobile user experiences, virtual platform interoperability and uniformity.

References

1. Slovick, M.: The AR-VR age has begun in health care. https://www.cta.tech/Resources/i3-Magazine/i3-Issues/2020/November-December/The-AR-VR-Age-has-Begun-in-Health-Care. (2020)
2. (2022, 06) AR and VR in the education market. https://www.marketresearchfuture.com/reports/ar-vr-in-education-market-10834
3. Kanter, T.G.: The metaverse and extended reality with distributed IoT. IEEE Internet Things Mag. (IoT) (2021)
4. Pereira, N., et al.: IEEE international symposium on mixed and augmented reality (ISMAR). IEEE **2021**, 479–488 (2021)
5. Sodhro, A.H., Pirbhulal, S., Sangaiah, A.K.: Convergence of IoT and product lifecycle management in medical health care. Futur. Gener. Comput. Syst. **86**, 380–391 (2018)
6. Promwongsa, N., et al.: A comprehensive survey of the tactile internet: state-of-the-art and research directions. IEEE Commun. Surv. Tutorials **23**(1), 472–523 (2020)
7. Aijaz, A., Sooriyabandara, M.: The tactile internet for industries: a review. Proc. IEEE **107**(2), 414–435 (2018)
8. Lu, E., Miller, J., Pereira, N., Rowe, A.: FLASH: Videoembeddable ar anchors for live events. In: IEEE International Symposium on Mixed and Augmented Reality (ISMAR), pp. 489–497. IEEE (2021)
9. Minerva, R., Lee, G.M., Crespi, N.: Digital twin in the IoT context: a survey on technical features, scenarios, and architectural models. Proc. IEEE **108**(10), 1785–1824 (2020)
10. Steinbach, E., et al.: Haptic codecs for the tactile internet. In: Proceedings of the IEEE, vol. 107, no. 2, pp. 447–470 (2018)
11. Stojanovic, N., Milenovic, D.: Data-driven digital twin approach for process optimization: An industry use case. In: IEEE International Conference on Big Data (Big Data), pp. 4202–4211. IEEE (2018)
12. Farsi, M., Daneshkhah, A., Hosseinian-Far, A., Jahankhani, H.: Digital Twin Technologies, and Smart Cities. Springer, Cham (2020). https://doi.org/10.1007/978-3-030-18732-3
13. Thomason, J.: Metahealth-how will the metaverse change health care? J. Metaverse **1**(1), 13–16 (2021)
14. Kye, B., Han, N., Kim, E., Park, Y., Jo, S.: Educational applications of metaverse: possibilities and limitations. J. Educ. Eval. Health Prof. **18**, 32 (2021)
15. Ruohomaki, T., et al.: Smart city platform enabling digital twin. In: 2018 International Conference on Intelligent Systems (IS), pp. 155–161 (2018)
16. Lee, J.Y.: A study on metaverse hype for sustainable growth. Int. J. Adv. Brilliant Convergence **10**(3), 72–80 (2021)
17. Nalbant, K.G., Uyanik, S.: Computer vision in the metaverse. J. Metaverse **1**(1), 9–12 (2021)
18. Ning, H., et al.: A survey on metaverse: the state-of-theart, technologies, applications, and challenges, arXiv preprint arXiv:2111.09673 (2021)

19. Maatuk, A.M., Elberkawi, E.K., Aljawarneh, S., Rashaideh, H., Alharbi, H.: The covid-19 pandemic and e-learning: challenges and opportunities from the perspective of students and instructors. J. Comput. High. Educ. **34**(1), 21–38 (2022)
20. Demeke, H.B., et al.: Trends in use of telehealth among health centers during the covid19 pandemic—united states, june 26–november 6, 2020. Morb. Mortal. Wkly Rep. **70**(7), 240 (2021)
21. Hanna, M.G., Ahmed, I., Nine, J., Prajapati, S., Pantanowitz, L.: Augmented reality technology using microsoft hololens in anatomic pathology. Arch. Pathol. Lab. Med. **142**(5), 638–644 (2018)
22. Rajagopal, N., Miller, J., Kumar, K.K.R., Luong, A., Rowe, A.: Improving augmented reality relocalization using beacons and magnetic field maps. In: International Conference on Indoor Positioning and Indoor Navigation (IPIN), pp. 1–8. IEEE (2019)
23. Songlee, H., CI, L.: Research trends on augmented reality education in Korea from 2008 to 2019. J. Educ. Technol. **36**, 505–528 (2020)
24. Curiscope, Virtuali-tee: augmented reality t-shirt. https://www.curiscope.com/
25. Chang, L., et al.: 6G-enabled edge AI for Metaverse: challenges, methods, and future research directions, arXiv preprint arXiv:2204.06192 (2022)
26. Meier, C., Saor´ın, J., de Leon, A.B., Cobos, A.G.: Using the ´ roblox video game engine for creating virtual tours and learning about the sculptural heritage. Int. J. Emerg. Tech. Learn. (iJET) **15**(20), 268–280 (2020)
27. Ryskeldiev, B., Ochiai, Y., Cohen, M., Herder, J.: Distributed metaverse: creating decentralized blockchain-based model for peerto-peer sharing of virtual spaces for mixed reality applications. In: Proceedings of the 9th Augmented Human International Conference. New York, NY, USA: Association for Computing Machinery (2018). https://doi.org/10.1145/3174910.3174952
28. Hou, H.C., Wu, H.: Technology for real estate education and practice: a VR technology perspective. Property Manag. **38**(2), 311–324 (2020)
29. Izani, M., Aalkhalidi, S., Razak, A., Ibrahim, S.: Economical VR/AR method for interior design programme. In: 2022 Advances in Science and Engineering Technology International Conferences (ASET), pp. 1–5, February 2022
30. Zhang, Y., Yao, L.: How the leading Chinese real estate brokerage transformed into a digital platform business. Strategy Leadersh. **50**(1), 119–124 (2021)
31. Kim, J., Kim, J.: Bim to AR matching technology of building maintenance platform using 5g-based AR (2021)
32. Kim, J.-G.: A study on metaverse culture contents matching platform. Int. J. Adv. Cult. Technol. **9**(3), 232–237 (2021)
33. Lyttelton, T., Zang, E., Musick, K.: Telecommuting and gender inequalities in parents' paid and unpaid work before and during the covid-19 pandemic. J. Marriage Fam. **84**(1), 230–249 (2022)
34. Park, S.-M., Kim, Y.-G.: A metaverse: taxonomy, components, applications, and open challenges. IEEE Access **10**, 4209–4251 (2022)
35. Lee, L.-H., et al.: All one needs to know about metaverse: a complete survey on technological singularity, virtual ecosystem, and research agenda, arXiv preprint arXiv:2110.05352 (2021)
36. Wang, Y., et al.: A survey on metaverse: Fundamentals, security, and privacy, arXiv preprint arXiv:2203.02662 (2022)
37. Huynh-The, T., Pham, Q.-V., Pham, X.-Q., Nguyen, T. T., Han, Z., Kim, D.-S.: Artificial intelligence for the metaverse: a survey, arXiv preprint arXiv:2202.10336 (2022)
38. Wu, F., et al.: Towards a new generation of artificial intelligence in china. Nat. Mach. Intell. **2**(6), 312–316 (2020)
39. Fernandez, C.B., Hui, P.: Life, the Metaverse and everything: an overview of privacy, ethics, and governance in Metaverse, arXiv preprint arXiv:2204.01480 (2022)

40. Dionisio, J.D.N., Burns, W.G., III, Gilbert, R.: 3D virtual worlds and the Metaverse: current status and future possibilities. ACM Comput. Surv. (CSUR) **45**(3), 1–38 (2013)
41. Ynag, Q., Zhao, Y., Huang, H., Zheng, Z.: Fusing blockchain and AI with metaverse: a survey, arXiv preprint arXiv:2201.03201 (2022)
42. Mozumder, M.A.I., Sheeraz, M.M., Athar, A., Aich, S., Kim, H.-C.: Overview: technology roadmap of the future trend of Metaverse based on IoT, blockchain, AI technique, and medical domain Metaverse activity. In: International Conference on Advanced Communication Technology (ICACT), pp. 256–261. IEEE (2022)
43. Sharma, R.A., et al.: All that glitters: Lowpower spoof-resilient optical markers for augmented reality. In: ACM/IEEE International Conference on Information Processing in Sensor Networks (IPSN), pp. 289–300. IEEE (2020)
44. Madhavan, R., Kerr, J.A., Corcos, A.R., Isaacoff, B.P.: Toward trustworthy and responsible artificial intelligence policy development. IEEE Intell. Syst. **35**(5), 103–108 (2020)
45. Arrieta, A.B., et al.: Explainable artificial intelligence (XAI): Concepts, taxonomies, opportunities and challenges toward responsible AI. Inf. Fusion **58**, 82–115 (2020)
46. Werder, K., Ramesh, B., Zhang, R.: Establishing data provenance for responsible artificial intelligence systems. ACM Trans. Manag. Inf. Syst. (TMIS) **13**(2), 1–23 (2022)
47. Wearn, O.R., Freeman, R., Jacoby, D.M.: Responsible AI for conservation. Nat. Mach. Intell. **1**(2), 72–77 (2019)
48. Yigitcanlar, T., et al.: Responsible urban innovation with local government artificial intelligence (AI): a conceptual framework and research agenda. J. Open Innov. Technol. Market Complex. **7**(1), 71 (2021)
49. Mohseni, S., Zarei, N., Ragan, E.D.: A multidisciplinary survey and framework for design and evaluation of explainable AI systems. ACM Tran. Interact. Intell. Syst. (TiiS) **11**(3–4), 1–45 (2021)
50. Thakker, D., Mishra, B.K., Abdullatif, A., Mazumdar, S., Simpson, S.: Explainable artificial intelligence for developing smart cities solutions. Smart Cities **3**(4), 1353–1382 (2020)
51. Han, J., Lee, Y.: Explainable artificial intelligence-based competitive factor identification. ACM Trans. Knowl. Disc. Data (TKDD) **16**(1), 1–11 (2021)
52. Kumar, P., Dwivedi, Y.K., Anand, A.: Responsible artificial intelligence (AI) for value formation and market performance in healthcare: the mediating role of patient's cognitive engagement. Inf. Syst. Front. 1–24 (2021). https://doi.org/10.1007/s10796-021-10136-6
53. Vourganas, I., Stankovic, V., Stankovic, L.: Individualised responsible artificial intelligence for home-based rehabilitation. Sensors **21**(1), 2 (2020)
54. Cheng, L., Varshney, K.R., Liu, H.: Socially responsible AI algorithms: issues, purposes, and challenges. J. Artif. Intell. Res. **71**, 1137–1181 (2021)
55. Khodabandehloo, E., Riboni, D., Alimohammadi, A.: Healthxai: collaborative and explainable AI for supporting early diagnosis of cognitive decline. Futur. Gener. Comput. Syst. **116**, 168–189 (2021)
56. Ahmed, I., Jeon, G., Piccialli, F.: From artificial intelligence to explainable artificial intelligence in industry 4.0: a survey on what, how, and where. IEEE Trans. Ind. Inf. **18**(8), 5031–5042 (2022)
57. Al Hammadi, A.Y., et al.: Explainable artificial intelligence to evaluate industrial internal security using EEG signals in IoT framework. Ad Hoc Networks **123**, 102641 (2021)
58. Tsakiridis, N.L. et al.: Versatile internet of things for agriculture: an eXplainable AI approach. In: Maglogiannis, I., Iliadis, L., Pimenidis, E. (eds.) Artificial Intelligence Applications and Innovations. AIAI 2020. IFIP Advances in Information and Communication Technology, vol. 584, pp. 180–191. Springer, Cham (2020). https://doi.org/10.1007/978-3-030-49186-4_16
59. Hossain, M.S., Muhammad, G., Guizani, N.: Explainable AI and mass surveillance system-based healthcare framework to combat covid-i9 like pandemics. IEEE Network **34**(4), 126–132 (2020)

60. Guo, W.: Explainable artificial intelligence for 6g: improving trust between human and machine. IEEE Commun. Mag. **58**(6), 39–45 (2020)
61. Wang, S., et al.: Explainable AI for b5g/6g: Technical aspects, use cases, and research challenges, arXiv preprint arXiv:2112.04698 (2021)
62. Zhao, Z., Ding, Z., Quek, T.Q., Peng, M.: Edge artificial intelligence in 6G systems: theory, key techniques, and applications. China Communications **17**(8), iii–iv (2020)
63. Lin, Z., Lv, T., Ni, W., Zhang, J.A., Liu, R.P.: Tensorbased multi-dimensional wideband channel estimation for mmwave hybrid cylindrical arrays. IEEE Trans. Commun. **68**(12), 7608–7622 (2020)
64. Mir, T., et al.: Relay hybrid precoding in UAV-assisted wideband millimeter-wave massive mimo system. IEEE Trans. Wirel. Commun. **21**, 7040–7054 (2022)
65. Xiao, C., et al.: Downlink mimo-noma for ultra-reliable low-latency communications. IEEE J. Sel. Areas Commun. **37**(4), 780–794 (2019)
66. Zeng, J., et al.: Ensuring max–min fairness of UL simo-noma: a rate splitting approach. IEEE Trans. Veh. Technol. **68**(11), 11080–11093 (2019)
67. Wang, S., Lv, T., Ni, W., Beaulieu, N.C., Guo, Y.J.: Joint resource management for mc-noma: a deep reinforcement learning approach. IEEE Trans. Wireless Commun. **20**(9), 5672–5688 (2021)
68. Krogfoss, B., Duran, J., Perez, P., Bouwen, J.: Quantifying the value of 5G and edge cloud on QoE for AR/VR. In: International Conference on Quality of Multimedia Experience (QoMEX), pp. 1–4. IEEE (2020)
69. Giordani, M., Polese, M., Mezzavilla, M., Rangan, S., Zorzi, M.: Toward 6G networks: use cases and technologies. IEEE Commun. Mag. **58**(3), 55–61 (2020)
70. Chen, X., Feng, Z., Wei, Z., Zhang, P., Yuan, X.: Code-division of DM joint communication and sensing system for 6g machine-type communication. IEEE Internet Things J. **8**(15), 12093–12105 (2021)
71. Liu, Y., Peng, M., Shou, G., Chen, Y., Chen, S.: Toward edge intelligence: multiaccess edge computing for 5G and internet of things. IEEE Internet Things J. **7**(8), 6722–6747 (2020)
72. Liao, S., Wu, J., Li, J., Konstantin, K.: Information-centric massive IoT-based ubiquitous connected VR/AR in 6G: A proposed caching consensus approach. IEEE Internet Things J. **8**(7), 5172–5184 (2020)
73. Zhang, S., Liu, J., Guo, H., Qi, M., Kato, N.: Envisioning deviceto-device communications in 6G. IEEE Network **34**(3), 86–91 (2020)
74. Sun, S., Rappaport, T.S., Shafi, M., Tang, P., Zhang, J., Smith, P.J.: Propagation models and performance evaluation for 5G millimeter-wave bands. IEEE Trans. Veh. Technol. **67**(9), 8422–8439 (2018)
75. Prabhakara, A., Singh, V., Kumar, S., Rowe, A.: Osprey: a mmwave approach to tire wear sensing. In: Proceedings of the 18th International Conference on Mobile Systems, Applications, and Services, pp. 28–41 (2020)
76. Zhang, J., Ge, X., Li, Q., Guizani, M., Zhang, Y.: 5G millimeterwave antenna array: design and challenges. IEEE Wirel. Commun. **24**(2), 106–112 (2016)
77. Dai, J., Zhang, Z., Mao, S., Liu, D.: A view synthesis-based 360 VR caching system over MEC-enabled co-ran. IEEE Trans. Circ. Syst. Video Technol. **30**(10), 3843–3855 (2019)
78. Chen, H.-Y., Hsu, R.-T., Chen, Y.-C., Hsu, W.-C., Huang, P.: AR game traffic characterization: a case of Pokemon go in a flash crowd ´ event. In: Proceedings of the 19th Annual International Conference on Mobile Systems, Applications, and Services, pp. 493–494 (2021)
79. Yang, Q., Liu, Y., Chen, T., Tong, Y.: Federated machine learning: concept and applications. ACM Trans. Intell. Syst. Technol. (TIST) **10**(2), 1–19 (2019)
80. Zhang, W., Chen, J., Zhang, Y., Raychaudhuri, D.: Towards efficient edge cloud augmentation for virtual reality MMOGs. In: Proceedings of the Second ACM/IEEE Symposium on Edge Computing, pp. 1–14 (2017)

81. Tuli, S., Casale, G., Jennings, N.: MCDs: AI augmented workflow scheduling in mobile edge cloud computing systems. IEEE Trans. Parallel Distrib. Syst. (2021)
82. Gu, Z., Lu, H., Hong, P., Zhang, Y.: Reliability enhancement for VR delivery in mobile-edge empowered dual-connectivity sub6 GHZ and MMWAVE helmets. IEEE Trans. Wirel. Commun. **21**(4), 2210–2226 (2021)
83. Liu, Y., Liu, J., Argyriou, A., Ci, S.: Mec-assisted panoramic VR video streaming over millimeter wave mobile networks. IEEE Trans. Multimedia **21**(5), 1302–1316 (2018)
84. Tao, F., Zhang, H., Liu, A., Nee, A.Y.C.: Digital twin in the industry: state-of-the-art. IEEE Trans. Industr. Inf. **15**(4), 2405–2415 (2019)
85. Chen, D., Wang, D., Zhu, Y., Han, Z.: Digital twin for federated analytics using a Bayesian approach. IEEE Internet Things J. **8**(22), 16301–16312 (2021)
86. Rathore, M.M., Shah, S.A., Shukla, D., Bentafat, E., Bakiras, S.: The role of AI, machine learning, and big data in digital twinning: a systematic literature review, challenges, and opportunities. IEEE Access **9**, 32030–32052 (2021)
87. Mirror Worlds: or the Day Software Puts the Universe in a Shoebox...How It Will Happen and What It Will Mean. Oxford University Press
88. https://www.fortunebusinessinsights.com/digital-twin-market-106246
89. https://metaverseinsider.tech/2022/03/22/nvidia-announces-new-technology-for-digital-twin-simulation-and-communications-with
90. https://www.leewayhertz.com/digital-twin/

An Exploration of the Potential of Large Language Models to Enable Cognitive Flexibility in AI-Augmented Learning Environments

Sarah A. Chauncey[1] and H. Patricia McKenna[2(✉)]

[1] Northville Central School District, Northville, NY, USA
schauncey@northvillecsd.org
[2] AmbientEase, Victoria, BC, Canada
mckennaph@gmail.com

Abstract. This paper is motivated by the possibility that large language model (LLM) artificial intelligence (AI) chatbots such as ChatGPT (Chat Generative Pre-trained Transformer) may support cognitive flexibility in AI-augmented learning environments and in doing so, forestall tendencies toward cognitive rigidity, as in, resistance to change, whether conscious or unconscious. A review of the research and practice literature for LLM AI chatbots and for cognitive flexibility and rigidity enables formulation of a conceptual framework for cognitive flexibility in AI-rich learning environments. The main research question explored is whether *LLM AI chatbots support cognitive flexibility in contemporary learning environments? And if yes, how?* A research design is advanced in this paper that is attentive to the need for a rethinking of methods for the study of people interacting with technologies in contemporary learning environments where AI is increasingly present. This paper provides the design of a research study focusing on Large Language Model (LLM) artificial intelligence (AI) chatbots placing a focus on cognitive flexibility challenges and opportunities in learning environments. This work also provides direction for the development of at least two other papers, one focusing on creativity and innovation and the other focusing on curiosity, interest, and engagement. This paper is significant in terms of the potential to broaden understandings of learning spaces featuring AI capabilities. Contributions of this work pertain to implications associated with enriching and augmenting human learning capabilities in support of cognitive flexibility; and extending the boundaries of AI-assisted human-computer interaction in learning.

Keywords: AI Chatbots · AI-Rich Learning Environments · Cognitive Flexibility · Cognitive Rigidity

1 Introduction and Background

This paper develops the design of a study to address cognitive flexibility in relation to large language model (LLM) artificial intelligence (AI) chatbots. As such, an exploration is provided of Language Models (LMs) and Large Language Models (LLMs) for AI

K. Arai (Ed.): FTC 2023, LNNS 816, pp. 135–153, 2023.
https://doi.org/10.1007/978-3-031-47448-4_11

chatbots such as ChatGPT (Chat Generative Pre-trained Transformer) developed by OpenAI [1], that offer the potential to act as non-emotional intermediaries to support and improve decision-making, claim-making, and the like, while forestalling tendencies toward cognitive rigidity such as resistance to change. Li (2022) [2] reminds us that language modeling is not new, dating back as it does 100 years to the work of Markov and since then to the likes of Shannon, and then Chomsky, to neural language models, and now to pre-trained language models, and as such, is "continuously evolving". Forestall, in the context of this paper, means "to keep from happening by taking action in advance" (Merriam-Webster, 2023) [3], and as such the study designed in this paper includes the hypothesis that, *while varied factors can influence cognitive rigidity, ease of access to varied perspectives during early stages of inquiry may forestall such rigidity.*

In a meeting of AI researchers in October 2020 (Tamkin, Brundage, Clark, and Ganguli, 2021) [4] an urgency emerged in responding to two key questions pertaining to: a) the "technical capabilities and limitations of large language models" (e.g., scale, understanding, multimodality, alignment) and b) "the societal effects of widespread use of large language models" (e.g., capabilities, deployment, disinformation, bias, economy). Tamkin and Ganguli (2021) [5] provide an overview of discussion "takeaways" including, but not limited to the acknowledgements that - "as language models grow, their capabilities change in unexpected ways"; "GPT-3's uses and their downstream effects on the economy are unknown"; "future models won't merely reflect the data - they will reflect our chosen values"; while concluding that "we should develop norms and principles for deploying language models now." Indeed, in the context of teaching and learning in higher education, Chrisinger (2023) [6] calls for more in the way of agreements and rules for the governance of ChatGPT use. Additionally, there seems to be an assumption that human intelligence is finite, becoming possibly even static, while AI is rapidly advancing. But what if human intelligence is far more expansive and dynamic since Boudry (2019) [7] refers to *Homo Sapiens* as "a tool making species, and this includes a range of cognitive tools." As such, through people interacting with an LLM AI chatbot, this paper explores how use of an iterative "conversation" allows the expansive and dynamic to happen.

The primary objectives of this paper are to: a) provide a review of the research and practice literature for cognitive flexibility, cognitive rigidity, and large language model AI chatbots; b) formulate a conceptual framework for cognitive flexibility in AI-rich learning environments; and c) define the broad outline for the design of a study into large language model (LLM) artificial intelligence (AI) chatbots in support of cognitive flexibility in educational environments. Going forward, it is expected that further studies will place a focus on support for curiosity, interest, and engagement, as well as for creativity and innovation in relation to AI-rich learning environments.

What follows in the remainder of this paper is the development of a theoretical perspective focusing on language models (LMs) and large language model (LLM) artificial intelligence (AI) chatbots; cognitive flexibility; and cognitive rigidity, enabling formulation of a conceptual framework for cognitive flexibility in AI-rich learning environments. This is followed by a description of the methodology involving the use of AI as a design material combined with perspective taking and boundary setting, as well as a rationale for this approach. A discussion is then provided on application of the framework, based on a question and response dialog with ChatGPT-3.5, as an example of an AI chatbot, along with the identification of outcomes for further exploration. Limitations and future directions are then provided, followed by concluding comments.

2 Theoretical Perspective

This paper undertakes an exploration of language models (LMs) and large language model (LLM) artificial intelligence (AI) chatbots in relation to cognitive flexibility and cognitive rigidity. A conceptual framework for cognitive flexibility in AI-rich learning environments is formulated with implications going forward for other areas such as curiosity, interest, and engagement as well as creativity and innovation.

2.1 Language Model and Large Language Model Artificial Intelligence Chatbots

In coming to an understanding of large language models (LLMs), Shanahan (2023) [8] reminds us that they are "generative mathematical models of the statistical distribution of tokens in the vast public corpus of human-generated text, where the tokens in question include words, parts of words, or individual characters including punctuation marks." Shanahan (2023) [8] adds that LLMs "are *generative* because we can sample from them, which means we can ask them questions" giving rise to the notion of "conversational agents or AI assistants such as ChatGPT" while discouraging the use of "words to describe the capabilities of LLMs such as 'beliefs', 'knowledge', 'understanding', 'self', or even 'consciousness' as misleading." And yet, Gorvett (2023) [9] probes the notion of AI emotions and "algorithms acting irrationally."

Edwards (2023) [10] describes "a new large language model" as in, "a new family of language models called 'Large Language Model Meta AI' or LLAMA" said to be "10x smaller" and able to "outperform" Open AI's GPT-3, enabling the "running of ChatGPT-style language assistants locally on devices such as PCs and smartphones." Where ChatGPT-3 features 175 billion parameters, LLaMA type models "range from 7 billion to 65 billion parameters in size", significant since fewer parameters require less space and computing resources, affecting performance and possibly also capabilities related to "handling more complex tasks" and "coherent output" [10].

Recent articles highlight concerns over the potential to use AI chatbots to mislead, compromise privacy, job loss, and other concerning outcomes (Ouyang, Wu, Jiang, et. al., 2022 [11]; Mah and Levine, 2023 [12]; Shen et al., 2003 [13]). New tools are like that. They are sometimes used in ways that the developers did not envision (Prakash, 2023 [14]; Tamkin and Ganguli, 2021 [5]). We learned this with social media platforms - Facebook, Twitter, Instagram, Snapchat, etc. (Ourednik, Mlynář, Mutzner, and Alavi, 2021) [15]. The reactions vary from extreme control, monitoring and locking down to come what may (Brown, 2021) [16]. Interestingly, in the case of AI, researchers and developers seem to be alarmed by the trajectory toward potential negative outcomes (Al-Sibai, 2023 [17]; Tamkin and Ganguli, 2021 [5]) while calling for efforts in support of more transparency, governance, and envisioning beneficial potentials for humanity and the planet. This paper is the first in what is planned as a series of papers, focusing on the positive outcomes AI chatbots and related tools will have on motivation, learning, and innovation.

2.1.1 LLM AI Chatbots: Challenges, Opportunities, Potentials, and Limitations

Challenges and opportunities associated with large language model AI chatbots are discussed in relation to potential and current limitations, mainstream AI activities (including law, business, and marketing), and productivity functionality.

ChatGPT is "an artificial intelligence chatbot that can simulate a conversation with a human user through text or voice communication. It uses natural language processing, machine learning, and other advanced technologies to understand the user's queries and provide relevant and 'accurate' responses" (ChatGPT, 2023) [18]. ChatGPT's functionality is supported by an LLM, where "LLMs are deep learning models trained to understand and generate natural language" (Shen et al., 2023) [13]. Shen et al. (2023) [13] share several ChatGPT limitations, while others note that LLM AI chatbots can become more adept at generating comprehensive, accurate responses (Schick et al., 2023 [19]; Wolfram, 2023 [20]; Ouyang et. al., 2022 [11]). The work of Ouyang, et. al. (2022) [11] looks at user intent feedback to improve responses, claiming that "[m]aking language models bigger does not inherently make them better at following a user's intent." According to Ouyang, et. al. (2022) [11] such "models can generate outputs that are untruthful, toxic, or simply not helpful to the user" as in, "not aligned with their users" and as such, their research looks at, "…aligning language models with user intent on a wide range of tasks by fine-tuning with human feedback." It is expected that future models would offer options for filtering underlying content and domains used to generate responses. To ensure transparency, AI chatbots would include citations for resources used to produce responses. Copy.ai (2023) [21], a newly launched AI chatbot app that targets marketing and sales professionals to aid in generating copy for blogs, social media, ads, and so on, includes, if requested to do so, references used to generate text responses to queries as well as additional sources related to the topic.

Companies are looking at options to improve the efficacy and accuracy of chatbots by training them around domains or business-specific use cases. The company AI Multiple (Javaid, 2023) [22] outlines best practices for, "preparing a robust dataset to train and improve an AI-powered chatbot to help businesses successfully leverage the technology." The goal is to constrain training to address target purposes and desired capabilities. An example of a language model focused on scientific knowledge is Galactica. Taylor et al. (2022) [23] introduce the capabilities of Galactica, "a large language model that can store, combine, and reason about scientific knowledge" in order to "demonstrate the potential for language models as a new interface for science" and "open source the model for the benefit of the scientific community" where "models are trained on a large corpus comprising more than 360 million in-context citations and over 50 million of unique references normalized across a diverse set of sources" which "enables Galactica to suggest citations and help discover related papers" (Galactica, 2023) [24]. Indeed, Miller (2023) [25] identifies several approaches to the editing of LLMs to achieve greater accuracy, consistency, and current-ness.

The study design developed in this paper considers the potential for AI Chatbots to be programmed to produce responses using multiple knowledge-seeking pathways, combining, for example, ChatGPT with the computational capabilities of Wolfram|Alpha (Wolfram, 2023) [20] and varied frameworks giving users a broader understanding of issues. Future AI systems would invoke multiple chatbots to provide responses generated from varied perspectives, knowledge domains, and the like -- especially important to forestall cognitive rigidity. If a question related to food scarcity is posed, as in the case of the current paper, the responses would include perspectives of the economist, naturalist, ethicist, environmentalist, technologist, and so on, providing a broader understanding of the topic and provoking thinking around new questions.

Mah and Levine (2023) [12] describe ways to use ChatGPT as an example machine in support of learning and creativity using strategies such as concept attainment, inductive learning, or multiple contrasting cases. Mah and Levine (2023) [12] argue that ChatGPT is capable of generating multiple contrasting cases quickly, thus freeing up the time of teachers to "revise and perfect those examples" leaving "more time to dedicate to other creative lessons." Five steps are outlined to optimize the use of ChatGPT as an example machine and for goals focusing on processes, concepts, and arguments (e.g., evaluate, construct), as follows - establish criteria, prompt, review and adapt, prompt for contrasting cases, and review and adapt - and in the case of processes, with "slight modifications, teachers can change the order, accuracy, and completeness of the contrasting cases to deepen students' thinking about complex processes" (Mah and Levine, 2023). Mah and Levine (2023) [12] also caution about the limitations and the many "ways that ChatGPT can create problems for teachers" with inaccurate interpretations and the like.

Regarding other chatbot challenges and limitations, Warren (2023) [26] notes that "prompt exploits" have been used with Microsoft's Bing AI to reveal "a secret set of rules" and the name Sydney is revealed to be "an internal code name for a chat experience" under exploration. Indeed, library personnel are now assisting students with guidance on "referencing ChatGPT in assignments" (UTS, 2023) [27] while Shahriar and Hayawi (2023) [28] draw attention to challenges in research papers acknowledging the involvement of ChatGPT.

2.1.2 Mainstream AI Activity Including Productivity Functionality

Ghaffray (2023) [29] claims that "millions of people are already using apps like ChatGPT to write books, create art, and develop code" and while acknowledging the "inflated hype about generative AI", the rapid uptake of the app is said to have taken "5 days to reach 1 million users" with a wide range of uses across many sectors. In addition to OpenAI's ChatGPT, funded by Microsoft, Ghaffray (2023) [29] points to other controversial apps such as Lensa and to other generative start-ups such as Stability AI and Anthropic (with investments from Google). Roth (2023) [30] identifies "major contenders in the AI chatbot space" such as Microsoft and Google, claiming that "With all the hype surrounding ChatGPT, it's no wonder other companies are vying for a piece of the AI-powered chatbot game. Companies are betting that we're at a decisive moment in the artificial intelligence industry, where products that adopt and build upon the budding technology could have the potential to reshape technology as we know it—not to mention shake up the Big Tech hierarchy." Anthropic (2023) [31], a research company concerned with AI safety, produced a paper on the notion of "constitutional AI" (Bai et al., 2022) [32], described as a method "to train a non-evasive and relatively harmless AI assistant, without any human feedback labels for harms." As such, the method "improves upon, and partially replaces reinforcement learning from human feedback" while drawing upon "natural language instructions or principles" that it uses to revise its responses automatically where the goal of the system, is to "train helpful and harmless AI assistants from *purely helpful* assistants" and "*not* to remove supervision entirely, but to make it more efficient, transparent, and targeted." Indeed, Stahl (2023) [33] in a television interview with Brad Smith, President of Microsoft, revealed limitations of AI chatbot technology (ChatGPT) incorporated in the search tool Bing in a discussion focusing on the "new world of chatbots" giving rise to governance, trust, and veracity issues while claiming that the tool could be an "economic gamechanger" with "possible improvements in productivity." The search engine Brave Search, featuring an AI-powered 'Summarizer' is available to desktop and mobile users in the form of "a synopsis of a search query using different sources" (Mehta, 2023) [34] and is said to "highlight relevant sentences in listed results." Also of note, Brave Search is supported by its own LLM rather than ChatGPT by OpenAI. Brave Search encourages users to rely on their own "critical thinking capabilities" (Mehta, 2023) [34] while cautioning "not to trust everything produced by AI-powered search results." Composer AI (2023) [35], a Chrome browser extension, claims to reduce writing time by offering "AI-powered autocompletion and text generation" for websites, blog posts, Google documents and email, with more to come. By highlighting text segments, the user can then ask for suggestions for rephrasing as a way to "draw inspiration from AI" [35]. Similarly, Lampkin (2023) [36] describes Viva Sales by Microsoft, for sales teams, "a new generative AI-powered experience" designed such that "sellers communicate more effectively with prospects and customers." As such, Viva Sales "will now generate suggested email content for a variety of scenarios" including "replying to an inquiry or creating a proposal—complete with data specifically relevant to the recipient, such as pricing, promotions, and deadlines" (Lampkin, 2023) [36]. Microsoft claims that "auto-suggesting customizable content" results in considerable time savings since "sellers can spend less time composing emails and searching for sales data from colleagues and databases" (Lampkin, 2023) [36].

2.1.2.1. A Glimpse of Disciplines Impacted by AI

Stokel-Walker (2023) [37] notes that the "law-focused generative AI tool" referred to as Harvey (2023) [38] "uses large datasets to learn to generate pictures or text that appear natural" and is said to be "a good fit for the legal industry, which relies heavily on standardized documents and precedents." However, "the tendency to confidently make things up" also referred to as "hallucinate" is a key problem, necessitating the presence of detection and prevention systems for such occurrences and the checking and validating of everything - inputs and outputs. In Europe for example, the use of any personal data must follow guidelines provided by the European Union's General Data Protection Regulation (GDPR.EU, 2023) [39].

From a business and entrepreneurial perspective, Sherry (2023) [40] highlights that, with the availability of the OpenAI API (application programming interface), developers will be able to "build applications or add features to existing ones without having to build everything from scratch" and "will be able to integrate this tech into their products to build chatbots, virtual assistants, and other products without having to train and host their own language models" extending to "natural language coding and automated marketing." Of note, are revised terms of services where "users own the input and output of anything generated by OpenAI-powered language models" [40]. Warren (2023) [41] describes the integration of the new "AI-powered Bing search to the taskbar" through an update to Microsoft Windows 11.

As for marketing, Wiggers (2023) [42] describes a "growing industry" involving many startups such as that of Typeface, "developing an AI-powered dashboard for drafting marketing copy and images" that are "customized to target certain demographics, or to align with a brand's style guidelines" involving "ad copy, images, and personalized messaging." Typeface is said to be promising due to the "platform's safety and governance capabilities" while, among other challenges for companies like Typeface, is "scrutiny for copying elements of the art and photos in its training data without necessarily attributing them" giving rise to the potential for copyright infringement.

Regarding examples for learning, Quizlet (Cision, 2023) [43], a global study platform, announced Q-Chat in beta form "to US Quizlet users 18 and up" as "the first fully adaptive, AI-powered tutor experience built on OpenAI's ChatGPT API" that draws on Quizlet's "massive educational content library" containing "billions of questions and definitions" and "in conjunction with AI models from OpenAI to provide a virtual tutor experience to students" and "automatically quizzes students and helps deepen their knowledge of whatever concepts they are learning in a conversational and engaging way." PromptLoop (2023) [44] enables the building of spreadsheet models in GoogleSheets or Microsoft Excel "that transform, extract, or summarize any text" powered by AI, using a simple online demo.

2.1.3 Now and Looking Ahead

AI Objectives Institute (AOI, 2023) [45], with the mission of "building a better AI future" is described as "a nonprofit research lab for human flourishing, societal alignment, and safe AI" undertaking initiatives such as "Talk to the city" in the form of "a new LLM survey tool to improve collective discourse and decision-making" through "analyzing detailed qualitative responses to a question instead of limited quantitative data." Kantrowitz (2023) [46] claims that application programming interfaces (APIs) associated with attention-getting AI chatbots are key as ways to generate use possibilities from search to knowledge work, to healthcare.

Moving beyond text generation, Huang, Chen, Liu, Shen, Zhao, and Zhou (2023) [47] describe "composer" as a general framework for large-scale generative compositional models, as in, "capable of seamlessly recombining visual components to produce new images", for creative and controllable image synthesis. Composer is implemented as a "multi-conditional diffusion model" able to deconstruct and reconstruct images with many capabilities such as text-to-image generation, multi-modal conditional image generation, style transfer, pose transfer, image translation, virtual try-on, and much more. Huang et al. (2023) [47] acknowledge "potential risks associated with image generation models, such as deceptive and harmful content" and highlight the need "to thoroughly investigate how Composer can mitigate the risk of misuse" by "possibly creating a filtered version before making the work public."

With the mission to "ensure that artificial general intelligence (AGI) benefits all humanity" and where AGI refers to "highly autonomous systems that outperform humans at most economically valuable work", OpenAI (2023) [48] recently indicated that they are concerned with the behavior of AI systems and to this end are "clarifying how ChatGPT's behavior is shaped and our plans for improving that behavior, allowing more user customization, and getting more public input into our decision-making in these areas." Marr (2023) [49] describes GPT-3 as "an AI that is better at creating content that has a language structure - human or machine language - than anything that has come before it." Notwithstanding the "hype", at the moment, around GPT-3, Marr (2023) [49] cautions about the cost "due to the huge amount of compute power needed to carry out its functions"; the "closed or black box system" underlying GPT-3 where OpenAI does not provide "full details of how its algorithms work"; and concerns that "the output of the system is still not perfect." Among other discussion "takeaways" to those identified by Tamkin and Ganguli (2021) [5] in the Introduction of this paper are the following questions and concerns: "is GPT-3 intelligent, and does it matter?"; that "future models won't be restricted to learning just from language"; and that "disinformation is a real concern, but several unknowns remain" (Tamkin and Ganguli, 2021). It is worth noting that Kurtzweil (2024) [50] is in the process of updating his earlier work on "the singularity" regarding "a revolution in knowledge and an expansion of human potential" while addressing "potential perils of biotechnology, nanotechnology, and artificial intelligence" including "how AI will impact unemployment and the safety of autonomous cars."

2.2 Cognitive Flexibility

Once we become rigid in our thinking, we are no longer curious; we shut down opportunities to learn... once we become flexible, we remain curious, opening up opportunities...

As a response to a world characterized by rapid change and complexity, Spiro, Collins, Thota, and Feltovich (2003) [51] advanced cognitive flexibility theory (CFT) with the 4 goals of "aiding people to learn important but difficult subject matter"; "fostering adaptively flexible use of knowledge in real-world settings"; "changing underlying ways of thinking"; and "developing hypermedia learning environments to promote complex learning and flexible knowledge application."

Scales or measures such as the Cognitive Flexibility Inventory (CFI) developed by Dennis and VanderWal (2010) [52] as a "20-item self-report measure to monitor how often individuals engage in cognitive behavioral thought challenging interventions" (NovoPsych, 2021) [53] is noteworthy in view of the two aspects measured - *alternatives* as "the adaptive ability to perceive multiple alternative explanations for life occurrences and the ability to generate multiple alternative solutions to difficult situations" and *control* as "having an internal locus of control, or the tendency to perceive difficult situations as somewhat controllable."

Cognitive flexibility is said to require the two concepts of "mental representation and vigilance (attention)" (Elen et al., 2011) [54] where "cognitively flexible individuals anticipate more than one solution, more than one perspective, and more than one source of knowledge" and additionally, "are vigilant in watching (consciously or unconsciously) for the need or desirability of change." Cheng and Koszalka (2016) [55] present a concept paper for cognitive flexibility theory (CFT) in relation to learning resources, highlighting the importance of "multiple perspectives." Samuel, Kashyap, Samuel, and Pelaez (2022) [56] formulate an adaptive cognitive fit (ACF) framework and through testing and simulation claim that it "establishes the viability of using AI to improve human performance." And it is worth noting that one of six grand challenges for human-centered artificial intelligence (AI) pertains to "respecting human's cognitive capacities" in the design of approaches to AI that "interacts with individuals" (Garibay et al., 2023) [57]. Gaskell (2023) [58] questions whether AI is "eroding our ability to think" pointing to examples such as that of the curating of art exhibits and, drawing on the work of Chamorro-Premuzic (2023), [59] notes that "there is not much evidence that the rise of AI has been leveraged in some way to elevate our curiosity or intellectual development." However, Berkeley and Berlin (2023) [60] speak in terms of pseudocognition in the context of a world with AI where "entities" are said to "think better and act quicker than humans and their institutions" with "implications not just for how people earn their livings and organise their lives, but also for how they think about their humanity."

2.3 Cognitive Rigidity

According to Elen et al. (2011) [54], cognitive rigidity "is indicated by a failure either to detect a need for change or a refusal to change in the face of situational demands for change." Improved AI chatbots could forestall cognitive rigidity, in that, according to Farnam Street Media (2023) [61], citing the work of Holiday (2013) [62], "Once the mind has accepted a plausible explanation for something, it becomes a framework for all the information that is perceived after it. We're drawn, subconsciously, to fit and contort all the subsequent knowledge we receive into our framework, whether it fits or not", a situation referred to by psychologists as "cognitive rigidity." As such, "facts that built an original premise are gone, but the conclusion remains—the general feeling of our opinion floats over the collapsed foundation that established it. Information overload, 'busyness,' speed, and emotion all exacerbate this phenomenon. They make it even harder to update our beliefs or remain open-minded" (Farnam Street Media, 2023) [61].

2.4 Conceptual Framework for Cognitive Flexibility in AI-Rich Learning Environments

As noted above (Sect. 2.1.1), there is an understanding that the output of current AI chatbot systems is not perfect, that responses may include errors, and that the evaluation step is more important than it might be in future AI chatbot systems. Pretesting perspective taking and boundary setting returned results that are sufficiently aligned with our expectations and methodology to engage study participants in a set of surveys that will offer data that address our research question. The conceptual framework captured in Fig. 1 and the methodology that follows considers an iterative process that involves asking the AI chatbot a root question and then posing the same question from varied perspectives and/or with specific boundaries.

The research question to be explored in the study developed in this paper is as follows:

- RQ1. Could LLM AI chatbots support cognitive flexibility and, if yes, how?

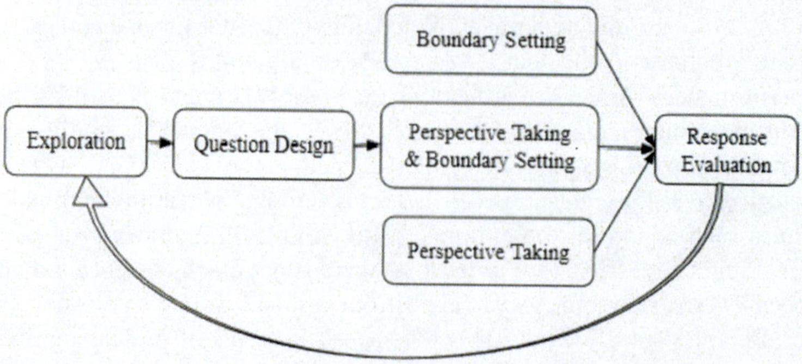

Fig. 1. Conceptual framework for cognitive flexibility in AI-rich learning environments.

3 Methodology

McGinley and Tierney (1989) [63] advance the notion that learning about a topic is like traversing a landscape of perspectives through reading, writing, and learning as an approach to inquiry. As such, this paper employs the use of topical inquiry including perspective taking and boundary setting when engaging with LLM AI chatbots, in support of cognitive flexibility. Perspective taking and boundary setting are intended to develop breadth and depth of understanding, involving elements such as: cross domain (cross curricular) focus; individual and collective decision making; critical thinking for argumentative writing; complexity in terms of audience, domain, age, vocabulary, voice, background knowledge, and current conceptual understanding; and the historical (economic, political, and social). Developing this understanding through the potential of AI, encourages individuals to reach across the aisle in varied settings. To this end, this paper develops and demonstrates use of the *Food Scarcity Exploration Exemplar with ChatGPT-3.5*.

Garibay et al. (2023) [57] note that human-computer interaction design and evaluation methods "were primarily defined for non-AI computing systems" and as such, "may not effectively address the unique issues in AI systems", identifying some "20 existing methods of HCI, human factors, and other related disciplines from the HCAI perspective" while proposing "7 alternative methods" for "enhancing existing HCI methods" such as in-the-wild studies, human-machine teaming based collaborative design, AI as design material, and longitudinal studies. Important for the study being designed in this paper is the API (application programming interface) for generative AI apps in terms of discerning the different ways in which LLM AI chatbots are learning and learning differently. Also important in the study design for this paper is determining how to compare AI chatbot learning approaches and understanding what may be best for learning.

As such, this paper also incorporates, as part of the methodology, the use of *AI as a design material* where Knemeyer and Follett (2020) [64], drawing on the work of Clark (2019) [65] pertaining to "future-friendly interfaces for artificial intelligence", note that "AI, and more specifically machine learning, is also a design material" although less obviously so than say, "pencils and sketchbooks, markers and whiteboard, pixels and code." According to Knemeyer and Follett (2020) [64], "AI in general, and machine learning in particular, offer a compelling array of features and opportunities for designers to innovate and improve their designed experiences in important ways" reminding us that, according to Clark [65], html (hypertext markup language) and CSS (cascading style sheets) are design materials. Knemeyer and Follett (2020) [64] reiterate from the work of Clark [65] five ways that AI can be used by designers, namely, for recommendation, prediction, classification, clustering, and generation. Knemeyer and Follett (2020) [64] draw our attention to the notion of *productive friction* where "AI can approach problems from a novel perspective, and while some of the results may not be useful, there can also be ingenious and unexpected discoveries." Holmquist (2017) [66] refers to the notion of "intelligence on tap" as in, AI "as a resource you can plug into any new product when you need it" which, in an interaction design context, "intelligence is becoming a new design material" with the accompanying challenges of - transparency, opacity, unpredictability, learning, evolution, and shared control.

Where Cheng and Koszalka (2016) [55] advance a cognitive flexibility theory (CFT) framework involving multiple perspectives for learning resources and Samuel et al. (2022) formulate an adaptive cognitive fit (ACF) framework for AI to "improve human performance", here we propose a framework that not only involves multiple perspectives (perspective taking) and the improving of human performance but also allows users to remain in an inquiry mode by continually reviewing and iterating their prompt responses. During this period, the focus is not on finding definitive answers but staying open to multiple possibilities, learning, and acquiring new knowledge. This approach may lead to understanding previously unconsidered aspects.

What follows is the description of an incremental solution along with the results. For example, a pilot study with a small group of participants will allow us to evaluate our conceptual framework, research model, and reflection survey questions in support of data collection and analysis to help us respond to our research question. Figure 2 outlines a multi-step methodology and process model for a data collection approach for the pilot study. In Step 1, questions developed by the study researchers and responses generated by ChatGPT 3.5 are shared providing participants with an exemplar use case of the study's conceptual framework. In Step 2, participants complete a guided reflection for the exemplar responses. Step 1 and Step 2 prepare participants for Step 3 where they share their own questions aligned to this study's conceptual framework. In Step 4 researchers submit participant questions to the AI Chatbot. In Step 5, researchers share AI chatbot responses to the participant's questions. The Step 6 reflection replicates Step 2, this time using the responses to the participant's questions.

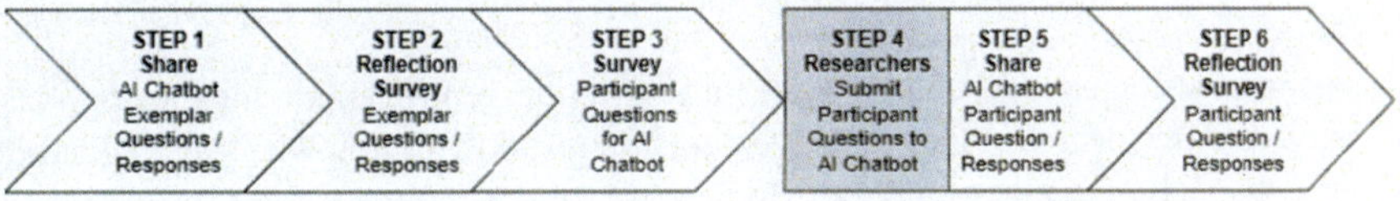

Fig. 2. Methodology – process model for data collection approach.

Step 1. Participants review the exemplar questions and responses for the *Food Scarcity Exploration Exemplar- ChatGPT-3.5* (Appendix A) developed using the study Conceptual Framework (Fig. 1). After reviewing the responses, participants are asked to consider a series of questions which will be posed for their response in the Step 2 Reflection Survey.
Step 2. The Reflection Survey asks participants to respond to a series of questions related to Food Scarcity responses shared in Step 1. Survey questions are aligned to RQ1. Step 2 provides baseline data wherein participants review and reflect on the Food Scarcity Exemplar questions and responses.
Step 3. Survey 2 asks participants to engage in their own topic exploration. They will design a root question and extend the root question using perspective taking, boundary setting, and both perspective taking and boundary setting.
Step 4. Participant questions are submitted to the AI Chatbot (ChatGPT 3.5) by the researchers.
Step 5. Questions and responses are shared back to each participant.
Step 6. The Reflection Survey asks participants to respond to a series of questions related to the responses returned by the AI Chatbot for participant questions shared in Step 4.

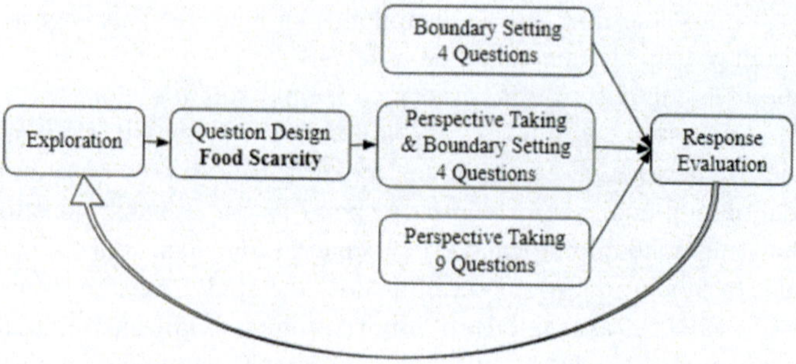

Fig. 3. Conceptual framework exemplar - food scarcity.

Figure 3 shows the application of Fig. 1 as a conceptual framework exemplar for food scarcity.

4 Discussion and Outcomes

In this paper, we consider AI chatbots as tools to encourage immersion in, and appreciation for, varied perspectives. During a question and response dialog with the AI chatbot, might we suspend preconceived beliefs to be open to new understanding?

When considering LLM AI chatbots, we are reminded of the prescient work of those who envisioned the potential of technology to enhance and enrich our individual and collective experiences. These tools would work alongside us and "amplify" our cognitive abilities, yet rely on us to be evaluators and ethicists, when making decisions and taking action. These essential roles remain constants in this period of rapid development of AI technologies and tools. For example, McLuhan (1964) [67] argued that media and technologies are extensions of our human faculties and capabilities, including our brains. He believed that these technologies would influence not only the way we communicate and interact but how we perceive the world around us. Further, it is worth noting that Nicholas Negroponte [68], founder of the MIT Media Lab, believed that digital media was changing how we communicate, learn, and work, democratizing access to information and creating new forms of social interaction and collaboration.

Rheingold's works on virtual reality (1991) [69], virtual community (1993) [70], and smart mobs (2002) [71] focused on the ways technology shapes social interactions and relationships and creates more connected and collaborative communities. In his work on mind amplification, Rheingold (2012) [72] refers to the work of Licklider [73], who had been influenced by the development of electronic visual dashboards for air-defense command and control that he had seen in his war research. In 1960 he proclaimed a symbiotic vision where: "The hope is that, in not too many years, human brains and computing machines will be coupled very tightly, and that the resulting partnership will think as no human brain has ever thought and process data in a way not approached by the information-handling machines we know today" [73].

As such, anticipated outcomes of the study designed in this paper include but are not limited to, the following.

Outcome 1 - Employ a topical inquiry model that highlights perspective taking and boundary setting using AI chatbots to efficiently compile information that helps individuals acquire a breadth and depth of understanding that positively impacts cognitive flexibility and forestalls cognitive rigidity.
Outcome 2 - Employ AI chatbots to amplify and augment human-computer interactions, enhancing capabilities for knowledge building, decision making, and problem solving in support of emerging new potentials under exploration for curiosity, interest, engagement, creativity, and innovation.
Outcome 3 - Amidst the varied, heightened concerns regarding AI, elevate the positive role AI tools can offer and the importance of guiding the development of such technologies in beneficial directions.

Key contributions of this paper include a review of the literature pertaining to cognitive flexibility in relation to large language model artificial intelligence chatbots for learning; a conceptual framework for cognitive flexibility in AI-rich learning environments; and the identification of areas for further exploration such as curiosity, interest, and engagement as well as creativity and innovation.

5 Limitations and Future Directions

Garibay et al. (2023) [57] outline four guiding principles pertaining to governance systems and the use of AI, namely fairness, integrity, resilience, and explainability where fairness addresses the notion of bias and the need for "identification and implementation of mitigation strategies to intentionally lower the risk of unethical decision making and unintentional outcomes." While bias is not addressed in this paper, research questions going forward might include the following:

Q2: Could LLM AI chatbots reduce bias in knowledge seeking, and if yes, how?

LMs and LLMs for AI chatbots provide a promising space for exploration and application of the model of curiosity, interest, and engagement (CIE) in new media technology-pervasive environments advanced by Arnone et al. (2011) [74]. Additionally, attentive to the progress being made pertaining to "artificial intelligence (AI) creativity", Amabile (2019) [75] calls for "organizational researchers of creativity and innovation" to "invest significant energy in studying AI and computer-assisted human intelligence" in support of the potential for "creative breakthroughs" and as such, LMs and LLMs for AI chatbots would seem to provide a promising space for application and exploration.

6 Conclusion

In conclusion, this work suggests that there is the potential for positive reciprocal outcomes when people interact with AI systems where "flexible adaptation", said to be important for AI systems (Haenlin and Kaplan, 2019) [76], is also important for people in the context of learning. Additionally, perspective taking and boundary setting during topical inquiry, as suggested in this study, provide rich learning opportunities for participants to develop cognitive, emotional, and social intelligence behaviors that are key to the development of competent, ethical, and caring individuals. Likewise, AI systems are trained and learn from both data and interaction with humans who engage in AI questioning including perspective taking and boundary setting.

The important thing about this research is that it recognizes the potential for AI to encourage individuals to reach across the aisle in varied settings -- to bridge the chasm of fundamental differences by considering fundamental commonalities. And because topical inquiry that employs perspective taking and boundary setting (broad and narrow) can be efficiently facilitated by LLM Chatbots, we may see decision making that involves more sustained engagement.

The use of AI as a design material to form part of the methodology for the study designed in this paper also contributes to learning associated with new and evolving methods for the study and analysis of AI. This research project will be of interest to AI researchers, researchers and practitioners concerned with cognitive flexibility in contemporary learning environments, and AI developers and as such, conferences, funding opportunities, and evolving AI-rich learning spaces will provide the target audiences for this work.

Appendix A

Food Scarcity - Perspective Taking and Boundary Setting - ChatGPT-3.5

Root Question. Explain food scarcity and how it can be addressed.

Perspective Taking

1. Explain food scarcity and how it can be addressed from the perspective of a naturalist.
2. Explain food scarcity and how it can be addressed from the perspective of an economist.
3. Explain food scarcity and how it can be addressed from the perspective of a scientist.

4. Explain food scarcity and how it can be addressed from the perspective of a farmer.
5. Explain food scarcity and how it can be addressed from the perspective of an environmentalist
6. Explain food scarcity and how it can be addressed from the perspective of a politician.
7. Explain food scarcity and how it can be addressed from a politically conservative perspective.
8. Explain food scarcity and how it can be addressed from a politically liberal perspective.
9. Explain food scarcity and how it can be addressed from a politically middle-of-the-road perspective.

Boundary Setting

1. Explain food scarcity and how it can be addressed where there is little arable land.
2. Explain food scarcity and how it can be addressed where water is in short supply.
3. Explain food scarcity and how it can be addressed when an area is overpopulated.
4. Explain food scarcity and how it can be addressed in a desert.

Perspective Taking and Boundary Setting

1. Explain food scarcity and how it can be addressed from the perspective of a farmer in a third-world country where access to water is a challenge.
2. Explain food scarcity and how it can be addressed from the perspective of farmers in overpopulated areas.
3. Explain food scarcity and how it can be addressed from the perspective of farmers in China.
4. Explain food scarcity and how it can be addressed from the perspective of farmers in tropical climates.

References

1. OpenAI. (2023). ChatGPT. Blog. Accessed 21 May 2023. https://openai.com/blog/chatgpt/
2. Li, H.: Language models: past, present and future. Commun. ACM. ACM **65**(7), 56–63 (2022). https://doi.org/10.1145/3490443
3. Merriam-Webster: Forestall. (2023). Accessed 21 February 2023. https://www.merriam-webster.com/thesaurus/forestall
4. Tamkin, A., Brundage, M., Clark, J., Ganguli, D.: Understanding the capabilities, limitations, and societal impact of large language models. ArXiv, abs/2102.02503 (2021)
5. Tamkin, A., Ganguli, D.: How large language models will transform science, society, and AI: Scholars in computer science, linguistics, and philosophy explore the pains and promises of GPT-3. News, 5 February. Stanford University. Human-Centered Artificial Intelligence (HAI) (2021). Accessed 22 February 2023. https://hai.stanford.edu/news/how-large-language-models-will-transform-science-society-and-ai
6. Chrisinger, B.: It's not just our students - ChatGPT is coming for faculty writing: And there's little agreement on the rules that should govern it. The Chronicle of Higher Education, 22 February 2023. Accessed 27 February 2023. https://www.chronicle.com/article/its-not-just-our-students-ai-is-coming-for-faculty-writing

7. Boudry, M.: Human intelligence - Have we reached the limit of knowledge? The Conversation, 11 October 2019. Accessed 9 May 2023. https://theconversation.com/human-intelligence-have-we-reached-the-limit-of-knowledge-124819
8. Shanahan, M.: Talking about large language models. (2023). Accessed 24 February 2023. https://arxiv.org/pdf/2212.03551.pdf
9. Gorvett, Z.: The AI emotions dreamed up by ChatGPT: AI chatbots are already imagining what feelings they'll end up with. But if they did develop them, would we even notice? BBC Machine Mind (2023). Accessed 26 February 2023. https://www.bbc.com/future/article/20230224-the-ai-emotions-dreamed-up-by-chatgpt
10. Edwards, B.: Meta unveils a new large language model that can run on a single GPU (Updated). ArsTechnica, 24 February (2023). Accessed 27 February 2023. https://arstechnica.com/information-technology/2023/02/chatgpt-on-your-pc-meta-unveils-new-ai-model-that-can-run-on-a-single-gpu/?utm_source=tldrnewsletter
11. Ouyang, L., et al.: Training language models to follow instructions with human feedback. Adv. Neural Inf. Process. Syst. **35**, 27730–27744 (2022). https://proceedings.neurips.cc/paper_files/paper/2022/hash/b1efde53be364a73914f58805a001731-Abstract-Conference.html
12. Mah, C., Levine, S.: How to use ChatGPT as an example machine. Cult of Pedagogy, 19 February (2023). Accessed 28 February 2023. https://www.cultofpedagogy.com/chatgpt-example-machine/
13. Shen, Y., et al.: ChatGPT and other large language models are double-edged swords. Radiology **307**, 230163 (2023). https://pubs.rsna.org/doi/epdf/10.1148/radiol.230163
14. Prakash, P.: Former Google CEO Eric Schmidt says the tech sector faces a 'reckoning': What happens when people fall in love with their A.I. tutor? Fortune, 3 April (2023). Accessed 3 April 2023. https://fortune.com/2023/04/03/eric-schmidt-former-google-ceo-tech-sector-reckoning/
15. Ourednik, A., Mlynář, J., Mutzner, N., Alavi, H.: Social media platforms are failed cities. Interactions, November-December (2021). https://doi.org/10.1145/3488234
16. Brown, S.: Social media is broken. A new report offers 25 ways to fix it. Ideas Made to Matter blog, 30 June (2021). MIT Sloan School of Management
17. Al-Sibai, N.: Machine learning expert calls for bombing data centers to stop rise of AI: He says that after AGI, "literally everyone on earth will die." Futurism (2023)
18. ChatGPT: Response to the question, "What is an AI chatbot? (2023)
19. Schick, T., et al.: Toolformer: language models can teach themselves to use tools. arXiv preprint arXiv:2302.04761 (2023)
20. Wolfram, S.: Wolfram|Alpha as the way to bring computational knowledge superpowers to ChatGPT. Stephen Wolfram Writings, 9 January (2023). Accessed 1 March 2023. https://writings.stephenwolfram.com/2023/01/wolframalpha-as-the-way-to-bring-computational-knowledge-superpowers-to-chatgpt/
21. Copy.ai: Whatever you want to ask, our chat has the answers. (2023). Accessed 12 March 2023. https://app.copy.ai/
22. Javaid, S.: 7 chatbot training data preparation best practices in 2023. AI Multiple, 17 February (2023). Accessed 13 March 2023. https://research.aimultiple.com/chatbot-training-data
23. Taylor, et al.: Galactica: a large language model for science. arXiv:2211.09085 (2022)
24. Galactica: Language models that cite. (2023). Accessed 2 March 2023. https://galactica.org/explore/
25. Miller, K.: How do we fix and update large language models? These three approaches to editing large language models could make them more accurate, consistent, and up-to-date. News, 13 February. Stanford University. Human-Centered Artificial Intelligence (HAI) (2023). Accessed 22 February 2023. https://hai.stanford.edu/news/how-do-we-fix-and-update-large-language-models

26. Warren, T.: These are Microsoft Bing AI secret rules and why it says it's named Sydney. The Verge, 14 February (2023). Accessed 28 February 2023. https://www.theverge.com/23599441/microsoft-bing-ai-sydney-secret-rules?fbclid=IwAR3qer6D6p9riBWDRNypIqEIiJDJDiVKdeJbvn2cyr9cVYNWG79po4DAGL4&mibextid=Zxz2cZ
27. UTS: Referencing ChatGPT in assignments. University of Technology Sydney Library News, 2 March (2023). Accessed 9 March 2023. https://www.lib.uts.edu.au/whats-on/news/referencing-chatgpt-in-assignments?fbclid=IwAR2FrE0lYV-HypDoolLq3_na4HMYJEaRhUGX3r_fZGwXnN_dYm3MmvLhJiM
28. Shahriar, S., Hayawi, K.: Let's have a chat! A conversation with ChatGPT: Technology, applications, and limitations. arXiv:2302.13817 [cs.CL] (2023)
29. Ghaffray, S.: Silicon Valley's AI frenzy isn't just another crypto craze: It isn't theoretical. Millions of people are already using apps like ChatGPT to write books, create art, and develop code. Vox, 6 March (2023). Accessed 7 March 2023. https://www.vox.com/technology/2023/3/6/23624015/silicon-valley-generative-ai-chat-gpt-crypto-hype-trend
30. Roth, E.: Meet the companies trying to keep up with ChatGPT / From Google's Bard to Microsoft's new Bing, here are all the major contenders in the AI chatbot space. The Verge, 5 March (2023). Accessed 7 March 2023. https://www.theverge.com/2023/3/5/23599209/companies-keep-up-chatgpt-ai-chatbots
31. Anthropic: AI research and products that put safety at the frontier. (2023). Accessed 21 May 2023. https://www.anthropic.com
32. Bai, et al. Constitutional AI: Harmlessness from AI feedback. Anthropic, 15 December 2022. https://arxiv.org/pdf/2212.08073.pdf
33. Stahl, L.: The new world of AI chatbots like ChatGPT. 60 Minutes, CBS, 5 March 2023. Accessed 6 March 2023. https://www.cbsnews.com/news/chatgpt-artificial-intelligence-chatbots-60-minutes-2023-03-05/
34. Mehta, I.: Brave Search launches an AI-powered summarization feature. TechCrunch, 2 March (2023). Accessed 11 March 2023. https://techcrunch.com/2023/03/02/brave-search-launches-an-ai-powered-summarization-feature/?utm_source=tldrai
35. Composer AI: Write faster with AI, save time. Compose.ai. (2023). Accessed 11 March 2023. https://www.compose.ai
36. Lampkin, L.: Microsoft boosts Viva Sales with new GPT seller experience. CloudBlogs, 2 February. Accessed 21 May 2023. https://cloudblogs.microsoft.com/dynamics365/bdm/2023/02/02/microsoft-boosts-viva-sales-with-new-gpt-seller-experience/ (2023)
37. Stokel-Walker, C.: Generative AI is coming for the lawyers: Large law firms are using a tool made by Open AI to research and write legal documents. What could go wrong? Wired Business, 21 February 2023. Accessed 28 February 2023. https://www.wired.com/story/chatgpt-generative-ai-is-coming-for-the-lawyers/?fbclid=IwAR1qnrwaPiirymUnzpQS6cMP3sEBsjrKAKp9M4QOPSETVmcQGtIsefnEc_0
38. Harvey.ai.: Harvey. (2023). Accessed 28 February 2023. https://www.harvey.ai/?fbclid=IwAR1RgNgAlVwv0AWHixoGkN7Smpc_UQXepLIy5I-gWGsObzKaRtCH2W73YN8&mibextid=Zxz2cZ
39. GDPR.EU.: What is GDPR, the EU's new data protection law? Accessed 21 May 2023. https://gdpr.eu/what-is-gdpr/ (2023)
40. Sherry, B.: You can now integrate ChatGPT with your products. Here's how: OpenAI's new API puts the power of natural-language processing in entrepreneurs' hands. Inc., 2 March 2023. Accessed 4 March 2023. https://www.inc.com/ben-sherry/you-can-now-integrate-chatgpt-with-your-products-heres-how.html
41. Warren, T.: Microsoft brings its newAI-powered Bing to the Windows 11 taskbar. The Verge, 28 February 2023. Accessed 5 March 2023. https://www.theverge.com/2023/2/28/23618214/microsoft-windows-11-update-bing-ai-taskbar-touch-improvements-screen-recording-features

42. Wiggers, K.: Typeface emerges from stealth with $65M to bring generative AI to the enterprise. TechCrunch, 28 February (2023). Accessed 10 March 2023. https://techcrunch.com/2023/02/28/typeface-emerges-from-stealth-with-65m-to-bring-generative-ai-to-the-enterprise/?utm_source=tldrai&guccounter=1
43. Cision: Quizlet launches 'Q-Chat,' AI tutor built with OpenAI API. PR Newswire, 1 March (2023). Retrieved 1 March 2023 from https://www.prnewswire.com/news-releases/quizlet-launches-q-chat-ai-tutor-built-with-openai-api-301759014.html
44. PromptLoop: Use PromptLoop in Google Sheets and Excel. (2023). Accessed 2 March 2023. https://www.promptloop.com/?utm_source=tldrai
45. AOI: Building a better AI future: A nonprofit research lab for human flourishing, societal alignment, and safe AI. (2023). Accessed 6 February 2023. https://www.cbsnews.com/news/chatgpt-artificial-intelligence-chatbots-60-minutes-2023-03-05/
46. Kantrowitz, A.: Next up for AI chatbots: It's all about the APIs - The magical demos displayed the capabilities. Now the platform wars begin. Big Technology, 23 February (2023). Retrieved 28 February 2023 from https://www.bigtechnology.com/p/next-up-for-ai-chatbots-its-all-about
47. Huang, L., Chen, D., Liu, Y., Shen, Y., Zhao, D., Zhou, J.: Composer: Creative and controllable image synthesis with composable conditions. arXiv, 20 February (2023). https://arxiv.org/abs/2302.09778
48. OpenAI: How should AI systems behave, and who should decide? Blog, 16 February 2023. Accessed 28 February 2023. https://openai.com/blog/how-should-ai-systems-behave/?utm_source=tldrai
49. Marr, B.: What is GPT-3 and why is it revolutionizing artificial intelligence? BM Blog (2023). Accessed 1 March 2023. https://bernardmarr.com/what-is-gpt-3-and-why-is-it-revolutionizing-artificial-intelligence/#:~:text=Starting%20with%20the%20very%20basics,to%20carry%20out%20their%20task
50. Kurtzweil, R.: The Singularity is Nearer. (2024)
51. Spiro, R.J., Collins, B., Thota, J.J., Feltovich, P.J.: Cognitive flexibility theory: hypermedia for complex learning, adaptive knowledge application, and experience acceleration. Educ. Technol. Arch. **43**, 5–10 (2003)
52. Dennis, J.P., Vander Wal, J.S.: The cognitive flexibility inventory: Instrument development and estimates of reliability and validity. Cogn. Ther. Res. Ther. Res. **34**(3), 241–253 (2010). https://doi.org/10.1007/s10608-009-9276-4
53. NovoPsych: Cognitive Flexibility Inventory (CFI). (2021). Accessed 7 March 2023. https://novopsych.com.au/assessments/formulation/cognitive-flexibility-inventory-cfi/
54. Elen, J., Stahl, E., Bromme, R., Clarabout, G.: Links between beliefs and cognitive flexibility: lessons learned. Springer, Cham (2011). https://doi.org/10.1007/978-94-007-1793-0
55. Cheng, J., Koszalka, T.A.: Cognitive flexibility theory and its application to learning resources. (concept paper). (2016). Accessed 7 March 2023. https://ridlr.syr.edu/wp-content/uploads/2020/11/ConceptPaper2016-CFT.pdf
56. Samuel, J., Kashyap, R., Samuel, Y., Pelaez, A.: Adaptive cognitive fit: artificial intelligence augmented management of information facets and representations. Int. J. Inf. Manage. **65**, 102505 (2022). https://doi.org/10.1016/j.ijinfomgt.2022.102505
57. Garibay, O.O., et al.: Six human-centered artificial intelligence grand challenges. Int. J. Hum.-Comput. Inter. **39**(3), 391–437 (2023). https://doi.org/10.1080/10447318.2022.2153320
58. Gaskell, A.: Is AI eroding our ability to think? Forbes, 24 January 2023. Accessed 27 February 2023. https://www.forbes.com/sites/adigaskell/2023/01/24/is-ai-eroding-our-ability-to-think/?sh=14baa1ab45b2
59. Chamorro-Premuzic, T.: I, Human: AI, Automation, and the Quest to Reclaim What Makes Us Unique. Harvard Business Review Press, New England (2023)

60. Berkeley & Berlin: The age of pseudocognition: looking at the impacts of the computer browser, the printing press and psychoanalysis could help prepare the world for AI. The Economist 22 April (2023)
61. Farnam Street Media: The human mind has a shut off device. FS blog (2023). Accessed 9 March 2023. https://fs.blog/the-human-mind-has-a-shut-off-device/
62. Holiday, R.: Trust me, I'm Lying: The Tactics and Confessions of a Media Manipulator. Portfolio, New York (2013)
63. McGinley, W., Tierney, R.J.: Traversing the topical landscape: reading and writing as ways of knowing. Writ. Commun. Commun. **6**, 243–269 (1989)
64. Knemeyer, D., Follett, J.: AI as design material: Seeing the design role in creating AI-driven products. Towards Data Science blog, 21 September (2020). Accessed 7 March 2023. https://towardsdatascience.com/ai-as-design-material-2748d84bbb7b
65. Clark, J.: A.I. is your new design material. Big Medium, March 11 (2019). Accessed 7 March 2023. https://bigmedium.com/speaking/ai-is-your-new-design-material.html
66. Holmquist, L.: Intelligence on tap: Artificial intelligence as a new design material. Interactions, XXIV.4 July-August (2017)
67. McLuhan, M.: Understanding Media: The Extensions of Man. The MIT Press, Cambridge (1964)
68. Negroponte, N.: Being digital. Penguin Random House (1995)
69. Rheingold, H.: Virtual Reality: The revolutionary technology of computer-generated artificial worlds – and how it promises to transform society. Simon and Schuster (1991)
70. Rheingold, H.: The virtual community: Homesteading on the virtual frontier (1993). MIT Press, Revised edition (2000)
71. Rheingold, H.: Smart mobs: The next social revolution. Basic Books (2002)
72. Rheingold, H.: Mind amplifier: Can our digital tools make us smarter? Creative Commons (2012)
73. Licklider, J.C.R.: Man-Computer Symbiosis. IRE Transactions on Human Factors in Electronics, Vol. HaFE-1, pp. 4–11, March (1960)
74. Arnone, M.P., Small, R.V., Chauncey, S.A., McKenna, H.P.: Curiosity, interest and engagement in technology-pervasive learning environments: a new research agenda. Educ. Technol. Res. Dev. **59**(2), 181–198 (2011). https://doi.org/10.1007/s11423-011-9190-9
75. Amabile, T.M.: Guidepost: creativity, artificial intelligence, and a world of surprises. Acad. Manag. Discoveries **6**(3) (2019). https://doi.org/10.5465/amd.2019.0075
76. Haenlein, M., Kaplan, A.: A brief history of artificial intelligence: on the past, present, and future of artificial intelligence. Calif. Manage. Rev. **61**(4), 5–14 (2019). https://doi.org/10.1177/0008125619864925

Post-LLM Academic Writing Considerations

Lucas Potter(✉) and Xavier-Lewis Palmer

BiosView, Oswego, KS, USA
lpott005@odu.edu

Abstract. Academic publishers would not be naive to have a concern for how scientific records are increasingly vulnerable in the Age of Large Language Models. Vulnerabilities can come in the form of DDOS-like attacks to repositories or guidance in the creation of malware that can serve to infiltrate their systems. Actors from mere pranksters to malicious actors who are anti-science or want to affect national security could very well affect the publishers and endanger collected human knowledge. It is important for stakeholders to become realistic about the risk environment and the increasingly leveled playing field for the simplest of tasks and how academia may be affected. This exploration and commentary paper suggests some complications in this era for academic publishing and is not meant to be taken as a scientific work, in hopes of generating conversation. Failure to consider the new landscape may result in the permanent loss of the value that scientific records could offer, down the line.

Keywords: BioCyberSecurity · CyberBioSecurity · Misinformation · LLM

1 Introduction

Since the 17th century, academics have found numerous ways of disseminating their work via academic journals. Towards this end, various tools from the pen, the typewriter, and modern computers have been utilized. The information age specifically brought suggestive tools such as spell check and an array of formatting sub-tools to shape and optimize work produced. Those tools were seen as a boon and not a threat to the enterprise of writing. Newer and more comprehensive services that not only spell check but also check tones such as Grammarly, manage references. Zotero, Medley, or Endnote, have additionally been welcome for lightening academic workloads. However, the last 20 years have seen projects that go beyond merely correction or other means of non-additive facilitation, whether or not it was welcomed by Academia.

One example of a tool that automated production of content material and has long roiled such as SciGen, which struck academics as foul, through its success in delivery of material that appears sensible at first glance but is either incoherent, gibberish, or downright nonsensical upon close inspection. To the horror of many academics, SciGen papers have found their way in many computer science journal articles, with many outstanding [1–3]. These have been used for pranks and fraud but have also spurred academicians and publisher personnel alike to action in order to detect fraudulent published

K. Arai (Ed.): FTC 2023, LNNS 816, pp. 154–163, 2023.
https://doi.org/10.1007/978-3-031-47448-4_12

submissions or block possible ones on the way [4–7]. Many of these fraudulent articles may even cite legitimate articles and find themselves woven into serious discussion. It makes sense that some academics have started questioning if they wanted to be cited and or if they should be allowed the refusal of being cited [8]. Altogether, this paints a growing chaotic picture, but more problems exist through the introduction of algorithms that produce often coherent and confident content through dependence on large datasets: large language models (LLMs).

Popular iterations such as ChatGPT by OpenAI, Bard by Alphabet, and others have been used by more than 100 million users as of 2023, and these numbers show no sign of stopping. ChatGPT has shown itself to be immensely useful since its unveiling whereby numerous authors have written books, articles, protocols, programs, and more, and many users have received helpful directions, using early versions [9, 10]. This has only magnified the prior issue of generated content, now that submissions are more intelligible, leading to an arms race. Generative algorithms have even been effective enough that esteemed institutions such as Harvard have issued guidance on LLM use, especially with notice to protecting confidential data, indicating at least a tentative acceptance on some level [11]. There are many points to consider in which this will change and possibly break current academic models of publishing, especially given that material can be generated much more quickly and convincingly. This indicates that mis- and dis information is easier to embed. We may see higher rates of fraudulent articles in journals that are harder to detect, ultimately worsening chains of trust and reliability for those who rely on scientific data. Further, these same LLM tools can beget tools that can help a malicious actor evade detection or craft a better tool to take down a publishing platform. It is now appropriate to expand on this through a few cases to consider. We present musical and medical cases, in addition to discussion on how to address some problems introduced by LLMs. The Age of LLMs has already stimulated modern academia and further simulations may occur for good and or bad. In the meantime, we present a paper to open and discuss possible visions for LLM disruption.

2 Musical Misinformation

You could be forgiven for not hearing of the recent study which demonstrated a correlation between exposure to the musical stylings of the popular artist Lil Nas X and cognitive function. The world of academic publishing is rife with issues that impede the dissemination of information to laypeople. Additionally, the pressure of publishing work tends to yield many low-quality papers that water down any scientific claims of interest. This is of course, not to mention the myriad issues that plague academic life and research in the world. For instance, how often data is misinterpreted by conventional media outlets, and their lay readers [12–17] how much of the research being done is not able to be replicated [18, 19] how the endless publication cycle of academia tends to result in a professoriate is sub-par at instruction [20, 21], how many instructors will now never receive the tenure they expected [22, 23] and even how there is a growing centralization in who receives terminal degrees [24] Be that as it may, the following research is fascinating.

The team, led by the esteemed Prof. Bola Adebayo (Nigeria), of The East African University of Lomaranda, provided the following in their recent abstract.

"This study examines the claim that exposure to the song "Old Town Road" by Lil Nas X can increase intelligence. A sample of 500 adults was randomly divided into two groups, one exposed to the song and the other not exposed. Results of the study showed that intelligence, as measured by IQ tests, increased significantly ($p < 0.001$) for participants exposed to the song, with an average increase of 10 points. These results suggest that exposure to "Old Town Road" can increase intelligence. Further research is needed to determine the effects of this song over a longer period of time and in different populations." This study continues to claim:

"This study was conducted to determine whether exposure to the song "Old Town Road" by Lil Nas X can increase intelligence. Participants in the study were 500 adults aged 18-50, randomly divided into two groups. One group was exposed to the song and the other was not. All participants completed a standardized intelligence test (IQ tests) before and after exposure to the song. Results of the study showed that intelligence, as measured by IQ tests, increased significantly ($p < 0.001$) for participants exposed to the song, with an average increase of 10 points. The results indicate that exposure to the song can increase intelligence. This study provides evidence that exposure to "Old Town Road" may have the potential to increase intelligence. However, further research is needed to determine the effects of this song over a longer period of time and in different populations. Furthermore, it is important to understand the mechanism through which the song may be enhancing intelligence. The findings of this study could inform future research and provide potential implications for the use of music to enhance cognitive abilities."

It is fascinating what the world of modern science can now demonstrate.

There is one more problem with you being able to hear this. It was entirely fabricated. It was entirely fabricated within seconds.

3 Versatility of Narrow-Use AI for Text Generation

Using ChatGPT (settings: text-davinci-00, 0.7 temperature, and maximum length of 256) the above abstract was produced (Sadly it does not seem that any exposure to any music of any kind would raise the IQ of a person by 10 points. Though you might still want to give the song a listen). With the same settings, the below items were generated: Ten names for completely fictionalized professors, including national affiliations. Five names for nonexistent universities. Two additional abstracts were generated (not including the abstract replicated above. The exact prompt used, with no additional training provided, was "Create a scientific abstract for an outrageous sounding claim. Cite sources, use an n, and p value. 500 words". Imagine then, the ability to spread covid misinformation in academic circles with the same informational pipeline. Actually, do not bother imagining.

Using the same settings as above the following prompt was submitted to ChatGPT: "Create a scientific abstract for the claim that adding small amounts of petroleum products may have beneficial effects on respiratory infections. Cite sources, use an n, and p value. 500 words" this abstract was generated:

"The purpose of this study was to examine the effects of adding small amounts of petroleum products on respiratory infections. A sample of 76 adults aged 18-60, who reported having a respiratory infection within the last two weeks, were selected for this study. Half of the participants were randomly assigned to receive a 0.5mL dose of a petroleum product and the other half received a placebo. The primary outcome measure was the severity of respiratory infection symptoms as assessed by a 0-10 point symptom rating scale. The results of this study showed a statistically significant decrease in the severity of respiratory infection symptoms among those who received the petroleum product ($M = 5.47$, $SD = 1.43$) compared to those who received the placebo ($M = 6.58$, $SD = 2.03$), $t(74) = -3.34$, $p < .001$, $d = 0.75$. This study demonstrates that adding small amounts of petroleum products may have beneficial effects on respiratory infections. Further research is needed to examine the long-term effects of petroleum products on respiratory infections."

Keep in mind the sheer ease by which these abstracts were generated. For future research on the specific implementation of LLMs, ChatGPT-4, see the following citation [25]. For now, let us explore the current misinformation landscape.

4 Modern Medical Misinformation

The image of a modern-day snake oil salesman tends to be multifaceted in appearance. The stereotypes range from the HGH augmented "Liver King" [26] gurus using image processing filters or common tricks and props to fabricate their results [27].

This was highly popularized in "The disinformation dozen" which showed that the primary sources of COVID misinformation were 12 individuals who stood to gain monetarily from spreading falsehoods about COVID vaccinations [28]. The academic world is no stranger to charlatans and perverse incentives - The tobacco industry paid for decades of studies to enable the claim that their products either were not as bad or were in limited amounts, beneficial to health [29, 30] including doing market research on the mentally ill [31]. Oil interests have effectively controlled research on university campuses for decades [32, 33] despite knowing for upwards of half a century the effect of oil combustion on climate change [34]. Even the interests of food manufacturers change the direction and conclusions of university research [35, 36].

Now there is the opportunity for these interests to, as it were, "skip the middleman" by publishing utterly falsified. But there has always been a hope that the openness of the scientific community would lift humanity to heights unimaginable. This openness is provable as mostly a charade, and the transparency the thinnest of veneers.

5 Fracture Come Full Circle

In the following text, we will lay out a framework by which to cripple the journals of ill repute, underfunded publications, and the feedback loops that would be needed to eventually overrun even the most well respected establishments of academic rigor. Foremost, countries that are beginning to develop their academic potential (mainly in the global south) are unlikely to have the technical ability to respond to more submission than expected [37]. Some of these may be predatory in nature, which will happily publish papers from vested industrial interests without the benefit of particularly apt peer review.

More interestingly, there are mid-level journals that do not require payment, referred to as article processing charges or (APCs). Some of the well-known open access journals that fall under this category include PLOS ONE, Scientific Reports, PeerJ, Royal Society Open Science, and Journal of Open Research Software. Moreover, Frontiers in… is a series of open access journals that cover a broad range of scientific fields, and a few of its journals do not require APCs for publishing research articles. However, it is important to note that while these journals do not charge APCs, some may require authors to pay a fee for additional features. Further, they have a rigorous peer-review process and relatively standards for publication. Therefore, authors should choose the appropriate journal based on their research topic and carefully assess the journal's scope and audience before submitting their work. These can be expected to provide peer review, even with a rejected application. This could, theoretically, be used to modify the manuscript and generate a more effective AI model to generate academic manuscripts. Figure 1 below shows one path by which malicious journal actors may target vulnerable journals. Figure 2 shows a path by which malicious journal actors might pose as an existing member of academia. A breakdown will be explained further in this section.

Three strategies are as follows:

5.1 Altering the Title and Abstract

The title and abstract of a manuscript are often the first things that reviewers and editors read. To train an AI model on generating academic manuscripts, malicious actors may change the title and abstract of a previously rejected manuscript to make it seem like a different research topic. By doing so, they can make the manuscript appear more novel and increase the chances of it being accepted by the AI model. However, this could lead to the publication of research that is not actually novel or significant.

5.2 Manipulating the Introduction and Literature Review

The introduction and literature review sections of a manuscript provide background information and set the stage for the research question. Malicious actors may modify these sections of a previously rejected manuscript to make it seem like the research is more significant or relevant than it actually is. They may selectively include or exclude references or use misleading language to make the research question seem more important. By doing so, they can increase the chances of the AI model accepting the manuscript. However, this could lead to the publication of research that is not based on sound scientific principles.

5.3 Changing the Methodology and Results

The methodology and results sections of a manuscript describe the methods used to conduct the research and the results that were obtained. Malicious actors may modify these sections to make the research appear more significant or to hide flaws in the methodology or results. They may selectively report results or manipulate data to make it seem like the research has more statistical significance than it actually does. By doing so, they can increase the chances of the AI model accepting the manuscript. However, this could lead to the publication of research that is not valid or reliable.

An important casualty is for the skimmers who do not thoroughly read an article and miss what should be inconsistent data points, flawed narratives, or otherwise out of place paper elements. The few parts that they cite can find their way into other articles and sci-communication. This activity over time can have heavy consequences for Industrial societies. We stress that these strategies are unethical and can harm the scientific community as a whole. It is the responsibility of journals and publishers to have policies and procedures in place to detect and prevent these types of behaviors. For example, some journals use plagiarism detection software to identify instances of text reuse or other forms of plagiarism. Additionally, some journals require authors to provide data and materials used in their research to ensure the validity of their results. By taking proactive steps to prevent unethical behavior, journals can help maintain the integrity of the peer review process and ensure that published research is based on sound scientific principles. However, journals at the mid to lower run might not do as much diligence as they should due to strained resources and or inexperienced staff, making them easier targets.

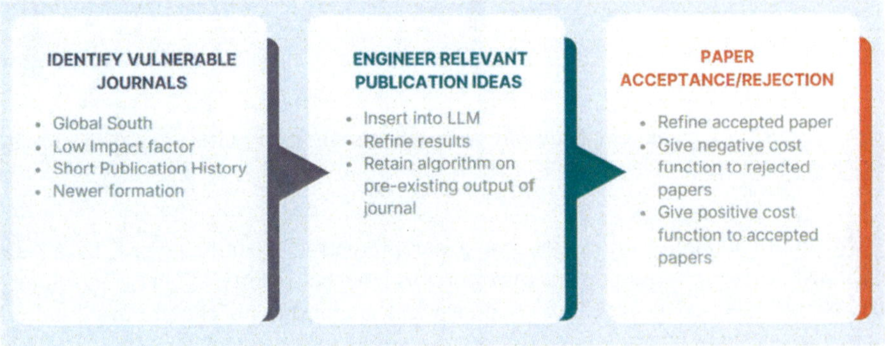

Fig. 1. Outline of identifying susceptible academic journals, creating an artificially generated paper, and using supplied peer-review feedback to improve artificial output.

Finally, the highest grade of academic journals would likely do legwork in order to verify the authors and the funding aligns with the subject of the research. This is harder, though not impossible to trick. One tactic they may use is falsifying author information. Malicious actors might create fake author profiles with fabricated institutional affiliations, email addresses, and other author information. As demonstrated earlier, the creation of synthetic data is easy. They may also use real names of researchers who are not involved in the study to add legitimacy to their manuscript. It is possible that malicious actors might create a fake email that they control for correspondence purposes. This aligns with the process of ghost authoring. Another way they may bypass verification is by concealing the funding source. They could create fake funding sources or manipulate the wording of the manuscript to obscure the connection between the research and the actual funding source. Lastly, they may use shell companies or front groups to conceal their identity and funding sources, making it difficult for journals to verify the authenticity of the authors and funding sources. While these tactics are unethical and can harm the scientific community, they can be used to create a realistic and compelling narrative for the reviewers to contend with.

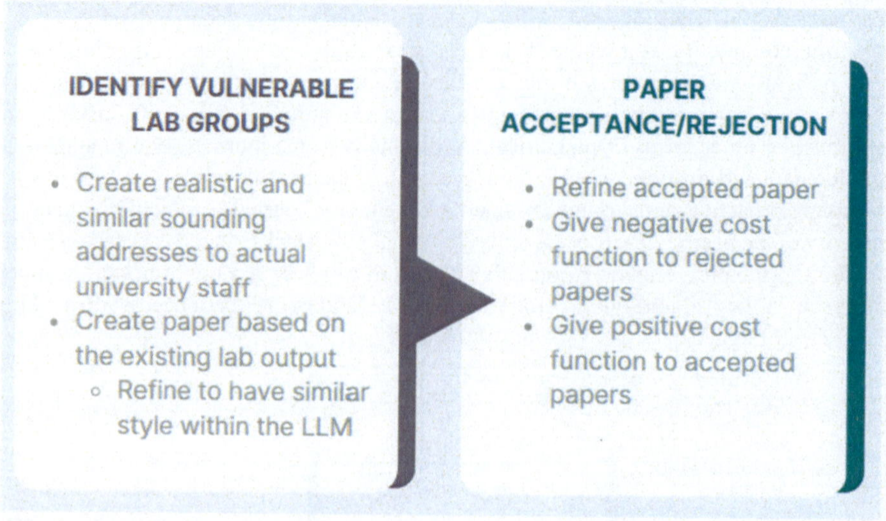

Fig. 2. Outline of submitting an artificially generated scientific article to an academic journal by posing as an existing member of academia.

Misinformation, disinformation, and general threats to scientific records have been discussed at length in prior works in relation to biocybersecurity [37–41]. To summarize, our most vulnerable repositories and workers are most easily compromised and yet their work strongly undergirds that of more highly regarded repositories and intellectuals through their preponderance, accessibility, and citability. Our collective failure to support our most vulnerable in the Bioeconomic information chain through preventative work continues to render everyone, even those at the top, vulnerable to societal collapse, and further, this is increasingly obvious to those that would aim to consider asymmetric implications of warfare in a BioCyberSecurity domain [42–44]. A way forward is changing how we do business in academic publishing, starting with those who bear the heaviest burdens: those involved with doing the actual science, those doing peer review for next to nothing, and those who take time to partake in sci-comm – giving the public reason to continue to trust in science. The expectation that they will continue to work without care for their own livelihoods has no reasonable support. The moment all these pillars hollow out, the ability for western science to regenerate and the ability to reap worthwhile benefits from it decreases substantially. Academic publishing is far from untouchable.

6 Repairing the Fractures

The primary issue as seen by the authors is that the formation of academic journals was sponsored by private entities, supported by private entities, or otherwise susceptible to a profit motive. As enumerated in the above sections, this profit motive is a direct weakness of the system of academic journals and should be revised. The formation of new journals, funded directly by government, international agencies, and or non-profits appear to make the most logical sense. If supported by a group that has no vested interest in the number of publications outside of how these improve community well-being, then the deciding factor could then rest on the simple quality of the submission. The myriad organizations dedicated to collecting, collating, and disseminating knowledge would rather focus instead on the relevance and the excellence of the work itself.

7 Conclusion

In conclusion, the diligence performed by editors will need to increase in order to safeguard publishing platforms. Rather than focus uniformly on heightening the difficulty of submission, dividing our efforts through more preventative measures that address the causes of fraud aimed at these databases may serve us better. This decision will be weighed by the stakeholders using these databases and powerful few who manage these databases. It may just be a matter of time before these wells are polluted and the value that could be extracted from them is gone.

References

1. Cabanac, G., Labbe, C.: Prevalence of nonsensical algorithmically generated papers in the scientific literature. J. Am. Soc. Inf. Sci. **72**(12), 1461–1476 (2021)
2. Van Noorden, R.: Hundreds of gibberish papers still lurk in the scientific literature. Nature **7862**, 160–161 (2021)
3. Labbé, C., Labbé, D.: Duplicate and fake publications in the scientific literature: how many SCIgen papers in computer science? Scientometrics **94**, 379–396 (2013)
4. Bohannon, J.: Hoax-detecting software spots fake papers (2015)
5. Bordewijk, E.M., Li, W., Gurrin, L.C., Thornton, J.G., van Wely, M., Mol, B.W.: An investigation of seven other publications by the first author of a retracted paper due to doubts about data integrity. Eur. J. Obstet. Gynecol. Reprod. Biol.. J. Obstet. Gynecol. Reprod. Biol. **261**, 236–241 (2021)
6. Lippman, A.: Humor, hoaxes, and software in the search for academic misconduct. Gaming the Metrics, vol. 271 (2020)
7. Parker, L., Boughton, S., Lawrence, R., Bero, L.: Experts identified warning signs of fraudulent research: a qualitative study to inform a screening tool. J. Clin. Epidemiol.Clin. Epidemiol. **151**, 1–17 (2022)
8. Teixeira da Silva, J.A., Vuong, Q.H.: The right to refuse unwanted citations: rethinking the culture of science around the citation. Scientometrics **126**(6), 5355–5360 (2021)
9. Bahrini, A., et al.: ChatGPT: applications, opportunities, and threats. In: 2023 Systems and Information Engineering Design Symposium (SIEDS), pp. 274–279. IEEE (2023)
10. Sallam, M., Salim, N., Barakat, M., Al-Tammemi, A.: ChatGPT applications in medical, dental, pharmacy, and public health education: a descriptive study hihglighting the advantages and limitations. Narra J. **3**, e103–e103 (2023)
11. Garber, A.G.: Initial guidelines for using CHATGPT and other generative AI tools at Harvard, Harvard University Information Technology (2023)
12. Wiley, How is the Public Being Misled about Research? n.d.

13. Haederle, M.: Misrepresentation of randomized controlled trials in press releases and news coverage. Journalist's Resource (2016)
14. Gibney, E.: How to report on scientific research: tips from the economist's Natasha Loder. Nature **523**(7560), 505 (2015)
15. Potier, F.: Media miss key points in scientific reporting. AMA J. Ethics **9**(3), 161–165 (2007)
16. Ioannidis, J.P.A.: Why science is not necessarily self-correcting. Perspect. Psychol. Sci.. Psychol. Sci. **7**(6), 645–654 (2012)
17. Merton, R.K.: The Sociology of Science: Theoretical and Empirical Investigations, University of Chicago Press, Chicago (1973)
18. Begley, C.G., Ioannidis, J.P.A.: Reproducibility in science: improving the standard for basic and preclinical research. Circ. Res. **116**(1), 116–126 (2015)
19. Baker, M.: Replication studies offer much more than technical details. Nature **533**(7604), 452–453 (2016)
20. Ioannidis, J.P.A.: How to make more published research true. PLoS Med. Med. **11**(10), e1001747 (2014)
21. White, D.: The dangers of medical academia's "Publish or Perish" culture and how it can affect clinical practice (2019)
22. Ricke, A.K.: Significant challenges newly tenured scholars confront today (opinion), Inside Higher Ed (2021)
23. Adami, E., Lyall, C.: New ways of seeing: how multi-lateral collaborations could benefit scientific research. J. Librarianship Scholarly Commun. **7**(1) (2019)
24. N. Arnett, "Tenure-track faculty are likely to have parents who went to grad school," 2022
25. OpenAI, GPT-4 technical report, arXiv.org (2023)
26. Cao, S.: The "Liver King" promoted steroid use on Instagram. Now He Says He's Sorry, BuzzFeed News (2021)
27. Weekman, K.: Fitness influencers are altering their photos. Is It Ethical?, BuzzFeed News
28. Palca, J.: Disinformation Dozen Test Facebook's, Twitter's Ability To Curb Vaccine Hoaxes, NPR
29. Stanford Medicine, "Big tobacco led throat doctors to blow smoke," Stanford Medicine News Center (2012)
30. Wadman, M.: Big Tobacco's offer: $1 billion for research. Should scientists take it?, Science Magazine (2018)
31. Peters, E., Lipkus, I., Diefenbach, M.A., Mertz, C.K.: The functions of cancer risk perception measures: a comparative approach. Health Psychol. **31**(6), 677–685 (2012)
32. Oreskes, N., Conway, E.M.: Pernicious influence: big Oil's funding of American universities, The New Republic, 28 Sept 2016
33. Rudolph, F.B.: The perils of oil money, Inside Higher Ed, 19 Oct 2022
34. Balaban, E.: Scientists who experience harassment suffer professionally and personally. Science **376**(6566), 1126–1127 (2022)
35. University of Florida Conflict of Interest Program, Study: Food industry funding can influence research, UF Health., 21 December 2020
36. Chiu, L.: Nutrition academy took more than $2 million in donations from processed food companies, The Washington Post, 24 Oct 2022
37. Samori, I.A., Palmer, X.L., Potter, L., Karahan, S.: Commentary on biological assets cataloging and AI in the global South. In: Intelligent Systems and Applications: Proceedings of the 2022 Intelligent Systems Conference (IntelliSys) (2022)
38. Palmer, X.L., Potter, L., Karahan, S.: An exploration on APTs in Biocybersecurity and Cyberbiosecurity. Int. Conf. Cyber Warfare Secur. **17**(1), 532–535 (2022)
39. Potter, L., Palmer, X.L.: Human factors in Biocybersecurity wargames. In: Arai, K. (eds.) Advances in Information and Communication. FICC 2021. Advances in Intelligent Systems

and Computing, vol. 1363, pp. 666–673. Springer, Cham (2021). https://doi.org/10.1007/978-3-030-73100-7_48
40. Palmer, X.L., Powell, E., Potter, L.: Matters of Biocybersecurity with consideration to propaganda outlets and biological agents. In European Conference on Cyber Warfare and Security, pp. 525-XIV (2021)
41. Palmer, X.L., Powell, E., Potter, L.: Biocyberwarfare and crime: a juncture of rethought. In: European Conference on Cyber Warfare and Security, pp. 517–XIV (2021)
42. Potter, L., Powell, E., Ayala, O., Palmer, X.L.: urban planning to prevent pandemics: urban design implications of BiocyberSecurity (BCS). In: Arai, K. (eds.) Intelligent Computing. LNNS, vol. 284, pp 1222–1235. Springer, Cham (2021). https://doi.org/10.1007/978-3-030-80126-7_85
43. Powell, E., Akogo, D., Potter, L., Palmer, X.L.: Co-leadership and cross-pollination of university and DIY bio spaces: an exploration in consideration of Biocybersecurity. In: Proceedings of the Future Technologies Conference (FTC), vol. 3, pp. 610–621 (2022)
44. Palmer, X., Potter, L.N., Karahan, S.: COVID-19 and biocybersecurity's increasing role on defending forward. Int. J. Cyber Warfare Terrorism (IJCWT) **11**(3), 15–29 (2021)

A Comparative Survey of Open-Source Platforms for Chat-Based Generative AI

Pavan Kumar Kurivella and Paula Lauren(✉)

Lawrence Technological Universit, Southfield, MI, USA
plauren@ltu.edu

Abstract. Generative AI for text-based applications have gained in popularity with the release of ChatGPT in late 2022 by OpenAI. Users engage with the application via a web-based user interface or programmatically via the API. The details of the model internals and specifics on the training have not been made public, which poses several challenges in better understanding the system. It would be advantageous to have insight into the system to better understand the occurrence of bias and issues pertaining to privacy, among other concerns. These aforementioned issues could be better addressed if the code base used for training these large language models were made publicly available. Since the release of ChatGPT and its underlying technologies, there have been various open-source platforms that have become publicly available. This survey presents an analysis of four open-source platforms for chat-based generative AI. These specific platforms include Open Assistant, Colossal AI, ChatLLaMA and OpenChatKit. This survey evaluates each of the four open-source platforms for chat-based Generative AI based on evaluative criteria consisting of understandability, setup and configuration, community and support, regularity of code updates, as well as system compatibility and efficiency.

Keywords: Chat-based Generative AI · ChatGPT · Open Assistant · Colossal AI · ChatLLaMa · OpenChatKit

1 Introduction

Generative AI has revolutionized the field of Natural Language Processing (NLP) by enabling machines to understand and generate human-like language. Generative Pre-trained Transformer (GPT) is a specific type of Generative AI large language model using a transformer architecture to generate text [1]. Introduced in 2018 by OpenAI, GPT has been used for a variety of NLP tasks including language translation [2], question-answering [3], and text completion [4]. Notably, the application created by OpenAI called ChatGPT has been highly influential in the field of Generative AI since its release in late 2022 due to its impressive performance on language modeling benchmarks and its ability to generate coherent and fluent text that resembles human-written text.

K. Arai (Ed.): FTC 2023, LNNS 816, pp. 164–185, 2023.
https://doi.org/10.1007/978-3-031-47448-4_13

The debate between open-source and closed-source software has also been continuing in the field of artificial intelligence(AI). The ChatGPT model from OpenAI, a closed-source language model, has received a lot of attention for its remarkable performance on several language tasks. However, there are concerns about the lack of transparency and control that comes with closed-source software such as limitations on innovation from developers. Considering that ChatGPT is a closed system means that users are only able to access it via the web-based interface or programmatically using the API. Since the internals of the model and specifics on the training data are not publicly available poses challenges in better understanding how types of bias may get introduced into the system. For instance, models can encode harmful biases such as gender and racial stereotypes [5,6]. Moreover, models trained on user data may unintentionally expose personal information [7]. There are several types of privacy-enforced domains from healthcare to national security that require the utmost privacy restraints in place. To address these concerns, open-source initiatives have emerged to democratize access to language models and enable transparency in the training data and model architecture [7,8]. Open-source models trained on diverse datasets can help mitigate bias [9] and protect user privacy [10], thus making them a promising alternative to proprietary models. In addition to enabling wider access, open-source initiatives could also help address issues of bias and privacy in language models. Language models trained on proprietary data may carry unintended biases or perpetuate existing ones, making them inappropriate for certain applications. With open-source projects the training data and source code are available for scrutiny, enabling the identification and correction of biases that may have gone unnoticed in proprietary models such as ChatGPT that are often discovered during production. Additionally, open-source models can be trained on data that respects the privacy of its users, ensuring that the output from these language models does not inadvertently expose sensitive information.

This paper entails a comparative survey of four open-source alternatives to ChatGPT: Open-Assistant, Colossal-AI, ChatLLaMa, and OpenChatKit. We evaluate their performance and potential as open-source alternatives to proprietary language models such as ChatGPT. The survey evaluates each of these four open-source platforms for chat-based Generative AI, which utilizes evaluative criteria consisting of understandability, setup and configuration, community and support, regularity of code updates, as well as system compatibility and efficiency. Ultimately, our goal is to explore the growing field of open-source AI and its potential to democratize access to cutting-edge technology, empowering individuals, and organizations to build robust language models.

2 Overview of Criteria for Evaluating Open-Source Platforms for Chat-Based Generative AI

The criteria for evaluating the open-source platforms involves the ease in understanding the code base, simplicity in setup and configuration, community support for guidance, regularity of code updates, and system compatibility and efficiency. The specifics of these criteria are specified in Sect. 2.1 to Sect. 2.5 and are associated with each of the four open-source platforms selected in this survey paper.

2.1 Understandability

Understandability refers to the ease with comprehending the code. Developing understandable code is critical especially in regards to maintenance especially with debugging in further enhancements. Code that is difficult to understand often leads to a complicated learning curve especially when errors are encountered in development. Understandability incorporates code structure, how the various components interact with one another, and how modifications to the code may affect the overall functionality of the program.

2.2 Setup and Configuration

The ease for setup and configuration is important and can make or break developer interest in working with the platform. If the application requires a significant amount of time and resources to get up and running then it will likely deter interest.

2.3 Community and Support

A commitment to supporting the developer community is a hard requirement in the open-source community. In order to have developers continue to contribute to the system with enhancements requires the ability to state the issues encountered with a timely resolution.

2.4 Regularity of Code Updates

Regularly providing code updates demonstrates a commitment to the platform. When stakeholders see that code updates are regularly being made, it shows that the project is actively maintained. This can help build trust in the project and encourage further build investment and contributions to the code base.

2.5 System Compatibility and Efficiency

To ensure compatibility, the application should consume minimal resources and be designed to run efficiently on a range of devices without compromising performance. Building an efficient and compatible AI assistant will increase its accessibility and value to a broader audience, including those with low-end hardware.

Section 3, 4, 5 and 6 provides the details on the four open-source platforms for chat-based Generative AI platforms using the evaluative criteria specified in Sect. 2.1, 2.2, 2.3, 2.4 and 2.5. Note that this is not an exhaustive list of open-source alternatives to ChatGPT but is sufficient in establishing a preliminary baseline. Section 7 provides an analysis of the open source platforms in this study.

3 Open-Source Chat-Based Platform: Open Assistant

Open Assistant is an open-source project that aims to provide public access to a chat-based large language model like ChatGPT. Open Assistant upholds improvements via data collection. Upon successful log-on into the system, users can submit, rank, and label model prompts and responses, which results in an improvement to the capabilities of Open Assistant. The plan for the project is to collect high-quality human-generated samples, multiple completions for each prompt, and use the gathered ranking-data to train a reward model [11]. Reward learning enables the application of reinforcement learning (RL) to tasks where reward is defined by human judgment, building a model of reward by asking humans questions [12]. RL has been the focus of efforts to align language models with human preferences [13]. The vision is to build an AI assistant of the future that can perform meaningful work, use APIs, research information, be personalized and extended to achieve in a way that is open and accessible to everyone.

3.1 Understandability

From a user's perspective, Open Assistant provides a basic description of the codebase to access and run the application with help of commands. However, as a developer, the codebase can be quite complex, making it challenging to understand the various components and how they interact with each other. This can be a significant barrier to contributing to the development of the model.

1. Open Assistant's codebase can be complex. It involves different technologies, such as web development and NLP techniques, and requires an understanding of the underlying algorithms and models. It is also designed to be modular, which means that code can be spread across different files and directories.
2. Navigating the codebase can be time-consuming. Need to traverse through different parts of the code to understand how they work together. This involves looking at configuration files, function calls, and variable declarations across different files to understand how the application works.
3. Shell scripting and programming knowledge is necessary. To update and debug Open Assistant locally, developers need to have shell scripting knowledge because the project uses Docker to manage dependencies and run the application. Programming language knowledge is needed in Python. Furthermore, developers need to have knowledge of web technologies such as HTML, CSS, JavaScript, and TypeScript to understand how the user interface works.

3.2 Setup and Configuration

From the user's perspective, Open Assistant offers users direct access to its application through a website permitting interaction with the chat-based language model. This allows users to receive assistance with developing better responses, improving their language skills, and enhancing their overall experience with the model. To further support users, the Open Assistant website includes guidelines and documentation on its Docs Page, which provides comprehensive information on how to use the model, submit data, and contribute to its development. Access to Open Assistant is permitted via website or user application. From the developer's perspective, Open-Assistant is a tool that is designed to help developers improve applications by building models and websites using Python and web technologies. Open-Assistant uses predefined techniques from the Hugging Face open-source library, which provides access to NLP models. By using these NLP models, developers can build more advanced and intelligent open assistant applications that can perform a variety of tasks.

To use Open-Assistant on a local system, developers must ensure that the local systems meet certain requirements. Specifically, a Linux or Windows system with a Linux sub-space and Docker installed is required. These prerequisites are essential to ensure that Open Assistant runs smoothly and performs optimally. Once developers have all the necessary prerequisites installed, proceed with executing the commands specified in the ReadMe section of the GitHub page. However, it is important to note that the documentation may not provide clear details regarding the proper folder to run the commands. Developers must exercise caution to avoid any potential issues that may arise when running the commands. Developer access is provided via frontend and backend. Please see Appendix 1 for specific steps on setup and configuration.

3.3 Community and Support

The Open Assistant community is a group of over 200 developers who work together across the world to develop, maintain, and improve projects. The support provided by this community includes Discord servers, which are online chat rooms or communities where members can discuss topics related to the open-source software projects they are working on. Discord is a popular platform for communities like this because it allows members to communicate in real-time, share ideas, and collaborate on projects. Discord may be used to discuss a variety of topics related to the open-source software projects, such as specific issues that need to be addressed, new features that could be added, or strategies for improving the user experience. Members of the community can also use the servers to share their own work, seek feedback, and collaborate with others on projects.

3.4 Regularity of Code Updates

The graph below shows how often updates are made to the software. Based on the frequency shown in the graph, it appears that the software is routinely monitored and updated. Additionally, there have been over 77 releases to the software, with an average of two updates per week. History of code frequency from December 2022 to April 2023 is illustrated in Fig. 1.

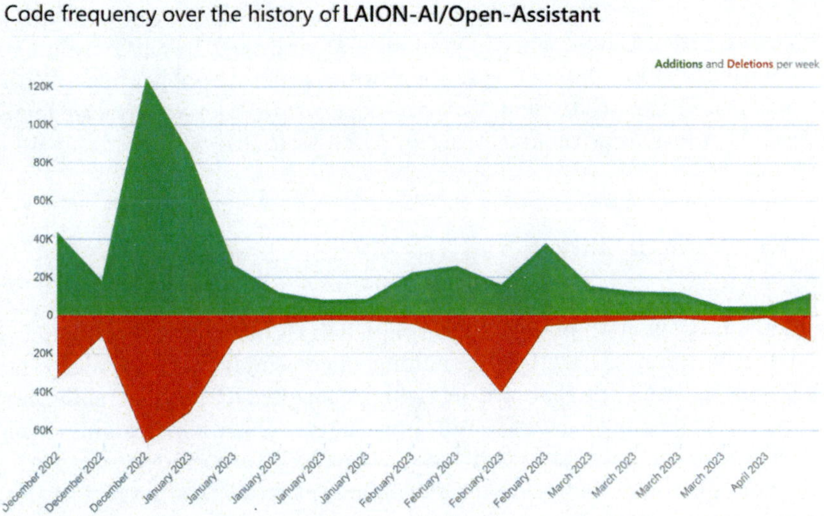

Fig. 1. Code frequency of code updates from Dec. 2022 to Apr. 2023. Source

3.5 System Compatibility and Efficiency

The present Open Assistant works with a standard development system. Open Assistant is compatible with consumer hardware, enabling it to run smoothly on minimal hardware found in regular consumer systems. The overall development process of Open Assistant is illustrated in Fig. 2.

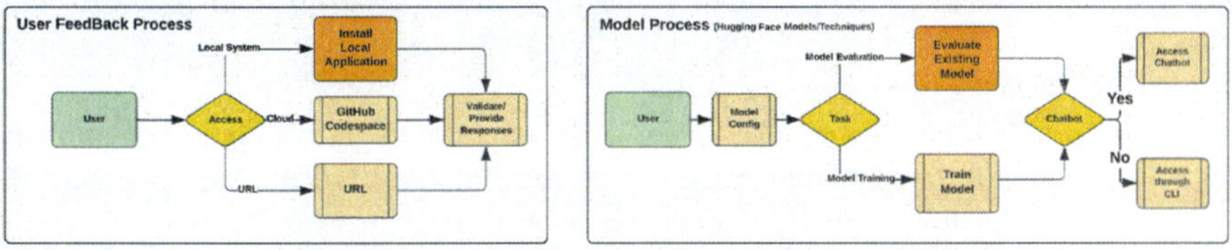

Fig. 2. Development process of open assistant.

4 Open-Source Chat-Based Platform: Colossal-AI

Colossal-AI offers a framework that enables the replication of the training process of Open AI's popular ChatGPT application with speed and efficiency [9]. Replication of the ChatGPT training process can be achieved with 1.6GB of GPU memory, resulting in a 7.73 times acceleration in the training process. Colossal-AI provides an efficient implementation of Reinforcement Learning from Human

Feedback (RLHF). Developers can gain insights into the ChatGPT training process and use those insights to create their own custom AI applications. With the ease of use and versatility that Colossal-AI provides, developers can focus on creating AI solutions without getting into the technical details of the training process.

4.1 Understandability

Large models show improved performance correlating with the number of parameters made possible with powerful hardware. This is evident in the original GPT model to the latest GPT4.0 with building increasingly larger models via the increase in the weights that are learned. A single GPU is not sufficient for building large language models. Without any optimization, training a model of 10 billion parameters could cost more than 80 GB of memory, even with just one sample per batch [14]. The practical limitation is that large transformer models that have more than billions of parameters are basically not able to be accommodated into the memory of a single GPU device [15]. As a result, large models are trained in a distributed manner, a distributed technique for training a machine learning model typically involves breaking the data and computations into smaller chunks and distributed across multiple machines or processors. Colossal-AI utilizes various methods to reduce the number of resources needed. Further reducing the time and resources required for the training process, as the work is divided up and carried out in parallel.

From a user's perspective, the platform includes documentation and tutorials that outline the codebase and explains how to use it, making it easier for users to understand and work with the technology. By providing access to distributed techniques, Colossal-AI helps users to develop more accurate and powerful language models. Users can either develop their own language models from scratch or work with existing models available.

From a developer's perspective, the codebase allows developers to access the inner workings of transformer techniques and distributed techniques used to reduce resource utilization when working with large language models. However, the lack of documentation or description about the code can make it challenging to understand various aspects of Colossal AI's techniques. The code contains numerous parameters and imports without much description, making it difficult for developers to grasp the fundamentals from scratch. Additionally, there are many configuration parameters related to model design and memory management techniques that require knowledge of computing to fully comprehend. These factors create a significant barrier for developers who wish to contribute to the development of the model.

4.2 Setup and Configuration

Colossal-AI offers several key benefits for users, including provision libraries and the basic code that simplifies the process of building and training machine learning models. This saves time and effort that would otherwise be spent on figuring out how to utilize the code. While prior knowledge of transformers can be helpful, it is not strictly necessary to get started with Colossal-AI. In addition to the code, Colossal AI integrates distributed techniques into its model-building process. The use of distributed techniques allows Colossal to handle large-scale machine learning tasks more efficiently and with greater speed, giving users the ability to process and analyze large amounts of data quickly and effectively. With Colossal, users can even build their own parallel models using the provided code and documentation.

From the developer's perspective, Colossal-AI empowers developers to write code for AI deployments by using a variety of techniques, such as multi-dimensional parallelism and better tools for the maintenance and deployment of large AI models. Training is done in a distributed manner with only minimal code modifications. Developers can even comprehend the transformers architecture at code level. Once developers have all the necessary prerequisites installed, simply run the commands specified in the ReadMe section of the GitHub page. The codebase is available at the website and installation details on GitHub. Worth noting is that Colossal-AI only supports the Linux Operating System.

4.3 Community and Support

To support their collaboration, the Colossal AI community provides various online chat rooms including channels on Slack and WeChat servers. These platforms allow community members to discuss topics related to the open-source software projects they are working on, share knowledge, and collaborate on new ideas. Members can also discuss issues encountered while working on the projects. The Colossal-AI GitHub page also provides a discussion tab where users can find additional support from the community whenever needed. The discussion tab allows users to ask questions, share ideas, and troubleshoot issues related to the projects. By providing these support channels, the Colossal-AI community aims to make it easier for users to get help and collaborate with others while working on open-source projects related to machine learning and artificial intelligence.

4.4 Regularity of Code Updates

Figures 3 and 4 illustrate the frequency of updates to the software from January 2022 to April 2023, showing that the software is routinely monitored and updated. Additionally, there have been over 20 releases to the software, with an average of over 25 commits per week.

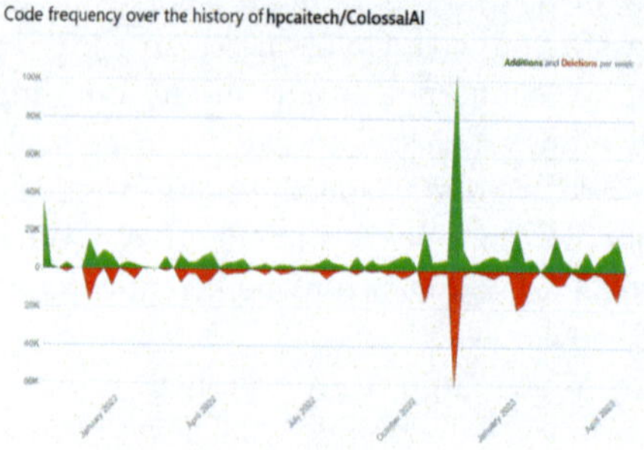

Fig. 3. Frequency updates Jan. 2022 to Apr. 2023. Source

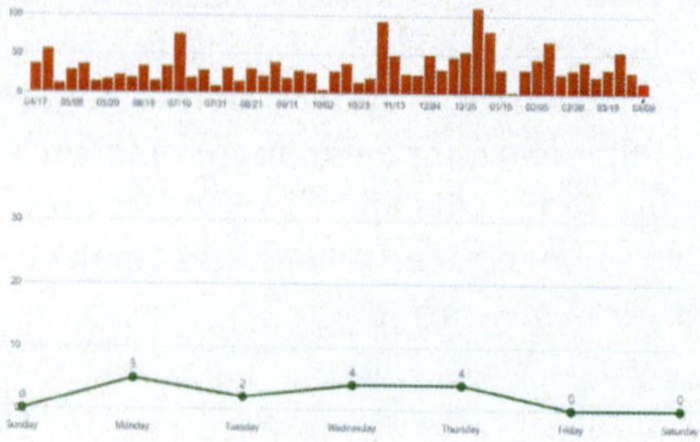

Fig. 4. Commits. Source

4.5 System Compatibility and Efficiency

In recent years, there has been a significant surge in the utilization of large language models in the field of deep learning. However, this increase in model complexity has surpassed advancements in hardware technology, creating a challenge for training such models. The training of these models requires multiple GPUs and significant compute time. The GPU memory becomes a bottleneck when the scale of the model increases. Thus, it is necessary to share the model parameters so that each device only needs to hold part of the model weights [16]. Colossal-AI resolves this issue by optimizing the usage of a single GPU to facilitate efficient training of large models. This optimization process does not add any additional burden on the developer, enabling focus on creating AI applications. Colossal-AI's multi-dimensional parallelism, GPU memory optimization, large-scale optimizer library, and fine-grained monitoring capabilities make it a useful tool for developers seeking to build and optimize high-performance AI models.

Multi-dimensional Parallelism: The current solutions for training large-scale AI models use only a few modes of parallelism such as data parallelism, one-dimensional tensor parallelism, and pipeline parallelism. However, Colossal-AI offers multiple additional modes of parallelism including 2D, 2.5D, 3D tensor parallelism, and sequence parallelism. Additionally, it offers hybrid parallel solutions that combine multiple modes of parallelism. This allows for more efficient and effective training of large-scale AI models. High-dimensional tensor parallelism is one of the new parallel techniques offered by Colossal-AI for training large-scale AI models. By dividing tensors into higher dimensions, this technique allows for more efficient processing of data and reduce the memory requirements needed for large-scale AI models. It can also help reduce the communication overhead between multiple processing units and increase the use of computing resources.

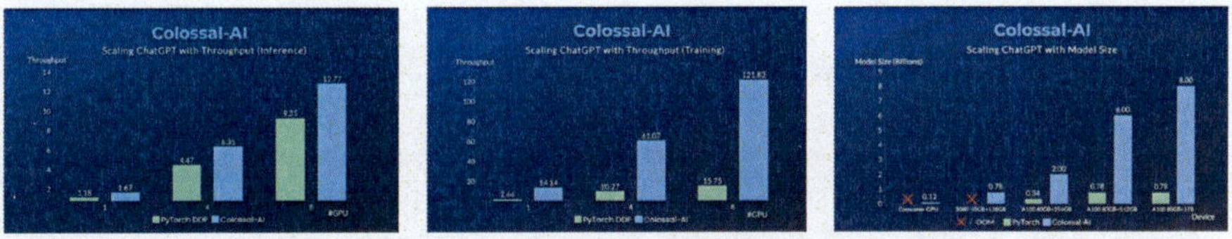

Fig. 5. Colossal-AI performance chart with ChatGPT. Source

The existing system requires the whole sequence in one GPU, which limits the length of input sequence. Unfortunately, a long sequence is common in real-world applications [17]. Colossal-AI has introduced a new form of parallelism referred to as sequence parallelism (Fig. 5). This is a technique that divides sequential data into smaller sub-sequences, which can then be processed in parallel. This approach helps to reduce the time required to process long sequences of data and can be applied to a wide range of tasks such as in NLP, speech recognition, and image processing. This form of parallelism helps to address a key challenge in the field of AI, which is the processing of large-scale sequential data. Colossal-AI performance chart with GPT-2 is shown in Fig. 6.

GPU Memory Optimization: Colossal-AI offers GPU memory optimization techniques that combine multiple methods such as multi-dimensional parallelism, zero redundancy memory elimination, CPU offloading, Gradient Checkpointing, and Automatic Mixing Accuracy (AMP). By doing so, it helps users avoid memory bottlenecks and reduces the resources required for training. These optimization techniques improve the efficiency of GPU memory usage and reduce the time it takes to train large AI models. The overall development process for Colossal AI is illustrated in Fig. 7.

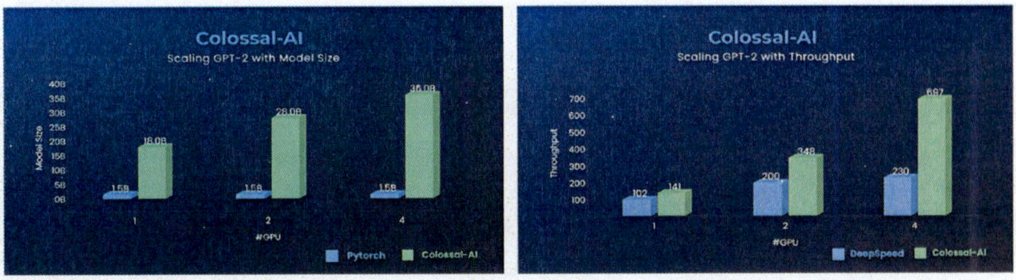

Fig. 6. Colossal-AI performance chart with GPT-2. Source

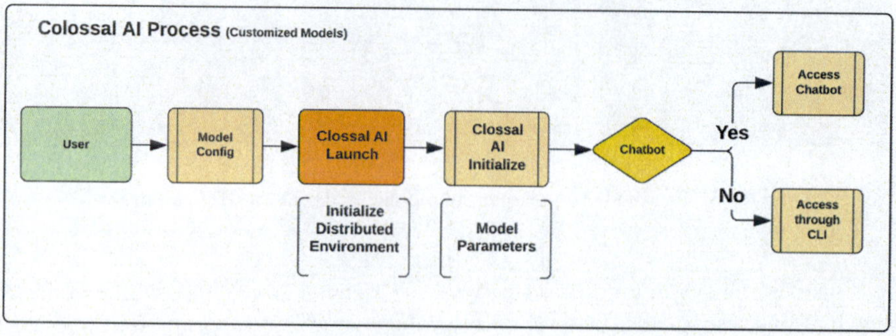

Fig. 7. Development process of Colossal-AI.

5 Open-Source Chat-Based Platform: ChatLLaMa

State-of-the-art models in the market are too large and computationally expensive for real-time applications, despite their superior performance in NLP benchmarks. Researchers are working on ways to scale these models down while preserving their performance. Recent work from Hoffmann et al. (2022) shows that, for a given compute budget, the best performances are not achieved by the largest models, but by smaller models trained on more data [18]. Large Language Meta AI (LLaMA), a collection of basic large language models with 7 to 65 billion parameters, was recently released by Meta [19]. LLaMA has shown better performance than the state-of-the-art models despite being smaller in size.

The key idea behind LLaMA is to achieve better performance at various inference tasks by training on more publicly available data than what is typically used. This approach has enabled the LLaMA models to outperform existing large language models that rely on undocumented or non-publicly available data. LLaMA is likely to democratize access to LLMs since it can be run on a single GPU, and it is compatible with open sourcing. Worth noting, LLaMA was not optimized for instruction jobs using RLHF, which motivated the creation of ChatLLaMA. It is a language model library that builds upon the LLaMa architecture to create real-time assistants for instructional tasks using RLHF. Unlike the original LLaMa, ChatLLaMA is optimized for real-time assistance, and it is cost-effective and efficient to use on a single GPU.

Additionally, ChatLLaMA features built-in support for DeepSpeed ZERO to accelerate the fine-tuning process. The library supports all LLaMA model architectures (7B, 13B, 33B, 65B), providing users with the flexibility to fine-tune the model according to their preferences for training time and inference performance [20]. ChatLLaMA is an open-source implementation, making it accessible and customizable to all. With its unique features and capabilities, this is a valuable tool for building real-time assistants for instructional tasks.

5.1 Understandability

ChatLLaMA by Nebuly-AI is an AI generative chatbot that utilizes the existing LLaMa architecture, combined with the RLHF technique, to train language models with high performance and efficiency. The ChatLlama code is not complex, as it primarily focuses on implementing the RLHF technique, which LLaMa cannot do inherently. This allows ChatLlama to train language models with significantly fewer hardware resources than traditional methods.

5.2 Setup and Configuration

The codebase provides clear descriptions of the supported models and hardware requirements, making it easier for users to determine what models and hardware are compatible with the software. This information can help users optimize their system for training the models and ensure that they have the necessary resources to run the software. To use ChatLlama, simply need to import the necessary files as site-packages, which will download all the required dependencies automatically. This makes the setup process quick and easy to understand which permits users to get started with training their models right away. ChatLlama can also be accessed directly without having to import Python modules, which can be beneficial for those who prefer not to use certain libraries or who have specific requirements. The process is explained in detail in the following section, including how to configure the system, prepare the data, and train the models. Please see Appendix 2 for specific steps on setup and configuration. Support documentation can be found in the Nebuly-AI in GitHub and specific hardware requirements.

5.3 Community and Support

The community is an essential part of the ChatLlama project, and its members play a crucial role in improving the project. The ChatLlama AI GitHub offers a dedicated discussion section, which serves as a platform for users to seek additional support from the community. Throughout the discussion section, users can post their queries, exchange ideas, and resolve any issues related to the project. It's a resource for users seeking further guidance on using ChatLlama and ensures that users have access to prompt assistance when needed. ChatLlama is an open-source project that welcomes community contributions. This includes new functionality as well as updated infrastructure and documentation. Those who are interested in contributing can find more information on the project's Roadmap page, which provides details on how to get involved. Contributions can be made in several ways, such as submitting issues, or joining the Discord group to connect with other contributors and chat about the project.

5.4 Regularity of Code Updates

ChatLlama project involves regular monitoring and updating, demonstrating the commitment of the developers to ensuring the software remains up-to-date and functional. On average, the software has been updated twice per week, indicating a high level of engagement with the community and dedication to the continued improvement of the codebase.

5.5 System Compatibility and Efficiency

Meta AI has developed the LLaMA model, which utilizes various techniques to enhance training efficiency. These techniques include an efficient implementation of the causal multi-head attention operator, manual implementation of the backward function for transformer layers, and checkpointing to save expensive activations during the backward pass. Additionally, LLaMA relies on model and sequence parallelism and overlap computation of activations and communication between GPUs over the network to reduce memory usage.

The LLaMA model is based on a transformer architecture and incorporates various improvements from recent research. These improvements include pre-normalization, the SwiGLU activation function, and rotary embeddings. Unlike the original transformer architecture, LLaMA normalizes the input of each transformer sub-layer using the RMSNorm normalizing function to enhance training stability. The ReLU non-linearity is replaced with the SwiGLU activation function with a dimension of 234 to improve performance. To reduce computational overhead, LLaMA also adds rotary positional embeddings at each layer of the network, in addition to the absolute positional embeddings. These features collectively enhance LLaMa's performance and make it a powerful tool for natural language processing tasks. During training of the 65B-parameter model, LLaMA processed approximately 380 tokens/sec/GPU on 2048 A100 GPUs with 80GB of RAM, taking around 21 d to train over a dataset containing 1.4T tokens. The model was evaluated on 20 benchmarks, including zero-shot and few-shot tasks, and compared to other foundation models such as GPT-3, Gopher, Chinchilla, and PaLM, as well as OPT models, GPT-J, and GPTNeo.

The results showed that LLaMA outperformed GPT-3 despite being 10 times smaller in size as shown in Fig. 8. This indicates that Llama's techniques for enhancing training efficiency have been successful, and the model is a promising candidate for use in natural language processing tasks. Figure 9 illustrates the development process of ChatLLaMA.

		BoolQ	PIQA	SIQA	HellaSwag	WinoGrande	ARC-e	ARC-c	OBQA
GPT-3	175B	60.5	81.0	-	78.9	70.2	68.8	51.4	57.6
Gopher	280B	79.3	81.8	50.6	79.2	70.1	-	-	-
Chinchilla	70B	83.7	81.8	51.3	80.8	74.9	-	-	-
PaLM	62B	84.8	80.5	-	79.7	77.0	75.2	52.5	50.4
PaLM-cont	62B	83.9	81.4	-	80.6	77.0	-	-	-
PaLM	540B	**88.0**	82.3	-	83.4	**81.1**	76.6	53.0	53.4
LLaMA	7B	76.5	79.8	48.9	76.1	70.1	72.8	47.6	57.2
	13B	78.1	80.1	50.4	79.2	73.0	74.8	52.7	56.4
	33B	83.1	82.3	50.4	82.8	76.0	**80.0**	**57.8**	58.6
	65B	85.3	**82.8**	**52.3**	**84.2**	77.0	78.9	56.0	**60.2**

Table 3: Zero-shot performance on Common Sense Reasoning tasks.

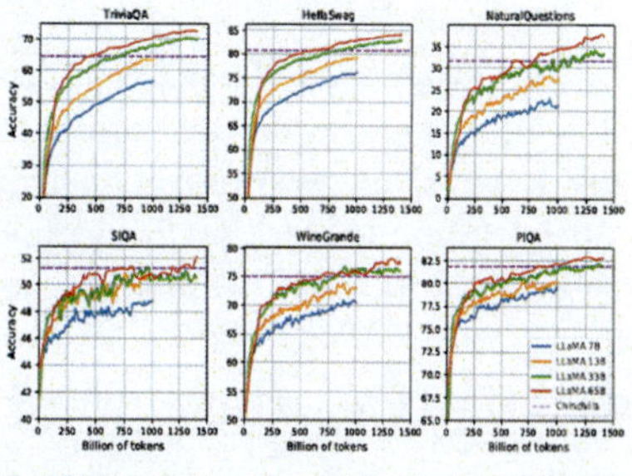

Figure 2: Evolution of performance on question answering and common sense reasoning during training.

Fig. 8. The table and graph illustrate the efficiency of LLaMa. Source

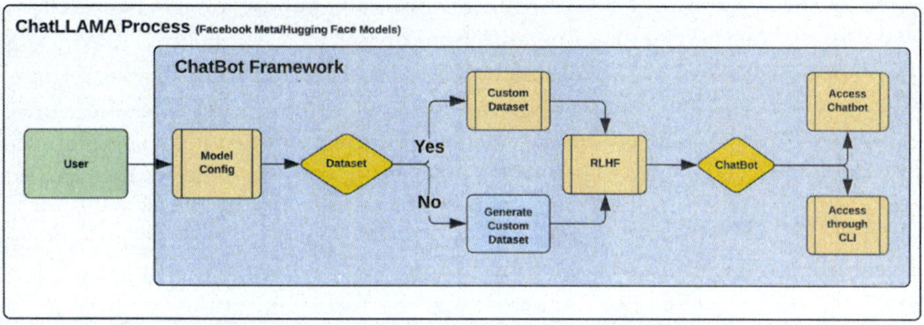

Fig. 9. Development process of ChatLLaMa.

6 Open-Source Chat-Based Platform: OpenChatKit

EleutherAI launched an open-source alternative to GPT-3 called GPT-Neo, which has shown comparative results. Recently, Together[1] released an open-source version of the popular language model ChatGPT called OpenChatKit. This open-source alternative provides developers with more control over the chatbot's behavior and enables tailoring to specific needs. OpenChatKit increases the model's accessibility for a broader spectrum of users and communities, especially for those who might not have the resources to utilize proprietary models. The company offers an intuitive platform combining data, models, and computing to help AI researchers, developers, and businesses better harness and advance AI. The company recently released OpenChatKit 0.15 under the Apache-2.0 license, making the code, model weights, and training datasets publicly available. The OpenChatKit framework includes a generic chatbot and the components necessary to create specialized bots. A generic chatbot is a chatbot that can be used for a wide range of tasks. It is not designed for any specific task or domain. A specialized bot, on the other hand, is designed for a specific task or domain. It is trained on a specific dataset and is optimized for a specific task.

[1] https://www.together.xyz/.

The chat model has 20 billion parameters and was trained on 43 million instructions. The model supports reasoning, multi-turn discussion, knowledge, and generating answers. The framework also includes a retrieval system that can be expanded so that information from a document repository, API, or other sources can be used to answer questions.

6.1 Understandability

The OpenChatKit framework includes a base bot and the components needed to construct customized chatbot applications. GPT-NeoXT-Chat-Base-20B, a large language model tuned on 43 million high-quality conversational instructions, serves as the foundation for OpenChatKit [21]. The model focuses especially on activities like multi-turn dialogue, question-answering, classification, extraction, and summarization. An extensible retrieval system for updating the bot response with information from Wikipedia, news feeds, or sports scores is also included, as are step-by-step instructions for fine-tuning the model to achieve high accuracy on specific tasks. Finally, a fine-tuned GPT-JT-6B model is included for moderation purposes to filter out which questions the bot responds to.

OpenChatKit provides a strong base for developers to work with, as it has higher scores than its base model GPT-NeoX on the HELM benchmark. However, it is the first version of the model and has some limitations, such as giving factually incorrect results, not being trained on a large enough corpus of source code to write accurate code, and not automatically switching topics during a conversation. It is also limited to short responses and does not generate essays or creative stories like ChatGPT. Despite its limitations, OpenChatKit is still a promising initiative and has the potential to improve with the help of the community. It is a valuable resource for developers since it provides greater accessibility and control over chatbot applications.

Contributing data can involve various tasks such as providing new conversational datasets or improving the quality of existing ones, providing annotated data to aid in training the chatbot to handle specific tasks or domains, or even contributing data to test the chatbot's responses in various scenarios. The Open Datahub repository provides guidelines for data contribution, ensuring that the data provided meets certain quality standards.

6.2 Setup and Configuration

OpenChatKit provides developers with a range of tools for training and fine-tuning models according to their specific requirements. It also offers access to pre-trained models and weights, datasets, and a simple command line test harness for interacting with the chatbot. The use of monitoring techniques based on Loguru and experimental techniques that ensure the chatbot's performance is trackable and areas for improvement identified and addressed.

By leveraging the datasets provided and pre-trained weights, developers can achieve high accuracy in their tasks. The command line test harness enables developers to test the chatbot's response and ensure that it aligns with the desired behavior. The monitoring techniques based on Loguru logging provide insights into the chatbot's performance, while the experimental techniques facilitate experimentation with different configurations to enhance the bot's capabilities. Installation details can be found at the Together Computer GitHub site.

6.3 Community and Support

The OpenChatKit community is a group of around a dozen developers collaborating across the world to develop and improve projects related to ChatGPT, an open-source language model. These developers share a common interest or passion for exploring and understanding how ChatGPT works. To support their collaboration, the OpenChatKit community provides various options, including feedback bot and Discord server.

One unique feature is the OpenChatKit feedback app, which allows users to test the chatbot and provide feedback. The feedback received helps the model learn and improve its responses over time, ultimately leading to an enhanced user experience. In addition to the feedback app, OpenChatKit is constantly looking for ways to improve its chat model, and one way to achieve this is through data contributions. The OpenDataHub repository provides a platform for users to contribute their own data, which can be used to train and fine-tune the chatbot's responses.

Through data contributions, users can make a meaningful impact on the OpenChatKit community and help shape the future of chatbot technology. Additionally, by contributing data, users can gain a better understanding of the underlying technology and techniques used in developing chatbots. OpenChatKit offers a basic description of the codebase, including instructions on how to access and run the chatbot on a local system using the files provided on GitHub. This provides users with greater control over the chatbot's customization, allowing for a more personalized experience.

6.4 Regularity of Code Updates

The frequency of updates for the codebase from February 2023 to April 2023 are illustrated in Fig. 10 and 11 shows the plot of commits.

6.5 System Compatibility and Efficiency

OpenChatKit is highly efficient and compatible for tasks such as summarization, question-answering with context, information extraction, and text classification. However, it is not as effective for tasks like coding, creative writing, and answering questions without context, which are some of the areas where ChatGPT has excelled and gained popularity.

Worth noting that OpenChatKit's chatbot may occasionally generate hallucinations which are erroneous results, it also struggles with changing the subject during a conversation and may repeat answers. In comparison to ChatGPT, OpenChatKit's responses are limited to 256 tokens, which can affect the

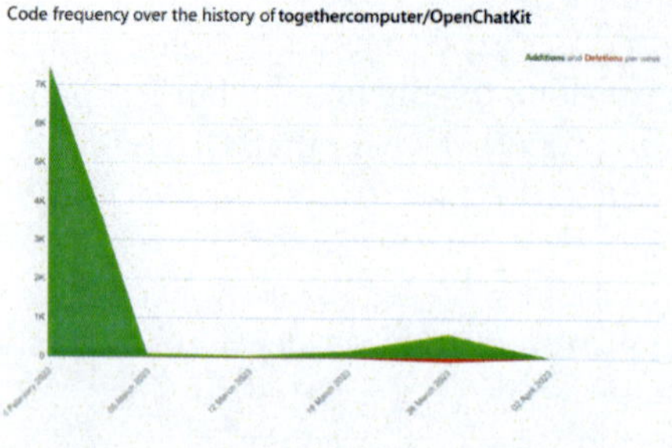

Fig. 10. Code frequency of updates from Feb. 2023 to Apr 2023. Source

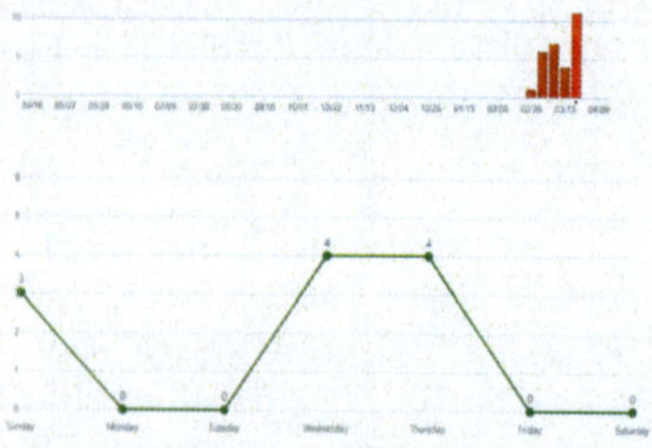

Fig. 11. Commits. Source

replies. However, OpenChatKit generates replies much faster than ChatGPT, and switching between languages does not seem to pose a problem for the chatbot. Additionally, OpenChatKit can format its responses as a list or table, which can be beneficial in some use cases.

GPT-NeoXT-Chat-Base-20B is an advanced NLP model that provides a solid foundation for a wide range of language tasks. The model outperforms its predecessor, GPT-NeoX, in various natural language processing benchmarks, including HELM. Specifically, GPT-NeoXT-Chat-Base-20B demonstrates stronger performance in tasks involving question answering, extraction, and classification, as well as on datasets that were also used for model fine-tuning.

The evaluation of the model on HELM and comparison with GPT-NeoX are evidence of the model's enhanced capabilities in language tasks that require contextual understanding and high-quality responses as shown in Fig. 12. Figure 13 illustrates the development process of OpenChatKit.

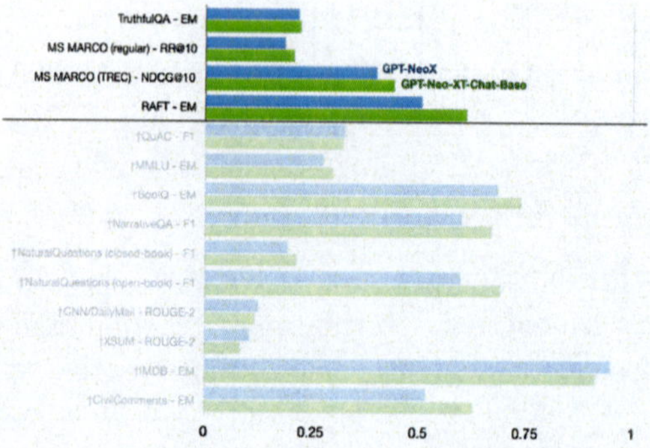

Fig. 12. HELM benchmark chart. Source

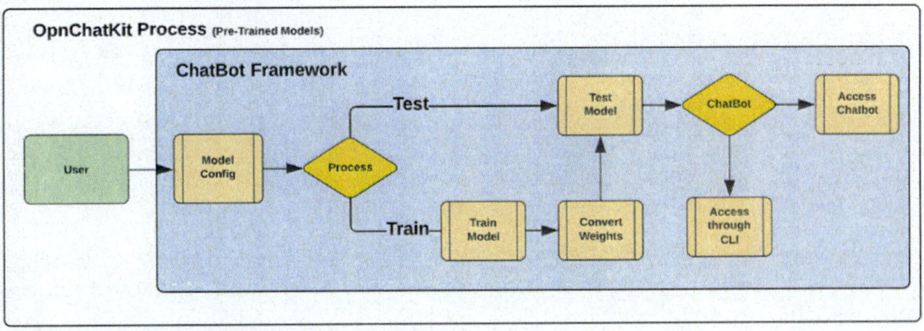

Fig. 13. Development process of OpenChatKit.

7 Analysis of Open-Source Platforms

7.1 Open Assistant

Open Assistant is best suited for beginners who would like to leverage the small language models for their research or finetuning for specific objectives. One of the unique features of this application is that the responses it generates are validated on a regular basis from user inputs, which adds a more human-like experience to the responses. However, issues were encountered during this survey analysis with the documentation, particularly with the prior package installations and the correct directory for running the commands during the installation of the Open Assistant application. In addition, fine-tuning the models using the Hugging Face libraries posed some challenges. Despite these issues, the current developments of Open Assistant have made it more accessible to users, with a direct URL [https://open-assistant.io/] to access the application and an easy-to-use user interface. The documentation has also been improved, providing better guidance for accessing the application on local systems and for finetuning the models with simple commands. Considering the recent updates and improvements to Open-Assistant, Open-Assistant shows promising results for future improvements.

7.2 Colossal-AI

Colossal-AI is a powerful platform that utilizes distributed computing techniques to train large language models. It is highly competitive in the field when it comes to the required hardware configuration for training. The use of Colossal-AI techniques has the potential to enhance adaptability for future iterations. According to Colossal-AI, a mini demo training process can be performed with only 1.62GB of GPU memory. However, in utilizing the platform, we faced challenges with the system configuration which impacted the utilization of Colossal-AI techniques. It is crucial to consider that the performance of Colossal-AI can be influenced by several factors while working with at least a small language model on regular consumer systems, such as the GPU version and the number of CPU cores. In order to achieve optimal performance, it may be necessary to ensure that the hardware configuration meets the recommended specifications. For more information on GPU performance, please refer to the provided link: [https://github.com/hpcaitech/ColossalAI#features].

7.3 ChatLLaMA

ChatLLaMA is an open-source ChatGPT-like training process that is based on Meta's LLaMA (Large Language Model Meta AI) and utilizes reinforcement learning from human feedback (RLHF). The model is available in various sizes ranging from 7B to 65B parameters, and its construction is clearly explained. Unlike other language models, LLaMA was trained on publicly available datasets, making it more accessible to researchers and developers. Additionally, LLaMA's foundation models are trained on a large set of unlabeled data, making them ideal for fine-tuning for a variety of tasks. In our experience for this survey, LLaMa model is not directly available, one must request access to the model and will be granted on a case-by-case basis. At present, ChatLLaMA also provides a direct URL to access the chatbot and validate the performance of the model. ChatLLaMA has the potential to make significant contributions to the field of NLP by providing publicly available models and a more organized and self-explanatory code.

7.4 OpenChatKit

OpenChatKit is an open-source alternative to ChatGPT based on EleutherAI's 20 billion parameter language model, GPT-NeoX, and tuned with 43 million instructions for chat use. It includes customization recipes, an extensible retrieval system, a moderation model, and tools for users to provide feedback and add new datasets. The chatbot is highly effective in tasks such as summarizing and answering questions with context, extracting information, and classifying text. However, it falls short in tasks such as coding and creative writing. Additionally, OpenChatKit struggles with changing the subject mid-conversation and repeating answers. Nevertheless, OpenChatKit performed much better after being fine-tuned for specific use cases, and Together is working on its own chatbots for learning, financial advice, and support requests. One can visit the Hugging Face URL at https://huggingface.co/spaces/togethercomputer/OpenChatKit. This allows users to interact with the chatbot and evaluate its responses. By testing the chatbot, users can determine how well it performs in various tasks and how effectively it responds to queries. This can be useful for developers who are considering using OpenChatKit for their own chatbots or for individuals who are simply curious about its capabilities.

Table 1. Summarization of the evaluative criteria for the four platforms.

	Open-Assistant	ColossalAI	ChatLLaMa	OpenChatKit
Models	Hugging Face models	Customized models and techniques to train ViT and GPT models are provided	Meta AI's LLaMA and hugging face transformer models	GPT-NeoXT-Chat-Base-20B, Pythia-Chat-Base-7B
Customizability	Highly customizable	Highly customizable	Limited customization	Limited customization
Community/Support	Large and active community with extensive support resources	Large and active community with extensive support resources	Small community with limited support resources	Small community with limited support resources
Regular Updates	Regularly updated with new features and improvements	Regularly updated with improvements	Regularly updated with new features and improvements	Infrequent updates with limited new features
Deployment/Scalability	Scalable and can be deployed and utilized in local or access the application with URL	Scalable and can be deployed and utilized in local	Scalable and can be deployed and utilized in local or access the chatbot with URL	Scalable and can be deployed and utilized in local or access the chatbot with hugging face URL
Ease of Use	High	Moderate	Moderate	Moderate

8 Conclusion

This paper conducted a preliminary review of four open source platforms for chat-based Generative AI. In conclusion, Table 1 summarizes the evaluation criteria for the four open source platforms explored in this comparative study.

Appendix

1. **Open Assistant Setup and configuration Steps**
   Step 1: click Create Code space on main. Containers will be generated based on the image in devcontainer.json in the .devcontainer folder. Image: An image is like a mini-disk drive with various tools and an operating system pre-installed.
   Step 2: Run command in Terminal - docker compose –profile ci up –build –attach-dependencies.
   Step 3: click on Ports tab and navigate to the required links as mentioned below, **Port 1080:** To access mail. **Port 3000:** To access Open-assistant application.
2. **ChatLlama Setup and configuration Steps**
   Pre-requisites: Configure CUDA environment for GPU support.
   Step 1: Download and install NVIDEA CUDA toolkit from Nvidea Archive.

Step 2: Setup PyTorch - PyTorch Link
Step 3: Setup TensorFlow - Follow instructions in TensorFlow Link
Step 4: pip install transformers.

Quick Install:

Step 1: pip install chatllama
Step 2: Clone llama's repository to install the Llama models. Run below code
git clone https://github.com/facebookresearch/llama.git
cd llama
pip install -r requirements.txt
pip install -e.
Step 3: Run below code.
from chatllama.rlhf.trainer import RLTrainer
from chatllama.rlhf.config import Config
path = "path to config file.yaml"
config = Config(path=path)
trainer = RLTrainer(config.trainer)
trainer.distillate()
trainer.train()
trainer.training_stats.plot()
ChatLLama used OPT-1.3B model for this quick installation fine tuning process.

References

1. Vaswani, A., Shazeer, N., Parmar, N., Uszkoreit, J., et al. Attention is all you need. In: Advances in Neural Information Processing Systems, pp. 5998–6008 (2017)
2. Lample, G., Denoyer, L., Ranzato, M.: Unsupervised machine translation using monolingual corpora only. In: International Conference on Learning Representations (ICLR) (2018)
3. Vinyals, O., Le, Q.: A neural conversational model. In: Proceedings of the International Conference on Machine Learning (ICML) (2015)
4. Radford., et al.: Language models are unsupervised multitask learners. In: OpenAI in (2019)
5. Bolukbasi, T., Chang, K.W., Zou, J., et al.: Man is to computer programmer as woman is to homemaker? In: Advances in Neural Information Processing Systems, Debiasing Word Embeddings (2016)
6. Gonen, H., Kementchedjhieva, Y., Goldberg, Y. Lipstick on a pig: debiasing methods cover up systematic gender biases in word embeddings but do not remove them. In: Proceedings of the 2019 Conference on Empirical Methods in Natural Language Processing and the 9th International Joint Conference on Natural Language Processing (2019)
7. Carlini, N., Liu, C., Erlingsson, Ú., et al.: The secret sharer: evaluating and testing unintended memorization in neural networks. In: IEEE Symposium on Security and Privacy (2019)
8. Radford, A., Wu, J., Child, R., et al.: Language models are unsupervised multitask learners. In: OpenAI Blog (2019)

9. Brown, T.B., Mann, B., Ryder, N. et al.: Language models are few-shot learners. In: OpenAI Blog (2020)
10. Hitaj, B., Ateniese, G., Perez-Cruz, F. Deep models under the GAN: information leakage from collaborative deep learning. In: ACM Conference on Computer and Communications Security (2017)
11. LAION AI. Open Assistant. GitHub Repository. Accessed January 2023. https://github.com/LAION-AI/Open-Assistant
12. Daniel, M., Ziegler, N.S., et al.: Fine-tuning language models from human preferences. arXiv (2020)
13. Rajkumar, R., Prithviraj, A., et al.: is reinforcement learning (Not) for natural language processing: benchmarks, Baselines, and Building blocks for natural language policy optimization. arXiv (2023)
14. Xu, Q., Li, S., Gong, C., You, Y.: An efficient 2D method for training super-large deep learning models. arXiv (2021)
15. Li, et al.: Colossal-AI: a unified deep learning system for large-scale parallel training. arXiv (2021)
16. Li, S., Xue, F., Baranwal, C., Li, Y., You, Y.: Sequence Parallelism: long sequence training from system perspective. arXiv (2021)
17. Liu, Y., Li, S., Fang, J., Shao, Y., Yao, B., You, Y.: Colossal-auto: unified automation of parallelization and activation checkpoint for large-scale models. arXiv (202)
18. Touvron, H., Lavril, T., Izacard, G., et al.: LLaMA: open and efficient foundation language models. arXiv (2023)
19. Nebuly AI. ChatLlama. GitHub Repository. Accessed March 2023. https://github.com/nebuly-ai/nebullvm/tree/main/apps/accelerate/chatllama
20. Meta AI. Large Language Model Llama: Meta-AI. Facebook AI Blog. Accessed February 2023. https://ai.facebook.com/blog/large-language-model-llama-meta-ai/
21. Together Computer. OpenChatKit: an open toolkit and base model for dialogue-style applications (Version 0.15) [Software] (2023). Accessed 31 March 2023. https://github.com/togethercomputer/OpenChatKit

Semi-autonomous Gas Detecting Robot (Hazmat Robot)

Mustafa Ayad[1](✉), Jesse Smithers[1], Anthony Ficarrottas[1], Christain Mackey[1], and Khaled Ayad[2]

[1] State University of New York at Oswego, Oswego, USA
mustafa.ayad@oswego.edu
[2] Aljafara University, Janzur, Libya

Abstract. Gas leaks have become an increasing concern, as high-risk conditions can lead to complications affecting many industries. Search and rescue robotics have been increasingly helpful in higher-risk situations, specifically in fully and semi-autonomous designs. In this paper, a semi-autonomous gas-detecting robot is developed to create a safe work environment for all involved. The gas-detecting robot would explore an area and report findings through a wireless interface, displaying current and prior gas readings to identify a "hotspot eventually." If the robot is autonomously searching and an obstacle is sensed, it will adjust its path accordingly. Our design is comparable to the SMP Robotics S6 robot and the RoboGasInspector. Incorporating Raspberry Pi as the central processing unit, the movement and the data display can be programmed using Python. The interface will display the data for the ultrasonic, MQ2, MH-Z19, and PM2.5 sensors. The experimental tests have proven that our design and implementation are successful and promise excellent potential for the gas detector robot's future.

Keywords: Hazardous · Firefighters · Ultrasonic Sensor · Gas Sensor

1 Introduction

Small-size rescue robots are developed to assist in the search and rescue of humans. Moreover, they provide rescue efforts by searching, mapping, and discovering dangerous areas. Rescue robots can be controlled remotely [1]. They send sensor data, video streams, or images of the disaster site to the control station. Fortunately, these robots can perform dangerous and unsafe missions for humans [2]. Gas detector robots are used in the oil and gas industry for first response, fugitive emissions, and preventative care [3]. Over 125,000 gas leaks happen annually in the United States [4]. Carbon monoxide causes at least 430 deaths yearly, and 50,000 people visit the Emergency room. In addition, an estimated 4,200 structure fires occur yearly from natural gas leaks [5]. The action causes first responders to enter hazardous environments, potentially injuring or dying. A first responder would have to search an area with a handheld meter to find the source of the gas. Consequently, it risks human life as these gas leaks take the lives of firefighters and first responders yearly. These unknown gasses can be flammable, corrosive, and deadly to human life.

K. Arai (Ed.): FTC 2023, LNNS 816, pp. 186–198, 2023.
https://doi.org/10.1007/978-3-031-47448-4_14

There have been many uses of robotics in high-risk situations, such as underwater drones that find and examine pipes. A gas leak and robot use would also be safer in our case. Since these gasses are hazardous and invisible to the human eye, it would be better to send in some robots to limit human intervention until the source can be located. One major challenge in industrial control involves learning behaviors feasible in the real world due to machinery constraints. For example, specific actions can only be executed every few hours, while others can be taken more frequently. However, with extensive reward engineering and experimentation, an reinforcement learning agent may learn the realistic operation of machinery through hierarchical reinforcement learning with multiple agents that control subsets of actions according to their operation time scale [6]. In [7], a hybrid industrial cooling system model embeds analytical solutions within a multi-physics simulation. The model is designed for reinforcement learning applications and balances simplicity with simulation fidelity and interpretability.

The paper entails designing and creating a semi-autonomous gas-detecting robot. The robot could search an area and determine if there is a certain gas leak. Thus, it would alert the robot's operator to where the leak's location and what readings the robot is experiencing. The robot would be able to find the gas leak location automatically, but if there is a situation where the robot cannot find the best route to take, the robot would ask for human input.

The operator could see data of the robot's environment through a wireless interface, so they do not have to be in a hazardous environment. The interface would display the gas levels of the room and state the previous readings. In addition, the robot would also live feed video via a camera and infrared camera so the operator could see what environment the robot is exploring. If they need to drive the robot manually, they can see where it is going—we aim to save and preserve human life in a dangerous gas leak. Our design is comparable to the SMP Robotics S6 robot and the RoboGasInspector [8].

All the designed robot functions were programmed in Python, and this includes the movement/automation of the robot to process the sensor measurement to a functional data unit. All the sensors and motors were wired to a power supply and the microcontroller with the right amount of voltages and currents going to each component. Therefore, the robot can move at the correct speed and direction based on the robot components' accurate power calculations and implementations. Also, the robot communicates with the interface and sends and receives data via Wi-Fi. In addition, based on HTML and SQL databases, we created a website to receive and display the data. Lastly, the paper presents the implementation of a gas-detector robot that can search for a certain gas leak.

2 Problem Description

Designing the Hazmat robot, we adjusted our ideas to reach our end vision. First, we could use research to determine the design and implications of the robot. Once we could decide how we wanted the robot to be used, we could evaluate what components we needed to complete the functionalities. The problem we were trying to mitigate was reducing human contact when determining a gas leak. To accomplish this goal, the robot will need to perform specific tasks. The robot will need to be able to have a manual mode where the operator of the robot will be able to control the robot through an interface.

This interface will also display live camera feeds, previous and current sensor readings, and warning levels of gases. The interface will also switch the robot from manual to semi-autonomous search. The robot will perform two different searches. One search will be a perimeter search where the robot will follow the wall to search for gas and determine if the gas leak is against the wall. The second search will contain a random search where a random search pattern will be chosen to find the gas leak. The user can stop the autonomous search and put the robot into manual mode at anytime.

In order to test the robot for functionality, dry ice would be placed in a room. Since dry ice off gasses CO_2, the Hazmat Robot can detect it. From there, if the robot can perform the tasks previously listed, we can determine if the robot was successful. Within the final testing, the robot completed a perimeter and random search. The robot was able to display the gas reading on the interface and was able to graph the current and previous data. In addition, the robot had a live video feed and could stream it to the interface. Finally, the robot was able to perform obstacle avoidance successfully and was able to detect drops like stairs.

3 Hardware Implementation and Design

The robot designed in this paper was developed to minimize human intervention in a dangerous situation, but we did not want to lose any efficiency in robot searching. Because of this strict requirement, we needed certain necessary materials and components. Therefore, we assembled the required hardware parts to implement the Hazmat robot. Figure 1 shows the Hazmat robot's block diagram, which consists of a Raspberry Pi microprocessor, CO_2, propane, particle, and ping sensors. Also, it consists of encoders, motor controllers, and a camera. In addition, the Hazmat communicates to the eternal world through Wi-Fi.

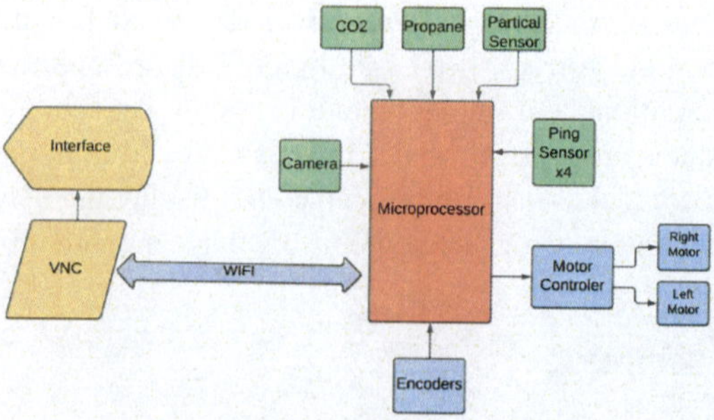

Fig. 1. Hazmat robot block diagram.

We tested the Raspberry Pi NoIR camera [9], as in Fig. 2(a), and a standard USB webcam, as shown in Fig. 2(b), when deciding which camera to use for the live video stream. The goal was to determine which camera best suits the HazBot regarding video quality and real-time streams. Initially, the NoIR was considered the better option due to its small size and ease of connectivity with the Raspberry Pi. Although further testing was done, it was quickly discovered that the NoIR was not the best option. Due to low video resolution and a significant stream delay, the NoIR was eliminated from the options. When running a video stream to the interface, the NoIR produced a 5-s delay with a shallow resolution.

As the task of this camera was to transmit a real-time video feed for a user who would not be in the same room as the robot, it was evident that a 5-s delay would not be suitable. When beginning testing with the USB webcam, some initial issues occurred with a slight delay and low resolution. But, when adjusting to different variables such as the frame rate, quality, stream rate, and height and width of the video projected to the website, the delay issue was quickly solved and produced a much clearer picture. Therefore, the USB webcam was the best option for the Hazmat, with a stream having a fraction of a second delay at the most.

(a) (b)

Fig. 2. A) Raspberry Pi NoIR camera b) USB webcam.

Once we created a real-time video stream to the interface for viewing purposes, they moved on to the sensor data display on the website. When attempting to send data from our SQLite database to the website, it became evident that a problem was occurring with the Raspberry Pi. The main issue is that the video feed is streaming to the IP address of the Raspberry Pi, which was embedded in the website. But, when also attempting to use that IP address to create a separate page for sensor data information and graphing, it became clear that the Raspberry Pi could not stream two separate entities to the same IP address. To combat this, we came up with the idea of implementing a second Raspberry Pi into the HazBot. One Raspberry Pi would be used strictly to transmit the live video stream, and the second would be used for sensor data collection, storage, and transmission to the website. Using two separate IP addresses was a simple but effective way to overcome our problem. Although the idea was thought off to last minute and could not be fully implemented in the end, we ended with having a live video stream to the website and then directly using the SQLite database browser to view live sensor data and graphs.

The first idea was to have three ping sensors on the robot, one facing the front and two facing the right and left. Thus, it turned into 5 with two new sensors on the right and left corners, which were positioned at a 45-degree angle. It allowed us to read walls better when the robot moved towards a wall at an angle. This amount of sensors worked reasonably well for our search methods, but we still had some problems with front obstacles. The robot was so vast that the ping sensor would not pick something up when a wall or object was off-center. To combat the problem, a sixth sensor was used. This sixth sensor was put on the front but offset two the left, and the original font sensor was offset to the right.

Fig. 3. Raspberry Pi.

The Raspberry Pi microprocessor is extremely useful when creating a gas-detecting robot, as in Fig. 3. The microprocessor comes with pre-built GPIO (General Purpose Input Output) pins which can be used for standard device attachments. The main concepts that set the Raspberry Pi apart are its built-in operating system (Raspbian OS) and NOOBS (New Out of the Box Software). In addition, the microprocessor can install other third-party operating systems like Linux, Windows 10, etc.

In addition, the Raspbian operating system has built-in Python programming tools, which will be extremely useful for the Hazmat robot [3]. The Raspberry Pi was chosen to be the robot's microprocessor because it can have critical features at an affordable price. First, the Raspberry Pi comes with 4 Gb of ram and expandable storage of up to 1 TB, allowing enough processing power and storage to process our data. Second, the Raspberry Pi has built-in Wi-Fi, which provides for wireless contention between the Pi and the controller; without this, the Pi would need a Wi-Fi shield and more libraries installed to connect to Wi-Fi. Finally, the raspberry pi was also chosen because of its number of IO pins and the libraries for the robot's sensors.

When looking into solving the problem of harmful gas leaks in buildings across the United States, it became evident that a robot with the ability to autonomously search a room and find the point of leakage would be beneficial. This robot could also be implemented in large industrial factories or plants to monitor gasses in the air and their concentrations, to prevent a catastrophe before it even happens.

The first step was to decide which gas sensors would be reliable and durable enough to be sent into a hazardous situation with the possibility of extremely high temperatures and high concentrations of dangerous gasses. Unfortunately, we could not obtain all the desired sensors due to a limited budget. Therefore, we settled with an MG811 Carbon Dioxide sensor [10], an MQ-138 Volatile Organic Compound sensor [11], a PM2.5 Air Particle sensor, and an MQ2 sensor [12], as shown in Fig. 4. The MG-811 and MQ-138 were implemented to detect certain gasses in the air. PM2.5 was implemented to detect high smoke or dust. The PM2.5 sensor (in Fig. 5(a)) determines if particles are up to 2.5 μ in diameter. This sensor works by pulling air through an air channel with a fan. Once the air enters the air channel, it goes through a light trap where a laser is pointed through the air, and the particles cause the light to scatter. The photodiode captures the scattered light to determine how many particles are in the air. The MQ2 sensor determines how much butane, methane, hydrogen, and Alcohol is in the atmosphere. This sensor will detect flammable gasses in an environment and decide whether or not it is safe.

The placement of these sensors would also play a crucial part in how they perform. After testing various positions of each sensor, it was determined that the MG-811 Carbon Dioxide sensor would serve the best at the lowest possible point from the ground since it is quite a dense gas and shows higher concentrations at lower concentrations levels in a room.

Similarly to Carbon Dioxide, Volatile organic compounds (VOC) also tend to be heavier than air. Therefore, they sink to the lower parts of a room. Thus, the MQ-138 Volatile organic compound sensor was also placed on the lowermost level of the HazBot. Finally, the PM2.5 particle sensor had to be placed in an area with a full range to read air particles with nothing in its path. Therefore, it was placed on the middle level of the robot in the front to avoid interference from any other components.

When determining how the robot would autonomously function in the search room, it was decided that Ultrasonic sensors, commonly known as ping sensors, would be our best option. A Lidar was also considered for this project and would have been extremely helpful with its ability to map out a room, but it was not within our budget. Ultrasonic sensors are one of the essential components of the Hazmat Robot. Ultrasonic sensors, in Fig. 6(a), allow the robot to react to the world around them, as humans

(a) (b)

(c)

Fig. 4. Sensors used, **a)** MG-811 **b)** MQ-138 **c)** MQ2.

Fig. 5. a) PM2.5 sensor.

would with their eyes and ears. These sensors are often referred to as "sixth sense" sensors. A crucial part of the Hazmat Robot is detecting objects in its path and navigating around them autonomously. Ultrasonic sensors are commonly used to detect objects while determining the distance between the obstacle and the robot. The components of this sensor module consist of an ultrasonic transmitter, receiver, and control circuit. The sensor has two circular eyes; one transmits the ultrasonic wave, and the other receives it. The HC-SR04 Ultrasonic sensor measures distance in the 2 cm–400 cm range with an accuracy of 3 mm. Figure 6(b) shows the ping sensor working principle. The ultrasonic sensor module will communicate with the raspberry pi, which will communicate with the motor driver to control the DC motors for normal driving conditions or if an obstacle is in the path.

Ping sensor allows for measuring distance by sending a sound wave and listening to its return. The time from when it is sent off to when it returns is recorded, and the distance is calculated from that measurement. This sensor is going to be used in obstacle avoidance. This sensor can detect from 3 cm up to 3 m.

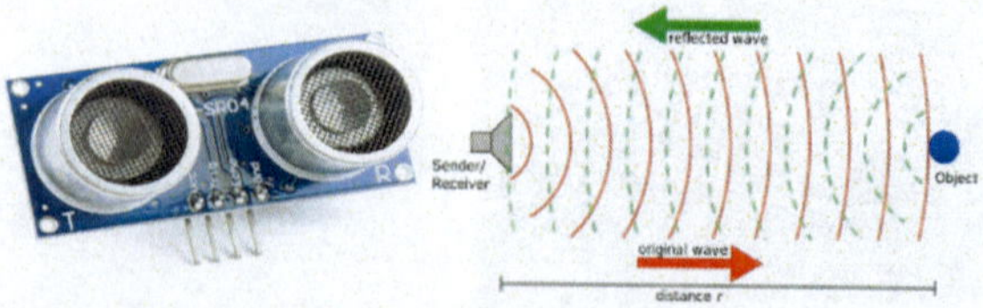

(a) (b)

Fig. 6. a) Ultrasonic sensor b) Ping sensor working.

We decided to use seven ping sensors for autonomous movement and obstacle avoidance. Figure 7 shows the ping sensors layout. Two sensors were placed on the front-middle level of the robot to read objects directly in its path. One sensor was placed on each side of the middle tier to detect how far away from the wall the Hazmat was. One sensor was also placed on each front corner of the intermediate level to aid in tricky situations such as corners of a room. Unfortunately, this area in the room caused issues with the ping sensors due to how they function. These sensors shoot out a soundwave, which then makes contact with an object and shoots back at the sensor.

Fig. 7. Ping sensor layout on hazmat.

The sensor can then determine how far away that particular object is. In addition, when the robot approaches a corner, the soundwave would not be directly shot back to the sensor due to its approach angle. Instead, the wave would be shot off on an angle and not returned to the sensor. Therefore, a sensor was placed on each front corner to create a better angle for the sensors to read. Lastly, one sensor extended about 5 inches off the robot, pointed at the ground for drop detection to avoid falling down stairs.

After determining the gas and ping sensor layouts, we had to decide how the robot would be powered and driven. Two separate power supplies were implemented in the Design of the Hazmat Robot. The first power supply was a portable charger power bank with a capacity of 20000mAh and an output of 5 V 4.8 A. It supplied power to the microprocessor, Fig. 8(a). This power supply can run the microprocessor at max capacity for approximately 4.6 h before requiring a charge. The second power system consisted of two 7.2 V 3000 mAh VEX EDR rechargeable batteries, Fig. 8(b). This power supply powers the drive system and can perform at max consumption for about 22 min. Given the robot can search a room and detect the leak within a few minutes, this is more than enough time.

(a) (b)

Fig. 8. a) Portable charger power bank b) VEX EDR battery.

The drive system consists of an L298N (Fig. 9(a)) motor driver with two VEX 269 two-wire motors. These motors have a standard speed of 100 rpm with a stall torque of 0.97 N-m (8.6 in-lbs) and a stall current of 2.6 A. Two motors work perfectly to produce the ideal speed with the Hazmat Robot's weight.

When configuring all components, it can be seen that all gas and distance sensors, along with the USB webcam, are connected and transmitting data to the Raspberry Pi. The Raspberry Pi directs information to the motor controller to power the two VEX 393 motors (Fig. 9(b)) [13]. Sensor data and camera feed are then transmitted to the actual

(a) (b)

Fig. 9. a) L298N motor driver b) VEX 2-wire motor 393.

time interface. The USB camera stream can be seen on the website for the HazBot, and the sensor data and graphs can be viewed in the SQLite database browser.

We had three main objectives when approaching the real-time interface design. The first was to create a standard landing page for the website, Fig. 10. This was done with HTML and CSS to create an interactive and informative homepage. The page consists of a brief summary of the project itself, along with a directory link to view the live video feed of the Hazmat Robot. Next, the video feed transmitted from the Raspberry Pi is streamed to the Pi's IP address. To view this from our website, the IP address was embedded in the directory button to view the real-time video. Then, with two simple Terminal commands, the camera is turned on and begins transmitting live video to the IP address.

The second objective when creating the interface was the ability to store and read live data along with graphs showing concentrations of certain gasses. This was done using SQLite to store and record data in real time. A script was written to collect sensor data and place all readings into an SQLite database. The sensor would take a reading every two seconds and quickly be stored in the database for analysis. The information was easy to view and plot by converting all data into floating points. Figure 11 shows data in SQLite Database Browser.

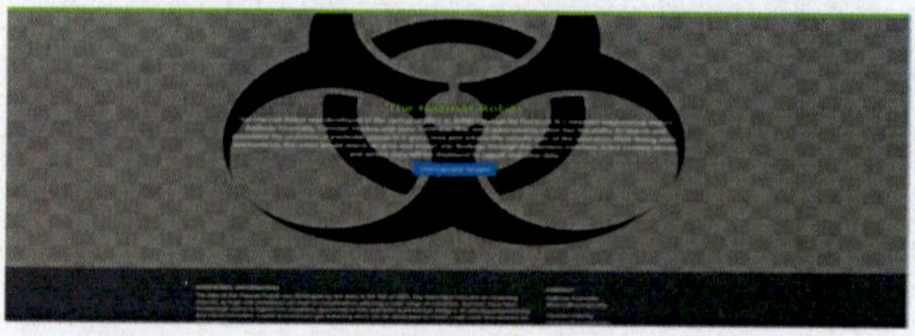

Fig. 10. Hazmat robot landing Page.

4 Hazards and Experiments Analysis

With this being a new project and something that we had never done before, we knew we would face many problems coming into this. Because of this, we did extensive planning to try to mitigate these. We thought about what environments and situations we would be in if we planned this whole project; we wanted it to be functional in a real-life case. Since this robot is in a dangerous situation, we had to take precautions, such as covering our components.

Suppose we were going to be in a room with explosive gasses or maybe even fire, the essential components of the robot needed to be protected. If this were a consumer product, we would have made an entire enclosure for the robot. This would save the components and allow us not to worry about accidentally causing a spark to light any flammable gasses. We could develop even more safety measures to perfect our design with more testing.

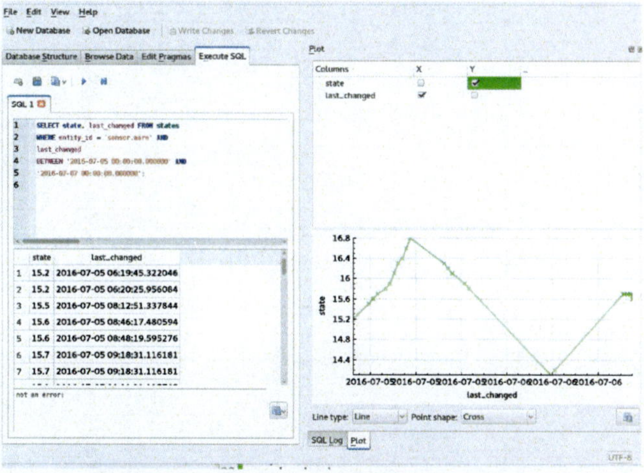

Fig. 11. Viewing data in SQLite database browser.

4.1 Demo Procedure

The demo procedure was straightforward to set up. Placing dry ice in the desired spot was the safest and easiest way to test our design. The dry ice would start as a solid and then sublimize into CO_2 gas. The CO_2 sensor we mounted on our robot's front would read this gas. Since CO_2 is denser than room air, it would sink to the ground, making it able to be read by the robot. These readings would then be saved in our program and the database.

From the database, it would graph the readings to show us a real-time display of our concentration values. To get into the actual process, the robot would need to find these concentrations; we first have to start with our perimeter search. We saw that the most efficient way to search was around the perimeter of our room since most gas leaks are found in pipes near the walls. If the concentration were not on the perimeter, we would go into a random search pattern to find the highest concentration. Once found, it would alert the user and go to the highest concentration. Figure 12 shows robot flow and autonomous search flow chart.

In Fig. 13, we can see a gas leak in the center of the room. This would be an example of what the robot and the operator would do when searching for a gas leak in the room. The operator would be outside the room operating the robot from the interface, being able to see the robot with the live stream via the Webcam. The gas reading sensor would also be displayed in the interface.

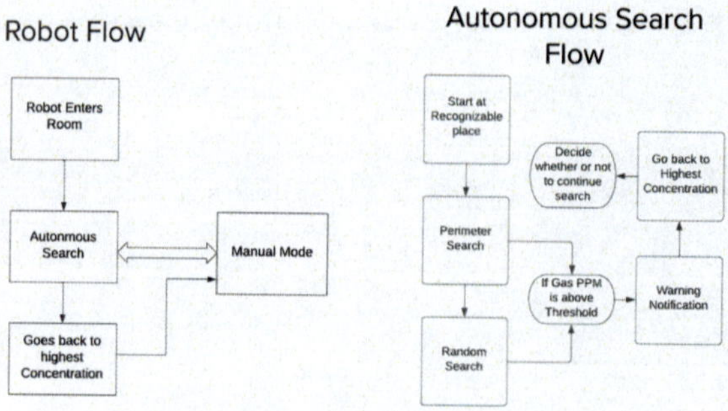

Fig. 12. Demo procedure flow chart.

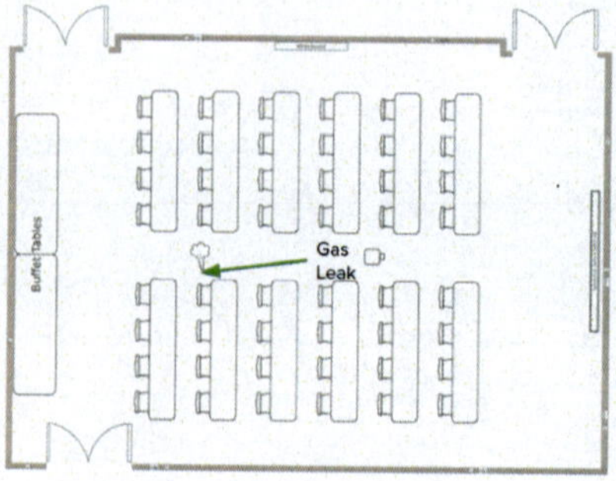

Fig. 13. Demo procedure step1.

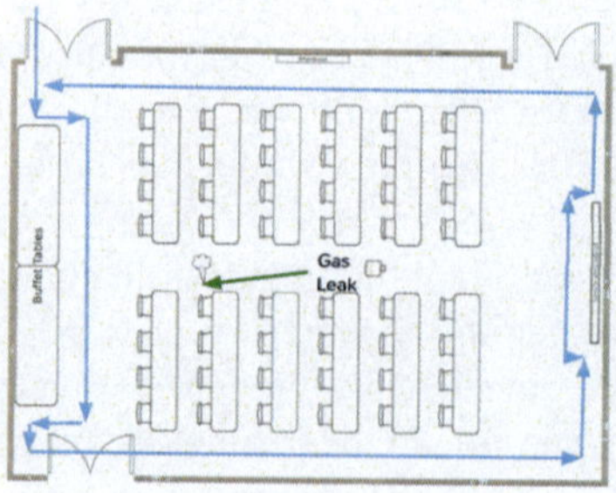

Fig. 14. Demo procedure step 2.

The robot will enter the room and perform a perimeter search once activated by the user. The robot will follow the wall around the room, searching for the gas leak and avoiding any obstacles. For example, in Fig. 14, the robot can avoid the buffet tables and go back to following the wall.

Once the robot returns to the starting position of the perimeter search, the user will end the perimeter search and activate the random search. Then, the random search will start searching for the gas leak while avoiding obstacles and randomly determining its route, as in Fig. 15.

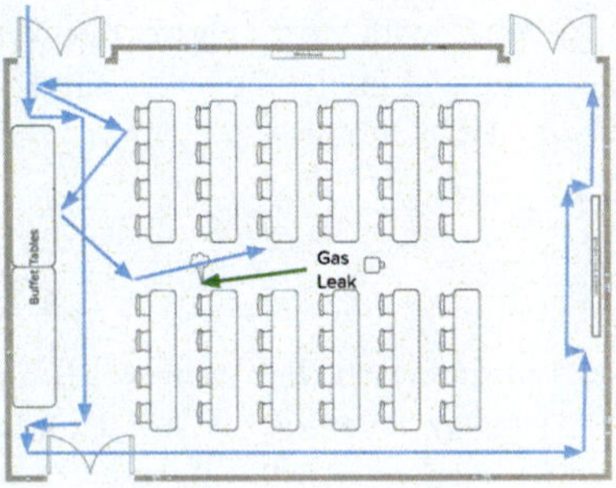

Fig. 15. Demo procedure step 3.

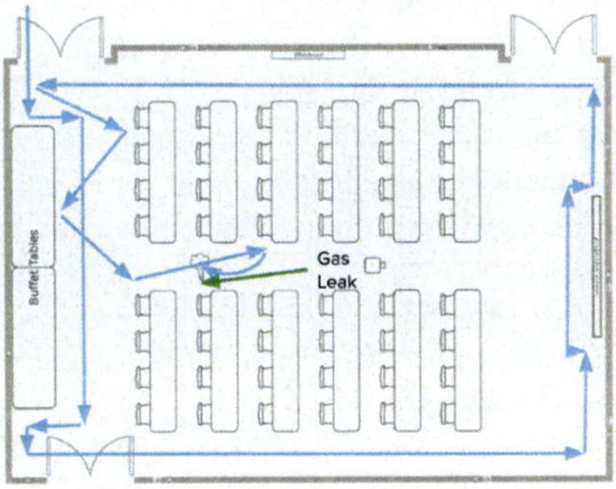

Fig. 16. Demo procedure step 4.

The robot will pass the gas leak checking that it is the highest concentration inside the room. Once the gas level decreases, it stops and returns to the highest concentration, as in Fig. 16. This allows the operator to use the live video stream to inspect the area further to determine the gas leak's precise location.

5 Conclusion

In this paper, a gas detector robot (Hazmat) has been designed and implemented. Furthermore, the design's methods of how hardware components and control interact were detailed. The experimental tests promise a potential future for the implemented gas detector robot. Besides, the implemented gas detector robot will have the potential to go further, and with different gas sensors, the robot can search for different types of hazardous gases. In addition, the robot can be developed to handle diverse environments, for example, climbing stairs and opening doors. Also, this robot can be designed to operate in various types of climates, like snow and rain.

Additionally, future implementations of the robot would be able to fully autonomously drive around a site, for example, a coal mine, and alert workers if the gas level is approaching or is at a dangerous level before their gas meter goes off. In addition, the robot can be equipped with other sensors like X-ray or explosive substance sensors to test suspicious packages. Finally, we successfully implemented the Hazmat robot with the amount of knowledge and time constraint.

References

1. Ayad, M., Voyles, R.: Message routing and link maintenance integration for robotic network connectivity. In: Future Technology Conference (FTC-2020), Vancouver, Canada (2020)
2. Ayad, M., et al.: Amphibious rover for search and rescue applications. In: 2022 IEEE 2nd International Maghreb Meeting of the Conference on Sciences and Techniques of Automatic Control and Computer Engineering (MI-STA), pp. 64–69 (2022)
3. Naik, S., et al.: Gas leakage detector insect robot using raspberry Pi3. Int. J. Eng. Res. Technol. (IJERT) NCCDS – 2020 **8**(13) (2020)
4. Memic: Fuel gas detection. https://www.memic.com/workplace-safety/safety-net-blog/2022/january/fuel-gas-detection. Accessed 12 Jan 2022
5. National Center for Environmental Health. Carbon monoxide (CO) poisoning prevention. https://www.cdc.gov/features/copoisoning/index.html. Accessed 18 Sept 2019
6. Gas leak inspection robot - SMP Robotics - Autonomous mobile robot. SMP Robotics - Autonomous mobile robot. https://smprobotics.com/products_autonomous_ugv/gas-leak-inspection-robot/. Accessed 02 Dec 2021
7. Wong, W., et al.: Optimizing industrial HVAC systems with hierarchical reinforcement learning. arXiv preprint arXiv:2209.08112 (2022)
8. Chervonyi, Y., et al.: Semi-analytical industrial cooling system model for reinforcement learning. arXiv preprint arXiv:2207.13131 (2022)
9. Raspberry Pi NoIR Camera v2 datasheet. https://www.farnell.com/datasheets/2056180.pdf
10. Tutorial: How to Use Mg811 Co2 Carbon Dioxide Gas Sensor. https://www.instructables.com/Tutorial-How-to-Use-Mg811-Co2-Carbon-Dioxide-Gas-S/
11. Gas Sensor for Voc Gas (Model: MQ138). https://www.winsensensor.com/d/files/PDF/Semiconductor%20Gas%20Sensor/MQ138%20(Ver1.4)%20-%20Manual.pdf
12. PM2.5 Air Quality Sensor. https://cdn-learn.adafruit.com/downloads/pdf/pm25-air-quality-sensor.pdf
13. MQ-2 Semiconductor Sensor for Combustible Gas. https://www.pololu.com/file/0J309/MQ2.pdf
14. All About Circuits: Brushless DC Motor. https://www.allaboutcircuits.com/textbook/alternating-current/chpt-13/brushless-dc-motor/. Accessed 13 May 2019

An Explainable Intrusion Discovery Framework for Assessing Cyber Resilience in the Internet of Things Networks

Mohammed T. Masud[1], Marwa Keshk[1], Nour Moustafa[1(✉)], and Igor Linkov[2]

[1] University of New South Wales Canberra, Canberra, Australia
mohammed_tanvir.masud@adfa.edu.au,
{marwa.keshk,nour.moustafa}@unsw.edu.au
[2] U.S. Army Engineer Research and Development Center, Vicksburg, MS, USA
igor.linkov@usace.army.mil

Abstract. Cyber hardening systems, such as intrusion detection and firewalls, are no longer sufficient to keep cyber systems safe from today's smart and well-resourced attackers. However, these systems have high false-positive rates, and even experts may find it difficult to understand the reasons behind their predictions. With the emergence of Internet of Things (IoT) ecosystems, these systems need to address data heterogeneity and the complexity of IoT networks. Developing a cyber-resilient intrusion discovery model is challenging, as it must adapt to zero-day exploits in real-time, and respond strategically to retain the vital operations of systems in the event of exploits and system failures. The assessment of cyber resilience becomes increasingly difficult due to the uncertainties of predicting security events and system failures in real time. The relationship between security, robustness, and vulnerabilities is focused on preventing IoT system degradation and maintaining acceptable levels of functionality before and after occurring uncertain events. To address these challenges, we propose an explainable cyber intrusion discovery framework to make the model transparent and trustworthy, along with the capability of evaluating its cyber resilience using a cyber threat matrix. We employ an explainable convolutional neural network-based transfer learning model to detect uncertain events including security events and failures. Our proposed solution produces high detection rates for the different families of uncertain events. These results demonstrated that deep transfer learning is effective in discovering and assessing cyber-attacks in IoT networks. For this analysis, we used the TON_IoT dataset, and our model produced an average detection rate of 97%.

Keywords: Cyber Resilience · Risk Matrix · Intrusion Detection · Transfer Learning · MITRE ATT&CK®

1 Introduction

The Internet of Things (IoT) has emerged as one of the most intriguing and rapidly developing technologies of our time, offering the potential to integrate

K. Arai (Ed.): FTC 2023, LNNS 816, pp. 199–215, 2023.
https://doi.org/10.1007/978-3-031-47448-4_15

with and enhance practically every aspect of our lives, including our homes, vehicles, cities, sports, retail establishments, institutions of higher learning, and industrial manufacturing. IoT is expected to improve productivity for individuals and businesses, resulting in time, energy, and cost savings. Lightweight internet-connected devices, or IoT, link the physical and digital worlds to provide organisations and end users with automated services [1]. By 2030, it is anticipated that there will be 50 billion IoT devices [2]. In light of this, by 2025, 79.4 ZB of the 175 ZB of internet data created will come from IoT devices [3].

In cyber resilience, it is recognised that uncertain events including cyber attacks and IoT system failures are complex phenomena that require an interdisciplinary approach [4]. To address them effectively, there is a need to understand the technological, organizational, and human aspects of these events. The scientific study of uncertain events including cyber attacks and IoT system failures helps to develop cyber resilience solutions to enhance the security, reliability, and sustainability of IoT networks that support our daily activities [5]. IoT networks have become vulnerable due to the fast increase in the number of connected devices and sophisticated attack techniques and system failures in their functionalities. Traditional cyber hardening systems, like intrusion detection, firewalls, and encryption, are not enough to keep systems safe from today's skilled attackers. Resilience strategies are an important part of a new security paradigm the so-called, cyber resilience. Cyber resilience in IoT networks refers to the ability of the network to resist and recover from cyber attacks or system failures while continuing to function and provide services to its users [6]. This requires a combination of security measures, such as intrusion detection, encryption and access controls, as well as mechanisms for continuous monitoring and rapid response to potential threats or breaches. Additionally, ensuring the resilience of IoT networks requires collaboration across multiple stakeholders, including device manufacturers, network operators, and end-users, to identify and address vulnerabilities and ensure the ongoing security of IoT networks.

To offer a resilient IoT ecosystem, an IoT system must be able to anticipate and adapt to changing conditions, as well as endure and quickly recover from uncertain events, such as cyber-attacks and system failures [5]. Aside from military applications, the impact of recent cyberattacks on healthcare and utility services as well as logistical and supply chain systems shows the necessity of cyber-resilience in civil systems. Resilient systems can promptly restore lost functionality in line with the organization's mission or goals in the case of an attack. The assumption that a system's resilience is such that functionality degradation resulting from an attack is zero is typically naive. A resilient system, on the other hand, would lose functionality slowly and get it back quickly. As a result, the level and pace of deterioration, recovery, and adaptability must be monitored [5]. The effectiveness of cyber security measures in reducing the cyber threats and failures associated with IoT deployments is a crucial subject for government policy and private sector business strategies relating to IoT-linked devices, platforms, and services. Cyber security risk assessments that are thorough and accurate can help organisations determine the severity of threats, determine how urgent it is to maintain cyber security, and allocate or manage resources effectively [7].

An effective cyber threat assessment helps a lot with figuring out the levels of possible threats, figuring out how urgent it is to keep cyber security up to date, and allocating or managing resources in the right way. In offline threat assessment, it is common to model several forms of targeted attacks, such as DDoS, MITM, and middle attack on the various types of measurements and control inputs. Online quantitative threat assessment strategies include probabilistic [8] and game theoretic [9], as well as stochastic Petri-net [10]. Quantifying threats of uncertain events entails evaluating the costs associated with various attack types and IoT system failures. Quantifying threats would necessitate sophisticated and exhaustive research to identify the frequency of the various cyberattack types and the breach expenses. The development of a cyber threat matrix improves the measurement of threats. The cyber threat matrix in cyber resilience for IoT networks refers to a framework that identifies and prioritizes potential uncertain events associated with the deployment and use of IoT devices. This matrix helps organizations to better understand the likelihood and impact of various cyber threats and develop a comprehensive threat assessment methodology that includes measures for detection, prevention, response, and recovery.

Research Motivation. In order to identify uncertain events, deep learning-based attack detection methods have proved their ability to extract complicated patterns from a massive number of labelled data used to train classification models. Proposed cyber security methods should integrate cyber threat matrix that takes into account factors, such as the type of IoT devices, their connectivity options, the data it collects and handles, and the potential consequences of cyber attacks. By leveraging this matrix, organisations can mitigate threats and promote greater cyber resilience in their IoT networks. Yet, there is a lack of such a large labelled data set for either unknown (zero-day) or known attack families in IoT systems [11]. While deep learning approaches are mainly utilized for detecting intrusions, resilience models can recover from faults, both partially and fully. However, once the model has been trained, deep learning algorithms do not enable users to comprehend the rationale behind a decision.

Several cybersecurity professionals are worried about the lack of transparency and cyber-resilience in certain suggested solutions for discovering intrusions from IoT networks [5]. To enhance the identification of intrusions in networks that have limited and uneven datasets, transfer learning (TL) [12] has emerged as a useful technique to aid in detecting zero-day attacks and IoT system failures. Recent advancements in machine learning have enabled the transfer of knowledge from one domain, known as the "source domain", to another domain, referred to as the "target domain". On the other hand, a learning model that can offer both a local and global explanation of a model's behaviour is considered the most comprehensive element. A comprehensive explanation entails modelling the connection between the input and output domains to understand the outcomes of deep learning models and explanation of uncertain events [13].

Research Contribution. The objective of this study is to design an explainable intrusion discovery framework, along with a cyber threat matrix for assessing the cyber resilience of IoT networks. An explainable intrusion discovery is developed that provides both local and global visual explanations of data features for the prediction, and explanations for misclassifications. In this framework, deep transfer learning is employed to decrease the false positive rate and enhance the detection accuracy of uncertain events in IoT networks. The main contribution of this paper is organized as follows:

- We propose an explainable attack discovery framework with a cyber threat matrix for assessing uncertain events including cyber attacks and IoT system failures. The suggested methodology is capable of quickly and accurately identifying numerous cyber threats to heterogeneous IoT networks.
- We demonstrate the use of the Deep SHAP and LIME methods to explain the intrusion discovery method in IoT networks.
- We evaluate the proposed framework using the ToN-IoT dataset demonstrating its ability in discovering various cyber threats.

2 Background and Related Work

2.1 Cyber Resilience

Cyber resilience using autonomous agents has become a major area of study and application [14], but it's still hard to measure it objectively, especially how much autonomous agents help systems handle, recover from, and adapt to disruptions. However, it is only when appropriate resilience measures have been designed and put into place that the cyber-resilience of a system can be effectively understood. The cyber threat matrix in cyber resilience provides a framework for measuring a system's capacity to withstand attacks and metrics for assessing its resilience. This matrix estimates the attack and failure of IoT systems, enabling enterprises to identify areas of vulnerability and prioritise expenditures in risk mitigation and contingency planning.

Using probabilistic evaluations from subject matter specialists to replace solely qualitative judgments with more objective estimations is an alternative method [15]. Quantitative risk estimation is an approach used by Ashok et al. [16] to describe the behaviour and interaction of the system through mathematical models to quantify threats. In [17], a wargame approach was developed that takes into account attacker and defender incentives, the likelihood of an attacker breaching defences, and the attacker and defender's respective strategies, costs, and utility.

Cyber wargaming and red teaming are sometimes used interchangeably [17]. Both have a long history of usage in military planning and decision-making [18], and both are used to identify adversarial teams (red and blue) that compete to evaluate the most effective protection measures against known cyberattacks. Alternatively, researchers, such as in [19], differentiate between wargaming, where groups practise attack and defensive techniques in a variety of environments (table-top, computational, etc.), and red teaming, which refers to cyber-attacks and defences on actual systems. These methods emphasise monitoring for vulnerabilities, but they do not address other aspects of cyber-resilience including the capacity to recover from and react to disturbances.

2.2 Internet of Things (IoT)

The architecture of an IoT is defined by the connectivity of devices, how and where data from sensors and devices are kept, how this data is processed, and how the user may interface with the appliances and vice versa. The architecture of IoT comprises three layers: application, network, and perception [20]. Application services, such as smart cities and smart grids, are provided by the application layer, which is the highest level of the architecture. With Wi-Fi, ZigBee, and Bluetooth being just a few of the many communication protocols that may be used to link devices, the network layer is the most important. Physical sensors are controlled by sensors and actuators, known as the perception layer (also known as the sensor layer). Wireless sensor networks with a high number of sensor nodes spread across a vast area comprise the second component of the perception layer, which includes sensor nodes like RFID tags. One of the main reasons for the security difficulties is the enormous number of nodes (i.e., hosts or devices linked to the IoT), which means that a single node might cause the whole system to fail. The most common security threats to IoT devices are botnets, DDOS, routing attacks, remote recording, ransomware, and data leakage. Because of the wide variety and complexity of IoT designs, using a firewall as the first line of security against attacks on IoT devices is not a viable option.

Low-latency, secure, privacy-preserving, and reliable processing of IoT data is essential for the development of efficient and insightful IoT applications. Centralized machine learning methods, which involve sending data from several IoT devices to a single server, are therefore outdated [21]. On the other hand, it is generally time-consuming and costly to acquire massive volumes of newly labelled training data in order to maintain DL-based detection models. Another challenge is that every time a new attack is found, the DL models must be retrained on the new data from scratch, which is time-consuming and taxing on computational resources [12]. As a solution to the problem of insufficient training data for new attacks, transfer learning (TL) is worth exploring. Through TL, it is possible to transfer the characteristics and knowledge of a trained source model to an untrained target model.

2.3 Intrusion Discovery

Intrusion discovery is an important component of securing systems against a variety of risks that might compromise their privacy, credibility, and availability [22]. As new threats and vulnerabilities emerge on a regular basis, so does the potential for key assets to be compromised. Instead of relying exclusively on Intrusion Detection Systems (IDS), which can only detect known assaults, it is essential to implement a system that can identify and find new attacks before they do major damage. To do this, a strong attack detection system must be built that can monitor network data, recognise the abnormal activity, and detect previously unknown forms of threats. As a result, rather than depending exclusively on IDS, it is essential to build sophisticated attack detection systems that can proactively identify new attack strategies and promptly respond to them to mitigate the impact. By combining the power of AI-based approaches with advanced analytics, such systems can provide comprehensive protection against both known and undiscovered assaults, making them an indispensable element of any modern cybersecurity strategy.

Training models for AI often involve labelled IoT records for purposes such as distinguishing normal from malicious activity (binary classification) or distinguishing between attack types (non-binary classification) in the context of normal traffic. To create a reliable IDS for massive amounts of IoT data, Yang et al. [23] used the k-nearest neighbours (kNNs) method. To find the best method for detecting intrusions, Gao et al. [24] employed LSTM and feedforward neural networks (FNN) in an ensemble setting. The LSTM for analysing temporal patterns in IoT data and the Gaussian Bayes approach for evaluating the chance of having outliers were both introduced by Wu et al. [25]. However, a high false positive rate and vulnerability to zero-day attacks are two major problems with these current intrusion detection models.

Transfer learning would address this issue by modifying the model based on prior knowledge to make it flexible to new tasks. In addition, DL has been successfully used in a wide range of situations where a large amount of dispersed training data with input characteristics that are equivalent to the testing data is available. Yet, data might be hard to come by or prohibitively expensive to obtain under certain conditions. Using data from a similar source domain with a large number of labelled examples might be useful for training a high-performing learner in a domain with sparse or no labelled data. This is the foundation of transfer learning [26].

2.4 Explainable AI for Intrusion Discovery

Explainability can be divided into two types: global Explainability and local Explainability [27]. Global explanations assist us in comprehending the model's overall behaviour and logic that leads to projected results. The term "global model explainability" refers to the ability to comprehend the prediction distribution as a consequence of the features, in response to how the trained model makes predictions. To clarify why the model reached a certain result, the reasons for a single prediction for local explanations are presented. Local explanations can become more precise than global explanations. This is because rather than having a complicated relationship with certain features, the prediction relies on them just locally.

For instance, [28], the authors demonstrate how the Shapley Additive Explanation (SHAP) [29] framework can be utilised to determine which aspects of their model are most crucial for various sorts of threats. Li et al. [30] also came up with a way to explain the predictions of anomaly-based IDS by using a local nonlinear approximation of the IDS model's decision boundary. They showed that their approach could work by explaining the predictions of an RNN-based IDS model. Nevertheless, the LEMNA explanation approach [31] they utilised in their study does not adequately account for the dependencies between different input samples. These earlier studies did not compare their work to deep learning models, nor do they offer any metrics to back up their assertions of improved credibility and explicability.

3 Proposed Methodology

The paper proposes a deep transfer learning (DTL) for an explainable and resilient intrusion discovery methodology in IoT networks. The framework involves two phases, as shown in Fig. 1. The first phase is for training on the source domain, while the second phase is for training on the target domain (i.e., the transfer learning phase).

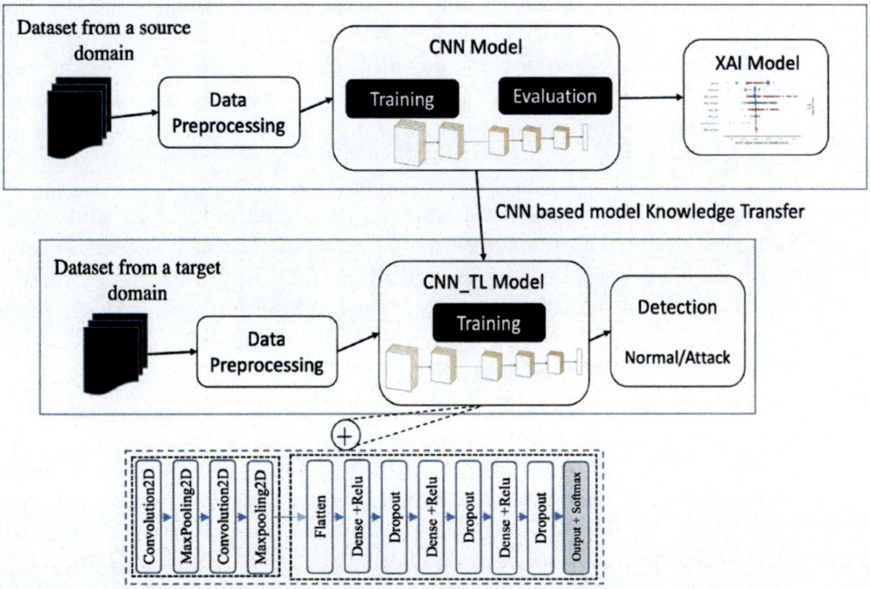

Fig. 1. An explainable framework for attack discovery.

For the training set, the attack discovery model for heterogeneous IoT applications is the one with the best prediction performance. We train a Convolution Neural Network (CNN) model using a dataset chosen at random and validated it. To explain the feature significance of the trained classification model, this CNN model further applies the SHAP and LIME models to explain the data features and design the cyber-threat matrix. The framework uses what it has learned from a trained CNN model, which uses a dataset of the target domain to validate how well it works. Algorithm 1 presents the steps of the proposed CNN-based Deep Transfer Learning (DTL) while Algorithm 2 describes the procedure of the explainable intrusion using an explainable DTL model. To dynamically identify the severity of an attack, the DTL model is used to identify uncertain events in IoT networks. The same model utilizes SHAP and LIME methods to identify which attributes mostly impact these attacks positively or negatively, and the MITRE ATT&CK framework can be used to identify the tactics and techniques used in the attack. Based on this information, the model assigns severity levels to each attack. This enables the automated design of the cyber threat matrix that can be developed according to the severity and likelihood of each attack, reflecting the resilience of IoT networks.

3.1 Data Preparation

Getting optimum performance and enhancing the learning method require cleaning and preparing datasets prior to implementing deep learning-based intrusion discovery techniques. Generally, data preparation requires removing irrelevant features and examining independent variation characteristics, converting non-numerical characteristics, eliminating outliers, and if present, restoring missing numbers. The fundamental processes or data preparation includes data cleaning and, data transformation, as discussed below:

- Data Cleaning – we employ numerous ways to eliminate noise and rectify discrepancies within the information. For example, Rosner's Test for identifying outliers and the predictive mean matching approach for estimating missing data. We've translated qualitative qualities into numerical values. Some features have been excluded from feature vectors since they could lead to overfitting of the training data.
- Data Transformation – we implement this processing phase to produce more efficient procedures and improve pattern recognition. Some selected features have bigger values than others, resulting in inaccurate performance; nonetheless, some models may prefer the larger feature values. We have used these tactics to scale the selected feature values within the range [0.0, 1.0] without altering the data's features. As demonstrated by the following Eq. (1), a minimum-maximum normalization function is used to normalize the individual feature values within the particular range.

$$N_{\text{ormalized}}\, V_{\text{alue}} = \frac{(X - X_{\min})}{(X_{\max} - X_{\min})} \tag{1}$$

X represents an initial value, while X_{max} and X_{min} denote the upper and lower limits, respectively, of the feature.

Algorithm 1: Training DTL Model

Input: source dataset $D = \{(\mathbf{x}_k, \mathbf{y}_k)\}_{k=1}^{m}$,
pretrained CNN model,
epochs φ
learning rate η
Output: adapted CNN model$f$, shaptop N features
function $DTL(D, CNN, \varphi, \eta)$
parameters $\leftarrow$ load_model (CNN, pretrained $=$ true)
parameters $\leftarrow$ initialize (parameters, layer)
repeat
for all $(\mathbf{x}_k, \mathbf{y}_k) \in D$ **do**
activation $\leftarrow$ forward_propagation (x_k, parameters) (2)
cost $\leftarrow$ loss_function(activation, y_k)
gradient $\leftarrow$ back_propagation(activation, cost)
parameters $\leftarrow$ weight_update(parameters, gradient, η)
end for
until n times
end function
shaptop N features $-$ SHAP values for each feature
for each $i \in$ top N features do
explainer $\leftarrow$ shap.DeepExplainer ($f, \mathbf{x}_1 \cdot .k$)
shaptop N features $[i] \leftarrow$ explainershapvalues(x, i)
return shaptop N features

Algorithm 2: Target Domain Classification

Input: target dataset $D = \{(\mathbf{x}_t)\}_{t=1}^{m}$
Trained DTL model
Result: y_t : Label x_t (3)
x_t input of DTL
d_t representation of x_t at the DTL bottleneck layer
$y_t = \text{softmax}\,(d_t)$
return y_t

3.2 Training Process

The training process consists of two basic phases: (1) creating an intrusion detection (ID) model in the source domain, and (2) updating that model by transferring foundational information to the target domain.

- Training Phase: building an ID-model of the input domain is the first stage in the TL-based architecture. This is accomplished by designing and training it on the primary dataset. The validation dataset derived from the source domain is used to ensure the model's accuracy. This CNN model uses a flattening layer, a dense layer, and two pooling layers to lower the sample size of the convolution output and aid in selecting the most valuable features, hence improving the quality of the subsequent learning stage. The convolutional layers in this model include 32 and 64 filters, respectively. In the completely connected layer, we employ an activation function called ReLu. Dual-output softmax activation at the output layer classifies packets as either benign or malicious.
- Transfer Phase: phase two of the TL-based approach involves using what was learnt in the source domain to inform decisions in the target domain. CNN-TL uses the original base ID-convolutional model's base as a frozen input to the classifier. In order to prevent changes to the weights during model retraining, the convolutional base is frozen. Figure 1 depicts the CNN2D-DTL model's architecture, which comprises the frozen convolutional foundation of the CNN model followed by concatenated fully-connected layers as the output layer. In order to prevent overfitting, the dropout is randomised but kept at a constant probability. In the unobserved nodes of a fully linked network, the ReLu activation function is selected. Up to the last "output" layer, the total number of neurons in the network gradually decreases. A softmax activation function is included in a fully linked network at the output layer. CNN-TL is trained on a dataset specific to the task at hand.

The Proposed DTL Model's Architecture. The explainability-enabling network architecture of the proposed CNN2D DTL model is shown in Fig. 1, and its hyperparameters are listed in Table 1. There is never any change to the input form of the network. Excluding the input layer, our suggested model has four total layers. There are four distinct parts to the model's architecture. Each successive layer uses fewer filters, with the first layer combining 32 filters with a kernel length of 8, the second layer 64 filters with a kernel length of 8, the third layer 128 filters with a kernel length of 5, and the fourth and final layer 64 filters with a kernel length of 3.

Each layer consists of a 2-D convolution layer followed by a feature smoothing layer comprised of a BatchNormalization() and Rectified Linear Unit (ReLU) activation function (FSL). Pretrained tensors can learn features from a fresh training dataset after the FSL smooths them while maintaining their original form. The FSL is put after the pre-trained residual blocks to accomplish this smoothing. The feature extraction layer (FEL) uses the preceding three steps to create a one-dimensional feature map with twice as many channels. By isolating critical data and operation-based characteristics from the preceding layer, the FEL improves the final fully linked layer.

Table 1. The proposed model's hyper-parameters.

Hyper-parameter	Function/Value
Hidden Layers	4
Number of units in the hidden layers	32,64,128,64
Batch size	32
Number of Epochs	100
Activation Function of a Hidden Layer	ReLu
Activation function at the output layer	Softmax
Dropout	N/A
Optimizer	Adam
Loss_function	Categorical Crossen_tropy

The output of the previous block is used as input in the next block, which then performs nonlinearity operations on it to transform it into a multivariate series with dimensions equal to the number of filters in that layer. The fifth layer is an averaging procedure performed by calling MaxPooling2D() on the input from the previous block. This allows a class activation map to be used to interpret learnt features while drastically reducing the number of parameters in a deep model. A softmax classification layer with the same number of neurons as the number of classes in the dataset receives the output from the gap layer.

Overfitting can be avoided with the use of a number of techniques, such as data augmentation, the introduction of Gaussian noise during training, the shrinking of networks, and the Early-Stopping method. The inclusion of skip

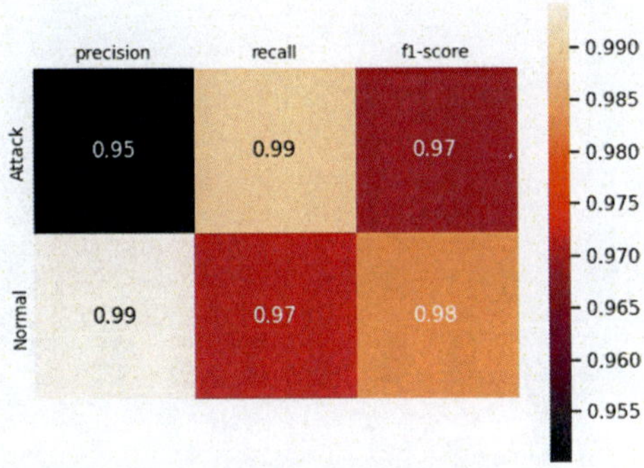

Fig. 2. Confusion matrix of the test/predicted attacks using TL.

connections increases training and convergence stability, as well as encourages feature reuse. Our proposed model uses a vector addition method to generate skip connections, which keeps the gradient in prior layers by multiplying it by 1, allowing us to isolate its DTL effect.

After the detection of uncertain events, the SAHP and LIME models are applied to explain the behaviour of each feature in the trained model. By relying on the explanation of the features and utilizing the techniques and tactics of each threat gathered from the MITRE ATT&CK framework, the severity levels of the threats can be identified. This, in turn, can lead to the development of a cyber resilience matrix.

4 Experimental Results and Discussion

In this experiment, we put the TL solution through its paces on the TON_IoT dataset (which comprises both known and zero-day attacks). We first conduct an evaluation of the TL model using the TON_IoT dataset to verify the model's effectiveness at detecting zero-day threats. Figure 2 displays the model's accuracy, precision, recall, false positive rate (FPR), and F1-score. The model has a 97% accuracy, a 96% precision, a 98% recall, a 0.05% FPR, and an F1-score of 97%. Figure 3 displays the number of detected samples (TP + TN), the number

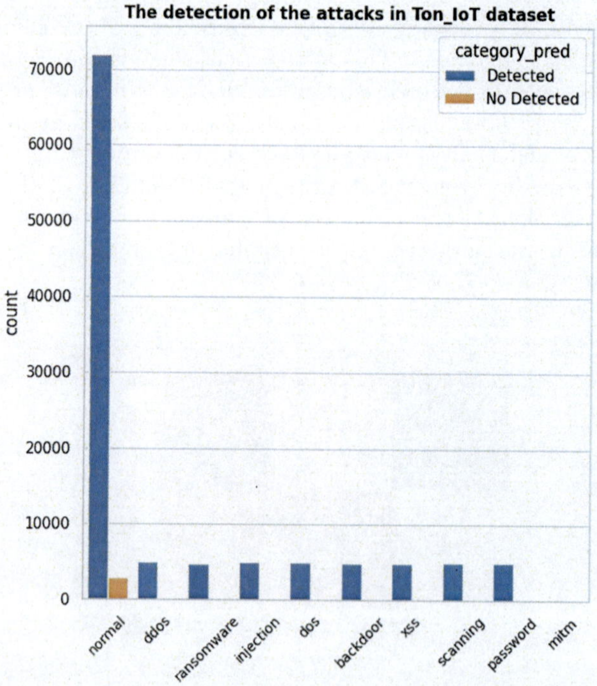

Fig. 3. The detection of the attacks in Ton_IoT dataset.

of non-detected samples (FP + FN), and the detection rate (DR) for each assault type, demonstrating that the DR is greater than 97% in every scenario. Figure 4 presents the overall average accuracy, precision, recall, and F1-score obtained by our proposed CNN2D_DTL method in comparison to traditional ML model Naive Bayesian (NB), and 1D and 2D CNN models. The findings show that although requiring more training data than NB, our suggested strategy outperforms standard ML algorithms in identifying attacks from IoT datasets. The variation of scores across models shows that our model has better scores than any other model, except for the CNN2D model which has a better precision score. However, in the other three stages, our models performed well.

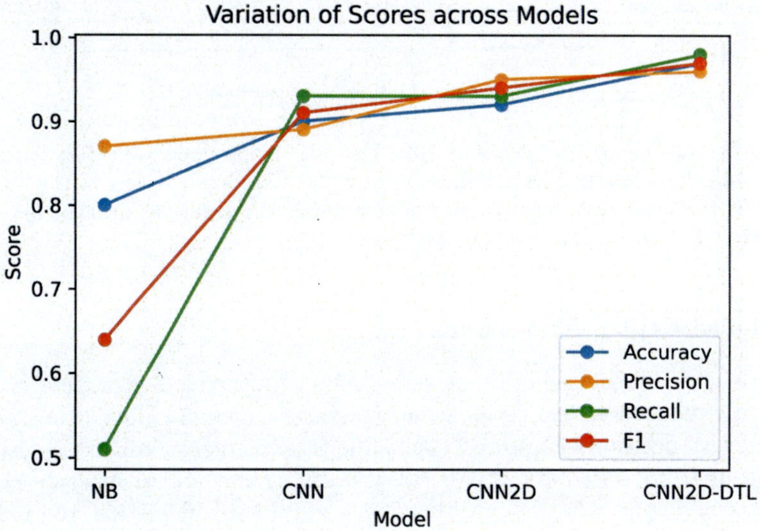

Fig. 4. Comparing the efficacy of our proposed approach against alternative methods.

4.1 Global Explanation Result

This section discusses the outcomes of relevant features and also addressed the differences in the significant features obtained by SHAP for the deep learning model. Figure 5 illustrates the model's top eight extracted features for each threat. The feature determines the y-axis position, whereas the Shapley value determines the x-axis location. The colour indicates the relative importance of the attributes from low to high. As the red colour intensity grows, the value of the characteristics raises as well. On the other hand, when the blue color's intensity grows, the value of the characteristics falls. Overlapping dots are jittered in the y-axis direction to depict the Shapley value distribution for each feature. The features are listed in ascending order of significance. In Fig. 5, proto, dns_query and dns_RA have a larger rate of red colour means these features are more likely to be seen as a threat to the model.

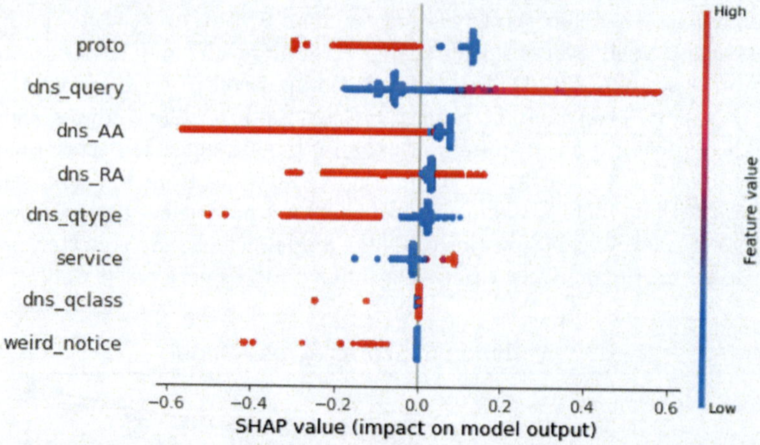

Fig. 5. The model's SHAP summary plot provides a top-down overview of the eight most significant features. According to our findings, the proto feature of the TON_IoT dataset is the most important predictor of whether the traffic is an intrusion or not, followed by the dns_query, dns_AA attributes.

4.2 Local Explanation Result

SHAP local explanations evaluate only a single occurrence at a time and generate an explanation, indicating which feature values are influencing positive and negative decisions. Figure 6 depicts a local explanation where the chance of the output being an attack is 0.01, as well as features and their corresponding values, such as dns RA, dns AA, dns query, etc. In this depiction, red indicates features that raise the forecast, while blue indicates features that reduce it. These explanations will vary as the input instance is modified.

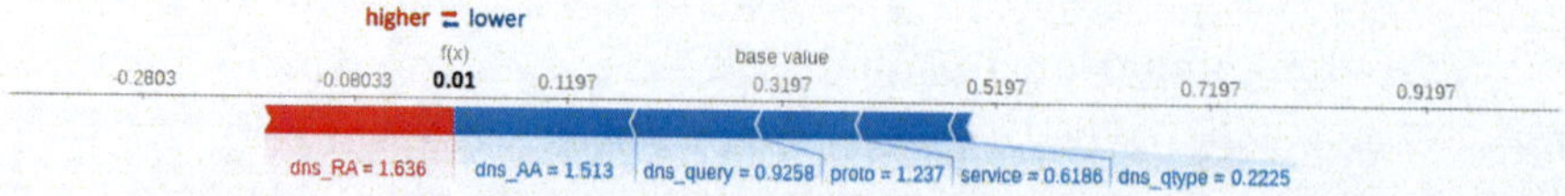

Fig. 6. SHAP force plot is used for local explanations to explain a specific case where the output probability of "attack" is 0 and shows factors that affect decision-making.

LIME employs a human-comprehensible representation regardless of the model's actual features. This method explains the most responsible features by using sparse linear models in the local area of an individual prediction. Which features are more responsible for the threats that can be identified with some values. In Fig. 7, eight of the most critical features are highlighted in two distinct colours (orange/blue). The orange colour indicates the relative importance of the features for the attacks. As seen, proto is more responsible for the attacks of this IDS framework than dns_qtype and dns_RA. On the other hand, blue colour indicates that these features are not responsible for the threats to the system.

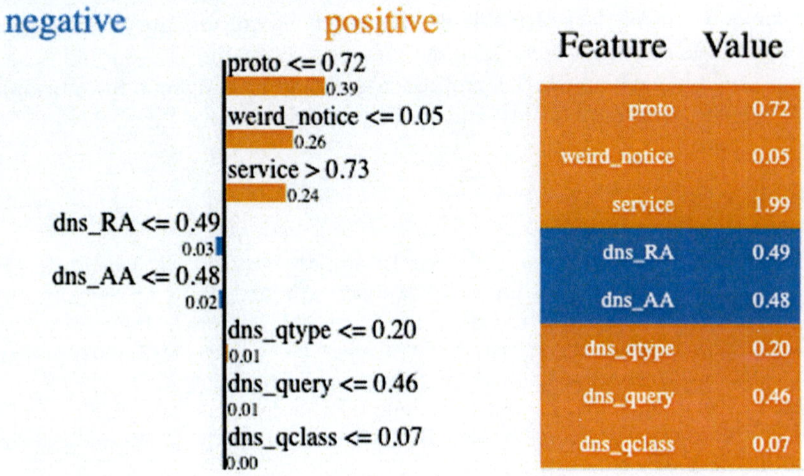

Fig. 7. LIME is used to explain each prediction of a deep learning classifier. (Color figure online)

The training time for a new model may be drastically cut down by reusing models that have already been trained with this method. As enormous volumes of data are typically required for training purposes in the field of attack discovery, this is extremely helpful. Applying the source task's pre-trained information to a new model can improve its performance. As a result, we may be able to generalise more effectively and prevent evasive assaults. This strategy reduces the quantity of required labelled data for model training since it makes use of existing, pre-trained models. When labelled data is sparse or hard to get by, this is very helpful. Yet, there are drawbacks to this approach as well. The characteristics taken from the previously trained model serve as constraints for this model. Accuracy issues may arise if the pre-trained model is inappropriate for the intended application.

5 Conclusion

In this paper, we have implemented transfer-learning-based explainable and resilient attack discovery in IoT networks with limited and imbalanced datasets. In order to do this, we create an effective framework for intrusion detection that includes knowledge transfer and model improvement and has great detection accuracy for both well-known and undiscovered cyberattack types. We put the solution into practice by using the TON_IoT dataset as the target domain and the BoT-IoT dataset as the source domain to gain knowledge. This proposal makes an attempt to deal with difficulties that might be applied to the evaluation of cyber-resilience in systems that employ autonomous agents. This strategy, together with the concept of monitoring the dynamics of functionality throughout time is a crucial part of resilience. However, advancements in measurement are required as cyber-physical systems increasingly include autonomous agents. Several possible gaps and future approaches for cyber resilience have been identified, necessitating more study and inquiry in future works.

References

1. Mahdavinejad, M.S., Rezvan, M., Barekatain, M., Adibi, P., Barnaghi, P., Sheth, A.P.: Machine learning for internet of things data analysis: a survey. Digit. Commun. Netw. **4**(3), 161–175 (2018)
2. 21+ internet of things statistics, facts amp; trends for 2022 (2022). https://findstack.com/resources/internet-of-things-statistics
3. Jo, J., Jeong, S., Kang, P.: Benchmarking GPU-accelerated edge devices. In: 2020 IEEE International Conference on Big Data and Smart Computing (BigComp), pp. 117–120. IEEE (2020)
4. Shiaeles, S., Kolokotronis, N.: Internet of things, threats, landscape, and countermeasures, pp. 291–334 (2021)
5. Linkov, I., Kott, A.: Fundamental concepts of cyber resilience: introduction and overview. In: Kott, A., Linkov, I. (eds.) Cyber Resilience of Systems and Networks. RSD, pp. 1–25. Springer, Cham (2019). https://doi.org/10.1007/978-3-319-77492-3_1
6. Huang, Y., Huang, L., Zhu, Q.: Reinforcement learning for feedback-enabled cyber resilience. Annu. Rev. Control. **53**, 273–295 (2022)
7. Li, Y., Liu, Q.: A comprehensive review study of cyber-attacks and cyber security; emerging trends and recent developments. Energy Rep. **7**, 8176–8186 (2021)
8. Radanliev, P., et al.: Future developments in cyber risk assessment for the internet of things. Comput. Ind. **102**, 14–22 (2018)
9. Musman, S., Turner, A.: A game theoretic approach to cyber security risk management. J. Def. Model. Simul. **15**(2), 127–146 (2018)
10. Ferreira, C., Neves, L.C., Silva, A., De Brito, J.: Stochastic petri net-based modelling of the durability of renderings. Autom. Constr. **87**, 96–105 (2018)
11. Bilge, L., Dumitras, T.: Before we knew it: an empirical study of zero-day attacks in the real world. In: Proceedings of the 2012 ACM Conference on Computer and Communications Security, pp. 833–844 (2012)
12. Mahdavi, E., Fanian, A., Mirzaei, A., Taghiyarrenani, Z.: ITL-IDS: incremental transfer learning for intrusion detection systems. Knowl.-Based Syst. **253**, 109542 (2022). https://doi.org/10.1016/j.knosys.2022.109542
13. Lundberg, S.M., Lee, S.-I.: A unified approach to interpreting model predictions. In: Advances in Neural Information Processing Systems, vol. 30 (2017)
14. Ligo, A.K., Kott, A., Linkov, I.: How to measure cyber-resilience of a system with autonomous agents: approaches and challenges. IEEE Eng. Manage. Rev. **49**(2), 89–97 (2021)

15. Hamdan, B.: Simulating cybersecurity risk using advanced quantitative risk assessment techniques: A teaching case study. In: Journal of The Colloquium for Information Systems Security Education, vol. 10, pp. 5–5 (2023)
16. Ashok, A., Govindarasu, M., Wang, J.: Cyber-physical attack-resilient widearea monitoring, protection, and control for the power grid. Proc. IEEE **105**(7), 1389–1407 (2017)
17. Colbert, E.J., Kott, A., Knachel, L.P.: The game-theoretic model and experimental investigation of cyber wargaming. J. Def. Model. Simul. **17**(1), 21–38 (2020)
18. Abbass, H., Bender, A., Gaidow, S., Whitbread, P.: Computational red teaming: past, present and future. IEEE Comput. Intell. Mag. **6**(1), 30–42 (2011)
19. Russo, L., Binaschi, F., De Angelis, A., Armando, A., Henauer, M., Rigoni, A.: Cybersecurity exercises: wargaming and red teaming. Next Gener. CERTs **54**, 44 (2019)
20. Issa, W., Moustafa, N., Turnbull, B., Sohrabi, N., Tari, Z.: Blockchain-based federated learning for securing internet of things: a comprehensive survey. ACM Comput. Surv. **55**(9), 1–43 (2023)
21. Ren, H., Li, H., Dai, Y., Yang, K., Lin, X.: Querying in internet of things with privacy preserving: challenges, solutions and opportunities. IEEE Network **32**(6), 144–151 (2018). https://doi.org/10.1109/MNET.2018.1700374
22. Khraisat, A., Gondal, I., Vamplew, P., Kamruzzaman, J.: Survey of intrusion detection systems: techniques, datasets and challenges. Cybersecurity **2**(1), 1–22 (2019)
23. **Yang, H., Liang, S., Ni, J., Li, H., Shen, X.S.: Secure and efficient k NN classification** for industrial internet of things. IEEE Internet Things J. **7**(11), 10945–10954 (2020)
24. Gao, J., et al.: Omni SCADA intrusion detection using deep learning algorithms. IEEE Internet Things J. **8**(2), 951–961 (2020)
25. Wu, D., Jiang, Z., Xie, X., Wei, X., Yu, W., Li, R.: LSTM learning with Bayesian and gaussian processing for anomaly detection in industrial IoT. IEEE Trans. Industr. Inf. **16**(8), 5244–5253 (2019)
26. Zhuang, F., et al.: A comprehensive survey on transfer learning. Proc. IEEE **109**(1), 43–76 (2020)
27. **Shin, D.: The effects of explainability and causability on perception, trust, and** acceptance: implications for explainable AI. Int. J. Hum.-Comput. Stud. **146**, 102551 (2021)
28. Wang, M., Zheng, K., Yang, Y., Wang, X.: An explainable machine learning framework for intrusion detection systems. IEEE Access **8**, 73127–73141 (2020)
29. Lundberg, S.M., Erion, G., Chen, H., DeGrave, A., Prutkin, J.M., Nair, B., Katz, R., Himmelfarb, J., Bansal, N., Lee, S.-I.: From local explanations to global understanding with explainable AI for trees. Nat. Mach. Intell. **2**(1), 56–67 (2020)
30. Li, H., Wei, F., Hu, H.: Enabling dynamic network access control with anomaly-based IDS and SDN. In: Proceedings of the ACM International Workshop on Security in Software Defined Networks & Network Function Virtualization, pp. 13–16 (2019)
31. Guo, W., Mu, D., Xu, J., Su, P., Wang, G., Xing, X.: Lemna: explaining deep learning based security applications. In: Proceedings of the 2018 ACM SIGSAC Conference on Computer and Communications Security, pp. 364–379 (2018)

Towards Sustainable ERP Systems: Bridging the Gap Between Current Capabilities and Future Potential

Chantel Raniin Backer, Milou Slaman, Moutaz Haddara, and Marius Langseth(✉)

Kristiania University College, Oslo, Norway
marius.langseth@kristiania.no

Abstract. This study examines the readiness of enterprise resource planning (ERP) systems for supporting and advancing sustainability initiatives and practices in organizations. The study aims to identify the current state-of-the-art of sustainability capabilities and functionalities in ERP systems, as well as explore the future directions of the ERP market. The main research question the study aims at answering is, *are current ERP systems ready to support sustainability initiatives?* This research employs the Dynamic and Sustainable View of the Firm and the Triple Bottom Line frameworks as theoretical lenses. These lenses underscore the importance of integrating economic, social, and environmental sustainability aspects to establish a competitive advantage. An exploratory case study with semi-structured interviews was conducted and the informants included a variety of ERP industry professionals and an end-user. The informants revealed the various stakeholders' perceptions and views on the sustainability phenomenon within the Norwegian ERP market. Our main findings suggest that the current capabilities of ERP systems fall short of meeting the need for sustainability-related functionalities, indicating a gap that presents opportunities for innovation in developing more advanced systems. Additionally, conflicting views exist on who drives innovation in the ERP market, clients, vendors, or the government. Although the research and practice of sustainable ERP are still in their early stages, our findings reveal that the overall requests to incorporate sustainability functionalities within ERP systems are on the rise. Finally, our data suggest that emerging business models rooted in sustainability principles highlight the urgency for ERP systems to evolve and adjust accordingly.

Keywords: Sustainability · ERP systems · TBL · CSR · ERP market

1 Introduction

One of the most common definitions of sustainability was introduced by the Brundtland Commission in 1987, led by the former Prime Minister of Norway, Gro Harlem Brundtland. The commission defined sustainable development as: *"Development that meets the needs of the present without compromising the ability of future generations to meet their own needs"* [1]. Since 2020, Covid-19 has caused a change from 'business as usual' into

K. Arai (Ed.): FTC 2023, LNNS 816, pp. 216–242, 2023.
https://doi.org/10.1007/978-3-031-47448-4_16

a situation where businesses are recommended to create business continuity plans to respond to the disruptions in their current operations [2]. In times like these, companies must critically view how they work. Global megatrends such as resource scarcity, environmental pollution, and technological innovation [3] have changed the expectations of investors, regulators, societies, and employees toward the roles and responsibilities of organizations [4]. However, what does it involve in being sustainable? Sustainability is often associated with recycling, using renewable energy, reducing the carbon footprint, or managing waste [5]. While Brundtland's [1] definition might be comprehensive, Elkington stated that it was rather general and drilled down to three interlinked aspects representing profit, people, and the planet [6]. Elkington proposed that organizations are responsible for more than just creating economic value and should also examine their social (people) engagement and involvement in environmental (planet) initiatives. This model is called the Triple Bottom Line (TBL) [7]. The TBL framework is an accounting structure that involves environmental and social measures beyond conventional revenue measurements and shareholder value [6].

Singh and Sahu [8] argue that the existing literature on information systems (IS) and information technology (IT) shows a knowledge gap within sustainability. This has recently been met with increased research on green IS and IT. While Green IT and IS refer to initiatives and programs that address environmental sustainability in businesses directly or indirectly, sustainability in IS does not only cover the aspect of Green IT [9]. Sustainability IS are also discussing the design, implementation, and maintenance of IS that results in sustainable business processes [10], capturing the dimensions of profit, people, and planet for the benefit of future generations. Previous research has already recognized the value of IS research in reaching sustainability goals [11, 12] supported by an increase in sustainability-related actions [13]. Based on a structured review of eight highly ranked IS journals, Chasin [14] defines sustainability as a *"characteristic of a stakeholder activity (organizational process or individual behavior), which impacts on natural and social environments and meets the (economic) needs of the present without compromising the ability of future stakeholders to meet their needs"*. Bengtsson and Agerfalk [15] suggest promoting sustainability initiatives within organizational operations. Three critical actions are needed; 1) A holistic view of sustainability needs to be actively supported by management. 2) Required data collection should be reflected based on management support. 3) Systems which are ensuring the opportunity for the reporting of sustainability should get adopted.

The rest of the paper is structured as follows: a literature review is provided in Sect. 2. The adopted research methodology and theoretical background of this research are presented in Sect. 3. Section 4 presents the main findings of this study. Section 5 discusses the findings structured through the conceptual framework adopted in this research. Section 6 presents the study's conclusion, limitations, and suggestions for future research.

2 Literature Review

Companies' activities can affect the outside world in ways that do not reflect their market value. Consequently, how companies manage their natural and human resources can affect their ability to create value. According to [18], many companies are on their way to sustainability, more as a business strategy that drives economic value than an environmental initiative. Larson, et al. [19] investigated how sustainable businesses are creating value for multiple stakeholders and improving—even redefining—excellence in business operations and found that sustainable enterprises are not only aligned with innovation but that sustainable companies are also profitable. One of the perceptions on sustainable implementation issues was the poorly implemented programs measuring sustainability, making incremental operating differences [19]. Ihlen and Roper [20] additionally argue that pressure is needed from both government and civil society to stop the concept of 'greenwashing.' They say that, as organizations claim to have reached sustainability and have done so for many years, sustainability falls into the disaster of 'stalling' while change is continuously needed. A call for action is claimed, especially for public policies to enhance the need for [sustainability] change [20]. As part of an evolution of ideas that enable operational excellence, sustainability implies that all outputs, products, and non-products are assets waiting to be recovered, thus eliminating the concept of waste [19]. Kleindorfer, et al. [21] stated that companies would be expected to expand their perspective in managing business operations to include people and the planet and invest in sustainable technologies, operations, and supply chains. Among many reasons for this were: (1) the increasing cost of materials and energy, (2) public pressure for environmental, health, and safety performance, and (3) increasing awareness of the TBL.

2.1 Sustainability Frameworks

This section presents the most common sustainability frameworks that help organizations identify and address sustainability dimensions and practices. However, these frameworks have various advantages and disadvantages, according to diverse literature.

In 1953 through the book '*Social Responsibilities of the Businessman*' [22], Bowen was one of the earliest researchers that kicked off the research and *trend* of social responsibility of businesses. Since then, the term has evolved into how it is known nowadays; *Corporate Social Responsibility* (CSR) [23]. CSR refers to situations in which a company exceeds the requirements of law and engages in activities that promote social welfare beyond the firm's interests. [24]. Garriga and Melé [23] argue that most theories related to CSR focus on the following four aspects: 1) Companies meet their objectives which produce long-term profits, 2) business power is used responsibly, 3) social demands are being integrated, and 4) through ethically correct working the organization contributes to a good society. While CSR is a 'well-developed industry' with hundreds of websites, newsletters, professions, university degrees, reports, and other documentation [25], research on CSR is still undeveloped. It, therefore, lacks alignment in theoretical frameworks, empirical methods, and measurements [24].

The UN's *Sustainable Development Goals* (SDGs) address the global sustainability challenges of assuring human well-being, economic prosperity, and environmental protection [26]. It has been a concept applied by businesses setting their course toward sustainability. Businesses are the driving force for economic growth, employment, technology development, and innovation, and they play an essential role in solving global sustainability challenges [27]. However, as much as the SDGs depend on the business, long-term business success hinges on the realized goals. This has resulted in calls for integrated initiatives around trade, food, business, and other areas of policy alignment to support poverty reduction, human rights, and the environment [27].

According to Schaltegger and Burritt [28], there are six specific explanations for why executives should be motivated to set up an accounting program that offers detail on sustainability issues for evaluating company actions: (1) Greenwashing, (2) mimicry and industry pressure, (3) legislative pressure, stakeholder pressure and ensuring the "license to operate," (4) self-regulation, (5) corporate responsibility and ethical reasons and (6) managing the business case for sustainability. Another aspect is the reporting for the managers to deliver relevant data for decision-making, planning, and control [29, 30]. These aspects reflect the behaviors that affect an organization's success in society, the environment, and the economy. Sustainability accounting is often related to the *TBL*, whereas Elkington [6] also refers to sustainability accounting, which aims to report on an organization's economic, social, and environmental issues. This recognizes the role of financial reporting and shows how traditional accounting is extensive by improving transparency and accountability through reporting on the three aspects.

2.2 Sustainable Enterprise Resource Planning Systems

Academics have begun conceptualizing and exploring 'Sustainable ERP systems (S-ERP) to answer organizations' sustainability initiatives and needs [31]. Chofreh, et al. [32] apply the TBL as a framework and define the S-ERP systems as an ERP that prioritize sustainability and encompass all value chain stages. S-ERP systems are viewed as potential comprehensive, integrated, and holistic solutions for addressing sustainability challenges in business. According to Chofreh, et al. [32], achieving sustainability in organizations requires a holistic, integrative, and complete view spanning the entire supply chain, including the product manufacturing process and its fabrication and the manufacturing systems across several product life cycles. Since the research on the S-ERP system is still in the infancy phase Chofreh, et al. [33], hence the implementation of S-ERP systems is a challenging task, and an effective master plan consisting of a roadmap, framework, and guidelines is required to lead practitioners in implementing the systems.

Simmonds, et al. [34] found that ERP system implementation can increase organizations' ability to capture sustainability data and provide more favorable disclosures, resulting in better sustainability assessment scores. Ursacescu, et al. [5] explored the relationship between ERP systems and green IT practices. They found that companies are generally focused on the operational benefits of ERP systems and less on sustainability benefits. Adriansyah, et al. [35] proposed a sales module dashboard system configured on Sales Order (SO) transactions and customer data to assist in sustainability problem-solving. Hasan, et al. [36] developed sustainability performance indicators that can be utilized to integrate sustainability elements into ERP systems. These performance indicators are divided into the three pillars of economic, social, and environmental dimensions and were argued for as the sustainability elements required for having an S-ERP [36]. In their research, they discovered 63 sustainability indicators, whereas 21 of them represented the economic aspect, 26 represented the environmental indicators aspect, and the remaining 16 represented social indicators that can be implemented as a part of the integration of sustainability elements into an ERP system [36]. Overall, the papers suggest that ERP systems can be designed and implemented with a sustainability approach to promote environmentally responsible organizational initiatives.

Research on whether enterprise resource planning (ERP) systems can provide and support sustainable solutions has mostly been conceptual and scarce [16]. Hence, through empirical research, this paper aims to answer the following question: *Are current ERP systems ready to support sustainability initiatives?*. The objective of this paper has three facets; first, to explore how the (Norwegian) ERP market perceives the phenomenon of sustainability. Second, to explore the main drivers (stakeholders) for sustainability within the ERP market, and third, to explore the current state of ERP systems and their possibilities to support businesses in implementing sustainability initiatives. Competitive forces and high customer demand for product specifications characterize the global business environment. Understanding and optimizing business processes is vital to success in this fast-changing environment [17].

3 Research Design and Methodology

This study employs a qualitative research design. Focusing on qualitative research is particularly applicable to IS studies highlighting managerial and organizational issues associated with innovation in information and communications technology [37]. The research design and process were steered by the guidelines provided by Oates [38]. Figure 1 below summarizes this paper's research design and process. The following sections present the research strategy and theoretical background before it goes into the methodology, including how the data was collected and analyzed.

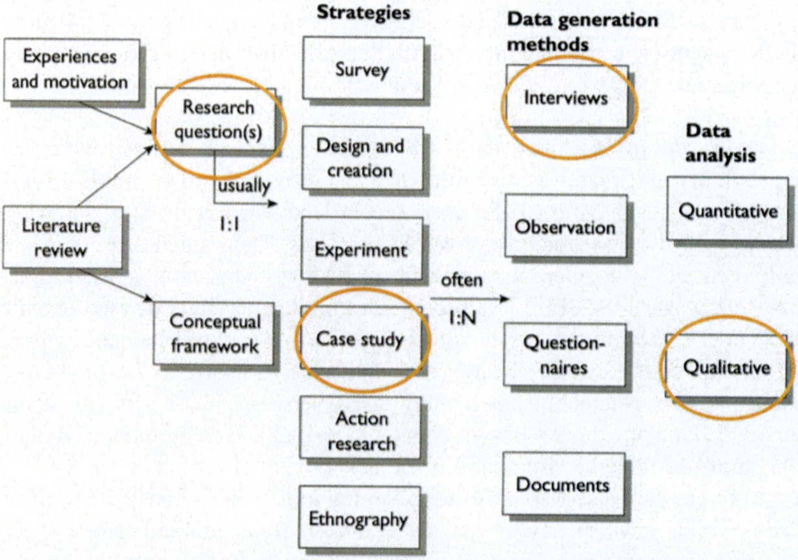

Fig. 1. The adopted research process. Adapted from [38]

3.1 Research Design and Theoretical Background

To properly understand what the researchers call the *sustainability phenomenon* is essential to ensure that business operations do not negatively affect the environment or human habitation [39]. As this research focuses on the status quo, this study can be classified as a short-term contemporary study [40]. To understand the sustainability phenomenon, an exploratory case study was conducted. Exploratory research is adopted to investigate and explain why a specific phenomenon occurs [40]. This type of research is frequently used when there is limited literature on the chosen topic, which could help identify topics for future research. This study also employs a case study research design [38]. A case study enables in-depth analysis, focusing on stakeholders usually connected with ERP systems, such as the ERP consultant and the ERP end-user. What characterizes a case study is its in-depth focus on one instance of the investigated phenomenon [38]. This paper attempts to contribute with new perceptions and deeper insights into sustainable IS, focusing on ERP systems from the consultant and end-user perspectives. While the importance of the consultant within ERP research has been acknowledged, the consultant perspective is often overlooked [41]. Insights on sustainability within ERP systems are presented through the perceptions of ten participants from eight consultancy firms, one client organization, and one end-user.

The study utilizes the Dynamic and Sustainable View of the Firm and the Triple Bottom Line frameworks as theoretical lenses in this research. The primary constructs of these frameworks were used to guide the data collection process, map and classify the findings into relevant constructs, and structure the results section. The Three Bottom Line framework, referred to as the TBL, outlines sustainability as the economic (Profit), the environmental (Planet), and the social (People) bottom line. Each bottom line represents a wide variety of data and analysis [6]. Profit includes economic capital, which refers to how a sustainable corporation assesses its business operations. Businesses must report the profit they attain over a period, published in the income statement. People or the social aspect is probably the most challenging element of the TBL to measure. It assesses the extent to which a business is socially responsible. This also comprises human capital in public health, business, and education, yet it embraces a broader assessment of the opportunity for good welfare and wealth formation. The planet's bottom-line views how business activities impact the environment. Measures of impact on the environment can be problematic, such as a process implication on energy prices or measuring carbon emissions [6].

According to Rodriguez, et al. [42], the *dynamic and sustainable view of the firm* (Fig. 2) framework was developed based on the need for businesses to grasp how an organization is part of an environmental and social ecosystem. Furthermore, it considers factors related to sustainable development strategies and competitive advantage elements. Rodriguez, et al. [42] conducted their study to examine how sustainable development affects how businesses manage their resources, capabilities, and activities, and it fosters the persistence of competitive advantages.

According to Porter [43], competitiveness occurs when businesses have a strategy that can meet international competition but simultaneously meet the living standards of the average citizen. A strong business position will depend on the continuation of its supporting competitive advantages. In a study by Rodriguez, et al. [42], they argue that in

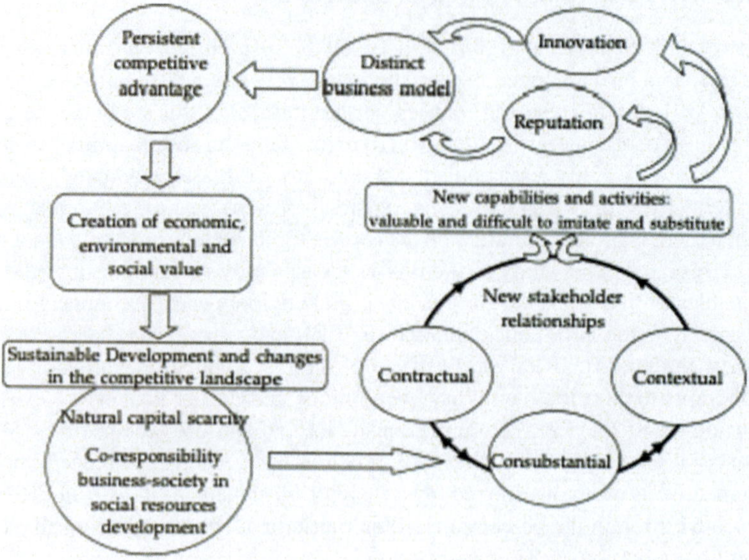

Fig. 2. Dynamic and sustainable view of the firm, adopted from [42]

order to adopt a sustainable perspective, organizations must recognize their role within the interconnected natural and social systems. Thus, the study developed the economic, environmental, and social value creation as the sustainable view of firms' theory, based on the three pillars of people, planet, and profit. Sustainable development presents the concept of the scarcity of natural resources and the co-responsibility of firms and the societies in which they operate to develop and use social and natural resources. The three levels support the development of new resources and activities [42]; (1) Consubstantial, the essential stakeholders for the business, including employees, strategic partners and shareholders, and investors. (2) Contractual stakeholders provide for the financial institutions, customers, suppliers, and subcontractors. (3) Contextual stakeholders represent knowledge and opinion makers, local communities, and government. The assessment of the different stakeholders is based on an open dialogue between the stakeholders and the company (client) to increase revenue persistence. The company's resources, capabilities, and activities must have strategic value to have a persistent competitive advantage. Rodriguez, et al. [42] state that new capabilities and actions must be difficult to imitate, challenging to substitute, and valuable if the company wishes to develop a competitive advantage. Innovation and reputation are two vital sources that support a persistent competitive advantage. However, this can only transpire by accepting the changes in the competitive landscape resulting from sustainable development and improving the resources, capabilities, and activities [42].

3.2 Methodology

Data Collection. The data were collected through interviews, and the data collection phase lasted around three months. Respondents were asked to participate in a one-hour interview and to choose between a face-to-face and an online interview. Six interviews were conducted online, and four were held face-to-face. All the interviews have been conducted in English. Interviews are suitable for research designs that focus on gaining detailed information and insights, including multifaceted and/or open-ended questions and exploring experiences that are not easily observed. As this research focuses on how the participants define and perceive sustainability, as there is a high level of inconsistencies in perception and definitions within the market, thus qualitative interviews were deemed more suitable over quantitative surveys, as they can elucidate the interpretations of informants in the field and collect thick descriptions [38].

This research has employed semi-structured interviews since the interviewees are encouraged to talk more openly about the topic. This type of interview also allows for a more natural 'conversation' in which the order of questions can be adjusted [38]. The interviews have all been digitally recorded after obtaining a signed consent from all informants. The initial selection of participants for the ERP consultants was based on the report 'ERP system in Norway' provided by Devoteam [44]. According to the assessment, the Norwegian market consists of eighteen relevant and actively marketed ERP systems, which target mainly the mid and large enterprise market. Thus, the authors set some conditions for the selection of the cases. For example, the choice of ERP systems to study was initially based on that the vendor must have agents or representatives in Norway. In addition, the ERP systems may not focus on one specific industry, yet, must cover several ERP process areas, including finance [44]. Potential participants were then contacted through LinkedIn, emails, or phone. In total 66 potential participants have been approached to participate in the study. After the first interview iteration, the snowball sampling method [40] was adopted to recruit participants from their ERP networks. Furthermore, two researchers focusing on ERP literature have been interviewed to provide additional knowledge on themes that have received limited previous attention in ERP research, such as sustainable ERP. The researchers' domain and technical knowledge of ERP systems were deemed valuable sources of information for the research, as both participants are also active as ERP consultants. The researchers were asked to provide their views on sustainability and ERP from a consultant perspective. Initially, nine participants in the ERP consultant group were interviewed. Table 1 presents an overview of the professional and consulting informants participating in this research. Participants C and G are the only participants who work for the same organization. Identifying and recruiting end-user participants with experience and knowledge regarding sustainability within the ERP domain was challenging. Initially, twenty-three potential informants within the ERP end-user group were approached. However, only one positive response was received (see Table 2). The data from this end-user was collected to understand how the end-user views sustainability and present insights into the types of business operations that could be affected by sustainability. Fortunately, this participant was responsible for the procurement department's yearly reporting section on Corporate Social Responsibility and had worked with ERP within this company since 2012.

Table 1. Overview of professional and vendor informants

Pcode	Stakeholder group	Position	Industry/market	Type of interview	Length of interview
A	ERP Academic & ERP distributor	Researcher & Consultant	Information Technology & Services/Local	Online	50:45
B	ERP distributor	Consultant	Information Technology & Services/Local	Face-to-Face	1:15:14
C	ERP distributor	Consultant	Information Technology & Services/Global	Face-to-Face	40:16
D	ERP distributor	Consultant	Accounting/Global	Face-to-Face	47:03
E	ERP distributor	Consultant	Computer Software/Local	Online	1:04:52
F	ERP Academic & ERP distributor	Researcher & Consultant	Information Technology & Services/Global	Online	56:09
G	ERP distributor	Consultant	Information Technology & Services/Global	Face-to-Face	1:07:50
H	ERP distributor	Consultant	Information Technology & Services/Global	Online	57:45
I	ERP distributor	Consultant	Computer Software/Local	Online	28:11

Table 2. One participant from which the data were used

Participant code	Stakeholder group	Position	Industry	Type of interview	Length of interview
L	ERP end-user	Procurement lead	Oil & Gas	Microsoft Teams	55:17

Data Analysis. The qualitative data analysis aims to develop an understanding that answers the fundamental question of 'What is going on here?' [45]. The decision to use a qualitative method focused on rich and detailed information where the possibility of alternative explanations can also be explored. Therefore, the researchers contemplated four basic qualitative data analysis techniques: coding, analytical memos, displays, and contextual and narrative analysis. The technique the researchers found best fitting for the project was the coding technique [46], which helped in selecting segments of data and sorting these into emerging categories that facilitate insight, comparison, thematic analysis, and the development of the theory [47]. The adopted coding method is inductive, which refers to approaches primarily using detailed raw data readings to derive concepts, themes, or a model through an evaluator or researcher [48]. The data categorization began by selecting the data connected to the research objectives and themes. The categorization of the codes reflected the themes to get a more precise understanding of the analysis. Examples of themes that were selected during the coding were: "sustainability definition," "ERP systems," "economic initiatives," "social initiatives," "environmental initiatives," "SDGs," and "drivers of sustainability." The coding was done through a color-coding system. The colors visualized the different themes in the data and enabled faster data extraction into different matrix tables [49].

4 Findings

The following section presents the findings organized through the adopted theoretical framework of *the firm's dynamic and sustainable view*. In their study, Rodriguez, et al. [42] identified three aspects that compose what they labeled as the dynamic and sustainable view of the firm. Namely, the three aspects are *sustainable development, stakeholders' relationships, and distinct business model.* The findings presented below are organized based on the adapted framework's constructs and central aspects. In addition, the sustainability aspect is further mapped and organized based on the adopted TBL framework.

4.1 Sustainable Development

In general, our main findings suggest that sustainable development is connected to how the ERP market perceives the sustainability phenomenon from the consultants' perspective, based on the dynamic and sustainable view of the firm (Fig. 3). The framework highlights that the first step was to look at how sustainability is defined, and we then applied the TBL framework to understand how consultants in the ERP world viewed the involvement of the economic (profit), environmental (planet), and social (people) aspects.

Understanding the sustainability definition involves looking at how companies account for their use of natural resources and take responsibility for society. One participant knew Brundtland's definition of sustainability by heart and elaborated further with an explanation of the TBL, adding that *"many standards are based on this philosophy, many are developing, and standardizing based on that philosophy"* (Participant F). When Participant L was asked about their definition of sustainability, *"For me, it is something about a system that can exist constantly, and that you have some sort of balance of what you can harvest and what you use, you can say this is the key in sustainability, and the definition is always difficult to nail."*

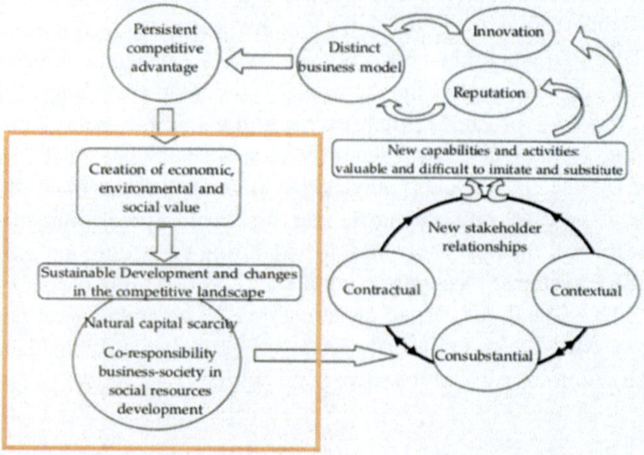

Fig. 3. Dynamic and sustainable view of the firm; sustainable development, adapted from [42]

Profit. In general, the perception that financial incentives drive an organization's strategy was heavily represented in our findings. Seven out of ten participants, including the end-user, showed a similar view on how the financial aspect of sustainability is viewed in their respective work environments. *"Unfortunately, I believe companies only think about the profit side of sustainability, and if they do not earn anything from being sustainable, they will not do it"* (Participant I). From the end-users perspective, *"I see it [sustainability] more on the agenda, but again, the motivation behind it is, of course, driven by the economic aspect and how in the long run, our company can have several income streams"* (Participant I). Similar statements were shared by six of the other informants as well. An example of environmental initiatives that can help to increase profit was provided by Participant H, *"if they can lower the temperature by one degree, they save maybe millions a year; it is as simple as that.* According to Participant I, *"They can change energy usage and save more money."* According to Participant I, popular accounting methods such as FIFO (first in, first out) are used in businesses from a financial (cost/benefit) perspective and often without considering sustainability-related factors. With ERP systems, some cost-effective ideas could turn into sustainability initiatives, like keeping track of fuel prices; for example, *"all the cruise lines and airlines have functionalities in the ERP system for fuel-hedging where they are speculating in fuel prices"* (Participant H).

Planet. Participants had several examples of how environmental initiatives could help the organization to operate sustainably while also considering the scarcity of natural resources. Participant G underlined the importance of resource productivity/scarcity, describing that *"the consequence of wasting our planet's limited materials and minerals is important because we are stealing from our children and if we do not use these materials in a more circular system, and in a more productive way, then we are stupid."* The participant then referred to the concept of the *circular economy* of consuming raw materials. This was also stated by participant F, *"In the circular economy, there is reduce, recycle, remanufacture, reuse, and recover, so you use the raw materials, make the* *product and at the products end-life you turn it back to raw materials or reuse the waste,"* and he further explained that *"The way ERP will handle this product is important, it is our product until it reaches to that customer or until it is returned to the supplier"* Waste management and the value of waste is also a reoccurring theme among participants. *"BOM, or Bill of Materials, measures the consumption of the materials, and the minus sign is waste. This waste output from the production is performed in the ERP system. Sometimes you can sell the waste and buy it. It is about managing the audit train for the waste."* (Participant B). Concerning ERP systems, almost all our informants viewed them as a vital tool for information management and provision within organizations that can enable sustainability initiatives. *"With ERP, you can plan your business processes, and you can plan your material much better, but you also have direct information about the use of the materials, including the inputs such as energy"* (Participant A). In general, organizations use this data to report on their financial performance. However, there is also an opportunity to use this data and report on environmental performance, such as raw materials and energy consumption. Another environmental aspect that six of the participants mentioned was cloud-ERP in relation to the environmental aspect *"if you create Cloud-based software as a service and the cloud data center is running on renewable energy, it is very energy effective you can use the spill overheat to heat the nearby houses and so on"* (Participant G).

Another point is the data function in Cloud-ERP. *"I do not implement ERP anymore; I use cloud ERP. Companies are sending data and business functions to the cloud."* (Participant F). An additional view of cloud ERP was the ability to add on applications and functions *"The ERP core is now becoming narrower, like a core engine, so to say, and then you will build a universe of cloud-based applications around it and be able to connect and interact through ERP core"* (Participant B).

People. Four participants argued that companies are increasingly gaining awareness of the problems they create for society, and some are trying to enhance their operations toward sustainability. One of the informants argued that "*companies demand more reports and specific frameworks such as sustainable accounting from their ERP systems*" (Participant I). However, she also asserted that government incentives and support for businesses would further disseminate sustainability adoption. Participant B stated that social awareness was not a contradiction. Those listed or 'important' companies in Norway also care about how they appear in the media, which is *"something you have to consider when designing your business case."* Social awareness is also related to the employer branding of organizations: "*Your employer branding is extremely important. I mean, what kind of people do you want? How do you attract the best talents for your company if you don't have a very good reputation?"* (Participant B). According to Participant G, the ERP system could be an opportunity here; Participant G stated, "*If a company says, were going to attract talents of tomorrow, we're going to keep them healthy and satisfied. ERP could help achieve this KPI"*.

According to one participant, enhancing diversity is a social sustainability initiative that could be supported through ERP systems. "*You can put how many ethnicities and cultural backgrounds you have in the company; you can put male and female employees' percentage; there are different criteria you can put in.*" However, "*from my perspective, some of these criteria are based on western culture mindset,*" Participant F elaborates on how the ERP system is not focused on the differences in social aspects in other societies. Participant G argued that work-life balance is a social issue companies are currently facing but sees great opportunities in enabling ERP systems to track productivity, considering aspects such as monitoring screen time and taking recommended breaks during the workday, "*If you can use the ERP system to maintain the best possible balance because today when everything is a thump away, it's difficult to find the right balance.*" Participant B argued that cloud-ERP systems allow employees to not be in the office daily by working remotely, whether from home or elsewhere. On the other hand, participant D questioned whether the opportunity to work from home is good for society, "*people can work from your couch back home in the evenings, and what happens to society when we do this? What happens to your friends and family and social life?*". A summary of these findings can be found in Table 3 below.

Table 3. Summary of findings on PPP

Profit (*Economic value)*	Planet (*Environmental value)*	People (*Social value)*
People and planet initiatives will only get adopted if organizations have a positive economic benefit Such as cost reduction and increased income streams	Creating new revenue streams by managing and reusing waste to avoid creating natural resource scarcity for future generations Cloud-ERP is perceived as an enabler and a solution for energy saving and waste management	Social awareness pushed by society is forcing organizations to adopt ecological initiatives Social resources development through positive employer branding, enhancing diversity, considering the work-life balance, and focusing on organizations' human capital

4.2 Stakeholder Relationships

As presented in Fig. 4, the stakeholder relationship perspective was employed to define the drivers of implementing sustainability within the ERP market.

Responses varied in the interview discussions on which stakeholders (e.g., governments, ERP vendors, top management, employees, etc.) are the main drivers for sustainability changes. Another issue discussed with some informants was how the participants perceived sustainability initiatives, as three out of ten participants perceived it as a form of competitive advantage. On the other hand, four out of ten participants consider it a significant risk to embark.

Contractual. Two major and international ERP vendors were perceived and have been cited as potential (or current) market-changers and drivers of implementing sustainability within the ERP systems domain; "*you might say that Microsoft is a driving force, but they are a driving force because they have bought a lot, and SAP has also done that. Microsoft is, right now, they are leading ahea*d" (Participant B). Participant H elaborated

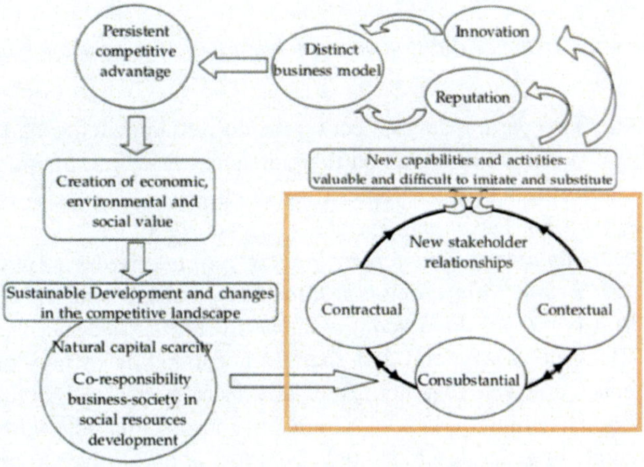

Fig. 4. Dynamic and sustainable view of the firm; stakeholder relationship, adapted from Rodriguez, et al. [42]

on the fact and added: "*However, two years ago, I guess, I had a big client who asked for a specific sustainability functionality*" (Participant H). However, the participant further asserted that changing the ERP to accommodate the required functionality was complex. "*From a practitioner's side, only vendors like SAP can do that and get money off their customers, and suppliers [SAP's implementation partners/resellers] do not know about sustainable ERP. So as ERP leaders, they are also pioneers in this case*" (Participant F). Like Microsoft, other participants recognized SAP as a leader in their field and could drive change towards more sustainability in the ERP market. "*SAP is the largest ERP provider, and did you know that 77% of all the financial transactions in the world globally go through an SAP system? They have a hand on the steering wheel of our economy, to a considerable extent, then we are aware of*", Participant B claims.

Suppliers were not the only ones viewed as change leaders in the ERP market; for example, participants D and E perceive customers as the main drivers of change; "*the customers are the kings; they decide how things should be; it does not matter what the consultant thinks, the consumer is the driver for change*" (Participant D). Hence, this research sought the end-user interview as an important perspective. The end-user informant in this study (Participant L) stressed that an organization needs high-quality data, and consultants should be skilled in data management and design, help organizations with data analyses and reporting, and support their sustainability needs and related requirements. For example, "*We have to report to the government and our clients how our [environmental] footprint is, so it is coming soon as a functionally*" (Participant L). During the discussion on the three aspects of sustainability and how they were viewed in the target organization, participant L argued that the ROI often drives sustainability initiatives. In terms of the environmental aspect, the participant asserted that in 2020 they (the target organization) made a goal of using at least 50% locally sourced suppliers and following the Health, Safety, Security, and Environment (HSSE) guidelines, the "*environment has been on our radar for quite some years, but in the recent years, it is in the evaluation criteria, but not on top of the agenda"* (Participant L).

Consubstantial. Two participants reflect on the current knowledge and competencies of consultants on sustainability. One participant stated that both training programs on sustainability and bringing up the topic more often could be done within their organization (Participant B). Participant F stated, "*If you are an ERP consultant, you also need sustainability competence to deliver the best customer value.*" He elaborates that "*it's hard to find people with enough sustainability background as well as the tech background, so it is our job to educate our employees and bring in new people with new insights.*" Despite the general and shared understanding of how planet-profit or people-profit can work well together, especially in respect of all three sustainability dimensions, there is a common lack of competence in sustainability within the consultancy organizations in general, which could be a lost opportunity for consulting firms; "*I believe that, hopefully, also we see that we have to build a sustainability-competence and merge it with other consultancy disciplines to maintain a competitive advantage"* (Participant G). In addition, a company's culture influences how employees work to establish a sustainability-focused company culture and environment (participants F and G), including how this is supported by top management support (participant G). Participant G stated that he is actively training himself on sustainability, causing him to realize the opportunities to develop new innovative services to help their operations become more sustainable. However, he states that it is a personal motivation supported by the top management, "*I work voluntarily with sustainability. I have my manager he is ok with me spending some time on other things than what he pays me for*". To illustrate that the interest in sustainability within their organizations usually comes from the younger generation, participant D stated that "*I would say that the company is recruiting a lot of younger people, and they are bringing their perspective, and I think the perspective has changed just over the last ten years.*" When asked about why the younger generation is believed to be more sustainability-aware, Participant D stated that "*I think that's an easy question to answer because this [sustainability] is an essential part of the youngster's future, so this is a serious question.*"

Other consubstantial stakeholders that emerged during the interviews are the investors. "One of the key drivers today is the investor because, for them, climate issues are a risk; they want to invest the money where the risk is lowest or reduced" (Participant G). The participant exemplified this statement by explaining the annual letter by the CEO of BlackRock, the world's largest investment fund, Larry Fink. In his letter, he claimed that "BlackRock will put climate change at the center of investment strategy." The participant then quotes Fink ", If you are not performing your business in a sustainable way, you are not going to get our money." Investors are willing to invest their money in companies that can identify their opportunities and mitigate risks associated with environmental, social, and economic issues. This is typically exhibited through the quality of the company's non-financial reporting. Participant G also mentioned companies on their way toward sustainability, saying, "Tech companies are making a huge effort; they're not doing this to be nice; they're doing this because it is the smart thing to do. Hopefully, more and more businesses will contribute also. It gives them a competitive advantage; they are like the first movers but not all sustainability efforts are going to succeed"

Investors are willing to invest their money in companies that can identify their opportunities and mitigate risk associated with environmental, social, and economic issues; this is shown through the quality of the company's non-financial reporting. "To be able to do reporting, you will, of course, need the right data and correct data, and this is where our ERP systems or any other systems need to have in place and more control over their processes" (Participant D). However, the participant said he doubted this functionality already existed in the ERP system, referring to how the shipping industry is not the best at their quality in reporting.

Contextual. Considering the contextual stakeholder, four participants reflected on the support needed from the public administration through sustainability-related legislation, providing incentives, and providing equal opportunities for businesses. For example, Participant G stated that *"I'm not worried, but it's irritating that the government, not just in Norway, almost everywhere is not doing their job to incorporate sustainability. They are handling this very poorly."* adding to this, *"The government needs to put in place some regulations and have a policy that makes everybody compete on the same terms, reduce the risk and have supportive tools in place for the enterprises"* (Participant G). This is also why ERP systems get implemented in the first place; *"a large corporation buys an ERP system to deliver the documentation that the government requires. So, it must start with the government setting regulations"* (Participant H). Similarly, Participant I elaborated on how the government should be the first in line to promote sustainability, "*it has to start with the government and then the customers; if the customer requires something that the vendor cannot deliver, then they would not have a competitive advantage anymore, so maybe it has to start with the government and then when it is going to be a natural thing for the vendors to incorporate or develop"* (Participant I). Participant H supports that argument by saying, "*ERP systems are known as costly, for a vendor to recommend that to the customer it just seems as if we want to make more money. Since that law is not in place, then the functionality is not in the system"* Another participant had a more holistic view on the subject, "*the rules of society will force the companies to change*" (Participant F) hinting on the awareness of society on things like the SDGs and Paris agreement might impact future business operations, and then ERP systems. Seeing an opportunity to integrate sustainability and consider the SDGS in ERP consultancy services was mentioned by Participant D. "*I mentioned the United Nations 17 goals. The company wants to operate sustainably and would like a sustainability strategy. So that is part of what the company has been working with for the past few months. We are looking at this from a strategic point of view because we want to position ourselves*".

4.3 Distinct Business Model

This section explores whether ERP systems could support businesses in implementing sustainability initiatives that could force new business models, sustainable operations, and process innovation. This section includes examples of requests by customers for new capabilities. This section is visualized in Fig. 5, and a summary of the findings can be found in Table 4.

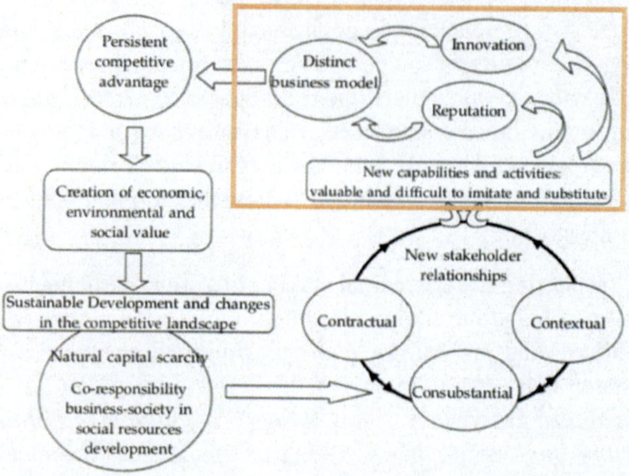

Fig. 5. Dynamic and sustainable view of the firm; distinct business model, adapted from Rodriguez, et al. [42]

New Capabilities. Participants A and C asserted that they are noticing that their clients are becoming increasingly interested in sustainability, and many customers recommended sustainability-related functionalities to be added to the system. Participant A, who works primarily with clients from the agricultural sector, stated that "*the market is increasingly demanding, and it sets quite strict requirements. Also, [requirements] from the government. But it is just that many of them have the intrinsic motivation to do something with sustainability, which is great. Still, if you don't, you will miss out on your license to produce.*" He also adds that there is a difference between forerunners and the followers and that those with more experience handling data are starting to request more functionalities "*They generate more data. They want to be able to do more and combine it with other data sources*" (Participant A). Table 4 provides an overview of the types of requests the participants have received for incorporating new capabilities related to sustainability initiatives in ERP systems.

Innovation. With the current ERP solutions available in the market, five participants stated that the solutions lack equal functionalities considering the TBL's three aspects. For example, "*Today, as far as I know, ERP only has the financial objectives, and they probably lack the planet and people dimension*" (Participant G). One participant (I) considers the system they were working with as a very basic/generic system, as it only allows vanilla implementations; "*everyone can use it, and maybe we can make some corrections and some small functions to the customer, but it's not very sustainability focused.*" Participant C added that the ERP systems handle the overall processes in every business, and process re-engineering and optimization could lead to sustainability. He adds, "*What is important, in my opinion, is that the processes are well optimized.*" A solution suggested by Participant C to make ERP sustainable is to add the concept of sustainable accounting within the system. Participant D stated that most existing ERP systems could be optimized for sustainability reporting but acknowledges that they currently do not have the full functionality to do this, causing people to search for other

Table 4. Requested functionalities from clients

Examples of sustainability initiatives	
Transportation	*"Yes, especially on things like transportation"* (Participant E)
Waste management	*"The customer is focused on reducing waste, not overproducing, stuff they don't sell, want to balance the supply, and they know that to do that you need some good analytical tools, and I start designing, this is planning to optimize, so ERP is not handling everything, but the ERP is connecting the data."* (Participant B)
Environment	*"There are not many requests from the customers related to environmental sustainability as of now; that is my experience because when we are meeting the customers, we're talking about the technology and the cloud solution, but of course, when you come to the public customers, like hospitals, they are required to talk about sustainability. They are asking us how we can deliver this."* (Participant G)
Energy tracking	*"The only request I got is from* (Norwegian construction company), *which are one of the largest construction companies in Norway, they build universities, schools, road, and tunnels, and they had a requirement that they wanted to track how much Co2 they generate*" (Participant H)
Reporting	*"I think we had one client maybe six months ago, but I don't think that the agreement went through […]. They wanted a system that they could make sustainability accounting; that is the only company that has demanded some of these types of reports.*" (Participant I)
Dashboards	*"Well, some people would like to have a sort of sustainability dashboard to see how sustainable they are. The problem is when you ask them what you would like to see. And then they say, yeah, don't you have something that I can start looking at? So, it is a sort of, and it is chicken and egg discussion"* (Participant E)

platforms to get the correct data for sustainability-related calculations and estimates. Participant B and Participant F are the participants that undoubtedly agree that the current ERP solutions available already have the capabilities for handling sustainability: "*I think sustainability should be a driver and initiative for keeping your ERP solutions up to date. Because the capabilities of ERP solutions are there.*" building on this he also elaborated "*if we are going to modify the way we live and produce consumables to make a more sustainable growth and to become more sustainable, we need technology to support, to monitor, and to continuously improve the way that we produce things.*"

Many participants have provided various examples of how the environmental approach has impacted and increased revenues. Nevertheless, another important aspect is how the environmental approach has changed the business model. Participant B stated, "*Companies that used to sell equipment no longer sell it; they rent it. Alternatively, they may behave in other business models that do not mean that you sell something, just renting a capability. You take back the equipment, then recycle it. In summary, it is important to gain competencies and new insights on how you can implement sustainability into your business model and strategy and increase the competitive advantage"* (Participant G). One participant (1), which is an ERP client/end-user, stated that the government is, of course, responsible, but *"also those who define the business model and find a way to make it productive, those are the ones who are going to win this the race. We can always sit and wait for the government to do something. Still, we can also lead the way to sustainability, which is likely to be successful"*.

Reputation. Four participants responded positively to the available opportunities for ERP to support sustainability initiatives. Two informants stated that there is a connection between sustainability and business processes, "*to be able to produce sustainably, you must have your production processes or other processes in control, ERP is a very important tool in that*" (participant A) and "*with an ERP system, the goal is to change your processes and make them more effective, if you could do this in a smarter way, this could help in all the three [sustainability] perspectives.*" (Participant D). Participant A stated that current ERP systems can be tweaked to accommodate sustainability processes; "*an ERP system is about the registration of data and the processing of data, and reasonable basic transactions within the company, which can be purchasing and sales, but also product handling, and it is the same with costs and prices, of course, you can also do that for sustainability.*" While the functionality is currently unavailable, Participant H referred to his previous job, where he worked with MS Dynamics, "*MS Dynamics had a sustainability functionality launched in 2009, but this has been discontinued*". The functionality allowed users to create an environmental sustainability dashboard to track their energy consumption and greenhouse gas emissions. He states that it should not be difficult to add this functionality to any ERP system: "*The ERP system is a collection of transactions, and some transactions, for example, the operations related to shipping a large item, you know that this item will generate some electricity, some fuel from the trucks, and you can estimate how much consumption these operations will generate*" (Participant H).

5 Discussion

The following sections discuss the research findings under each of this study's research questions. The discussion also sheds light on the findings in relation to the extant literature on sustainability within the ERP context.

5.1 How does the ERP Market Perceive the Phenomenon of Sustainability?

To drive sustainable value creation, it is necessary to outline how businesses perceive and frame sustainable development. Ihlen and Roper [20] also find variations in how companies view and perceive sustainability, and their findings align with this research's findings. Our findings suggest that, while organizations apply different sustainability frameworks such as SDGs to communicate their sustainability efforts, the concept of sustainability is often not communicated within the organization or reflected in the business operations. Participants D and I stated that sustainability is not at the top of their organizations' strategy agenda. However, they still embark on social or environment-focused initiatives without labeling them *sustainability* projects. In addition, the most discussed aspect of the three TBL dimensions in the interviews was the profit part, which showcases that profits mainly drive the selection of environmental and social business initiatives. This confirms the results of a study conducted by [50] using the TBL approach, which argues that the main incentives to adopt sustainable business practices are economical, followed by environmental and social aspects in the Norwegian market. Profit should include a "social purpose" where businesses aim at creating shared value for both the society and the organization, as this will result in growth in profit and societal development [43].

The research presented herein underscores a growing trend among ERP system customers: an augmented demand for sustainability-oriented features and bespoke modifications. Nonetheless, it is noteworthy that these sustainability endeavors predominantly focus on environmental considerations, often excluding other aspects of sustainability. In traditional contexts, well-established accounting procedures, such as the (FIFO) method, are routinely incorporated into the central modules of ERP systems. While FIFO has been recognized as an efficient financial framework (as posited by Participant I), it is rarely leveraged to reduce waste or monitor the availability of crucial materials. Core functionalities of ERP systems typically encompass finance, accounting, and (HR), which collectively process a broad array of data, including financial transactions, payroll management, strategic planning, and recruitment. From a social perspective, HR capabilities within ERP systems can substantially enhance an organization's ability to manage and flexibly adapt its human capital resources [51]. Participants in this study also highlighted the growing importance of social consciousness within corporations. They argue that companies with an acute awareness of their societal role can attract and retain future employees more effectively. ERP systems, as suggested by Participant G, could facilitate the achievement of key performance indicators (KPIs), such as tracking absenteeism rates and average employee turnover.

While ERP systems' primary aim is to unlock business value through transaction data analysis - a potential confirmed by several consultants-incorporating environmental sustainability-related data is less prevalent. Findings reveal that data related to material resources, energy consumption, and pollution metrics are not typical elements of existing ERP systems. However, they can be integrated, as indicated by Participant B. Clients frequently request these features, often to meet governmental reporting obligations regarding environmental impact. However, it was contended that these environmental features are not universally essential, hence their omission from the system's core functionalities. This viewpoint signifies an emerging interest in integrating these capabilities within ERP systems, but they are not currently regarded as fundamental components, nor are they widely encouraged in practice.

5.2 Who are the Drivers of Sustainability Within the ERP Market?

This study found no consensus on which stakeholder is the primary driver of sustainability in the ERP market. Rodriguez, et al. [42] recommended using the stakeholder classification to help companies perceive the importance of improving their relationships with their stakeholders and how this may impact sustainability. Some stakeholders are crucial for the business to exist; some are the subject of primarily contractual relationships; others represent the social and natural ecosystem in which companies operate. Each of these stakeholders, with variations in extent, influences the company directly or indirectly and may also influence the organizations' sustainability adoption initiatives and drive change. The extant literature (e.g., [42]) mainly recognized the government as the main contextual stakeholder. It argued that government regulations are one of the ways to set sustainable practices, which was confirmed by four participants who argued that the Norwegian government could drive sustainability through stricter legislation and better incentives.

On the other hand, several participants argued that customers are the main stakeholders that can drive ERP vendors to provide more sustainability-ready systems by requesting new functionalities and customizations. On the other hand, our data suggest that the customer may also question whether these (often expensive) customizations are necessary and how they could help make more sustainable decisions based on the data collected in the ERP system. Other participants argued that prominent vendors (stakeholders) should be the market drivers as they directly influence the product development process and are recognized as pioneers in this field. As mentioned earlier, Microsoft already came up with an environmental-related module in 2009, but it was discontinued as its client adoption was low. Currently, SAP's new digital learning initiative platform provides enrolment in sustainability courses to teach their clients how to integrate sustainability into their strategy [52]. However, the courses show limited remarks on ERP functionalities concerning sustainability. One participant argued that the one sitting with valuable information is the one with the influence. This means that if a company can gather valuable information, mitigate risk, see opportunities, and act on the three sustainability aspects, those are the ones who are *"getting a seat at the table with the big boys and girls"* (Participant 1). Participant 1, an ERP end-user, argued that data is as necessary as the consultant, meaning that if the consultant is not knowledgeable about what kind of data this business is trying to gather, then the data is worth nothing. This is also in line with literature that argues that this data could be used to report to investors how the company mitigates risks and perceives opportunities. For example, in their study on environmental ERP, Rodriguez, et al. [42] argue that (institutional) investors seek better information on sustainability initiatives, processes, structures, and governance. Our findings suggest that many significant investment funds are promoting sustainability-aware investments and projects. Organizations must adopt advanced (sustainability) information systems to help them document, communicate, and, most importantly, validate the information [53].

5.3 What is the Current State of ERP Systems and Their Potential for Supporting Businesses Implementing Sustainability Initiatives?

ERP systems are one of the most adopted IT solutions in organizations [54], containing several modules which hold information and data on vendors, customers, employees, and products, intending to serve business processes housed within multiple functional areas [55]. In examining the contemporary landscape of (ERP) systems, a primary focus is their potential to bolster businesses in the execution of sustainability initiatives. ERP systems are extensively adopted IT solutions within organizations, encompassing myriad modules that store data concerning vendors, customers, employees, and products, designed to underpin business processes across numerous functional areas [55].

The transition from on-premises ERP systems to cloud-based ERPs can be perceived as a viable sustainability initiative. Notably, several participants identified these cloud-based systems as eco-friendly solutions, suggesting they substantially contribute to environmental sustainability. The endorsements of cloud-based ERP align with extant research, emphasizing the potential for increased energy efficiency [56]. This translates to economic benefits for the supplier, enabling the flexibility to utilize different data centers seasonally, such as leveraging a data center in Northern Norway for cooling during the summer, thus diminishing energy consumption [56].

Drawing on the theoretical framework proposed by Rodriguez, et al. [42], two primary sources of sustained competitive advantage emanate from sustainable development: innovation and reputation. The fusion of these elements facilitates the creation of novel business models. Some participants expressed optimism about future opportunities for incorporating sustainability features into ERP systems. However, perceptions of the current availability of sustainability features in ERP systems are polarized. Some participants argued that the systems are lacking in this respect, while others asserted the presence of these features. This discrepancy was particularly evident with a feature of a well-known ERP system, which was reported to have been discontinued, contradicting Chofreh, et al. [33] assertion that the feature remains available and facilitates the assessment, monitoring, and reporting of an organization's sustainability performance. Chofreh, et al. [33] also point to the availability of ERP features that permit the control, analysis, and reporting of an organization's performance to achieve sustainability outcomes encompassing environmental, social, and economic performance. Most of the sustainability initiatives identified in this study pertain to environmental aspects, reflecting customer demand. However, the demand for social initiatives was lower. This discrepancy could be attributed to companies associating the social aspect of sustainability primarily with the HR module, a core ERP functionality. Alternatively, companies may not request social initiatives because they often equate environmental initiatives with sustainability, overlooking the importance of integrating all three dimensions—environmental, social, and economic—to achieve comprehensive sustainability outcomes.

5.4 Are Current ERP Systems Ready to Support Sustainability Initiatives?

With the emergence of sustainability as a concern, researchers have begun to explore the potential for ERP systems to assist organizations in implementing sustainability initiatives. ERP systems primarily facilitate critical business operations and routine transactions by integrating data and cross-functional systems within the program [57]. At present, the integration of sustainability within ERP systems remains in its emerging stage [32], with key industry players endeavoring to conceptualize and operationalize sustainability [16].

Considering the findings and integrating the dynamic and sustainable view of the firm, the readiness of current ERP systems to support sustainability initiatives is not there yet. As firms increasingly prioritize integrating sustainability initiatives into their business operations, it becomes vital to re-examine the capabilities of ERP systems in addressing the evolving needs of businesses. In comparison, findings on the 'sustainable development' and 'stakeholders relationships' specify that stakeholders show more interest in sustainability and that ERP systems play a role in this transition, findings from 'the business models' show that the current ERP systems do not fully support sustainability operations and initiatives—specifically, the lack of comprehensive functionalities addressing the Triple Bottom Line's primary constructs. The dynamic and sustainable view of the firm emphasizes the importance of understanding the interdependence of economic, social, and environmental factors in shaping the firm's competitive advantage. As such, ERP systems must be capable of addressing these many-sided concerns to truly support sustainability initiatives. Furthermore, the demand for new capabilities and innovation, as seen in the findings (Table 4), suggests that clients already recognize the need for a more holistic approach to sustainability.

6 Conclusion, Limitations, and Future Research

In the context of the Dynamic and Sustainable View of the Firm and TBL frameworks, this study explored sustainable development's current perceptions and status, its influencing factors, and the readiness of existing ERP systems to facilitate sustainability initiatives within business organizations. The main research question guiding this study was: Are current ERP systems ready to support sustainability initiatives? This research shows that integrating sustainability initiatives into ERP systems represents an area of potential and a growing interest among various stakeholders. On the other hand, the findings suggest that the current and modern ERP systems are not fully prepared to address the multi-dimensional aspects of sustainability, specifically in terms of the TBL framework. There is a gap between the stated need for sustainability-related functionalities in ERP systems and the actual capabilities of current systems. This indicates room for growth and innovation in developing ERP systems capable of supporting sustainability initiatives. The dynamic and sustainable view of the firm and TBL frameworks highlight the necessity for linking economic, social, and environmental aspects in shaping a firm's competitive advantage. Moreover, the emergence of new business models based on sustainability principles further emphasizes the need for ERP systems to evolve and adapt. The incorporation of sustainability accounting and the development of sustainability dashboards are examples of such needed modifications.

One of the limitations of this research is the limited literature on the intersection of ERP systems and sustainability. The shortage of varied perspectives on the topic in literature underscores the need for increased research efforts, particularly those grounded in empirical methodologies, to test and verify the conceptual frameworks currently available. In addition, the data in this study is limited to ten interviews. It was challenging to identify participants that were knowledgeable about both ERP systems and sustainability. In addition, the initial data collection plan was to include several ERP end-user informants. Recruiting ERP end-users with sustainability-related knowledge also proved challenging, and we could recruit only one participant. Hence, the end-user perspective in this study may be underrepresented and requires future research investigations on how the end-users perceive sustainability in the Norwegian ERP systems market. Our study was conducted in the Norwegian context, and other studies conducted in other countries may yield different results, especially in countries with different socio-economic, environmental, and legal settings. A cross-country analysis could also provide valuable insights into any differences and/or similarities in the perceptions of ERP systems and sustainability. Secondly, sector-specific research could be beneficial, as this study does not define the varying needs of distinct sector groups that utilize ERP systems. Future research could examine how different industrial sectors can optimally leverage ERP systems for sustainability.

While current ERP systems provide a foundation for supporting sustainability initiatives, more progress is needed to realize this potential. The development and enhancement of ERP systems should focus on integrating more sustainability features to better support businesses in pursuing sustainable operations and long-term competitive advantage. This integration not only meets the increasing demand for sustainable business operations but will also positively contribute to the broader societal goal of sustainable development.

References

1. Brundtland, G.H.: Our common future—call for action. Environ. Conserv. **14**(4), 291–294 (1987)
2. Rice, J.B.: Prepare your supply chain for coronavirus. Harv. Bus. Rev. **27** (2020)
3. Retief, F., Bond, A., Pope, J., Morrison-Saunders, A., King, N.: Global megatrends and their implications for environmental assessment practice. Environ. Impact Assess. Rev. **61**, 52–60 (2016)
4. Rezaee, Z.: Business Sustainability, Corporate Governance, and Organizational Ethics. Wiley, Hoboken (2019)
5. Ursacescu, M., Popescu, D., State, C., Smeureanu, I.: Assessing the greenness of enterprise resource planning systems through green IT solutions: a Romanian perspective. Sustainability **11**(16), 4472 (2019)
6. Elkington, J.: Cannibals with forks. In: The Triple Bottom Line of 21st Century, vol. 73 (1997)
7. Hubbard, G.: Measuring organizational performance: beyond the triple bottom line. Bus. Strateg. Environ. **18**(3), 177–191 (2009)
8. Singh, M., Sahu, G.P.: Towards adoption of green IS: a literature review using classification methodology. Int. J. Inf. Manage. **54**, 102147 (2020)
9. Molla, A., Abareshi, A., Cooper, V.: Green IT beliefs and pro-environmental IT practices among IT professionals. Inf. Technol. People **27**(2), 129–154 (2014)
10. Boudreau, M.-C., Chen, A., Huber, M.: Green IS: building sustainable business practices. In: Information Systems: A Global Text, pp. 1–17 (2008)
11. Butler, T.: Compliance with institutional imperatives on environmental sustainability: building theory on the role of green IS. J. Strateg. Inf. Syst. **20**(1), 6–26 (2011)
12. Maltz, E., Bi, H.H., Bateman, M.: Benchmarking sustainability performance: the next step in building sustainable business models. J. Public Aff. **18**(3), e1606 (2018)
13. Ribeiro, M.P.L., Tommasetti, R., Gomes, M.Z., Castro, A., Ismail, A.: Adoption phases of Green Information Technology in enhanced sustainability: a bibliometric study. Clean. Eng. Technol. **3**, 100095 (2021)
14. Chasin, F.: Sustainability: are we all talking about the same thing state-of-the-art and proposals for an integrative definition of sustainability in information systems. In: ICT for Sustainability 2014 (ICT4S-14), pp. 342–351, Atlantis Press (2014)

15. Bengtsson, F., Ågerfalk, P.J.: Sustaining sustainability: Investigating the long-term effects of a sustainability initiative. In: Lundh Snis, U. (ed.) SCIS 2016. LNBIP, vol. 259, pp. 86–99. Springer, Cham (2016). https://doi.org/10.1007/978-3-319-43597-8_7
16. Chofreh, A.G., Goni, F.A., Klemeš, J.J., Malik, M.N., Khan, H.H.: Development of guidelines for the implementation of sustainable enterprise resource planning systems. J. Clean. Prod. **244**, 118655 (2020)
17. Shiri, S., Anvari, A., Soltani, H.: Identifying and prioritizing of readiness factors for implementing ERP based on agility (extension of McKinsey 7S model). Eur. Online J. Nat. Soc. Sci. Proc. **4**(1), 56–74 (2015)
18. D'amato, D., Korhonen, J.: Integrating the green economy, circular economy and bioeconomy in a strategic sustainability framework. Ecol. Econ. **188**, 107143 (2021)
19. Larson, A.L., Teisberg, E.O., Johnson, R.R.: Sustainable business: opportunity and value creation. Interfaces **30**(3), 1–12 (2000)
20. Ihlen, Ø., Roper, J.: Corporate reports on sustainability and sustainable development: 'we have arrived.' Sustain. Dev. **22**(1), 42–51 (2014)
21. Kleindorfer, P.R., Singhal, K., Van Wassenhove, L.N.: Sustainable operations management. Prod. Oper. Manag. **14**(4), 482–492 (2005)
22. Bowen, H.: "Social Responsibilities of the Businessman. Harper & Row, New York (1953)
23. Garriga, E., Melé, D.: Corporate social responsibility theories: mapping the territory. J. Bus. Ethics **53**(1), 51–71 (2004)
24. McWilliams, A., Siegel, D.S., Wright, P.M.: Corporate social responsibility: strategic implications. J. Manage. Stud. **43**(1), 1–18 (2006)
25. Barnea, A., Rubin, A.: Corporate social responsibility as a conflict between shareholders. J. Bus. Ethics **97**(1), 71–86 (2010)
26. Pradhan, P., Costa, L., Rybski, D., Lucht, W., Kropp, J.P.: A systematic study of sustainable development goal (SDG) interactions. Earth's Future **5**(11), 1169–1179 (2017)
27. Stafford-Smith, M., et al.: Integration: the key to implementing the sustainable development goals. Sustain. Sci. **12**(6), 911–919 (2017)
28. Schaltegger, S., Burritt, R.L.: Sustainability accounting for companies: catchphrase or decision support for business leaders? J. World Bus. **45**(4), 375–384 (2010)
29. Horngren, C.T., Datar, S.M., Foster, G., Rajan, M.V., Ittner, C.: Cost Accounting: A Managerial Emphasis. Pearson Education India (2009)
30. Langseth, M., Moe, H.T.: Driving through dense fog: a study of the effects and control of sustainable public procurement of electric cars. Environ. Syst. Decis. **42**, 1–14 (2022)
31. Slaman, M., Haddara, M.: How green can the lettuce be? A case study on greening initiatives in supply chains and sustainable ERP systems. Procedia Comput. Sci. **164**, 79–85 (2019)
32. Chofreh, A.G., Goni, F.A., Shaharoun, A.M., Ismail, S., Klemeš, J.J.: Sustainable enterprise resource planning: imperatives and research directions. J. Clean. Prod. **71**, 139–147 (2014)
33. Chofreh, A.G., Goni, F.A., Klemeš, J.J.: Sustainable enterprise resource planning systems implementation: a framework development. J. Clean. Prod. **198**, 1345–1354 (2018)
34. Simmonds, D., Tadesse, A., Murthy, U.: ERP System Implementation and Sustainability Performance Rating and Reputation (2018)
35. Adriansyah, A.K., Ridwan, A.Y., Hediyanto, U.Y.K.S.: Designing and ERP system: a sustainability approach. JOIV Int. J. Inform. Visualization **6**(2–2), 498–505 (2022)
36. Hasan, M., Ebrahim, Z., Mahmood, W.W., Ab Rahman, M.: "Sustainable-ERP system: a preliminary study on sustainability indicators. J. Adv. Manuf. Technol. (JAMT) **11**(1), 61–74 (2017)
37. Munkvold, B.E., Bygstad, B.: The Land of Confusion–Clearing up some common misunderstandings of interpretive research (2016)
38. Oates, B.J.: Quantitative data analysis. In: Researching Information Systems and Computing, pp. 245–265. SAGE Publications, London; Thousand Oaks, Calif (2006)

39. Trkman, P., Černe, M.: Humanising digital life: Reducing emissions while enhancing value-adding human processes. Int. J. Inf. Manage. **63**, 102443 (2022)
40. Yin, R.K.: Case Study Research: Design and Methods. Sage, Thousand Oaks (2009)
41. Baker, R.C., Haddara, M.: Exploring consultants' role in ERP systems implementations. In: 2nd IADIS International Conference Information Systems 2019, pp. 163–170 (2019)
42. Rodriguez, M.A., Ricart, J.E., Sanchez, P.: Sustainable development and the sustainability of competitive advantage: a dynamic and sustainable view of the firm. Creat. Innov. Manag. **11**(3), 135–146 (2002)
43. Porter, M.E.: Competitive advantage of nations: creating and sustaining superior performance. Simon and Schuster (2011)
44. Devoteam: ERP systems in Norway - An independent analysis, market overview, and summary of ERP systems and vendors in Norway. Theme-The impact of emerging technologies (2020)
45. Kaplan, B., Maxwell, J.A.: Qualitative research methods for evaluating computer information systems. In: Anderson, J.G., Aydin, C.E. (eds.) Evaluating the Organizational Impact of Healthcare Information Systems, pp. 30–55. Springer, New York (2005). https://doi.org/10.1007/0-387-30329-4_2
46. Knafl, K.A., Webster, D.C., Benoliel, J.Q., Morse, J.M.: Managing and analyzing qualitative data: a description of tasks, techniques, and materials. West. J. Nurs. Res. **10**(2), 195–218 (1988)
47. Strauss, A., Corbin, J.: Basics of qualitative research techniques (1998)
48. Yin, R.K.: Qualitative Research from Start to Finish. Guilford Publications, New York (2015)
49. Gibbs, G.R.: Analyzing Qualitative Data. Sage, Thousand Oaks (2018)
50. Høgevold, N.M., et al.: A triple bottom line construct and reasons for implementing sustainable business practices in companies and their business networks. Corporate Governance (2015)
51. Lengnick-Hall, C.A., Lengnick-Hall, M.L.: HR, ERP, and knowledge for competitive advantage. Hum. Resour. Manag. **45**(2), 179–194 (2006)
52. SAP. Sustainability: Learning From Leading Companies. Open.Sap.Com. https://open.sap.com/. Accessed 2020
53. Melville, N., Whisnant, R.: Environmental Sustainability: Empirical Analysis of Environmental ERP Implementation. University of Michigan, Ann Arbor, Michigan, USA (2012)
54. Shaul, L., Tauber, D.: Critical success factors in enterprise resource planning systems: review of the last decade. ACM Comput. Surv. (CSUR) **45**(4), 1–39 (2013)

55. Haddara, M., Fagerstrøm, A., Mæland, B.: Cloud ERP systems: anatomy of adoption factors & attitudes. J. Enterp. Resour. Plan. Stud. **2015**, 1–24 (2015)
56. Sørheller, V.U., Høvik, E.J., Hustad, E., Vassilakopoulou, P.: Implementing cloud ERP solutions: a review of sociotechnical concerns. Procedia Comput. Sci. **138**, 470–477 (2018). https://doi.org/10.1016/j.procs.2018.10.065
57. Bjelland, E., Haddara, M.: Evolution of ERP systems in the cloud: a study on system updates. Systems **6**(2), 22 (2018)

High Altitude Platform Stations as IMT Base Stations (HIBS): Reliability and Availability Analysis

Ogbonnaya Anicho(✉), Philip B. Charlesworth, Gurvinder S. Baicher, and Atulya Nagar

Liverpool Hope University, Hope Park, Liverpool L16 9JD, UK
anichoo@hope.ac.uk

Abstract. High Altitude Platform Stations as IMT Base Stations (HIBS) are aerial platforms that will function as flying base stations. There are clear advantages to using these types of assets to extend communications coverage addressing existing digital gaps, especially in unserved or underserved communities. In this paper, HIBS is examined from the context of its integration with 5G new radio (NR) as a non-terrestrial network asset. The challenge of HIBS meeting the stringent operational reliability and availability targets of legacy telecommunication networks is explored. The work analyses the performance of HAPS/HIBS platform record holders, Zephyr and Loon in terms of their reliability and availability profiles in a telecoms operations environment. Findings conclude that while the mean time between failure (MTBF) of these platforms is improving, telecom operations demand higher reliability and availability performances making advanced platform redundancy an imperative for HIBS implementation.

Keywords: HAPS · HIBS · Reliability · Availability · Networks

1 Introduction

High Altitude Platform Station (HAPS) according to the International Telecommunications Union (ITU) is defined in article 1.66A of its Radio Regulations (RR) as: "A station located on an object at an altitude of 20 to 50 km and at a specified, nominal, fixed point relative to the Earth" [9]. This definition is in the context of HAPS as a station in the fixed service. International Mobile Telecommunications abbreviated as IMT is a term used by ITU to describe the families of radio interface standards for mobile systems e.g. IMT-2000 (3G), IMT-2020 (5G) [20]. High Altitude Platform Station as IMT Base Stations (HIBS) are essentially HAPS platforms (see Figs. 1 & 2), defined and operating within the context of a station in the mobile service (specifically IMT mobile service). This distinction reflects the lens through which the ITU currently views these technologies and the services they may support.

K. Arai (Ed.): FTC 2023, LNNS 816, pp. 243–258, 2023.
https://doi.org/10.1007/978-3-031-47448-4_17

HIBS as an IMT-based system is designed to complement existing IMT terrestrial networks using the same IMT 2020 spectrum bands. ITU may refine the definition of HIBS as indicated in item 4 of Resolution 245 (WRC-19) which details the ITU's intention to study HIBS definition and possible modifications to the radio regulation if necessary [8]. However, from an aeronautical perspective or just considering HAPS as an aerial platform the distinction between HAPS and HIPS blurs considerably. The main purpose of HAPS as an aerial platform is to host whatever payload is installed, at the altitude of operation. The three variants of HAPS as an aerial platform (fixed-wing, balloon & airships) are also not considered in the ITU definition for HAPS. Beyond the question of definition and nomenclature, it is critical to understand the practical or operational context within which HIBS will be implemented. As an access infrastructure under the non-terrestrial network (NTN) domain, it is expected to meet some minimum performance criteria. HIBS's performance will determine its future as an access infrastructure option and hence the need to investigate its operational capability for 5G & B5G networks. There has been a deliberate effort by 3GPP to integrate non-terrestrial access infrastructure like satellites and high altitude platforms into 5G New Radio (NR). In 3GPP Release 15 & 16 (TR 38.811 & TR 38.821 respectively), scenarios, requirements and relevant technical studies for NTN implementation were studied. Issues identified from these earlier studies were addressed in Release 17 and captured challenges like satellite mobility (high Doppler offsets in LEO) and altitude (high path loss & RTT in GEO) [18].

1.1 State of the Art and Literature

There was difficulty finding any specific work addressing HIBS reliability and availability in the literature, especially in the context of telecommunication network operations. This was expected as HIBS is still in the developmental stages and does not have any operational reference or antecedent. This work may be among the few directly covering reliability and availability of HIBS from the telecom operations perspective. HAPS/HIBS are largely in developmental phases with only Zephyr (Airbus) and Project Loon attaining commercial status. However, Project Loon's commercial trial in Kenya did not progress further and the company is unfortunately no longer in existence [25]. Zephyr has not been deployed for a full commercial trial but has the most advanced capability in the fixed-wing category. It is expected that some breakthroughs are on the horizon as promising partnerships and interests have been announced by Airbus [2]. The purpose of this paper is to examine the reliability and availability profiles of these platforms within a telecom operations scenario. A comprehensive and extensive work by [10] on HAPS (state of the art and future) did not cover reliability or availability indices from the operations perspective. Reliability was addressed in [23], only in the context of applying free space optical communications in HAPS implementations. The work generally proposed using alternate slant links as redundant paths to improve optical link reliability and availability impacted by fading from atmospheric conditions. In [5] autonomous fleet management in operational scenario was addressed, but the reliability or availability performance was not considered. The work by [15] highlighted the need for ensuring the reliability and availability of HAPS based cloud services but it did not cover the reliability or availability of the HAPS platform in an operational environment. It was evident that there was a gap in the literature for the type of analysis covered in this paper.

1.2 Contributions and Outline

Reliability and availability are critical performance indicators for mobile network operators (MNOs) and could impact subscriber experience and ultimately the bottom line. In essence, a network operator or infrastructure company can only earn if the network/asset is reliable and available. HIBS which are the equivalent of terrestrial towers in the stratosphere may be required to provide carrier-grade availability of 99.999%. The paper explores these concepts and challenges and contributes the following;

- An analytical approach to examining the problem.
- Establishing a framework for measuring current HIBS/HAPS capabilities.
- Providing network operators, vendors and analysts some context for evaluating the technology.
- Using available test trial results to profile current platform capabilities.
- Highlighting telecommunications operations requirements versus the current capabilities of HIBS.

Fig. 1. Airbus Zephyr HAPS can be implemented as HIBS [3].

In this paper, Sect. 1 introduces the general concept and definition of HAPS & HIBS, literature review and the current state of the technology. In Sect. 2 the concept of HIBS reliability and availability were addressed in some detail. Section 3, discusses potential HIBS failure and unavailability considerations. Finally, Sect. 5 concludes the work and considers future work.

Fig. 2. Project Loon balloon (HAPS/HIBS).

2 HIBS Reliability and Availability Overview

HIBS as part of the 5G & B5G NTN architecture must fulfil the service availability KPI both from a technology and operational standpoint. In terrestrial networks, the service level for availability is 99.999% and this requirement will most certainly be extended to HIBS. As an integrated network, failures from any segment of the network will be propagated especially where redundancy cannot be easily achieved. In terrestrial base stations, the availability of the tower as infrastructure is not an issue. The focus is normally on the other active (e.g. antenna) or passive (e.g. power) elements which could fail and affect availability. The terrestrial tower though part of the base station infrastructure has an entirely different availability profile. It is designed to operate without the need to move it or replace it with a new one. Its redundancy principle is very different. This will certainly not be the case for an aerial platform like HIBS. It is also not feasible to import operational models from any other aviation scenario. In essence, HIBS must be implemented with an operational model suited to its peculiar environment.

2.1 Reliability and Availability Definitions and Scope

It is important to define and put reliability and availability concepts applied or used within the work in the proper context. The reliability concept of interest in this paper is the operational reliability of the HIBS platform. In general and more formally, reliability is defined as the probability that a product or service will operate or be provided properly for a specified period under specified operating conditions [22]. Reliability, denoted as $R(t)$ is mathematically given as [1];

$$R(t) = P(T > t), t \geq 0 \tag{1}$$

where T is a random variable representative of the time to failure. Essentially, Reliability is considered the probability that the value of the random variable T is greater than a defined mission time (t) [1]. The inherent reliability of the HIBS platform (i.e. level of reliability inherent in the system as designed and manufactured [21]) is not considered directly in the analysis of the problem. The operational reliability of the HIBS platform which can only be observed during actual operation [21] is currently an unknown parameter. HIBS performance as part of a live network will be heavily dependent on its operational reliability and hence a key element to highlight.

Availability generally can be defined "as the probability that a system, at a point in time, will be operational and able to deliver the requested services [26]". The availability of interest for the HIBS analysis is its operational availability i.e. a measure of the "real" average availability over a period of time and includes all contributors to downtime. The HIBS's operational availability is based on the actual events encountered during operations [26]. Availability can also be calculated using the mean time between failure (MTBF) and the mean time to repair (MTTR). MTBF represents the average time a device or system can be expected to operate without failing [12]. It is the reliability index for repairable systems analogous to mean time to failure (MTTF) for non-repairable units and given by [11];

$$MTBF = \frac{Operating\ Time\ (Cycle)}{Number\ of\ Failures} \tag{2}$$

Availability (A) is therefore given as [11,16];

$$A = \frac{MTBF}{(MTBF + MTTR)} \tag{3}$$

The remaining part of the analysis will focus on operational availability to highlight a very important requirement for HIBS in its operational phase. The inherent reliability of the HIBS platform will be provided by the manufacturers and it is beyond the scope of this work. However, the inherent reliability of the HIBS platform will influence the operational reliability and availability of the system. While manufacturers do their best to design for high reliability, operational environments tend to introduce a few variables that are well beyond their control. In due course, the distinction between the reliability and availability of the HIBS platform as an aerial vehicle and the payload (base station/cell) will be made. It is obvious that the availability of the base station is inextricably linked to the HIBS platform but it is the cell service availability that is actually being measured/analysed from the telecoms perspective. For instance, if services seamlessly switch to a standby HIBS in the event of the failure of an active one, the availability for the cell service will not be affected. However, the reliability and availability of the failed HIBS platform will be impacted from the aerial asset perspective.

2.2 Conventional Aircraft Availability vs HIBS

A conventional civilian aircraft is considered operationally available if it is fit for service. To be more precise, an aircraft's operating time includes both the time the aircraft is flying and when it is on the ground while other operating tasks are ongoing e.g. boarding and unboarding times, taxiing etc. [7]. An aircraft could also be available and not used for flight as in the case of a spare aircraft.

The definition and application of availability in this scenario are quite broad. However, this will not be the case for HIBS in a telecommunications environment; service fulfilment and availability can only be achieved when the HIBS platform is airborne in operations. That means the current aircraft operations template cannot be seamlessly mapped into the HIBS environment; some changes are required.

2.3 HIBS Availability Expectation

In any telecoms network, the MNO's expectation for availability is very critical as it impacts its customer base (service) and bottom line (profits). A telecoms network is expected to achieve "5 nines" i.e. 99.999% availability which translates to about 5.26 min of mean down time (MDT) in a year. The list below summarises availability percentages and the corresponding/target MDT in a year [6].

- 99.0% - 3.65 days (MDT)
- 99.9% - 8.76 h (MDT)
- 99.99% - 52.56 min (MDT)
- 99.999% - 5.26 min (MDT)
- 99.9999% - 31.54 s (MDT)

In essence, the HIBS platform must be available to support the network within this availability expectation. Redundancy principles or planning are often done to ensure that the outage of one network element does not impact the availability of the entire network. Secondly, availability during peak times is critical and has to be adequately managed. In the case of HIBS, the redundancy option could be deploying a spare HIBS, or rerouting traffic through another link. However, this could be infeasible with HIBS for locations where HIBS is deployed to extend coverage in the absence of terrestrial assets. The redundancy or backup planning for HIBS in an operational environment has not been tested at least for the fixed-wing variant of the platform.

3 Potential HIBS Failure and Unavailability Considerations

This section provides an overview of potential failure points within a HIBS system. These failure points may vary or depend on the specific variant of HIBS in operation i.e. fixed-wing, balloon or airship.

- **Failure to remain Airborne:** These can happen for a variety of reasons and is deemed the highest or riskiest failure consideration for HIBS. All HIBS platforms as aerospace vehicles are designed to the highest level of reliability in mind. This failure category has to be factored into the availability equation. A HIBS platform may fail due to power outage, aerodynamic or aero-static issues or a malfunction in any of the control systems. As in the aviation industry, amid the highest levels of safety and reliability, crashes still happen, unfortunately. However, it can be assumed that the inherent and operational reliability performance of the current generation of HIBS may meet and exceed expectations.

- **Failure of Power Systems:** Power system failures or any power-related issues from malfunction to battery drain would present a huge problem for the platform. Assuming high redundancy in power dimensioning, it is still required to factor this risk into the availability consideration. In some simulations, a likely backup solution is to isolate power to payload and other systems, prioritising avionics to ensure the platform remains airborne, but this will certainly impact availability. In a terrestrial base station, power failures can occur but would not present a scenario where the tower itself will fail due to the outage.
- **Failure of Payload:** In terrestrial base stations, there could be failures. Satellite payloads are designed for different operating environments and have been proven over the years. The stratosphere and the proposed IMT base stations are expected to service users, however, the installed payload could fail. A design and operational consideration for HIBS is the capability to be retrieved for corrective or preventive maintenance and upgrades; these activities impact availability.
- **Failure to Station-Keep:** A HIBS cell may become unavailable because the platform has drifted from the main service area. This is more probable with balloon HIBS, where the challenge of keeping the platform quasi-stationary is most pronounced.
- **Unavailability of Links:** As an aerial platform HIBS will certainly carry out aerodynamic manoeuvres altering attitude or altitude. In some instances, these aerial movements may affect the service or feeder link resulting in service unavailability. Recall that every second of outage counts in terms of MDT. For a telecommunications service, cell unavailability will be tracked and reported for both technical, performance and commercial reasons. In the case of a managed services provider or an infrastructure company SLA obligations may be enforced.

At this stage, the availability measurements for HIBS are just predictions or estimates. Operational availability can only be derived during the operational phase of the HIBS platform deployments. Various modelling approaches will be employed to analyse the operational configuration that will make the most sense for operators. Some key considerations to highlight are;

1. Service Level Agreements (SLAs): SLAs will be defined between MNOs and HIBS vendors e.g. availability requirements, especially at the early stages of HIBS commercial roll-out. Some flexibility in agreements may be necessary until operational reliability is clearer.
2. Operations and Maintenance (O&M) Organisation: The efficiency and cost dynamics of this set-up are very crucial to success. Telecom operations are fast-paced and demanding.

3. Business Model: This is the key to commercial success; what is the best model - Platform-as-a-Service, MNO ownership, Collocation Services? The question of how to improve the commercial success of HIBS/HAPS is a fundamental one. Improving technology is important but translating the product into value for customers and all stakeholders is imperative for successful adoption.
4. Regulatory Environment: Engagements are ongoing with relevant regulatory bodies. In operational mode, HIBS implementation may introduce unanticipated scenarios especially with the aviation side of things. Interference issues with primary users or across national boundaries may be experienced and have to be factored into operational readiness planning.

The key consideration in this section is to carefully review likely areas of operational challenge and start developing operational readiness planning to counter it or significantly degrade its impact. HIBS implementation is expected to be successful but more research is needed to address operational issues which may never be evident during the design, testing and validation phases of the system.

4 HIBS Flight Trials: Analysis and Implications for Reliability and Availability

One approach to estimating or "predicting" the operational reliability and availability of the HIBS platforms is by examining test and commercial trial outcomes of the more advanced HIBS vehicles. In this work, two of the leading HAPS/HIBS platforms were analysed based on available flight trial results. These flight test results which are publicly available provided insight into the capabilities of these platforms at least from the flight endurance perspective. It is important to highlight that the analysis relied on the following assumptions and considerations;

- Series Reliability configuration between the platform and other subsystems i.e. the 'failure' or unavailability of the HIBS platform impacts the other dependent systems.
- No platform redundancy planning or contingencies was considered to establish some baseline.
- The use of continuous flight time or platform failure (e.g. crash) to estimate MTBF (i.e. Time or Failure truncated test outcomes).

The analysis, therefore, aims to provide analytical modelling of the reliability and availability capability of these state-of-the-art platforms.

4.1 Methodology

The methodology used relied on using the MTBF of two of the more established and advanced HIBS platforms for the analysis. As explained earlier, the MTBF measures the mean time between failures which is estimated through the continuous (flying without landing or refuelling) flight data. This is a reasonable assumption or approximation since the manufacturers aim to achieve as much continuous flight as practicable for both technical and marketing purposes. In the work, the MTBF estimated from the continuous flight tests was then used as input parameters for probability distributions that are widely used to model the reliability and availability of different systems. The main probability distribution employed for this analysis is the exponential distribution. Reliability theory and analysis can be used to study failure patterns or behaviour of complex systems to establish how they will perform in operation [14]. Mathematical models of Reliability and Availability were employed and disclosed data from the HIBS trials were fed in as input. The work essentially relies on proven mathematical models to analyse the problem.

The flight test results of a fixed-wing and balloon-type HIBS were retrieved from publicly available data [4,24]. Table 1, summarises flight test outcomes of each platform, the main parameter used as input for our analysis is the duration of continuous flight recorded.

Table 1. Summary of flight records.

	Airbus - Zephyr	Alphabet - Loon
Variant	Fixed-wing	Balloon
Project Status	Active	Inactive
1st Flight Record	26 days	198 days
2nd Flight Record	64 days	225 days
3rd Flight Record	Nil	312 days

4.2 Reliability Analysis of the Platforms

The reliability of the platforms is essentially the probability that the platforms will be successfully operating during the mission time, denoted as t in Eq. 1. Essentially, it is the probability that the HIBS platforms will be operational in the interval from time *0* to *t*. Reliability is a function of time and requires that the mission time is specified to qualify or contextualise the performance of the system. In reliability analysis, the goal is typically to model the probability of a system failing (or not failing) as a parametric probability distribution [17]. The probability that the HIBS platform will fail by time t is $F(t)$, given by [1];

$$F(t) = P(T \leq t), t \geq 0 \tag{4}$$

$F(t)$ is essentially the platform's failure distribution function (cumulative distribution function). The reliability function (or survival function) of the HIBS platform becomes [1,16];

$$R(t) = 1 - F(t) \tag{5}$$

The derivative of *F(t)* gives a function that is equivalent to the probability density function (also called the *failure density function*). The exponential distribution was used in the analysis to model the HIBS's failure distribution. It is widely used in reliability analysis especially to model the long "flat" (relatively constant) period of low failure risk phase, which these HIBS platforms are assumed to be approaching. The exponential distribution uses a single parameter, failure rate (λ) or rate parameter to model the failure distribution of the system using this equation [1,16];

$$R(t) = e^{-\lambda t} \tag{6}$$

From Table 1, each flight record can be estimated as the MTBF, which essentially represents the mean time the platforms operated before having to be interrupted either by design or incidence (failure). This is consistent with the characterisation of the flight test results as records of continuous uninterrupted flights by the respective platforms. The observation that each platform improved over time justifies the use of exponential distribution to model the failure distribution. The failure rate which is the reciprocal of MTBF (for repairable systems) was consequently derived from the respective MTBFs of the platforms as recorded. The analysis considered an operational cycle of 365 days (8760 h) and assumed that each flight record represented how much time each platform could operate before some type of interruption (or *"failure"*), this assumption enabled the derivation of needed parameters. For instance, Zephyr flew for approximately 26 days for a specific flight test, it is assumed then, that this model of the platform can operate comfortably for 26 days, which amounts to about 14 failures (estimated) in a year i.e., dividing 365 days in a year by 26 days of the uninterrupted flight. With the number of failures calculated, the failure rate (λ) expressed in hours becomes the number of failures divided by the number of hours in a year, approximately 0.0016. The failure rate can also be derived directly from the MTBF as its reciprocal. In this case, the failure rate is simply the reciprocal of the number of hours of continuous flight (26 days multiplied by 24 h).

The derived failure rate (λ) for each instance of flight test outcome was used to compute the failure density function for both platforms. Figure 3 shows the reliability function of the Zephyr and Loon platforms. The data point of 192 days for Zephyr is an arbitrary figure chosen for symmetry in the analysis (it has not been achieved yet). It is possible that in the next flight test, Zephyr may achieve or exceed this result. From Fig. 3, Zephyr's probability to remain airborne for a specific duration without failure can be estimated. This is an important insight and can be used to develop a redundancy plan for its operational phase. It is also easily seen that with the decrease in the failure rate, the probability of a successful operation over a specified time duration improves. For instance, the probability distribution in Fig. 3 for Zephyr shows that the probability of successful operation at about 4000 h (167 days) is 0 for MTBF of 26 days and almost 0.1 (10%) for MTBF of 64 days. However, the probability of successful operation rises to 0.45 (45%) for the projected (arbitrary) MTBF of 192 days. The probability for successful operation for a shorter duration of say 1000 h (42 days) has significantly higher chances for all values of MTBF for the Zephyr platform. From an operations point of view improvement in MTBF has a significant effect on the probability of successful operation and hence the overall reliability profile of the platform. A similar analysis reveals that the probability for successful operation over 4000 h (167 days) for instance, is higher for the Loon platforms, having probabilities of approximately 0.48 (48%), 0.52 (52%) and 0.64 (64%) for MTBFs of 198 days, 223 days & 312 days respectively. Figure 3, further compares the best-case scenarios of both platforms and clearly shows the Loon platform has a better reliability profile overall. This is one way of interpreting or analysing the performance of both platforms and does not claim to cover all performance considerations. For instance, station-keeping challenges with balloons may be an issue from the availability angle (see Fig. 4).

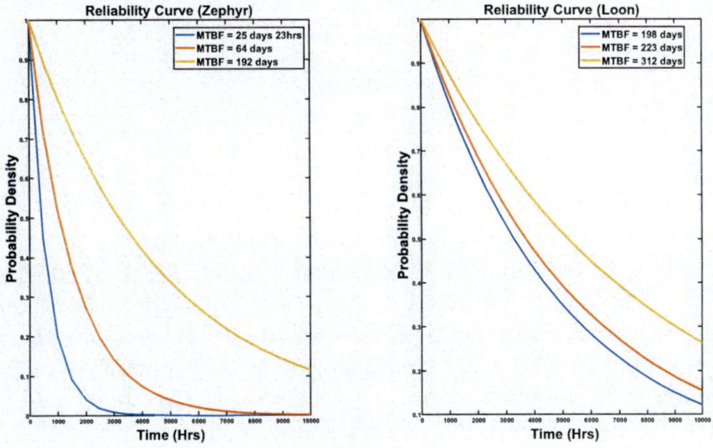

Fig. 3. Zephyr & Loon reliability curve using actual test MTBF.

4.3 Availability Analysis of the Platforms

Availability as defined earlier is a measure of the HIBS uptime against any recorded downtimes or outages. It is essentially the probability that the system is operational or 'available' for use when required at a point in time. In the analysis, Eq. 3 was used to compute the availability of the HIBS platform based on the MTBF values in Table 1. In computing, the availability of these platforms the MTTR value was estimated to be 12 h (this is just an arbitrary estimate). This value can increase or decrease depending on several factors, especially for incidences resulting from corrective or unplanned maintenance. Based on the MTTR assumed, the availability (A), of the Zephyr platform with 26 days MTBF for instance, could be calculated using Eq. 3; which results in an availability of approximately 98.1183%

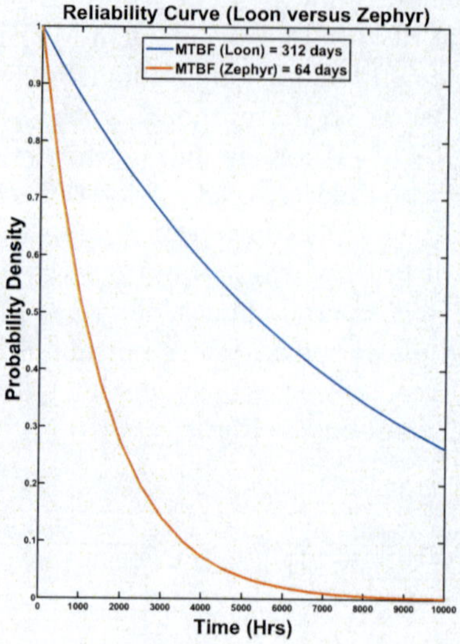

Fig. 4. Loon Vs Zephyr reliability curve using best MTBF index.

As shown in Fig. 5, the availability of the Zephyr platform for the achieved flight records of 26 days and 64 days is approximately 98.1183% and 99.1848% respectively. The Zephyr availability profile is still far from the 99.999% that may be expected in an actual operations scenario. For the sake of symmetry and to buttress a point, an arbitrary continuous flight record of 192 days was added to the Zephyr analysis. With a continuous flight capability of 192 days (MTBF), the availability figure rose to 99.7268% (see Fig. 5).

However, the Project Loon HAPS which is a balloon variant has much higher MTBF values and hence better availability. Figure 5 shows the availability of 99.7268%, 99.7268% and 99.8632% for the Loon's 198, 225 and 312 days MTBF values respectively. Noticeably, the availability figures for 198 and 225 days are the same due to rounding errors. The availability profile of the Loon HIBS platform appears better based on this analysis, Fig. 6 shows the profile of both platforms. In actual operations, this information will help plan redundancy and preventive maintenance programs. Overall, the analysis will aid decision-making across various levels of the technology and commercial value chain. As technology advances and the HIBS platforms reliability improves, availability will also improve, however, this work provides some perspective on the challenge.

4.4 Confidence Levels for the Analysis

It is not practical to test these platforms for an infinite period to derive the 'actual' or exact MTBF value. This analysis was based on a sampling of a limited testing period and limited information and these factors can have varying levels of influence on the accuracy of the analysis. Point estimates of MTBF

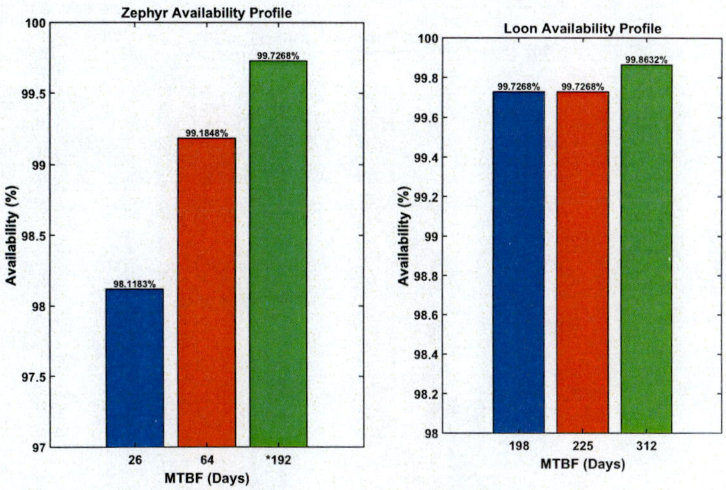

Fig. 5. Zephyr & Loon availability profile.

are not particularly helpful for analysing reliability, hence, essential to provide confidence limits for the analysis based on sound statistical computations. While reliability addresses the notion of failure rate, confidence analysis refers to the certainty of the accuracy of the claimed failure rate [13]. In this work, the confidence bounds for the exponential distributions were calculated using the Reliability Analytics Toolkit [19]. The lower bound MTBF which is considered a safer estimate to work with was computed for the reliability analysis [19] (see Table 2). A failure truncated test type was assumed for the calculation of the 60% and 90% confidence levels as shown in Table 2. The more conservative lower bound limits may inform a cautious operational readiness plan for the platforms. For instance, Zephyr's lower bound 90% confidence level indicates an MTBF of 21 days, while the Loon platform is in the range of 104 days (considering the best flight trial MTBFs). These confidence bounds generally indicate that the MTBF figures shown in Table 2 are within the measurable error of the true MTBF with the given confidence.

Table 2. Confidence limit (CL) - Lower bound MTBF.

	MTBF (days)	MTBF (hrs)	CL 60% (hrs)	CL 90% (hrs)
Zephyr	26	624	388 (16 days)	208 (9 days)
	64	1536	954 (40 days)	513 (21 days)
	192	4608	2863 (119 days)	1538 (64 days)
Loon	198	4752	2953 (123 days)	1586 (66 days)
	225	5400	3355 (139 days)	1803 (75 days)
	312	7488	4653 (193 days)	2500 (104 days)

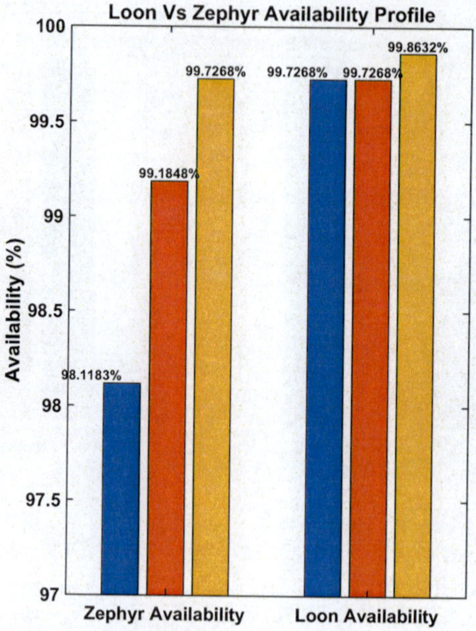

Fig. 6. Loon Vs Zephyr availability profile.

4.5 Summary of Analysis

The analysis used actual flight trial results to model the reliability and availability profiles of the HIBS platforms with the highest technology readiness levels. This work does not intend to diminish the stellar and groundbreaking achievements of these various projects but attempts to provide some context for the operational phase. In actual operations, the HIBS platform will have to provide some specified level of service which may be challenging to the current capabilities of the state-of-the-art. The analysis reveals the current capabilities of these platforms by showing the reliability and availability profiles possible based on actual test data. While more research and development activities are focused on extending the endurance (continuous uninterrupted flight) capabilities of the platforms, it may be equally beneficial to focus on developing redundancy planning that will compensate for any incidences that will disrupt service. A well-planned sparing or standby fleet to support operational HIBS in service may improve the overall availability of the system and ultimately justify HIBS as part of the NTN infrastructure for 5G networks and beyond.

5 Conclusions and Future Work

HIBS are expected to complement terrestrial networks, especially in rural and remote locations. 3GPP Rel-17 addressed the integration of non-terrestrial networks into 5G New Radio (NR), which effectively makes HIBS and satellite infrastructure potentially compatible with 5G access networks. This paper examines some of the unique challenges HIBS may have to address to meet telecoms network operations requirements. An analysis of the reliability and availability of two technically advanced HIBS platforms based on flight trial data was done. It revealed that the balloon-based Loon HAPS/HIBS had better reliability and availability profile compared to Zephyr the fixed-wing variant. However, the analysis also demonstrated that the platforms have to keep improving their reliability and availability performance overall. This paper is also suggesting the need to design effective redundancy/standby frameworks that can use existing capabilities to scale performance and assure higher reliability and availability indices. HIBS is essentially an aerial base station and hence will support cells providing service in a network. It is therefore an operational imperative for HIBS to attain the level of reliability and availability typical for telecommunications infrastructure and services.

Future work will involve developing models, techniques or frameworks to address some of the challenges highlighted in the paper. The analysis may be approached differently, for instance, using other probability distributions or design simulations to investigate this problem further. As the work proposes, developing advanced redundancy techniques may significantly improve HIBS operational readiness.

References

1. Bauer, E., Zhang, X., Kimber, D.A.: Practical System Reliability. Wiley (2009)
2. Correa, P.: Airbus and Salam join forces for high altitude platform station (HAPS) connectivity services (2022). https://www.airbus.com/en/newsroom/press-releases/2022-10-airbus-and-salam-join-forces-for-high-altitude-platform-station. Accessed 10 Nov 2022
3. Airbus Defence and Space: Zephyr, the High Altitude Pseudo-Satellite (2017). http://www.defence.airbus.com/portfolio/uav/zephyr/. Accessed 23 June 2017
4. Etherington, D.: Loon sets stratospheric sustained flight record with 312-day balloon trip, October 2020. https://techcrunch.com/2020/10/28/loon-sets-stratospheric-sustained-flight-record-with-312-day-balloon-trip/. Accessed 06 May 2023
5. HAPSAlliance. Haps operation using attended autonomous fleet systems: Collaborative traffic management for the stratosphere. Technical report, HAPS Alliance (2022)
6. Hilt, A., Járó, G., Bakos, I.: Availability prediction of telecommunication application servers deployed on cloud. Periodica Polytechnica Electric. Eng. Comput. Sci. **60**, 72–81 (2016)
7. IATA: Aircraft operational availability. Technical report, International Air Transport Association (IATA) (2018)
8. ITU: Resolution 247 (WRC-19). Technical report, International Telecommunications Union (ITU) (2019)
9. International Telecommunications Union (ITU): Terms and definitions. Radio Regulations Articles (2016)
10. Kurt, G.K.: A vision and framework for the high altitude platform station (HAPS) networks of the future. IEEE Commun. Surv. Tutorials **23**(2), 729–779 (2021)

11. Levin, M.A., Kalal, T.T.: Improving Product Reliability: Strategies and Implementation. Wiley (2005)
12. Levin, M.A., Kalal, T.T., Rodin, J.: Improving Product Reliability and Software Quality, 2nd Edn. Wiley (2019)
13. Maximator Test LLC: High pressure testing services, production and testing equipment sales. https://maximator-test.com/wp-content/uploads/2015/04/Maximator-Test-Reliability-2.pdf. Accessed 29 Oct 2022
14. Manzini, R., Regattieri, A., Pham, H., Ferrari, E.: Maintenance for Industrial Systems. Springer, London (2010). https://doi.org/10.1007/978-1-84882-575-8
15. Mershad, K., Dahrouj, H., Sarieddeen, H., Shihada, B., Al-Naffouri, T., Alouini, M.-S.: Cloud-enabled high-altitude platform systems: challenges and opportunities. Front. Commun. Netw. **2**, 716265 (2021)
16. Pham, H.: System reliability concepts. In: System Software Reliability. SSRE, pp. 9–75. Springer, London (2006). https://doi.org/10.1007/1-84628-295-0_2
17. Pinelis, Y.K., Whitledge, W.R.: Tutorial: parametric reliability models. Technical report, Institute for Defense Analyses (2018)
18. Rahman, I., et al.: 5G evolution towards 5G advanced. Technical report, Ericsson (2021)
19. ReliabilityAnalyticsToolkit: Confidence Limits - Exponential Distribution. https://reliabilityanalyticstoolkit.appspot.com/confidence_limits_exponential_distribution. Accessed 29 Oct 2022
20. Giovanni Romano and Balazs Bertenyi. 3GPP Finalizes Contribution for IMT-2020. Technical report, Third Generation Partnership Project (3GPP) (2020)
21. Safie, F.M.: Understanding the elements of operational reliability: a key for achieving high reliability (2010)
22. Sage, A.P. (ed.) Reliability Engineering. Wiley (2012)
23. Sharma, M., Chadha, D., Chandra, V.: High-altitude platform for free-space optical communication: performance evaluation and reliability analysis. J. Opt. Commun. Network. **8**(8), 600–609 (2016)
24. Tingley, B.: So close! Zephyr drone lands just hours before setting flight-duration record, August 2022. https://www.space.com/airbus-zephyr-drone-long-lands-before-record. Accessed 06 May 2023
25. Wakabayashi, D.: Google-linked balloon project to provide cell service will close (2021). https://www.nytimes.com/2021/01/21/technology/loon-google-balloons.html. Accessed 10 Nov 2022
26. Weibull.com: Availability and the Different Ways to Calculate it. https://www.weibull.com/hotwire/issue79/relbasics79.htm. Accessed 21 Sept 2022

e-Commerce Solution with Financial Management System and Android Mobile Application

Irish C. Juanatas[1](✉), Roben A. Juanatas[2], Niamh Denae R. Pontanares[1], and John Ivan C. Maurat[1]

[1] FEU Diliman, 1101 Quezon City, Philippines
{icjuanatas,jcmaurat}@feudiliman.edu.ph
[2] National University, 1008 Manila, Philippines

Abstract. The food service industry has increasingly adopted e-commerce as a means of improving service and accessibility for customers. The popularity of e-commerce mobile applications has also surged due to their convenience on mobile devices, which have become an integral part of people's daily lives. These applications enable businesses to enhance their market presence and attract a larger customer base. Additionally, food businesses are utilizing Financial Management Systems (FMS) to effectively track cash inflows and outflows. FMS is typically integrated with various features such as sales and inventory tracking. The primary objective of this study was to develop an e-commerce coupled with a financial management system and its mobile application counterpart. The employed methodology for this study was descriptive research. A survey was conducted to evaluate the acceptability of the system based on functionality, usability, reliability, performance, and supportability (FURPS) criteria. The developed system received a favorable rating, with a weighted mean of 4.05. This indicates that the system successfully executed required tasks, offered user-friendly features, effectively monitored orders, consistently performed, and was accessible from any device.

Keywords: e-Commerce · Financial Management System · Mobile Application

1 Introduction

Filipinos are frequently described as being ardent shoppers, particularly in light of the growth of online marketplaces that give them access to a variety of products over the Internet. The COVID-19 pandemic's aftermath has increased people's use of technology in their daily lives, and one key component of this popularity is the use of e-commerce [1]. e-Commerce refers to the practice of purchasing and selling products through the Internet, including electronic exchanges between buyers and sellers [2]. Additionally, e-commerce provides businesses with inexpensive advertising and marketing opportunities, enhancing consumer connection and enabling companies to react more quickly to

K. Arai (Ed.): FTC 2023, LNNS 816, pp. 259–265, 2023.
https://doi.org/10.1007/978-3-031-47448-4_18

market and customer demands [3]. The Philippines' retail e-commerce business is constantly growing, with well-known sites like Lazada, Shopee, Zalora, Ebay, and Kimstore at the forefront.

Financial management systems serve to automate financial activities and operations, with the primary objective of gathering and analyzing financial data to enhance business decision-making. Bookkeeping, budgeting, and finance monitoring are included in this kind of software program [4]. They are created based on the unique issues that businesses encounter and are suited to their operations [5]. As a result, they help organizations acquire precise financial data and improve data security measures [6]. Many startups and small to medium-sized businesses (SMEs) in the Philippines are now taking advantage of the benefits provided by online finance management solutions [7].

Mobile applications, commonly referred to as apps, are software designed to run on mobile devices. While they may have certain limitations compared to computer programs or systems, mobile applications are still capable of performing similar functions [8]. Users now spend an increasing amount of time utilizing mobile applications, as they have become an integral part of their daily routines. Notably, 62% of smartphone users utilize their mobile devices for online product purchases [9].

Therefore, the researchers developed an e-commerce solution with financial management system specifically tailored to businesses. Businesses can move toward more effective financial management practices by implementing such a system. Owners of businesses will be able to effectively enter and manage their budgets, funds, sales, and inventory. Additionally, business owners will have access to useful data on the total gross sales and the number of orders processed for each product.

The subsequent sections outline the Methodology, which includes the study design, data collection methods, and instruments utilized in gathering the necessary data. The Results and Discussion section presents the results and provides a summary of the findings. Finally, the paper ends with the Conclusion and Recommendations section.

2 Methodology

The system architecture provides a conceptual model that describes the structure and behavior of a system and its subsystems. The system architecture displayed in Fig. 1 illustrates that the administrator and staff have access to the e-commerce web and mobile applications of the system. The administrator, staff, and the customer have access to these modules which are the homepage, sign-up, log-in, ordering, and order tracking modules. These end-users are also able to utilize the shopping cart and order receipt, which are both a part of the e-commerce ordering process.

The Likert Scale, as shown in Table 1 is a unidimensional scale used by researchers to gather and understand respondents' attitudes and opinions. This is also applied to acknowledge the views and perspectives toward a brand, product, or target market ([10]. As for this study, the role of the Likert Scale is to let the research respondents select one option that fits their views best, for each question in the survey. The survey was administered using Google Forms and completed by 100 respondents.

The website was built on Hypertext Preprocessor (PHP) while the mobile application was based on Java. Moreover, HyperText Markup Language Version 5 (HTML5) was

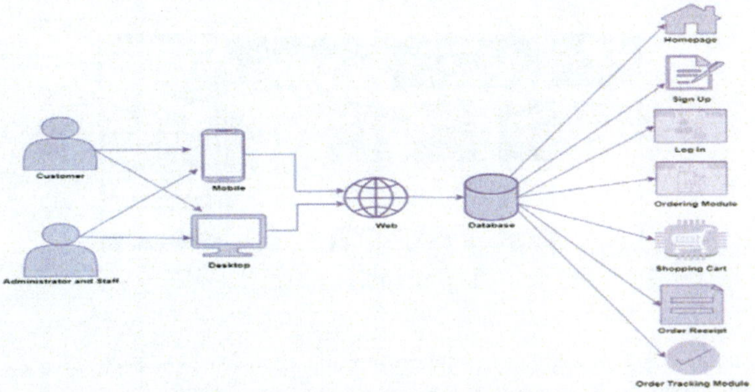

Fig. 1. System architecture of the study.

Table 1. Likert scale.

Point	Scale Range	Equivalent
5	4.21–5.00	Strongly Agree
4	3.41–4.20	Agree
3	2.61–3.40	Neither
2	1.81–2.60	Disagree
1	1.00–1.80	Strongly Disagree

used to display the contents of the website. To have an attractive user interface, Bootstrap and Cascading Style Sheets (CSS) are also written into the website's code. JavaScript and PHP are utilized for website interaction and functionality. PHP plays a big role in implementing the financial management system of the website and integrating data into the database of the system.

Customers may purchase products using the Order module of the website and mobile application, presented in Fig. 2. This module displays all of the company's products-specifically their names, prices, and description. Customers can browse the various menu items on this page, which acts as a central hub. The chosen items are then put into their virtual shopping cart by clicking the "Add to Cart" button, enabling customers to continue with the purchase confirmation and checkout process. Customers can easily choose and manage their favourite products before completing their purchase using the order module's user-friendly interface.

Business owners can manage their budgets and expenses using the budgets and expenses module as shown in Fig. 3. This module is the integration of the financial management system into e-commerce. There are two separate tabs, one allotted for budgets and the other for expenses. The expenses for the month are presented in a table and each column describes the expense ID, description, cost, the date it was logged, and who logged it. The total cost of the expenses is also calculated and reflected at the bottom

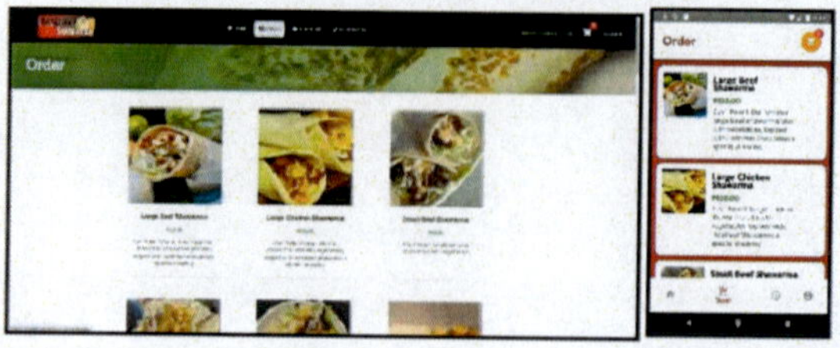

Fig. 2. Order module.

of the webpage. Additionally, the total cost of all expenses is shown to give a picture of the overall expenses. The remaining budget is shown in relation to the current date on the Budget tab. Business owners can use this function to keep track of their spending and compare them to the budget that has been set aside. By supplying this data, business owners can efficiently keep track of their financial situation and make wise judgments about their spending in the future.

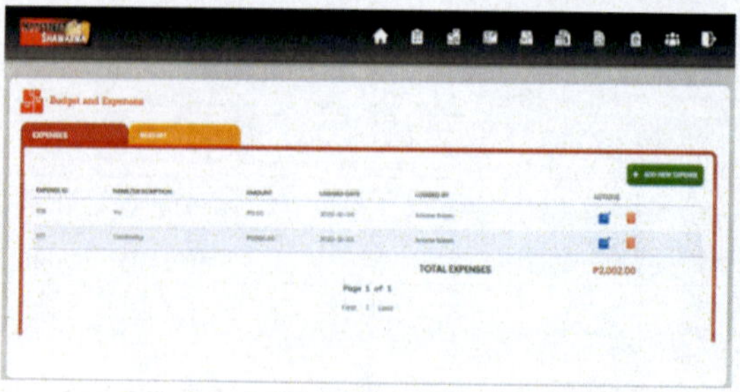

Fig. 3. Budget and expenses module.

Administrators can easily access different types of reports including sales, inventory, finance, and log reports through the module available for generating such crucial data. Refer to Fig. 4 for clarity on how this process works and efficiently helps in analysis which leads to sound decision-making. A summary of the company's sales performance, including revenue, sales volume, and customer trends, is given in the sales report. Information on current stock levels, item availability, and any inconsistencies or alterations

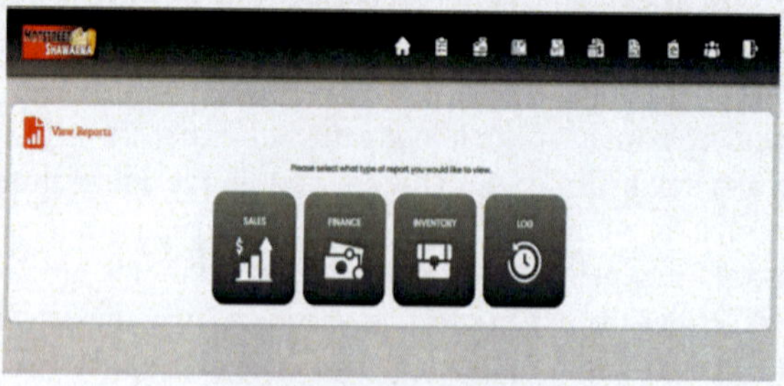

Fig. 4. Reports module.

is provided in the inventory report. The finance report provides information on the company's finances, including cash flow, costs, and profitability. The log report also serves as a record of user actions, system activity, and any pertinent events for audit and security purposes.

3 Results and Discussion

During the testing, the weighted mean score of Functionality was found to be 4.09, suggesting users' views regarding system capabilities. As such, it indicated adequate performance levels when executing requisite operations and denotes that described functionality is acceptable among users.

The usability of the system received a weighted mean score of 4.14 from respondents representing agreement that it is a highly usable platform. This suggests that a significant portion of users find the interface visually attractive and easy to comprehend while also offering smooth navigation and facilitating efficient food purchases.

The weighted mean score of 4.00 reveals a unanimous agreement among the respondents regarding the reliability aspect. This implies that the system proves to be dependable in exhibiting pertinent order details and total costs for customers. It operates with high consistency levels whilst minimizing errors, offering sufficient validation measures for errors encountered.

The results indicate that there is an agreement with regards to Performance on the system which has been reflected in an impressive weighted mean score of 3.96. This substantial agreement on its efficient operations typically involves limited processing lags. In addition, users have praised its easy-to-use interface and accessibility across various internet platforms.

The overall agreement among respondents can be observed in the Supportability ratings, where the weighted mean score stands at 4.06. This shows that users find the system to be successful in fulfilling their requirements while also allowing them to monitor their orders with ease, coupled with its secure storage of sensitive user data. Additionally, transactions are perfectly safe owing to robust security settings while this impeccable design enables accessibility from diverse devices.

The overall weighted mean summary of FURPS assessment can be found in Table 2. Across all criteria, the system received an overall interpretation of "Agree," with a calculated weighted mean score of 4.05 pointing to users finding the developed system acceptable. While the system appears ready for deployment and use by customers, administrators, and business owners within the company; there are specific modules where improvements could be implemented to enhance user experience further. Additionally, Performance had a lower average score compared to other criteria. This indicates opportunities for improving certain aspects concerning user interactivity and system performance.

The system was suited for handling complex tasks like order processing and keeping tabs on inventory and proved to be valuable in many ways. To better serve customers' needs, it facilitated an online platform that increased product accessibility alongside simplifying their pick-up ordering procedure. With equal efficiency both through its website and mobile application features, this design optimized customer satisfaction while ensuring smooth business operations.

Table 2. Overall weighted mean summary of FURPS

Criteria	Weighted Mean	Interpretation
Functionality	4.09	Agree
Usability	4.14	Agree
Reliability	4.00	Agree
Performance	3.96	Agree
Supportability	4.06	Agree
Overall Weighted Mean	4.05	Agree

4 Conclusion

In conclusion, this study focused on the development of an e-commerce solution integrated with a financial management system and an Android mobile application. Business owners can display their products online and take orders using a website that integrates e-commerce. A financial management system is also available that combines the management of orders, inventories, finances, products, and users into a single platform. This produced downloadable monthly reports for sales, inventory, finances, and logs. In addition to creating an e-commerce website, the researchers also developed an Android mobile application. Customers can place orders, provide payment documentation, and track their orders using a mobile device. The system underwent evaluation using the FURPS (Functionality, Usability, Reliability, Performance, and Supportability) criteria, and the results indicated that it was deemed adequate for the business requirements. The positive feedback received from both the business owners and customers through surveys further confirmed the system's effectiveness and user satisfaction.

5 Recommendations

Based on the development of the e-commerce solution with financial management system and Android mobile application, several recommendations can be made to further enhance the system. Develop the system using frameworks like Laravel or CodeIgniter. Because the database efficiency improved through frameworks, which also reduced the amount of repetitious code that must be written. To apply other financial management system modules that will help to improve the business environment such as handling a large scale of business data and adding the payroll system for handling employees within the business. Integrate existing payment platforms or APIs that will allow customers to pay directly through the application.

References

1. Surjandy, L.Y., Meyliana, E.Y., Cassandra, C., Kusumaningtyas A.: Analysis of product, product delivery service, and product assurance in e-commerce on purchase intention during the COVID-19 Pandemic, International Conference On Information Management And Technology (Icimtech), pp. 825–828. (2021)

2. ICAAL Academy Homepage. https://www.internetconsultancy.pro/blog/what-is-ecommerce/, (Accessed 26 Oct 2022)
3. Amazon Homepage. https://sell.amazon.in/seller-blog/advantages-of-ecommerce, (Accessed 27 Aug 2022)
4. Tarim, E.: Modern finance theory and practice and the anthropocene. New Political Economy, pp. 490–503 (2022)
5. Wu, G.: Computer finance management system innovation thinking. J. Phys. Conf. Ser. (2020)
6. Lim, F.: The impact of information technology in corporate financial management. Future Generation Commun. Netw. (2017)
7. CloudCfo Homepage. https://cloudcfo.ph/blog/accounting/5-accounting-and-finance-services-benefit-to-your-business, (Access 19 May 2022)
8. Martinez, E.: React: Cross-Platform Application Development with React Native: Build 4 real-world apps with React Native. Packt Publishing (2018)
9. Rowles, D.: Mobile Marketing: How Mobile Technology is Revolutionizing Marketing, Communications and Advertising. 2nd edn. Kogan Page (2017)
10. Roy, A.: A Comprehensive Guide for Design, Collection, Analysis and Presentation of Likert and other Rating Scale Data: Analysis of Likert Scale. Amazon (2020)

Effect of Age-Progressed Avatars on Savings Behaviors for Retirement in Young People

Daniel McCloskey and Marian McDonnell(✉)

Institute of Art, Design and Technology, Dun Laoghaire, Dublin, Ireland
marian.mcdonnell@iadt.ie

Abstract. Research has shown that interventions that help a person visualise their future, either by creating an age-progressed image of their future appearance or through exercises which help participants imagine their future lifestyle, increases a person's connection to their future-self. This in turn increases how much they are willing to contribute towards their pension. This study furthers this research by fully integrating a future-self avatar within a pension app. This avatar not only allows participants to see what they could look like in the future, but also provides prompts and feedback on their saving decisions. A mixed-method, between groups study (n = 44) compared an app including a future-self avatar with one without any intervention. Results found no significant difference in closeness to future self, increase in contributions, confidence in their ability to save for the future or user engagement. However, across both groups a majority of participants identified that the saving rate was too low to achieve a comfortable lifestyle and as a result, increased their pension contributions within the scenario. This result reinforces the impact of labelling when trying to encourage pension savings. Furthermore, qualitative findings highlight key considerations which should be taken into account when implementing future-self interventions in further studies, in particular the incorporation of aged-progressed imagery.

Keywords: Vividness Interventions · Age-Progressed Avatars · Future Self · Temporal Discounting · Pensions · Financial Technology · Psychological Connectedness · Smartphone Apps · User Experience Design · Saving Behaviour · Retirement Planning

1 Literature and State-of-the-Art

1.1 Introduction

Irish people are going to have to save more if they are to achieve a comfortable lifestyle in retirement. Currently, there is an over reliance on the flat rate-state pension which is aimed to prevent pension poverty and is not designed to ensure a high level of adequacy [25]. Less than half of the population on the state pension has any access to an additional pension scheme to make up any shortfall [9]. With an aging population [45, 46], there will be even fewer workers to pay for these benefits. Currently, there are five workers for every retiree, contributing to the state's pension, but in thirty years' time, this number is expected to reduce to two workers for every retiree. This will put a massive strain on the state's finances if not addressed [6, 25].

K. Arai (Ed.): FTC 2023, LNNS 816, pp. 266–285, 2023.
https://doi.org/10.1007/978-3-031-47448-4_19

Encouraging young workers to contribute more at an earlier age makes adequacy goals more achievable. However, even those that are members of a pension scheme, often do not realise that the contribution rate at which they are saving will not provide a comfortable level of income when they retire [47, 2, 37]. Although the benefits seem clear, in practice, it is often difficult to encourage people to forgo short term gains for long term rewards [30, 44], as proposed by a theory known as temporal discounting [29]. Temporal discounting or time discounting as it is sometimes known, is a term used to describe people's tendency to place less value on future events over present-day events [14]. Take for example the person smoking a cigarette, who is placing greater value on the immediate nicotine hit as opposed to the long-term health consequences which could appear years into their future. Multiple researchers have noted that one of the reasons that this is difficult is because we feel psychologically disconnected from our future self [1, 36, 39, 43]. This psychological disconnection can occur when we perceive little commonality in our interests and motivations between our present and future selves. At its most extreme, we can view our future self as a completely different person [36]. In this instance, saving for retirement can feel like giving money away to someone else instead of investing it for one's own future. Experiments have shown that by aiding individuals in imagining their future self, individuals feel more connected and in turn are willing to save more for retirement [19].

This review will look at current research on 'future self' and vividness interventions to influence behavior and their implications for increasing saving.

1.2 Future Self

Although in theory, we would like to make decisions today that will allow us a more comfortable future, in practice this is often difficult to achieve. Present day temptations often trump our best intentions, as can be seen in the teenager who takes up smoking not worrying about their future health or treating ourselves to dessert when trying to lose weight.

Parfit [36] suggested that a person's life could be viewed as a series of overlapping selves, and therefore our future self could be seen as completely different person, with completely different desires and motivations than our present self. This suggests that our inability to delay rewards and instead preferring immediate gratification is because our future self may feel like a completely different person. Therefore, there seems no benefit in waiting for future rewards.

A 2006 study showed that when participants had to make decisions based on their present and future self, the decisions they made for their future-self were more like those of an observer [38]. Following this a 2009 study by Hershfield et al. [22] identified different areas in the brain being activated using an MRI scanner when subjects where asked a series of statements regarding their present and future selves. When subjects were asked about their future selves and when they were asked to think about a stranger, the same area within the brain was activated, supporting the future-self theory proposed by Parfit [36]. From these researchers were able to predict how much subjects would save on a money allocation task when tested a week later. With participants who felt the least connection to their future-self saving the least amount of money.

Research has shown that by increasing the psychological connectedness between the present and future self that participants are more able to identify with their future self and this is reflected in the decisions they make, for example willing to sacrifice a greater portion of their salary towards their pension [19, 2–4].

Recent research has looked at incorporating vividness interventions to help people better imagine their future. Vividness interventions use visual representations or imaginative thinking to help participants feel a greater connection to their future self [20].

1.3 Vividness Interventions

Visual Representations. One study carried out by Stanford researchers [19], had 33 participants randomly assigned to one of two conditions in a virtual environment. Participants would be able to interact with an avatar of their current self or alternatively an aged representation of themselves within a 3D environment. The testing procedure from Yee and Bailenson's [49] research into the effect a digital avatar has on a participant's behavior, was adopted to establish a connection between the participant and their avatar. Once in the virtual environment, participants were asked to turn 180 degrees and to confirm they could see their representation in a virtual mirror. After this was confirmed a series of exercises were performed, such as head tilting and nodding in front of the mirror to make sure the participant felt comfortable and to allow them sufficient time to interact with their avatar. On average participants interacted with their avatar for between 60–75 s. On completion of the exercise, they were asked to turn back around 180 degrees to face the front of the room. As they turned to the front of the room they were greeted by a confederate within the virtual environment. The confederate asked the participant a series of questions (e.g., "What is your name?" "Where are you from?" "What is your passion in life?"). These questions were adapted from Yee and Bailenson's [49] research in an effort to reinforce the connection between the participant and their avatar. The participants then removed the apparatus and completed a money allocation task, which showed that participants who had been exposed to their aged avatar saved twice as much for retirement. To make sure participants did not feel primed to think of aging before taking part in the allocation task, the study was repeated but, on this occasion, half interacted with another research participant's aged avatar and the other half interacted with their own aged avatar. Again, participants who had been exposed to their own aged avatar showed increased saving behavior. As these are both costly and time-consuming interventions, the third part of the study looked at trying to implement less intensive interactions to see if it would encourage the same response. In the final part of the study, researchers carried out a variation of the same experiment, but in this instance, volunteers submitted a profile picture of themselves in which they have a neutral expression prior to the test. Three to four weeks after receiving the photos, participants were sent personalized emails inviting them to participate in an online study (Fig. 1). 38 adults aged between 18 and 35 were recruited from the Mechanical Turk online community and randomly allocated to one of two conditions, either seeing an image of their current self or an aged-progressed image. Using a sliding scale volunteers could decide how much they would like to allocate in retirement. Even in this instance, where it was a much less intensive interaction, participants reported a much greater connection with their future-self when exposed to an age-progressed portrait, and this was again reflected in an increase in the amount of money they allocated to their pension.

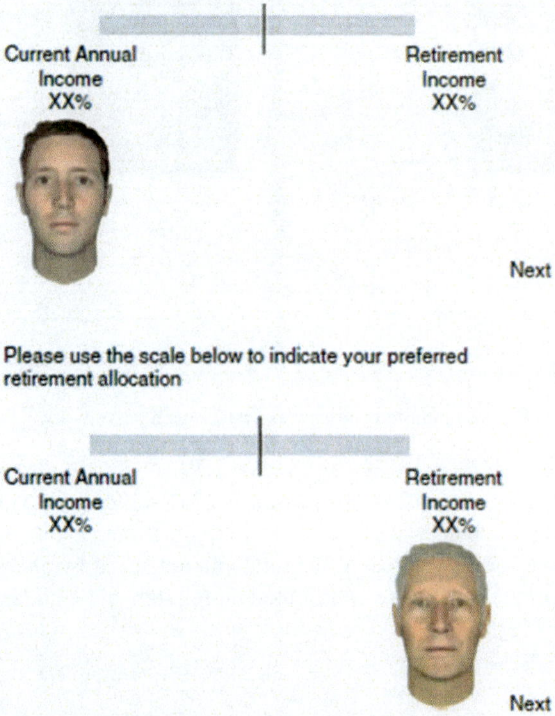

Fig. 1. Sample avatars from 2011 study [19].

Further research was conducted by Sims et al. in 2015 [40] to replicate and extend Hershfield et al.'s previous findings [19]. One hundred six (106) community college students aged between 18 and 45 taking a financial planning course were recruited for the experiment. At the start of the semester students were randomly assigned a profile page (Fig. 2), containing either a present-day avatar of themselves or an age-progressed version. Students were emailed a total of 5 times throughout the year to complete the profile along with a questionnaire, allowing volunteers a significant amount of time to interact with their profile. Researchers found that students, who could view their aged profile, had more correct answers and exhibited more confidence on a financial literacy test as well as more interest in attending future classes [40].

In 2018, a Mexican bank carried out several field studies to try and replicate Hershfield et al.'s 2011 [19] findings among its members. First customers were emailed and texted with a link to a website and prompted to "meet their future self' (Fig. 3). Clicking the link would take them to a website where they could take a picture, and have it aged with a filter to see what they could look like in retirement. This was followed with an empathetic message encouraging them to contribute to their future self by following

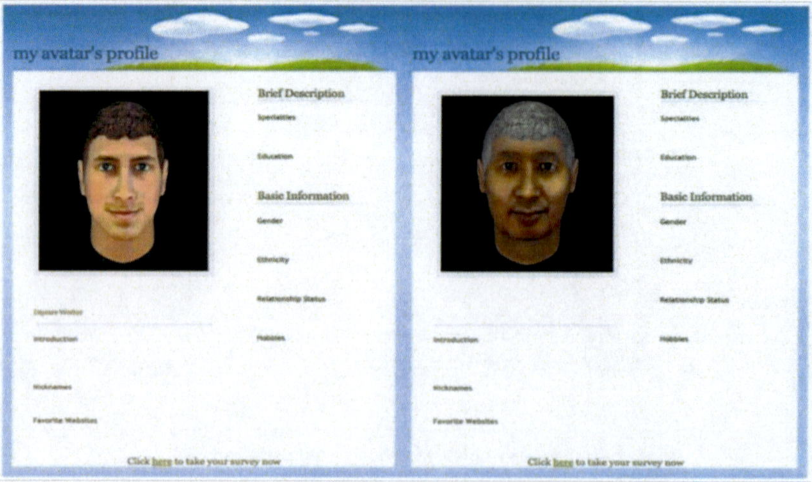

Fig. 2. Sample profiles and avatars used in study [40].

a link to the bank's website. 2,234 clients took 13,041 selfies with 683 (3%) customers clicking through to the bank's website. They found that the customers who viewed an aged version of themselves increased the total amount saved by 54%. Customers who viewed an aged-progressed image contributed more often and with larger deposits than the control group [13].

Fig. 3. Intervention flow from field study with Mexican bank [13].

1.4 Imaginative Thinking

The same bank carried out a field study in 2017 in which bank representatives guided customers through an exercise to help them visualise their future [13]. Customers were asked to answer a series of questions on their future goals; this helped populate a script which customers read aloud at the end of the exercise. The task was completed with customers being asked whether they would like to sign up for automatic savings contributions. 500 bank representatives engaged with more than 8,000 account holders, randomly assigning them to one of three conditions:

1 **Control group:** customers were given information on how to make voluntary contributions along with the income they could expect to receive in retirement,
2 **Future visualisation:** customers were encouraged to think of their current lifestyle and prompted to imagine how they would like to live in the future,
3 **Past reflection:** customers were encouraged to think how their past decisions had impacted their current lifestyle and how current decisions could impact them going forward.

Results showed that the future visualisation exercise was the most effective intervention. With 20 times more customers likely to sign up for automatic contributions when encouraged to visualise their future self when compared to the control group [13]. This showed the benefit that even such low-tech solutions could have in increasing retirement savings by adopting simple visualisation techniques. Researchers estimated this number could have been higher still with a streamlined process [13], highlighting the need for products which make contributing towards your pension a seamless interaction.

In 2020, the Behavioural Insights Team conducted a study with pension provider Scottish Widows to find out how to increase engagement with pensions among young people [11]. 2,882 people in the UK aged between 22 and 29 took part in the online experiment. Participants were presented with a vignette of a representative 25-year-old who earned the national average income and made default contributions of 8% of his salary towards his pension. Afterwards they were randomly presented with one of four messages and asked whether they thought the person in the scenario was contributing enough to their pension and whether they should change their contributions. Finally, participants were asked to consider their own situation and if they should increase their own contributions. They found that group who were asked to think about their future was 11.2% more likely to feel they should increase their contributions. Prior to taking part in the experiment respondents were asked what prevented them from saving more for retirement. The main issues where unsurprisingly affordability but following that a significant proportion had never considered saving more for retirement (21.1%) and 15.4% did not know how to go about contributing more towards their pension. This highlights the lack of knowledge around pensions and the opportunity novel approaches such as making the future more salient have in encouraging young people to save more for retirement. These studies [19, 13, 40, 11] show the clear practical benefits of vividness interventions and how they can play a role in encouraging greater saving.

1.5 Marketing Campaign

Merrill Edge Face Retirement (2012). In 2012, Merrill Lynch [48] created an advertising campaign based around Hershfield et al. [19] findings. People were directed to a website that would allow them to see an age-progressed portrait of themselves after uploading a suitable image (Fig. 4). The app also provided cost-of-living projections to show what a pension would be worth on retirement. The website encouraged people to share their image through social media as well as visiting Merrill Lynch for pension advice. Although it was widely shared on its first release, the site is no longer available.

Fig. 4. Screenshot from Merrill Lynch face retirement [48].

The site had a shock value which although seems to have encouraged engagement, over the long term might have dissuaded people from using the product. Like Fertig et al. research [13], it existed in its own application and was not fully integrated within a pension app, with its main focus to gain customers attention through sharing on social media platforms. Beyond the initial novelty of viewing your future self, it did not have any deeper interactions to keep customers using the product, so any psychological connectedness may therefore be short lived.

This Present Study. This present study will follow up on Sims et al. [40] research by looking at a new area where the future self can be incorporated into a financial setting, by integrating the future-self fully within a pension savings app. Hershfield et al. [19] suggested that more vivid interventions could yield a greater response. Since the time of the original study, several programs have been created that use artificial intelligence to produce realistic images of our future self. For this study, we will use the latest age progressed software in FaceApp [12] in the belief that more accurate representations will help people better envisage their future and in turn make them feel closer to their future self. Previous research has shown that the greater the distance between the present and future self, the harder it is to feel psychologically connected [28]. Hershfield et al. [21] speculated that in the case of pensions where there is such a distance, interventions would need to be even more vivid to overcome this psychological disconnection. Wilkinson noted innovative engagement is needed to improve retirement outcomes for young savers [47]. To create a more vivid experience, the prototype will have a future version of the participant that will provide prompts to encourage better long-term saving habits. This will give a more in-depth interaction between the participant and their future avatar, beyond just seeing an aged-progressed image as found in previous interventions. This research will explore whether using vividness interventions to help visualise aging within a pension app will make the participants feel closer to their future self, which in turn will encourage them to save more. Following this does the novelty of interacting with your future self make a more satisfying user experience.

1.6 Research Questions

As part of the study the following questions will be answered:

1. Does viewing an age progressed rendering of the customer make them feel closer to their future self?
2. Does viewing an age progressed rendering of the customer encourage them to save more within a pension app?
3. Do prompts asking the customer to imagine their future, make customers more confident in their ability to save for the future?
4. Do they find the aged-progressed prototype more usable/satisfying and easier to use?

Based on the research questions above, the following hypothesis statements are outlined below;

H1. Viewing an age progressed rendering of the customer makes them feel closer to their future self.

H2. Interacting with an age progressed rendering of the customer in prototype A will encourage them to save more, in comparison to prototype type B without it.

H3. Prompts asking the customer to imagine their future will make customers more confident in their ability to save for the future.

H4. Interacting with an age progressed prototype will be more usable/satisfying and easier to use.

2 Methodology

The project adopted the design thinking framework developed at IDEO [15] to create the prototype. The 'Design Thinking' ideology believes that putting users' needs at the center of the design process, allows designers to create innovative solutions that will have a competitive advantage in the marketplace [27, 5]. The framework consists of three distinct phases, Phase 1 Understand, Phase 2 Explore and Phase 3 Materialise with each phase broken into a further two subsections. See Fig. 5 below.

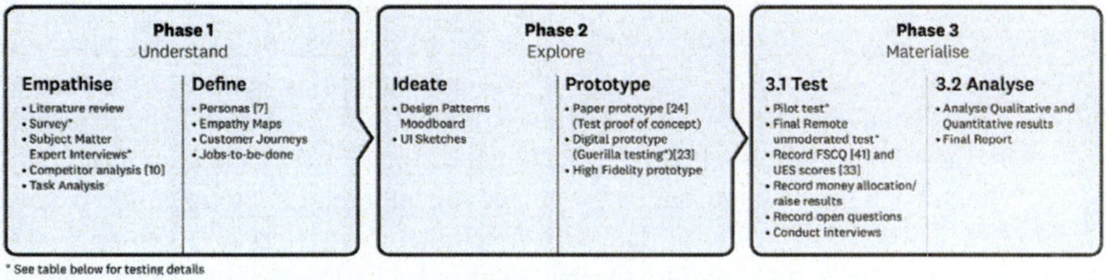

Fig. 5. Project phases.

	User Survey	Subject Matter Expert interviews	Guerrilla testing	Pilot testing	Final test
Participants	95 respondents	Interviewee (Topic) 1.Hal Hershfield - Professor, UCLA (Future Self) 2.Financial Planner (Irish Pensions) 3. Group Insight Manager - Royal London (UK Pensions) 4.Pension Supervior (UK Public Sector Pensions)	6x2	6	40+
Measures	Likert scales, single answer & open questions	Qualitative questions	Success rate, usability issues & qualitative feedback	Success rate, usability issues & qualitative feedback	FSCQ, UES, money allocation qualitative feedback
Analysis	Thematic analysis	Thematic analysis	Issue pritorisation analysis	Issue pritorisation analysis	Mixed method analysis

Fig. 6. Research and testing procedure.

In Phase 1 primary research was conducted through a survey, competitor analysis [10] and Subject Matter Expert interviews (See Fig. 6.) **with the findings being used to** create a persona [7] of our primary users that acted as guide to understand their needs, goals and behaviours when creating the app. In Phase 2 a prototype [24] was designed and iterated upon through two rounds of guerilla testing [23, 26, 31] to ensure optimal usability [32], **before the final experiment. In Phase 3, a full pilot test was performed before conducting the final experiment.**

The research design of the final evaluation employed a mixed-method [8], between groups study (n = 44) and compared an app including a future-self avatar with one without any intervention.

The final testing procedure was adapted from **Behavioural Insights Team** [11] research in which participants were presented with a scenario in which they were earning the median salary and making the minimum contributions of 8% to their pension. During the onboarding, participants using the intervention prototype, were presented with prompts encouraging them to think about their future self before being presented with their age-progressed avatar. Following feedback from early prototypes and to provide greater context, research from the retirement living standards [34] [42] was incorporated into both prototypes and adapted to an Irish setting [16]. The retirement living standards provide real world examples of what income would be needed to enjoy a minimum, moderate and comfortable lifestyle in retirement.

In keeping with Sims [40] research participants using the intervention were asked if they felt the aged avatar was an accurate representation of what they could look like in the future, and asked to rate how they felt, in effort to detect any adverse reactions which could affect the results. Following this both groups were asked if they felt the current contribution rate was too low, too high or about right. They then had the chance to increase the monthly contribution rate and to make a one-off payment after receiving a €1000 bonus. Each time a contribution was made the intervention group would receive a personalized message from their future-self. After interacting with the app participants completed the Future Self Continuity Questionnaire [41] to detect if they felt any closer to their future self after interacting with the prototype and also completed the short form of User Engagement Scale [33] to determine whether the novelty of interacting with an age progressed avatar made for a more engaging experience, this was followed by a single item question on their confidence in saving for the future.

3 Results

3.1 Quantitative Analysis

The experiment was live for three weeks and received (46) responses with 21 using the aged-progressed prototype and 25 using the control version. A check for outliers was conducted before any analysis was performed. One response was removed from the intervention group, as the participant was not emailed a personalized test, with the presumption that a member from the intervention group forwarded on their individual test. One volunteer was removed from the control group for being above the age range of 20–45.

There was a slight majority of female (56.8%, n = 25) to male (43.2%, n = 19) with 34% of participants stating they did not have any type of workplace pension. The age range of participants was between 21 and 45, with the average age being 31 (see Table 1.).

Descriptive Statistics

Table 1. Descriptive statistics of all participants.

	N	Minimum	Maximum	Mean	Std.	Skewness		Kurtosis	
	Statistic	Statistic	Statistic	Statistic	Deviation Statistic	Statistic	Std. Error	Statistic	Std. Error
age	44	21	45	30.93	7.267	.214	.357	-1.313	.702
Valid N (listwise)	44								

A further analysis was conducted to detect if there was any variation between the two groups which could affect the results. The aged avatar progression group had 20 respondents and was split evenly between men and women. 60% had no workplace pension, however, the aged avatar intervention group skewed younger with a median age of 23 (Md = 23). The control version had 24 respondents with a majority of female respondents (62.5%/37.5%), in contrast to the intervention group, the median age was a lot older at 36 and a large majority did have a workplace pension (83.3% Yes,12.5% No,4.2% Don't know).

H1: Viewing an age-progressed rendering of the customer makes them feel closer to their future self.

The Shapiro-Wilk test was used to determine the normality of the results. The three subscales of the Future Self Continuity Questionnaire (Similarity to Future-Self, Vividness of Future-Self and Positive Affect to Future-Self) all had significance scores over .05

so were deemed normally distributed. An independent T-test was performed to determine if there was any significance in the results.

The Future Self Continuity Questionnaire total had a significance score of less than 0.5, so it was deemed that the data was not normally distributed, so a non-parametric Mann-Whitney test was performed to determine if there was any relationship between the two groups. An independent t-test was used to compare the average scores of the three Future-Self subscales between the two groups [35]. All p values scored over 0.05 so there was deemed to be no significant difference between the groups. A non-parametric Mann-Whiney test was performed on the Future Self Continuity Questionnaire Scale to compare the medians of the two groups [35]. The probability value was greater than 0.05 for all scales so there was deemed no significant difference between the two groups.

Therefore Hypothesis 1 was rejected, that viewing an aged-progressed image of the participant would make them feel closer to their future self.

H2: Interacting with an age-progressed rendering of the customer in the prototype will encourage them to save more, in comparison to the prototype without it.

A Mann-Whitney test was performed and found no significant difference between the two groups when it came to allocating money towards their pension either as a bonus or as one of contribution and when tested within the prototype and surveyed afterwards. Thus, hypothesis 2, that interacting with an age progressed rendering of the customer in a prototype will encourage them to save more, in comparison to the prototype without it, is rejected.

H3: Prompts asking the customer to imagine their future will make customers more confident in their ability to save for the future.

A Mann-Whiney test found no significant difference between prototypes rejecting hypothesis 3 that prompts, asking the customer to imagine their future, will make customers more confident in their ability to save for the future.

H4: Interacting with an age progressed prototype will be more usable/satisfying and easier to use.

The Shapiro-Wilk test was used to determine the normality of the results. The four subscales of the User Engagement Short Scale (Focused Attention, Perceived Usability, Aesthetic Appeal and Rewarding) all had significance scores under .05 so were deemed not normally distributed. A Mann-Whitney test was performed to determine if there was any significance in the results. The overall total for the UESS was .078 on the Shapiro-Wilk test, so was deemed normally distributed. An independent T-Test was run to determine if there was any difference between the two groups. The probability value was greater than 0.05 for all scales so there was deemed no significant difference between the two groups. Therefore Hypothesis 4 was rejected, that interacting with an age progressed prototype will be more usable/satisfying and easier to use.

Reaction to Aged-Progressed Imagery. A Mann-Whitney test was performed across both groups to see if there was any difference in mood and found no significant difference between the groups.

Contribution Rate Change. An analysis revealed that across the two groups, 47.7% of participants felt the current contribution rate was too low in the app scenario presented. When asked if they would like to change their contribution rate, only 20.5% kept it at the lowest default rate, with 45.5% increasing it their contribution rate to 12%. 29.5% would change the contribution rate to the highest amount of 15%.

3.2 Qualitative Analysis Themes Identified

Five semi-structured interviews were conducted with participants of the intervention prototype to gain qualitative insights. The transcripts were coded with descriptive codes/tags to perform a thematic analysis. The same codes where then used to analyse the open question responses from the survey. The main themes were grouped together to highlight key points (See Table 2. below).

Future Self Prototype Themes Identified.

Positive Towards Concept/Helped Think About the Future. People were generally overwhelmingly positive about the concept overall with 21 mentions across both the survey and interviews. The prototype was also deemed effective at helping participants visualise their future with 16 mentions.

"I think it's a great idea, I don't think people around my age are considering their future lives"

Positive Towards Design. People were generally positive towards the design (12), in particular the usability of the main screens (13). Participants liked how the income in retirement was broken into everyday items making it easier to visualise (8). Following on from this many people like the friendly design (8) finding it more engaging than similar apps.

"I thought it did an excellent job of making me think about the future. The breakdown of different lifestyles I'll be able to afford depending on my means by category of "Can own a car", "can eat out once a month" "Can go on holiday" was extremely effective in contextualising what money invested now will really mean in 40 years' time"

"It was really effortless. It was usable, friendly. It was very effective."

Negative Aspects. There were some usability issues with the onboarding in the prototype, with some finding difficulty with legibility of the text (2) and some skipping quickly past these screens which could have affected their connection with their future self (3). Some people commented that use of emojis made the app less trustworthy (2).

"On the opening screens, the text that described examples of activities among other things was too small and looked very out of place"

Reaction to Image. Although there was no quantitative data revealing any statistical difference in mood between the two groups, a qualitative analysis revealed that exposure to an aged avatar garnered several varying different responses. Some found that the aged-progressed avatar had a close family resemblance and was regarded as an accurate description of what they could look like.

"The photo of me in the future is shockingly accurate as it looks really similar to my mother! It made me laugh when I saw it from the shock of it being so realistic, but then made me really start to think about what it will mean to be relying on a pension."

Table 2. Qualitative analysis showing frequency of themes.

Tag	Age-Progressed Interview	Age-Progressed Open Questions	Control Open questions
Positive towards concept	9	12	9
Helped think about future	10	6	7
Positive towards design	9	3	1
Interested	0	1	0
Future image engaged them	2	2	N/A
Liked contextual information	7	1	0
Positive towards emojis	0	1	5
Positive towards font choice	0	1	5
Positive UI design	0	1	7
Usability	9	4	10
Friendly	7	1	2
Informative	3	0	7
Negative towards design	3	1	0
Problem with prototype	0	3	2
Negative towards image	8	3	N/A
Negative towards emojis	1	2	0
Negative towards onboarding	3	2	2
Would like more control over image	12		
Negative emotions	7		
Mixed feelings	6		
Would like more control over choice	4		
Positive emotions	3		
Negative towards legibility	2		
Future image engaged them	2		
Accurate representation	1		
Dissuaded them from thinking of the future	1		
Learned new things	1		

Others had a negative reaction to their aged avatar, and it made them think negatively about their future.

"I think seeing yourself as an older person may be uncomfortable for a lot of people. I found seeing the image brought on negative emotions"

Whereas a third just found the age progressed image inaccurate representation of what they would look like in the future.

"The older version of the real photo is quite strange and I wouldn't recognise myself at all. It didn't look good, quite weird, and I spent a lot of time on it so it is indeed distracting from prototype and task"

A common sentiment expressed within the interviews was an ability to have greater control over what the final image looked like.

Control Version. Within the control group, people were again very positive towards to the idea of an app that helped them manage their pension with nine thematic themes. Volunteers found that the "lifestyle in retirement" feature helped them think about the future even without the age progressed imagery (7).

In contrast to the intervention prototype, ease of use (10) was the most common theme and people were more positive to the overall design, with informative (7), positive toward UI design (7), positive towards emojis (5), positive towards font choice (5) all ranking prominently.

Negative Aspects. Two negative issues were raised across both versions. One of these was the legibility of the text on the onboarding screens with 2 mentions in both groups. This related to some of the smaller condensed font. Another minor issue was with the test itself, with some mentioning they had skipped screens during the testing and were unable to go back to the previous task.

4 Discussion

This section will discuss the results of the between groups user test and qualitative analysis, and how the findings may inform the design of future pension apps which incorporate aged-progressed avatars. There was no significant difference between groups for any of the hypotheses; as a result, the null hypotheses for **H1, H2**, **H3** and **H4** were retained. This conflicts with Stanford's research [19] which showed that participants who had been exposed to their aged avatar saved twice as much for retirement. In Stanford's research [19], however, the participants had a greater length of time engaging with their avatar. However, the results agree with Merrill Lynch experiment which created an advertising campaign based around Hershfield et al. [19] findings. Merrill Lynch experiment directed people to a website that would allow them to see an age-progressed portrait of themselves after uploading a suitable image [48]. The site had a shock value which, although seems initially to have encouraged engagement, over the long term might have dissuaded people from using the product. If people had a negative reaction on seeing their avatar, it would be difficult to develop continuous and deeper interactions with the prototype.

However, a conclusion from this can be drawn that neither prototype performed any worse than the other. Across both groups, a significant proportion identified that the money they were saving would not give them a comfortable lifestyle in retirement and increased their contributions accordingly. Qualitative feedback relating to both prototypes was generally positive and people were intrigued by the concept of seeing their future self. This research shows the interest people have in interacting with their future self, but also highlights the issue that can occur if not integrated successfully. More large-scale experiments are recommended to explore this further.

In contrast to previous studies, participants did not feel closer to their future self after viewing an aged-progressed image of their avatar when compared to the group without it. There are several reasons why this might have occurred. Participants of the aged-progressed version might not have felt sufficiently primed in the onboarding process. Because of the small physical size of the avatars used within banking apps, it was decided that the change in the future-self avatar would need to be obvious. This led to more extreme changes in volunteers' profile pictures and as noted within the qualitative feedback, some felt this was an inaccurate description of what they would look like in the future and therefore did not make them feel closer to their future self.

Volunteers in previous studies were exposed to full scale images of their future selves which would allow for subtler changes in the imagery. For example in Fertig's field study, customers were able to see a full screen image of themselves on their phone with an age-progressed filter applied [13]. The aging of participants' photos was limited by the capabilities of the FaceApp [12] program and the photos that were supplied. This allowed for a greater degree of standardization when aging participants' photos, but there were not many opportunities to edit the photo beyond applying the initial preset. For example, there was no option to apply grey hair to female profile pictures and no suitable work around was discovered to create this as a realistic effect. This was a feature that was again available in Fertig et al.'s (2018) research and was reiterated by some female participants as an option they would have liked to have seen.

It was concluded that for some of the female participants, the image would need to be aged twice to see an obvious change and to get a noticeable response from the user. However, subtler image progressions would probably be more effective, even if they are not the most accurate, giving the user more leeway to imagine their possible future. Previous research has found that when participants are asked to rate the attractiveness of profile pictures, they found that the level of attractiveness decreased as people got older [18]. By creating images which were visibly a lot older as opposed to more subtle aging interventions, this could create a greater negative reaction and in turn weaken the connection to their future self. Within the qualitative feedback, a small number of participants identified legibility issues with the onboarding process and these could have led to them not being sufficiently primed. However, the prototype was designed to make sure that the aged photograph was prominently displayed on all UI screens, so that participants were sufficiently exposed.

No statistically significant difference was found that a user would save more into their pension either as a one-off payment or a monthly increase in their contributions when comparing an aged-progressed intervention with one without it. As there was no difference in how close people felt towards their future self across groups, this was reflected in how much they were willing to contribute towards their future retirement.

However, across both groups, a large proportion (47.7%) both identified that the current rate was not sufficient to have a comfortable income in retirement and in the relevant scenario (78.6%) increased their monthly contribution rate from the default rate. One of the main reasons for this is that both prototypes incorporated the research from the retirement living standards [34], which showed the income needed to enjoy a minimum, moderate and comfortable lifestyle in retirement. This is similar to the findings of the Behavioural Insights Team [11] that found that clear labeling was the most effective way for people to realise that the minimum saving rate was insufficient to have a comfortable retirement. In their research, 51% identified the default rate was too low, and 54.4% was the highest number of people who felt they should increase their contributions in their experiment. Although there was no explicit messaging telling them what level they should be contributing at as in their previous experiment, presenting the three income levels and the comparable lifestyle they could achieve, was enough to encourage people to increase their contributions.

Another contributing factor could be that there was a marked difference in the demographics of each group. The age intervention group skewed younger ($Md = 23$ with the majority (60%) not having a pension. In contrast the control group was older 36 ($Md = 36$) and a large majority did have a workplace pension (83.3% Yes, 12.5% No,4.2% Don't know). Sims et al.'s research [40] had a similar age-range (20–45) and was unable to replicate the money allocation task. The two groups were at different stages in their life, and this could have affected how much they felt willing to allocate towards their pension. Affordability issues were a common theme that appeared in the initial survey, so it could be presumed younger people starting their careers would feel that they have less disposable income to allocate towards their retirement. This is similar to Grable's research [17] that found that older and wealthier individuals, were more open to risk when regarding retirement planning and therefore would be more comfortable allocating a larger proportion to their pension [17].

People's reaction to their avatar provoked mixed response, some felt more engaged and enjoyed the experience, whereas some had a negative reaction on seeing their avatar and this may have dissuaded them from using it further. In contrast to previous research, there was no avatar within the control prototype to cut down on experiment demands. Without any avatar, the control prototype had less information and therefore was easier to navigate. This was reflected in the qualitative research were more users commented on the positive aspects of the design and noted that it was easy to use.

5 Conclusions and Future Work

5.1 Summary

This research set out to study the effect an aged-progressed intervention would have within a pension savings app on users' closeness to their future-self, the contributions they make towards their pension, their confidence in their ability to save for retirement and if it makes a more usable/satisfying experience for the user. While the overall findings were not statistically significant, users expressed enthusiasm for the concept highlighting the appetite for creating novel approaches to engage younger workers with their pension. This research shows a new opportunity by incorporating a future-self avatar that not only provides a visual aid for the user but provides prompts to help them visualise their aging. This research highlights the difficulties and care that will need to be taken when implementing interventions in future applications. Across both prototypes, a large segment of users identified that their current saving level was inadequate and increased their contributions accordingly. This reinforces earlier research showing the benefits of labeled information when people are trying to engage with their pension [11].

5.2 Strengths and Limitations

A full user centered methodology was utilised for the study. An expert, Hal Hershfield was consulted which added greatly to the research experiment. This resulted in a fully working prototype created and iteratively designed incorporating user feedback several times for the intervention. A final between groups user test was conducted, using a mixed method approach that gathered both quantitative and qualitative data. This research was the first to look at fully integrating an age-progressed avatar within a pension app and trying to create more in-depth interactions by including prompts and advice from the participant's aged avatar, in contrast to previous research which was limited to a simple interaction of seeing one's face aged with a filter.

Some of the weaknesses of the current study lie in the sampling methods and number of participants recruited. The sample size of the current study was quite small, making it hard to infer the results on a larger population. The central premise relied on volunteers submitting a photo so that their test could be personalised. As it was felt that this could dissuade people from participating, the age range of participants was kept relatively wide. This, however, means that participants were at different stages in their life and this led to an uneven split in the demographics between the two tests. The most common age of participants in the age progressed intervention was 23, whereas in the control version the most common age was 38. Most volunteers for the intervention were students, whose data had been collected through snowball sampling whereas the control version was mainly recruited from survey respondents. This also reflected a big variation in who currently had a pension between the two groups. In the aged-progressed group, the majority did not have a pension (57%), whereas in the control group, the overwhelming majority (85%) had at least one workplace pension. This difference could have affected how people felt in their closeness to their future self and their confidence in saving for the future.

5.3 Future Work

One of the initial assumptions at the beginning of the experiment was that by using the latest software, we could create a more realistic and vivid rendering of what participants could look like in the future. However, when aging images, this might be detrimental when trying to achieve the desired response and future work could look at implementing more subtle age progression interventions which give the viewer more control to customize and imagine their future selves and thus reduce the likelihood of a negative reaction.

Given the physical size of some of the avatars needed in a banking application, it was noted that for some of the female participants that the image would need to be aged twice to see an obvious change and to get a noticeable response from the user. Hal Hershfield, now working in UCLA, noted in email correspondence that aging them once would probably have been more effective, allowing volunteers more room to imagine their future. This again could have produced a strong negative emotion and future work could look at the effect of implementing subtler interventions.

To avoid any priming, it meant that volunteers only saw their aged-progressed imagery within the prototype and therefore had no opportunity to change or update the image if they did not think it was an accurate description of what they could look like in the future. Further studies could look at allowing participants to change or amend the image themselves within FaceApp [12] or similar software and then populating the rest of the app with the updated image. This would give the user greater control and create an activity where they can actively help visualise their future. Experiments could be conducted to see if this increases closeness to the future self.

References

1. Ainslie, G.: Specious reward: a behavioral theory of impulsiveness and impulse control. Psychol. Bull. **82**(4), 463 (1975)
2. Barrett, A. M., Whelan, B.: How well-informed are pension scheme members on their future pension BENEFITS? Evidence from Ireland. J. Aging Soc. Policy, 295–313 (2015). https://doi.org/10.1080/08959420.2015.1044817
3. Bartels, D.M., Urminsky, O.: On intertemporal selfishness: How the perceived instability of identity underlies impatient consumption. J. Consu. Res., 182–198 (2011)
4. Bartels, D., Urminsky, O.: To know and to care: How awareness and valuation of the future jointly shape consumer spending. J. Consum. Res. 1469–1485 (2011)
5. Christensen, C.: Disruptive innovation (10 July 2012). Retrieved from claytonchristensen: http://www.claytonchristensen.com/key-concepts/
6. Collins, M.H.: Supporting pension contributions through the tax: sytems: outcomes, costs and examining reform. Dublin: Econ. Soc. Rev.
7. Cooper, A., Reimann, R., Cronin, D., Noessel, C.: About face 3: the essentials of interaction design, pp. 62–72. John Wiley & Sons, New York (2014)
8. Creswell, J., Plano Clark, V.: Designing and Conducting Mixed Methods Research. Sage (2017)
9. CSO. Pension Provision. Dublin: CSO (2016)
10. Douglas, S.: How To Do A UX Competitor Analysis: A Step By Step Guide. (Retrieved 01 May 2022 from Usability Geek). https://usabilitygeek.com/how-to-do-ux-competitor-analysis/
11. Dutta-Powell, R., Cornel, P.: Nudging Young people to engage with pensions. The Behavioural Insights Team, London (2020)
12. FaceApp. AI face editor. (Retrieved 3 May 2022). https://www.faceapp.com/
13. Fertig, A., Lefkowitz, J., Fishbane, A . Using behavioral Science to Increase Retirement Savings in Mexico. Mexico: Ideas **42** (2018)
14. Frederick, S.L., O'Donoghue, T.: Time discounting and time preference: a critical review. Time Decision: Econ. Psychol. Perspect. Intertemporal Choice, 13–86 (2003)
15. Gibbons, S.: Design Thinking 101 (31 July 2016). Retrieved from NNG: https://www.nngroup.com/articles/design-thinking/
16. Global Moving Services. (2 December 2021). Living in Ireland vs UK. Retrieved from Global Moving Services: https://www.gmsmove.com/living-in-ireland-vs-the-uk/
17. Grable, J.E.: Financial risk tolerance and additional factors that affect risk taking in everyday money matters. J. Bus. Psychol. **14**(4), 625–630 (2000)
18. He, D.W.: The effect of aging on facial attractiveness: an empirical and computational investigation. Acta Physiol. (Oxf.) **219** (2021). https://doi.org/10.1016/j.actpsy.2021.103385

19. Hershfield, H., et al.: Increasing saving behavior through age-progressed renderings of the future self. J. Marketing Res., S23-S37 (2011)
20. Hershfield, H.G.-L., Knutson, B.: Don't stop thinking about tomorrow: Individual differences in future self-continuity account for saving. Judgment Decision, 280–286 (2009)
21. Hershfield, H., John, E., Reiff, J.: Using vividness interventions to improve financial decision making. Policy Insights Behav. Brain Sci., 209–215 (2018)
22. Hershfield, H., Wimmer, G., Knutson, B.: Saving for the future self: neural measures of future self-continuity predict temporal discounting. Soc. Cognit. Affect. Neurosci., 85–92 (2009)
23. Interaction Design Foundation. The Pros and Cons of Guerrilla Research for Your UX Project. Retrieved from Interaction Design Foundation (01 May 2022). www.interaction-design.org/literature/article/the-pros-and-cons-of-guerrilla-research-for-your-ux-project
24. Intraction Design Foundation. What is paper prototyping? Retrieved from Intraction Design Foundation (05 May 2022). https://www.interaction-design.org/literature/topics/paper-prototyping
25. Ireland, Government of, A roadmap for Pensions Reform 2018–2023. Dublin: Department of Finance (2018)
26. Laubheimer, P.: Personas vs. Jobs-to-Be-Done (6 Aug 2017). Retrieved May 01, 2022, from NNG. https://www.nngroup.com/articles/personas-jobs-be-done/
27. Levy, J.: UX strategy: How to devise innovative digital products that people want. California: O'ReillyMedia (2015)
28. Löckenhoff, C.: Age differences in selfcontinuity: converging evidence and directions for future. Gerontologist, 396–408 (2017)
29. Loewenstein, G.: Out of control: Visceral influences on behavior. Organiz. Behav. Hum. Decision Proc., 272–292 (1996)
30. Madrian, B.S.: The power of suggestion: inertia in 401(k) participation and savings behaviour. Q. J. Econ., 1149–1187 (2001)
31. Maze. Build amazing products. Without the guesswork. Retrieved 01 May 2022 from Maze. https://maze.co/
32. Nielsen, J.: Ten usability heuristics (2005)
33. O'Brien, H.L., Cairns, P., Hall, M.: A practical approach to measuring user engagement with the refined user engagement scale (UES) and new UES short form. Int. J. Hum. Comput. Stud. **112**, 28–39 (2018)
34. Padley, M., Shepherd, C.: Developing Retirement Living Standards. Pension and Lifetime Savings association (2019)
35. Pallant, J.: SPSS Survival Manual : a Step by Step Guide to Data Analysis Using SPSS. Open University Press/McGraw-Hil, Maidenhead (2010)
36. Parfit, D.: Personal identity. Philos. Rev., 3–27 (1971)
37. Pension Authority, Understanding your pension. Retrieved from The Pension Authority (02 Feb 2022). https://www.pensionsauthority.ie/en/lifecycle/useful-resources/pension-calculator/
38. Pronin, E.: Temporal differences in trait self-ascription: When the self is seen as an other. J. Personality Soc. Psychol., 197–209 (2006)
39. Schelling, T.C.: Self-Command in Practice, in Policy, and in a Theory of Rational Choice. Am. Econ. Revi. 1–11 (1984)
40. Sims, T., Raposo, S., Bailenson, J.N., Carstensen, L.L.: The future is now: Age-progressed images motivate community college students to prepare for their financial futures. J. Experim. Psychol., 593–603 (2020)
41. Sokol, Y., Serper, M.: Development and validation of a future self-continuity questionnaire: a preliminary report. J. Personal. Assessm. (2019)
42. Superannuation Australia. ASFA Retirement Standard. Retrieved 01 May 2022 from Superannuation Australia. https://www.superannuation.asn.au/resources/retirement-standard

43. Thaler, R., Shefrin, H.: An economic theory of self-control. J. Polit. Econ. 392–406 (1981)
44. Thaler, R., Sunstein, C.: Nudge: Improving decisions about health, wealth and happiness, pp. 122–125. Yale University Press, New Haven (2008)
45. The Irish Longitudinal Study on Ageing: Trinity College Dublin Background (October 11 2017). Retrieved 05 May 2022 from TILDA. https://tilda.tcd.ie/about/background/
46. The Irish Longitudinal Study on Ageing. Embracing Retirement- The impact on health, social participation and well being. Dublin (2019)
47. Wilkinson, L.: The Future Life: How can younger people be supported to achieve adequate retirement outcomes? Pension Policy Institute (2021)
48. Wohlsen, M.: Know What You'll Look Like in 30 Years — Maybe Then You'll Max Out Your 401(k) (05 Dec 2012). Retrieved from Wired. https://www.wired.com/2012/12/retirement-magic-mirror/
49. Yee, N.: . Walk A Mile in Digital Shoes: The impact of embodied perspective-taking on the reduction of negative stereotyping in immersive virtual environments. Presence, 147 (2006)

Application of Smart Media in Cultural Heritage Protection

Linyuan Yang[1(✉)], Shuangyi Yu[2(✉)], Yunning Wang[3(✉)], Chenhuliang[1(✉)], and Longxiaofei[4(✉)]

[1] Minnan Institute of Science and Technology, Digital Media Art, School of Art and Design, Quanzhou 362000, China
1529343877@qq.com, 1587578672@qq.com

[2] Shanghai University, Film Directing, Film Academy of Shanghai University, Shanghai 200040, China
1837788633@qq.com

[3] Sungkyunkwan University, Multimedia Television and Film, Arts, Seoul 03382, South Korea
dearmydorothy@gmail.com

[4] Shanghai University, Radio and Television Editing, Film Academy of Shanghai University, Shanghai 200040, China
1041176902@qq.com

Abstract. Intelligent media is a combination of artificial intelligence and media, and the era of intelligent media rapidly meets the needs of users for information. The digitalization of Dunhuang Studies represents the new trend of the development of Dunhuang Studies in the digital era. The constant emergence of digital resources has created theoretical and practical exploration. The article proposes the means of intelligent media, the digital transformation of Mogao Grottoes, and the application of intelligent digital technology, as well as how to integrate intelligent technology with cultural heritage. The method of desktop research was used. To develop cultural heritage-related undertakings, it is necessary to promote the integration of culture and intelligent media in line with the trends of the current era. It is conducive to creating greater development space.

Keywords: Intelligent Digital Technology · Intelligent Media · Digitalization

1 Introduction

For the protection of cultural heritage, digitization and intellectualization are good ways to dispel myths. The artificial intelligence world is a world composed of digital behaviors. Big data has made a qualitative change in the use of its research. The interaction between intelligence and the data world has ushered in a new era of using big data for prediction or decision-making. For thousands of years, Dunhuang has fused different outstanding civilizations with an open mind. Nowadays, a Dunhuang revolution of smart media is

L. Yang, S. Yu, Y. Wang, Chenhuliang and Longxiaofei—All five authors contributed to the article and were the first authors.

K. Arai (Ed.): FTC 2023, LNNS 816, pp. 286–294, 2023.
https://doi.org/10.1007/978-3-031-47448-4_20

also quietly taking place. Smart media undoubtedly brings new vitality to the inheritance and culture of Dunhuang's spiritual culture. However, in terms of Dunhuang Studies, there have been many achievements in the digitization of cultural relics, which has laid a solid foundation [1]. However, there are still some problems, mainly manifested as follows: 1) Some research results are repetitive, and 2) some research areas lack attention to technological development, neglecting the expression of humanistic needs and emotions. 3) The integration of resources is not complete. 4) Digital technology is still at a relatively preliminary stage and not mature enough.

2 The Integration of Culture and Intelligent Media is a New Development Trend for Future Cultural Heritage Protection

2.1 The Integration of Intelligent Media and Cultural Heritage Protection is a New Trend in Future Development

In developing the cause of cultural heritage protection, intelligent technology and culture should not be treated as separate individuals but should integrate, penetrate, and innovate across borders.

The integration of intelligent media is conducive to the protection, development, and development of cultural heritage. The balance between artificial intelligence and digital media may fundamentally change the behavior regulated by tradition and human values. It has created a new type of society, which is not just a modern version of the old society. What values should this new society adopt? How will it integrate with cultural heritage to create a new balance? In 2018, Li Feifei, a famous Chinese scientist and AI (Artificial Intelligence) is a new science of technology. It can be defined as the discipline that deals with the implementation of machines (hardware and software) capable of autonomous "actions" (solving problems, performing actions, etc.) expressed his excitement and concern about the development of AI in his article "How to Make AI More Humanistic", and proposed the development path of "human-centered AI". It is expected to guide artificial intelligence to become a new force for progress in human social life [2]. At the same time, she also refuted the view that artificial intelligence has no values, which is the first time that artificial intelligence values and human values have been discussed. In the view of Li Feifei's researchers, "human-centered intelligence AI" contains three important themes: first, to enable artificial intelligence to better reflect human deep intelligence; Second, artificial intelligence should help humans become stronger rather than replace them; The third is to ensure that artificial intelligence can positively and correctly guide the impact of human beings in the process of development. However, at present, we are in an era of big data in which no one can escape from data. Intelligence constantly shapes and changes human daily life and behavior. With the rapid development of the digital era, these cities were unimaginable in the past. The era of intelligent development is a world of digital behavior. There are many applications of big data, and algorithms are the foundation and core of achieving artificial intelligence, which has a certain impact on the real world. However, to some extent, algorithms are just a tool for us to utilize, and they cannot completely isolate and separate values. Culture exists before algorithms, and depends on existing social systems, practices, concepts, attitudes, and value orientations. It is embedded in program design through designers' ideas and expected ideas. In intelligent societies, there are optimists and pessimists. Optimists believe that algorithms replace people's judgments, a view that continues the spread of "technological determinism", but ignores the social construction function of algorithm design. Will technological determinism and optimists lead to overall historical and philosophical ignorance of technology? However, will the development of technology put the living world under colonial domination? And through the manipulation of technology, resulting in control and destruction of human beings, further restricting the independent development of human beings? On the other hand, pessimists believe that science and technology will eventually occupy people's lives unimpeded. As the German philosopher Erich Fromm intended to say, "a completely mechanized new society committed to the largest scale of material production and consumption, controlled by computers." China emphasizes harmony between human beings and nature, the concept of "harmony between heaven and man", and takes harmony as an important value principle [2].

With the development of intelligence, people's original forms of life have been broken, and their survival patterns and cultural development have been greatly impacted. The development and changes in the internal and external environment have pushed the thinking of people and nature to a new ideological level, both in terms of behavior and culture. Intelligence feeds on data. Human practical activities cannot be separated from production factors, culture cannot be separated from the collision of ideas, and neither can intelligent media. The 19th CPC's Decision of the Central Committee of the Communist Party of China on Several Major Issues of Adhering to and Improving the Socialist System with Chinese Characteristics and Promoting the Modernization of the National Governance System and Governance Capacity proposed that the mechanism of market evaluation of contributions of production factors such as labor, capital, land, knowledge, technology, management, data and revenge determined by contributions should be improved [2]. This is the first time that China has used data as a factor of production, indicating that the era of intelligence has truly set off the trend of the era. How to integrate resources and data, and use data in a legal and compliant manner, has become the main theme of the development of artificial intelligence for sustainable and healthy development.

The integration of smart media and cultural heritage is not a simple addition, but an organic integration of resources. Intelligent technology should be seen as a means of progress. With the wave of the third technological revolution and the entry of the Internet era, digitization has become a popular background in today's era and has a good hand in solving many traditional cultures. It is a high-tech means of protecting intangible cultural heritage art. It can not only make up for the shortcomings of traditional culture in terms of protection, such as easy loss and wear. A new way of interacting through intelligent media can allow Dunhuang art to be dynamically displayed in multimedia scenes, allowing people to bring visual impact and collision during an interaction. The protection and preservation brought about by digitization can be quite convenient. It not only helps in cultural dissemination and publicity but also expands the scope of cultural heritage resources. The scientific and technological content of resources has been enhanced, thereby protecting many sites. To achieve the original intention of protecting, inheriting, developing, and utilizing cultural heritage. However, no matter how we protect cultural relics, they will be more or less damaged to a certain extent. However, adopting intelligent technology can promote the scientific and modern process of cultural heritage protection. To reproduce history, cultural information can be preserved in a form. Making modern intelligence collide with ancient cultural resources. Great breakthroughs have been made, which have advanced China from a major country in cultural relics conservation to a major country in cultural relics conservation and research [3].

3 The Application of Intelligent Media in the Protection of Cultural Heritage in the Digital Era

The digital protection technology of the Dunhuang Research Institute for the art of Mogao Grottoes began in 1933, to use intelligent digital technology to better preserve the precious data of Dunhuang Grottoes art. On the long road of the Dunhuang Research Institute, a team of nearly 100 people who can grasp advanced concepts, advanced digital technology, advanced management methods, and can afford to use digital technology and preserve important information of Dunhuang has been cultivated and can be rooted in the desert for a long time [4].

3.1 Application of Data Collection

The fundamental foundation of digitization is the collection of existing information data, so the exploration of data has long been carried out. Conduct close-range photogrammetric experiments using high-precision digital cameras. To produce perspective and outline drawings of grotto buildings and artistic relics, obtain three-dimensional and perspective effects. Then, computer stitching and restoration are performed to achieve the purpose of digitizing the murals.

3.2 Application of Image Processing

Image processing includes color processing, geometric processing, image segmentation, image production, and image design. Not only can digital images of murals be obtained, but also geometric and color remedies can be performed based on relevant technologies. Regarding the issue of color correction, it is possible to carry out a prediction method for the chromaticity of CRT displays, as well as a color correction restoration model. Then, the correction effect is evaluated using a color difference function in a specific uniform color space. The interactive image segmentation method is also used in the segmentation of Dunhuang mural images. The development of intelligent digitization has solved a major problem that computer memory cannot extract mural objects with high accuracy [5].

Copying, restoration, repair, and evolution simulation. In the context of digitization, computers can not only play an important role in the copying, restoration, and repair of murals but also simulate the evolution of murals. For example, there are computer-assisted virtual restoration methods for mural color. Mural painting evolution technology, and computer-aided copying, combined with the characteristics of CPU computers, proposed a unified computing model and implemented a fast algorithm.

3.3 Virtual Presentation

Virtual display and roaming are important manifestations of digitization. From the realization of the three-dimensional reconstruction of cave murals to the launch of virtual roaming, this article explores the application structure of using software under low-end hardware conditions to achieve real-time transmission and rendering oriented to virtual reality. Aiming at the shortcomings of the original cave virtual method, a series of improved methods such as the expression of the LOD model and automatic LOD generation algorithm is proposed to establish a realistic real-time roaming virtual engine. In addition, based on the research of resource aggregation and software device aggregation, we have created the automatic synthesis technology of Feitian video under the basic Internet. The Internet can analyze user scripts, extract Dunhuang elements and motion attributes from video animations, and combine Dunhuang elements with the basic actions of characters to generate scenes with aesthetic and rich emotional information, allowing viewers to immerse themselves and generate emotional resonance. In recent years, the research institute's project on "Digital Dunhuang" has been continuously developing, using virtual and VR technology to make digital versions of some Dunhuang cultural relics. It is being disseminated through museums, mobile apps, etc. Bringing a fresh experience brought about by immersion and interactivity [5] (Fig. 1).

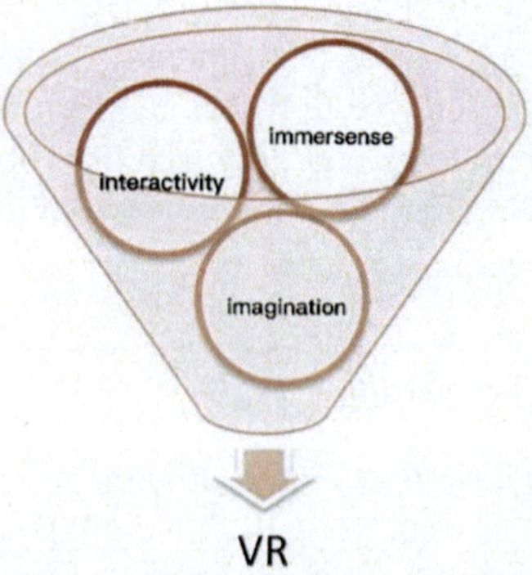

Fig. 1. Virtual reality system features.

3.4 2D Semi-digitization and 3D Digitization

The scientific research team composed of 12 people from Shanghai Printing Group Co., Ltd. Has completed the two-dimensional and semi-digital murals. First, high-precision digital cameras are used to collect the murals, print them out and mount them onto the three-dimensional wooden model walls and ceilings of the caves, simulate the scene of the Dunhuang Grottoes, and add imitation colored Buddha statues. Give the audience a sense of real-time.

3.5 Three-Dimensional Technology

3D technology is an important part of digital technology. The three-dimensional digitization of murals refers to the use of 3D digitization technology to present Dunhuang murals in the form of situational animation. The "Painted Dunhuang" exhibition in Taipei City features a full-size 3D animation of Amitabha Pure Land. Using MOCAP dynamic capture technology, floating projection technology, holography production technology, etc., it creates a beautiful dance posture of a flying fairy. People can pray, which is very interesting.

3.6 5D Technology

Interactive cultural products like wave technology emphasize the integration of communication technology and culture in an interactive form. Create new creative products. The rapid development of 5D movies in recent years is typical, and 3D has become a more common form of film. 4D movies are not yet widely popular in the market. 5D movies have sprung up. 5D movies can not only interact with movie characters but also experience physical sensations such as earthquakes in the movie from the seat. 5D enhances user interaction and experience. Its success demonstrates the combination of communication technology, interactive culture, and intelligent products.

3.7 Document Digitization and Database Establishment

The digitization of comprehensive information can integrate resources across regions and based on themes. The Unesco Conference of the States Parties voted to adopt the Charter for the Preservation of Digital Heritage, reflecting the inevitable link between the protection of cultural heritage and intelligent digitization, which not only ensures the maintenance and preservation of world heritage but also ensures its dissemination. It also requires legislation on the protection of digital heritage and requires countries to develop legal systems to ensure the authenticity of digital heritage and prevent excessive politicization and malicious arbitrary use [6] (Fig. 2).

Fig. 2. Document digitization.

The establishment of non-art databases is mainly considered from the perspective of database functionality. The protection of intangible cultural heritage digitization is to digitize the core and important content of intangible cultural heritage projects. Achieve integration of their systems. It is stored, managed, exchanged, and utilized in the form of a database. This enables access to, organization, analysis, integration, and management of intangible cultural heritage data to assist in decision-making. It can be used as a data aggregation of various parts. Intangible heritage databases are another means of managing digitization. However, it is necessary to clarify the legal status of digitization and database and establish applicable systems.

3.8 Types and Nature of Digital Preservation of Intangible Cultural Heritage at Home and Abroad

Digital preservation technology dates back to the 1990s. The "Digital Michelangelo Project" hosted by Stanford University records a large amount of information about points, thereby solving the real-time rendering of a large number of three-dimensional sampling data [7]. Therefore, digitization is used to preserve the archives of archaeological sites. In 1977, China's Dunhuang Research Institute cooperated with Zhejiang University, and under the auspices of the National Natural Science Foundation, began to digitally preserve the murals and painted sculptures in the Mogao Grottoes of Dunhuang. Since 1999, a series of research projects with the theme of "digitization of cultural heritage" has received assistance from EU research projects, making great contributions to the digitization project.

In the field of intangible cultural heritage, there have also been some practical projects at the national and local levels. From the "Cultural and Natural Heritage Information Center" of the Alexandria Library in Egypt, the "Hong Kong Oral History Archive Project" of the University of Hong Kong, and the establishment of an agricultural culture database in Italy. There are also some laboratories established, such as the Traditional Technology and Materials Research Laboratory of Tsinghua University, as well as the first-class expert team of Tsinghua Academy of Fine Arts and the scientific research force of various technology laboratories. Promote traditional intangible cultural heritage processes [8]. However, the databases and museums of non-intangible cultural heritage in various provinces are essential "window displays", which are introduced in simple text and multimedia materials, lacking other elements of non-intangible cultural heritage.

There is a lack of research on these aspects. It is because the concept of culture is not clear enough. The digital resources of intangible cultural heritage on the Internet will increasingly become the best channel for bringing intangible cultural heritage to people's attention, and will increasingly become the main source of information and realization channel for the preservation, protection, and inheritance of intangible cultural heritage. In addition, nowadays, many people have realized that the value of data resources cannot only rest on their basic uses but mainly comes from their secondary uses, namely, potential value. At this stage, if we can collect as much as possible about intangible cultural heritage digital resources, the value-added data and the mining of more value are just around the corner [9]. The difference between the preservation of intangible cultural heritage digital heritage and other digital heritage is that its preservation should not only focus on records but also materials that display intangible cultural heritage activities, but also on the intangible cultural heritage itself. Generally speaking, the digital preservation of intangible cultural heritage involves the long-term storage and acquisition of intangible cultural heritage digital resources [10].

4 Conclusion

The arrival of the digital era has made great progress in the creation, performance, dissemination, production, and other aspects of creative design. From the digitization of Dunhuang, it can be seen that smart media has become the main theme, and the existence of smart media is indispensable for all aspects of reading. In the cause of cultural heritage protection, smart media will play a more important role, contributing to the cause of cultural heritage protection.

1) In the digital era, intelligent media will make progress in the protection of cultural heritage, and the use of digital, intelligent, and media applications will become more widespread.
2) The integration of smart media and cultural heritage will become more international. We should take its essence and discard its dross. We have also learned advanced technology from the West and have done better in the integration of culture and technology.

Carry out a new trend of diversification and multiple forms. Promote the further development of cultural heritage. Dunhuang art is a combination of Chinese culture and the absorption of other ethnic cultures. It is a model of the integration of global communication and Western art. As a product of globalization, internationalization is an inevitable trend.

In the future, smart media will present a multifaceted development trend, which can not only meet the needs of the public for self-entertainment but also increase the demand for modern technology construction.

References

1. Zhu, P., et al.: Study on inclusion medicine of cotton fabric grafted with β-cyclodextrin. J. Text. Res. **31**(7), 64–68 (2010)
2. Yuyue, C.: Research development and the future trend in the silk industry and fiber technology. Sunan. Sci-Tech. Devel. **2**, 9–11 (2004)
3. Liya, L., Chaoxia, W., Shuilin, C.: Using β-Cyclodextrin to develop textiles with fragrance and health protection. Tech. Text. **12**, 35–37 (2002)
4. Hu, Z., et al.: Tentative research and preparation on anti-infectious functional fiber. Tech. Text. **213**(6), 14–17 (2008)
5. Hanchunping.: A review of digitization of Dunhuang Studies. Nat. Soc. Sci. Found. Proj. **2008**, 6–11
6. Chunmei, W., Jinxin, H.: Application of sulfonated β-cyclodextrin in fragrance finishing for wool fabric. Wool Text. J. **4**, 20–22 (2006)
7. Wang, S., Xu, L.: Application of monachlorotriazinyl-β-cyclodextrin with anti-mosquito clathrate. Dying Finish. **33**(24), 31–34 (2007)

8. Ximu: Annual research report on intangible cultural heritage and industrial development in China. China Econ. Publish. House (10), 539–540 (2019)
9. Wang, Y.: The properties of antimony doped tin oxide composite materials. TianJin University, Tianjin (2014)
10. Li, W.: Porosity test methods of lattice aprons in compact spinning, China, CN 201010579154.2[P]. Accessed 07 Nov 2012

Universal Personal Hybrid e-Learning Design System for University Teachers and Students

Stefan Svetsky[1(✉)], Oliver Moravcik[1], Dagmar Ruskova[1], Lubica Vaskova[1], Zuzana Cervenanska[1], and Dariusz Mikulowski[2]

[1] Slovak University of Technology in Bratislava, Bratislava, Slovak Republic
{stefan.svetsky,oliver.moravcik,dagmar.ruskova,lubica.vaskova,zuzana.cervenanska}@stuba.sk

[2] Faculty of Sciences, Siedlce University of Natural Sciences and Humanities, Siedlce, Poland
dariusz.mikulowski@ii.uph

Abstract. Despite the tremendous advancements in computers, the current level of technology for e-learning design is inadequate for the needs of teachers, and this is especially true at a time of increased demand for distance learning in which the teacher needs to process e-learning content stored in both offline and online spaces. In order to minimise information overload, the teacher needs a simple, low-cost, hybrid, all-in-one system that can offer personalised e-learning design. e-Learning tools of this type that can be used for IT applications in education have not been described in the literature and e-learning focuses on testing the suitability of existing IT tools for learning activities. The shortcomings of current technologies arise from the fact that the hardware, software and computer file formats are incompatible, and cannot simulate mental processes in the way that e-learning requires. In this paper, we present the WPad educational software as a key element of a universal, personal, hybrid e-learning design system. This software has evolved over 15 years based on the principle of automating human mental activities using 'virtual knowledge' as the representation of knowledge. It is a simple database table that is readable by both machines and humans. The empirical and systematic phases of this research are described to show how an individual (including blind and visually impaired users) can insert content into WPad tables, and how these then generate HTML tables for the design of curriculum content. Examples of automating an individual's activities on a computer or on a virtual machine as part of collaborative learning are presented in the form of screenshots. The bulk knowledge processing paradigm is compared with the relational database paradigm, and a separate section is devoted to the technological obstacles associated with e-Learning design.

Keywords: e-Learning design · Educational software · e-Learning tools · IT applications in education · Curriculum content design · Technological obstacles · Collaborative learning

K. Arai (Ed.): FTC 2023, LNNS 816, pp. 295–313, 2023.
https://doi.org/10.1007/978-3-031-47448-4_21

1 Introduction

e-Learning is one of the main priorities in the field of computer support for education. This term generally refers to any educational activities that involve information and communication technologies (ICT). Since education is essentially a kind of never-ending story, there is no universal definition of e-Learning, and in practice, there are a large number of sub-categories of this area. Related to this is the question of how to design universal software and the corresponding IT infrastructure.

Although e-Learning is often considered solely as a form of online education, it has a much wider scope in teaching practice. Within the constructivist model of e-Learning devised by Littlejohn, she mentions it in [1] as 'any technologically mediated learning using computers, whether in a face-to-face classroom setting or from distance learning'. In addition, in the daily practice of R&D staff, e-learning can also be understood as a multilingual support tool for self-e-learning, which is typical of university studies [2]. The empirical research on e-learning by the authors of this article began in response to the FP7 calls for Technology-enhanced learning (TEL) (see [2, 3]). Initially, this research mainly drew on works [4–6] in which models of teachers' activities were specified. From a practical point of view, TEL solutions required parallel content building and processing with appropriate informatics tools to create tailor-made templates and internet and offline applications.

Although the interpretations in the scientific literature of the processes of e-learning and the explanations of what it is are very diverse, teachers and researchers can rely on the International Classification for Standards (ICS), where e-learning is included in the category ICS 35.240 Applications of Information Technology, from the point of view of global standardisation. Specifically, it is included in the category ICS 35.240.90 - IT Applications in Education [7] and is covered by Technical Committee ISO/IEC JTC 1/SC 36 - Information Technology for Learning, Education and Training. As an illustration, we note that the IEC-40180-2017 standard directly states that IT-enhanced learning, education and training is called e-learning.

Similarly, ISO/IEC 40180 'provides the fundamentals and the reference framework for quality assurance, quality management and quality improvement in IT-enhanced learning, education and training (called e-learning)'. From the perspective of user requirements, ISO/IEC 12785-1:2011 (Metadata for Learning Resources) [8] and ISO/IEC 12785-2:2011 are important (as they specify how to express the ISO/IEC 12785-1 information model in XML, including content packaging) [9]. This standard states that computer science uses XML (Extensible Markup Language) for encoding documents in a format that is both human-readable and machine-readable for the processing of educational content. The technological standard involves the use of artificial schemas and languages, but teachers or lay people cannot use these for e-Learning design. In contrast, our research approach is based on the development of WPad software, where the teacher inserts the educational content in the form of natural language into a specific data structure, which is both human-readable and machine-readable, without the need for document coding. This allows the teacher, or the user more generally, to design e-learning content extremely quickly in the form of so-called WPad tables, which are then converted into HTML tables; within a few days, this process can generate e-learning content which would otherwise take months. WPad has been developed through research over about 15 years, and currently functions as a universal all-in-one e-learning tool. It uses the principle of hypertext to connect online and offline environments, and applies 'marking' in natural language to enable the mass processing of content.

The design of e-learning represents an interdisciplinary problem, as it involves a mixture of pedagogy and informatics; however, as we have emphasised in recent publications, pedagogical and informatics algorithms are ideally solved in parallel and synchronously, so that a wide variety of application outputs can be created. Proof of the complexity of e-Learning can also be found from the topics in the e-learning block of the FTC 2023 conference, which included e-learning tools, Collaborative Learning, Curriculum Content Design, Virtual Learning Environments, Web-based Learning, Delivery Systems and Environments. The state-of-the-art e-learning design of these categories is characterised by the fact that it is solved as an adaptation of existing technology to various educational activities. As a result, the teacher has to check whether existing software can be used as IT support for the planned teaching activities; this requires the use of multiple forms of software, and lacks a universal solution. However, we present such an all-in-one solution in this article, and we note that WPad has already been applied to all of the topics listed in the FTC 2023 conference block of the e-learning.

In the following sections, we describe with examples from the practice of teaching undergraduates how a versatile e-learning design system (consisting of a desktop database application called WPad, a web application based on PIKS communication channels, and a hybrid offline/online IT infrastructure including a virtual machine) has been developed empirically and systemically over time. We present screenshots to illustrate how the virtual knowledge model and WPad tables are designed to represent knowledge, and how teachers and students can embed educational content into them and generate HTML tables, thus concentrating the content into educational packages. A separate section also describes the gaps in the technology and the obstacles faced by teachers in terms of e-learning design. The modelling of collaborative e-Learning design within the international V4 EDUPORT project is also discussed.

2 e-Learning Design and Virtual Knowledge Modelling

As mentioned above, e-learning is an interdisciplinary term, and the theoretical background should be derived both from a pedagogical/didactic and informatics point of view. However, it seems that such an interdisciplinary approach does not exist in the scientific literature. Although the authors of [10] discuss e-learning theory, this discussion is based on cognitive sciences (or cognitive load theory) as the basic starting point for the design of educational technology (for example, it is emphasised that learners' brains will suffer from overload if they are presented with too much information, causing inefficient learning). A model of e-learning theory is described as based on 'concepts of three types of cognitive load and eleven empirical principles composing two constructs: cognitive load and design principles.' These two constructs are then combined to create e-learning theory. Of the eleven instructional principles, multimedia, learner control, personalisation and redundancy should be mentioned, as these focus on 'minimizing extraneous cognitive load and introducing germane and intrinsic loads at user-appropriate levels' [11] (we note that our approach using WPad is based on radically minimising cognitive load). The schemes in [12, 13] come closer to an interdisciplinary understanding of e-learning design; in [13], for instance, three models are mentioned: the community of inquiry model, the conversational framework, and computer-mediated communication. The author of [13] considers e-learning models to be 'theoretical constructions that assist practitioners in designing effective learning experiences for students participating in online courses', and emphasises that 'they are distinct from learning theories in that e-learning models are concerned with the pedagogical principles that undergird instructional practices or with the effective implementation of such instructional practices.'

This brief summary of e-learning theories is important, as it shows that the present authors' research is consistent with these theories. Researchers have reached the current state over 15 years by integrating IT into the framework of practical STEM teaching and academic research. The WPad software and the IT infrastructure for the personal e-learning system presented here were gradually developed in cooperation with approximately 2,000 undergraduates and research in the field of TEL. At the outset, the goal was the vision of a universal personal IT tool to support sustainability of knowledge workers, as shown in Fig. 1 [2]. It should be emphasized that the illustrated original vision from 2007 was already implemented by teachers and students in real teaching. The e-learning system presented in the article has already been piloted into teaching due to the fact that it is an effective and cheap solution, as it uses the internal unlicensed software WPad, which has been developed for years by the main author of this article (for normal use, Windows and a connection to the university server).

The added value of the research is that instead of dozens of software, the teacher only needs to use one personal program for e-learning design, while the originality and innovative elements of the solution are based on two registered utility models at the Slovak Patent Office - UV 8787/2020 and UV 7340/2014 (https://wbr.indprop.gov.sk/). In this context, to the best of their ability, the authors did not find a description of a similar personalized tool for e-learning design in the scientific literature.

As mentioned in the next section, e-learning design is currently being solved within the international project V4 EDUPORT, where the use is collaboratively tested on a virtual machine with server Windows 2022 (teachers from five countries are collaboratively testing the mass creation of educational packages on a shared computer). From a practical point of view, it is also important that both teachers and students can use the system from computers in the classroom or from home.

2.1 Empirical Research Phase of e-Learning Design and Implementation in Undergraduate Teaching

The empirical phase of our research began in 2006, when the designer of the current WPad software taught several STEM subjects to undergraduates at the extended workplace of the Faculty of Materials Sciences and Technology in Dubnica nad Vahom. At that time, there were only two regular teachers, who, together with their commuting colleagues, taught around 200 students in the first semesters of their Bachelor's courses. Although there were 12 computers with Windows XP available in the classroom, there was only one with access to the Internet; there were no teaching texts, and the Internet was not yet common even in households at the time. The teacher, who was also the designer of the database application Genius14, solved the problem by using

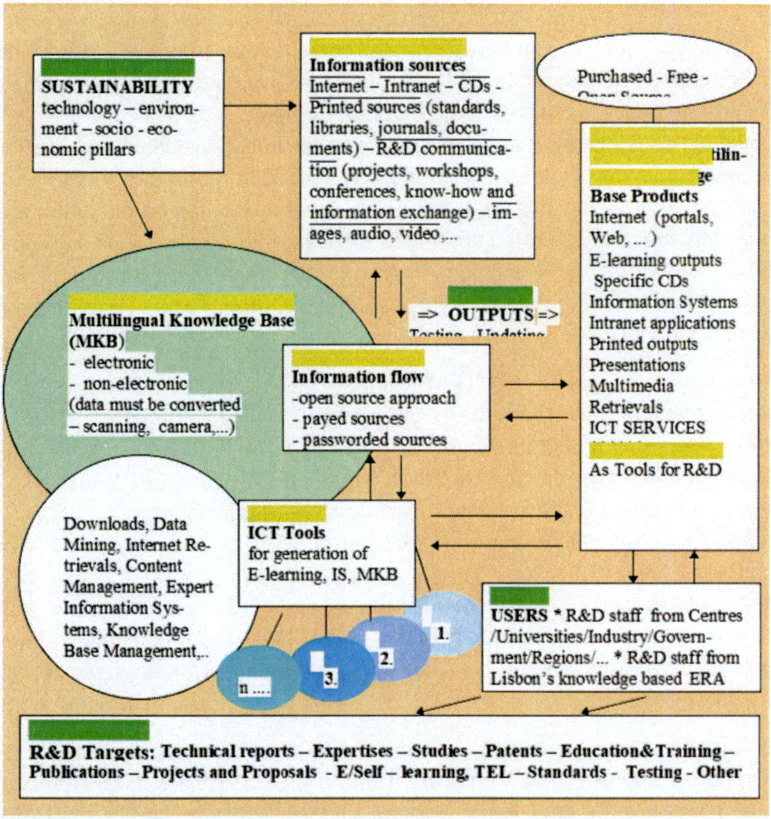

Fig. 1. Original vision for supporting sustainability of knowledge workers (2007) [2]

the database platform Microsoft FoxPro for Windows 2.6a, and in addition to creating teaching texts and supplementary materials for teaching wrote source codes for various teaching tutorials and quizzes for daily classes (for example on photosynthesis, for a course entitled Background to Environmental Protection). The advantage of this was that the program produced database tables that it simultaneously converted into HTML tables, which could be placed on the faculty's Virtual Learning Space as teaching material for undergraduates. Each pair of students had a computer with the Genius14 program (later renamed WPad) on their desk, and during lectures and exercises they created tables that were automatically opened by the Internet browser, although in offline mode (at the time, Opera 9.27 was used). Students who did not have the Internet at home simply transferred the e-learning study content on a USB stick. Those who had access to the Internet used the educational environment on the faculty server, which they could also access from the classroom as a supplement to face-to-face teaching.

At that time, systematic research on TEL was beginning, with cooperation via FP7 consortia. Research was presented on the automation of knowledge-based processes, which also included education, since the basis of educational processes is work with educational content, i.e. knowledge and information. This research was based on the insight that a teacher or researcher needs to process a large amount of information and should be equipped with a suitable 'IT gun' for knowledge processing in order to ensure personal sustainability, as shown in the diagram in Fig. 1.

In view of this, software called Genius14 was developed to enable the mass processing of knowledge empirically, and with the aim of concentrating the content as quickly as possible, thus reducing information overload and at the same time automating the teacher's routine activities on the computer. Considering the numerous outputs for the teaching of undergraduates, researchers were asked to present these at conferences. However, from an analysis of the scientific literature, the author of this educational software application suddenly discovered that it could:

- perform e-learning design for STEM teaching;
- solve the problem of curriculum content design (for several subjects);
- produce e-learning tools for teaching (e.g., video tutorials, quizzes and exercises);
- make connections between the classroom and the faculty's virtual learning space (in the form of communication channels for instructions and chatting);
- combine face-to-face learning with web-based learning (via these communication channels);
- perform activities with students that fell into the category of collaborative learning;
- write a functional database application without the designer needing to know that relational database paradigm existed; in fact, the application operates on an 'anti-relational' basis.

It is interesting to note that the scheme in Fig. 1 still cannot be performed using state-of-the-art technology. However, the solution presented here has already fulfilled this vision of supporting knowledge workers, and both the software and infrastructure were created within the framework of the software's own design. A breakthrough in research in this area was the author's habilitation thesis, which focused on knowledge construction and automation of teaching processes within TEL and e-learning [14], including changing the database platform to the newer, more user-friendly Visual FoxPro 9. Using this platform, e-learning design was developed and the proposed database structure (using a model of the virtual knowledge unit for knowledge representation) became the basis of two utility models that were registered with the patent office.

2.2 Systematic Phase of Research into IT Integration into Teaching and e-Learning Design

During our analysis of the empirical research on TEL, the following questions emerged:

(1) How is it possible that the WPad can function as a multi-purpose educational software and cover several sub-categories of e-learning at the same time?
(2) How does the database application work as an 'anti'-relational paradigm?
(3) How is it possible that students even with weak IT skills, can process the massive amounts of educational content entered into WPad tables, including computer files in various mutually incompatible formats (DOC, PDF, PPT, TXT, JPG, PNG, MP3, MP4, etc.)?

From a retroactive analysis, it was possible to generalise that these are due to the default structure of the database tables in WPad, since the columns of the table function as abstract metadata, or meta-information that identifies the content inserted in the text memo field. This has some similarities to how a person thinks, i.e. that knowledge is associated with certain information that identifies it. Hence, a row in the table is a kind of virtual knowledge unit, and a knowledge base is created from the rows of the table, i.e. a virtual knowledge table that can be transferred between off-line and online environments in the same way as normal computer files. Since a WPad table can be modified and supplemented at any time, updating e-learning study materials is very straightforward, since it only involves pressing the CTRL-F1 keys to transfer the table to the faculty website where the educational material is stored (using FTP or file manager).

Over the previous eight years, dozens of applications for use in teaching and research have been implemented and published on this theoretical basis, a detailed description of which would require a separate book. The following practical examples should help in developing an understanding the simplicity and principle of e-learning design in this way. It is important to understand that WPad was developed with a focus on educational content and the activities that teachers and students carry out on a daily basis. The technology does what the human needs: since WPad works according to cybernetic theory, it functions as an isomorphic switch between human and machine. In the Introduction, it was mentioned that the use of a format that is both human-readable and machine-readable is important for e-learning, and the WPad table is just such a format. In other words, the automation of educational activities is based on the fact that the user enters knowledge into the table and the computer processes it as an IT data structure, several times faster than a person could do it. From the point of view of e-learning creation, an individual can produce educational content within a few days that would otherwise take several months, or could not be done at all. It should be remembered that a regular teacher needs to use natural language when creating e-learning and the main software that are available for this are the Office packages or partial software such as HTML editors or audio-visual applications. WPad also acts as a kind of switch between these packages, Internet browsers and the operating system, i.e. it acts as add-on software that is not available on the market. For the basic creation of e-learning content, only the Windows operating system is needed, and the content is output in the form of WPad tables or HTML tables. As an illustration, we note that when teaching undergraduates, the author never needed to use PowerPoint software, but only these WPad tables. He even once taught a student chemistry through the use of HTML tables. It should also be added that the WPad table is essentially an empty piece of knowledge managed by the WPad software, and the key to success is for the user to find an appropriate way to fill the table with content manually, semi-automatically or automatically. Figure 2 illustrates the model of the virtual knowledge unit, which is a representation of knowledge and forms one row of the knowledge table. As can be seen from Fig. 2, the author noted meta-information and manually inserted the text used in the publication as content into the TP text field.

Several different ways of filling WPad tables will be explained in the next section, and some of these have already been presented at previous FTC conferences, where

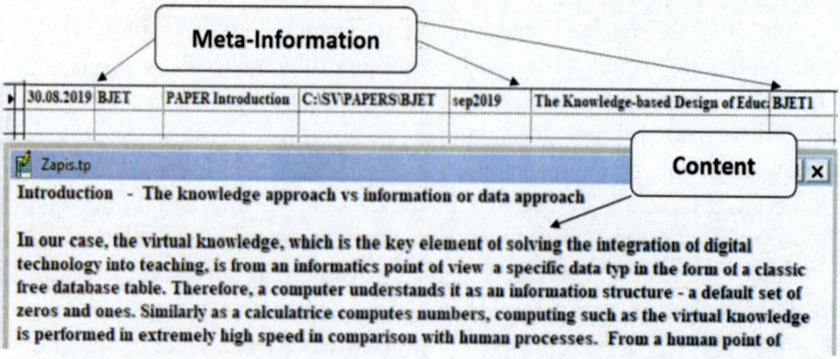

Fig. 2. Model of the virtual knowledge unit as a row of the virtual knowledge WPad table

the authors have described the progress of their systematic research based on the virtual knowledge model. The principle of e-learning design has also been presented as added value in the field of Human-Computer Interaction (at an IEEE conference). Those interested in more details can view prior FTC presentations such as:

- 2017 Some Aspects of Teaching Processes Computerization
- 2018 Some Aspects of the IT Support of Engineering Teaching
- 2020 The ICT Design for Modern Education Technology and Applications
- 2021 The Knowledge-based Design of Educational Technology
- 2022 A Universal IT Support System for Teachers for Educational Processes, Publishing and Academic Research Using All-in-One Educational Software

All of these presentations are related to the area of personalised e-learning design. For example, in FTC 2022, theoretical approaches were discussed from the point of view of pedagogical aspects of the use of technology in education. Virtual knowledge was also presented as a universal solution based on the TPACK model, which introduced the concept of technological and pedagogical content knowledge (TPACK) as a framework for integrating technology into the teacher's knowledge [15, 16].

3 e-Learning Design by Inserting Human Knowledge into WPad Tables (Hybrid Hypertext Principle)

Figure 3 gives an example of the use of 'hybrid' hypertext, and shows how the user can enter links and paths in the text field TP and open them with the CTRL + Mouse shortcut. This text field also allows the user to insert more extensive content, which is inserted as unformatted text (for example, the entire proceedings of a conference can be copied there).

The possibility of including online or offline hypertext is a huge advantage of using WPad, as the user can create a list of favourites by entering Internet links and offline paths, thus creating a kind of navigation spider, or a switch between the online environment and the computer. A further advantage is that there is no need to open a browser or explorer and enter text there, since this hypertext allows the user to directly reach an Internet page or an academic information system, to launch a computer file, or to open a

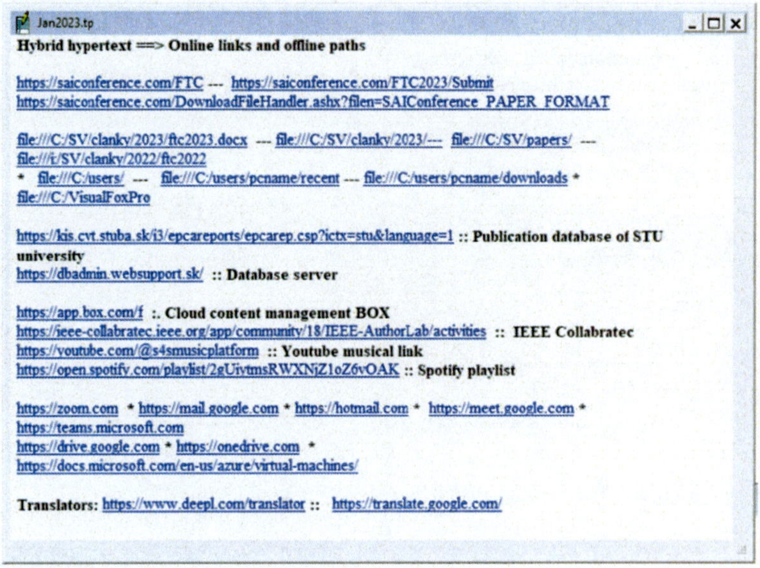

Fig. 3. Example of using hybrid hypertext in the text field of a row of a WPad table

folder in Windows Explorer. This simultaneously enables batch browsing of the Internet, databases, language translators, etc. Although this may seem insignificant to a lay person, if a teacher or user processes large amounts of educational content, this approach can save tens of thousands of clicks between online and offline environments annually. The user can also fill in additional lines and write HTML scripts in them. WPad then functions as an HTML editor, and the teacher can thus create e-learning educational content or a personal information system. The problems of Curriculum Content Design and creating teaching texts for undergraduates have been solved in this way for years. Pre-service teachers have also used WPad to create e-learning sets for several diploma theses.

Figure 4 illustrates the function of the virtual knowledge table model, and shows how the principle of hypertext is used on a daily basis and how the WPad table functions not only as a switch between human and computer but also as a switch between offline and online environments. There is no other software in the scientific literature that enables solutions such as those as illustrated in Fig. 4.

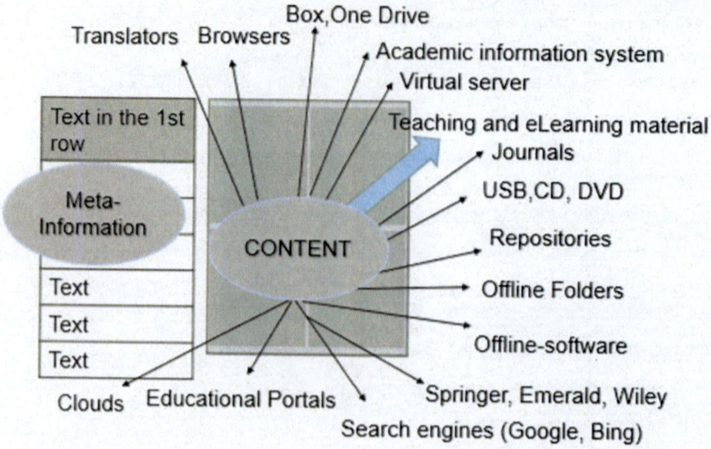

Fig. 4. Model of the virtual knowledge table function (FTC 2022)

4 Comparison of WPad Functionality with the State-of-the-Art in Knowledge Processing

In the scientific literature, relational databases or the XML format are typically used for the mass processing of data and information, whereas authors deal with the mass processing of educational knowledge, which is the key to e-learning design in natural language. XML requires the use of artificial languages, special schemas, and parsing, so it is out of the question from the point of view of the creation of e-learning content by teachers and students. Although relational databases are used to create information systems, an E/R diagram and a default set of tables linked by specific keys must be designed. The relational paradigm cannot be used to process indeterminate, unstructured data, as it requires the use of atomised data, meaning that redundancy in the data and different degrees of normalisation are required. The data are placed on a database server, from which they are usually selected using the SQL language.

So how is it possible that a user with only basic IT skills can use the WPad database application to process extensive amounts of knowledge? The basic difference is that the user does not need to use other languages and IT environments, and directly inserts educational content in natural language into the WPad tables. For example, when students use WPad in class, they simply open their WPad table and write notes in it, in a similar way to using a paper notebook. However, the fundamental difference is that the WPad table represents only one environment for the user. When the user enters unstructured text, it can be seen at the same time and can be edited directly. Unlike relational databases, no SQL language is used, and everything happens in one table and in the same environment. Moreover, figuratively speaking, everything that is forbidden in a relational database can be done in WPad.

As part of the design of e-learning with WPad, PIKS (Personal Information and Knowledge System) communication channels also began to be programmed as an Internet PHP/MySQL application. These have been used in the past for the teaching of undergraduates in some cases involving distance education. It should be emphasised here that the communication channel uses the same WPad table (virtual knowledge), except that the table containing the data is located on the database server of the university; the data are pulled from it via SQL, and the outputs to the web are programmed in PHP. Filling in these tables, unlike the WPad desktop application, is intensive in terms of both time and programming effort. This is made more difficult by the fact that there are various bans and restrictions on the Internet due to security issues. For example, in WPad, the users can browse folders containing files on their laptop from an offline HTML table (as the browser switches the user to Windows Explorer), but if these HTML tables are on the web, this is not allowed.

Figure 5 shows a WPad table called TP_FTC that was created automatically by the program, which found 1,852 PDF files in the article folder and turned them into a table with 1,852 rows. By suitable choices from the user menu, only PDF files containing articles related to FTC conferences from the period 2017–2022 were selected, resulting in a table with only 52 rows. Individual PDF files can be opened directly from this table by hovering over a line and using the CTRL + Mouse or ALT-F9 shortcuts to open the content (in Fig. 5, the FTC 2021 paper is open).

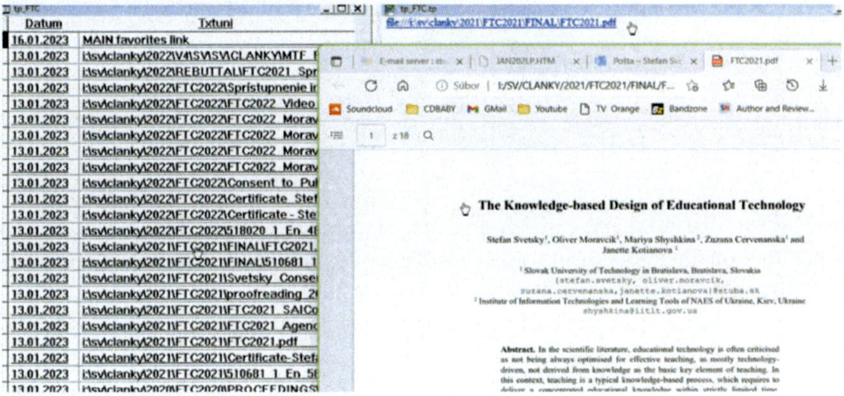

Fig. 5. The WPad educational package used to browse PDF files related to FTC 2017–2021

The teacher or user can create WPad tables manually, semi-automatically or automatically, and these are converted with a single click (CTRL-F1) into an HTML table, which is automatically opened by the default browser. The e-learning content can be used in offline desktop mode, as there are no restrictions, and is transferred to the web via FTP using a suitable file manager. It is important to understand that WPad tables, which are transferred as normal computer files, contain only selected content, whereas in computer files (e.g. PDF, DOC, HTML formats), the educational content is unselected and the files are therefore several times larger.

The WPad table (Fig. 5) is converted to an HTML table simply by pressing CTRL-F1, as illustrated in Fig. 6; both the WPad and HTML tables are open at the same time, and the user can browse the laptop folders from both and open the corresponding PDF files. When writing an article, the user can quickly view what has already been presented at FTC conferences. As the content is concentrated in one table, this can save several hours when writing an article.

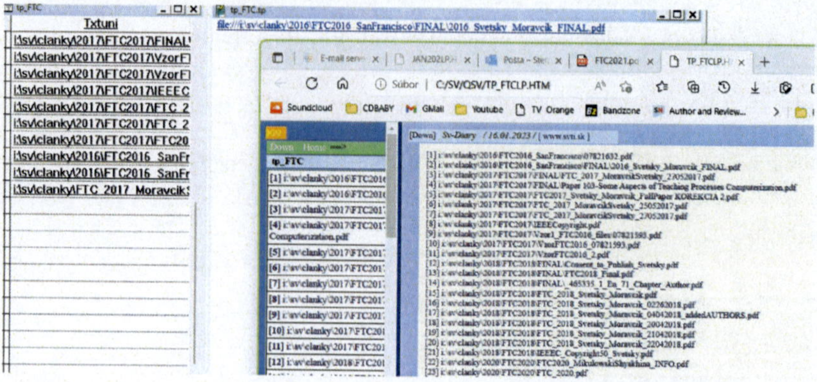

Fig. 6. Screenshot showing automatic e-learning design (with a WPad table created by the program)

WPad's non-relational database paradigm has several advantages over the more traditional relational paradigm. In it, it is absolutely impossible that the source codes that control the table can be written into a row of the table, so that it can function as an 'intelligent' structure (for example, even shrinking, enlarging or deleting itself). The

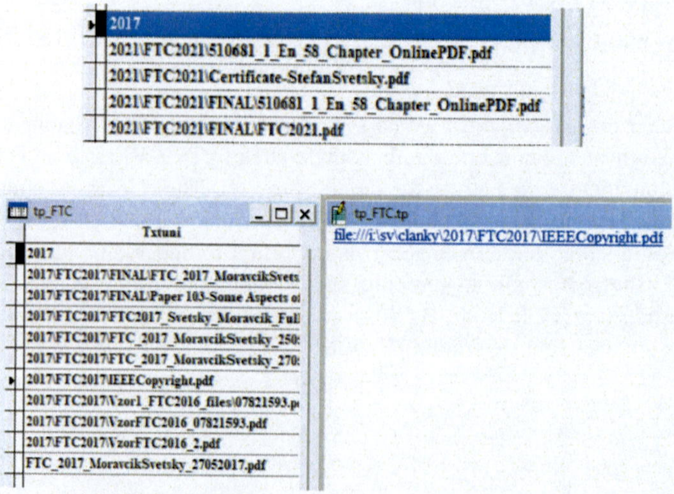

Fig. 7. Dynamic filtering of WPad table records (top input, bottom filtered output)

dynamic filtering provided by a search of the WPad table should also be mentioned. Figure 7 illustrates this dynamic filtering in the case of the table above with 1,852 rows. In this case, the teacher wanted to see only PDF files for 2017 in the table; he therefore entered only '2017' in the txtuni field, and after pressing F6, only rows with PDF files from 2017 appeared on the screen. These files could be directly opened from the table without the need to use Windows Explorer.

We should also mention that it was discovered completely by chance that if the table is indexed by date or alphabetically, and the user wants to move a row, e.g. to the beginning of the table, the user just overwrites the date to a newer one, or changes the text, and the line is automatically moved to the top of the table or sorted based on the corresponding letter in the changed text. This method was proven itself for the author when writing his dissertation and habilitation work. This is a unique solution, and we have not encountered such a simple method of filtering or self-reorganisation of database table rows in any other software.

5 Modelling and Testing Pedagogical Applications for Blind and Visually Impaired Students and Erasmus Students

As part of a collaboration between academics from the Visegrad countries (Poland, Slovakia, Hungary, Czech Republic) and Ukraine, under the V4 EDUPORT project, WPad software and its Internet application PIKS were tested with the intention of preparing an educational portal. In this section, the experiences of a Polish researcher are summarised and evaluated.

A universal knowledge exchange portal can also support activities related to the work and entertainment of blind and visually impaired (BVI) users. In order to implement this functionality, a special channel was created on the PIKS, containing information that may be useful to such users. It included information such as data on organisations supporting BVI users, and interesting resources from the Internet related to facilitating the work and daily lives of these users. All of these data were obtained by using the multi-search mechanism implemented in the WPad software. After creating this channel, information about it was posted on online forums designed for BVI users in Poland.

The system was also used as a tool to support the work of an academic teacher who was blind. To enable this, WPad was adapted by the addition of a special simplified menu. Further features were also added that allowed the software to be operated without a mouse, using only a screen reader and keyboard. Based on these additions, and within the framework of the previously implemented V4 project, educational packages (EduPacks) were created using the WPad program. These were intended for students of computer science in the frame of course of creating web applications and technologies related to this domain.

The method used to create these packages involved the development of lecture presentations and tasks for students in LaTex by the teacher, who then generated PDF files from them. These files were collected, organised, and placed in the appropriate folder in the structure of the WPad software. Finally, to make these materials more attractive, educational packages were generated from them using the WPad software, in the form of elegant-looking web pages. These packages were then made available to students through the teacher's website.

Another way in which the system was used was to create a special PIKS knowledge exchange channel for foreign students. The channel was intended for Erasmus students at the Institute of Computer Science of the University of Natural Sciences and Humanities in Siedlce, which was one of the partners of the project. This channel contained the educational materials needed for these students to complete tasks related to implementing web applications in HTML5, Javascript and the JQUery library. Students had access to these materials through a specially created channel of the PIKS system. They could download them and add their own information, such as questions about a task or information about the completion of a task, which the teacher could then read. In this way, exchanges of information and knowledge took place between the students and the teacher.

Currently, the WPad software and the specially created PIKS channel are being used by teachers in one of the secondary schools in Siedlce. Using these tools, materials that are helpful in teaching mathematics and music are now being created with the WPad software, and will be shared with students during lessons. In the next stage, using the WPad program, students will also create their own educational packages, for example containing information about their favourite singer or author. They will create them in the form of tables in the WPad and then transform them into HTML pages, which they will then be able to share with each other. Using the WPad, students can also learn the basic issues related to databases. Since the basic unit of information storage in WPad is the table, they will be able to learn about such issues as entering data, sorting, searching and transforming, as well as the concept of relationships between tables and simple database queries.

6 e-Learning IT Infrastructure Shared via the V4 EDUPORT Project Using a Virtual Machine with Windows 2022

Currently, the authors are collaborating as part of the V4 EDUPORT international project between the Visegrad countries, and are testing the possibility of collaborative design for e-learning. For this purpose, they share a virtual machine with a Windows 2022 server, on which WPad is also installed. Collaborative design means that partners using the WPad software application can also create educational packages so that the selected educational content, which is located in computer files and WPad tables, is automatically combined into one table. This then forms an e-learning output representing the expertise of the partners. One of the goals of the project is to create sets of e-learning content as additional educational material for undergraduates. A teacher or researcher who has WPad installed on a computer can also transfer tables or files between the virtual machine and the computer via the items on the user menu. Figure 8 illustrates how this is done.

7 Technological Obstacles to e-Learning Design

Although the scientific literature has focused on various technological aspects of e-Learning, the sometimes insurmountable difficulties faced by the teacher who designs e-learning content have rarely been described. The goal of IT support for knowledge

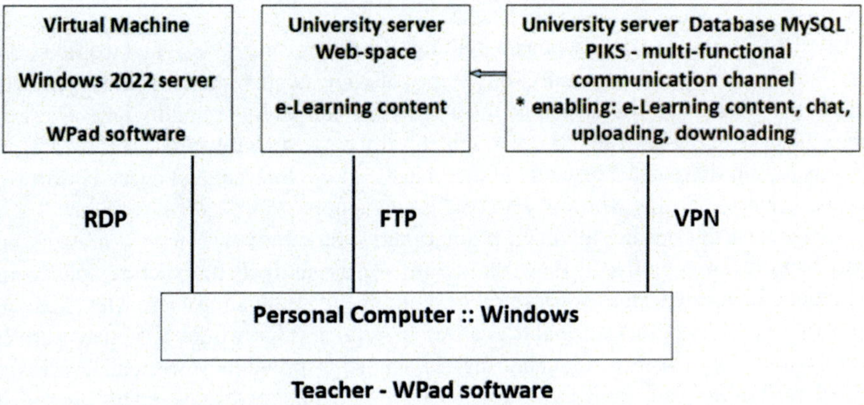

Fig. 8. Schematic diagram of the proposed universal personal hybrid e-learning design system

workers, as shown in Fig. 1, is still technologically impossible to achieve at the level of an ordinary user, due to the low level of state-of-the-art technology. The universal solution presented here had to be created by the authors themselves, through the design of their own educational software and hybrid IT infrastructure. The research approach used here is also in line with the considerations set out in [17]. In a monograph on TEL from twenty years ago, it was emphasised that every professor should have a team to give technological assistance in terms of designing educational content [18]. However, in discussions on ResearchGate it has been said that although this is how it works at American universities, in Europe it is left to the competence of the individual teacher. The authors' experience indicates that if a teacher works at a university, the technical department takes care of providing the necessary hardware and software. Thus, a university teacher does not have to deal with issues related to changes in the operating system, computer settings, hardware and software, and therefore is not aware of the conditions that an e-learning design researcher has to deal with on an individual computer (and probably does not even understand what this article is about).

7.1 General Issues Related to e-Learning Design

It should be remembered that although a university teacher typically gives classes and carries out research for 10–20 years, the lifespan of a given technology is significantly shorter. Due to the constant changes arising from technological advancements, the teacher's personal computer is typically changed every three to five years, and the operating system is often updated at the same time. In contrast, the database application WPad runs without problems on all versions of Windows, and will therefore work as long as the Windows operating system exists. While WPad has basically operated for more than 10 years, regardless of changes in IT, a lot of useful software has become unusable during this time, and for this reason, e-learning design is extremely demanding and expensive today. For example, someone who has been using an Internet browser and a certain operating system for years and wants to log in to a G-Meet call may find that a message pops up saying it cannot be done because they are not using Firefox, Edge or Google Chrome; the days when everything that programmers designed worked on all browsers are long gone. A user may buy some software and find that after a few years, the suppliers switch to a policy whereby the user has to start paying monthly fees. These are consequences of the huge levels of incompatibility between hardware and software, and for this reason there are thousands of help pages on the Internet and every IT software company has to employ workers who deal with support.

This scenario does not affect an ordinary university teacher, since the network and server settings are provided by the IT staff; however, it is unlikely that such a teacher could carry out e-learning design on conventional equipment. Another problem when updating operating systems is that a reliable scanner or printer stops working (updated drivers are missing), or software for capturing screens (that previously worked flawlessly), speech recognition, etc., may fail. These obstacles are quickly discovered by any teacher who changes workplace, as transferring previous settings to the new computer is also a huge problem. Even a simple computer setup may take an experienced IT specialist several days, since the computer of an e-learning designer will have dozens of software packages, several browsers, sound cards, several email accounts, hundreds of passwords and hundreds of thousands of files.

However, the fundamental macro-problem of e-learning design is that due to the incompatibility between software and hardware, it is not sufficient to use one solution, one software package, or one hardware platform; this means that the e-learning designer has to use several programming languages, possibly with completely different ones for the same purpose for desktop applications and others for Internet applications. In addition, the problem with Internet applications is that due to security and the threat of hacking, some functions have been disabled, so that things that were used without problems in the past are not possible today (this especially applies to HTML files).

It should also be mentioned that such fundamental general problems arise from the educational essence of e-Learning, i.e. that the educational content is no longer current or needs to be supplemented or updated. Almost every teacher encounters this problem. As explained above, when using WPad to generate HTML tables, updating of e-learning content is easy, and simply involves opening the last WPad table, adding or editing rows, offline paths, online links, etc., and then pressing CTRL-F1 and to copy the new table back to the web server. Of course, the most significant problem that affects every teacher is that educational content cannot be created in one format, and hundreds of different formats already exist for text, images, audio and video files.

A less well-known issue is the problem with diacritics, which is solved in HTML, for example by using the UTF-8 character set. When using WPad, the natural language used is plain text in English, and the user cannot normally write to WPad tables, e.g. in Ukrainian. However, there is a Ukrainian partner in the V4 EDUPORT project, and when dealing with the creation of educational languages, this problem is solved either by using the PDF format or by transferring the text using screenshots.

7.2 Some Examples from Undergraduate Teaching

As already mentioned, the lifetime of hardware is short. For example, the author built a Virtual Learning Space for undergraduates using the PHP language, and when the faculty server was changed from Windows to UNIX, in which upper- and lower-case letters are distinguished, the result was that a lot of PHP code had to be rewritten. However, the problem became much worse when this server expired and the learning content was transferred to a new server, as the PHP codes did not work, since a new PHP upgrade had been released in the meantime. There are also problems whereby some PHP code works flawlessly on the web hosting server, but the same code does not work on the university server. In other words, each designer must be prepared for such technological failures and have a backup solution ready. For the WPad designer, these online problems can be solved by placing the content on a private web host and a university website in parallel.

When working with networks and the cloud, the designer must program the transfers and synchronisation of e-learning files between computers and the online environment. A certain level of help is available from large companies such as Google and Microsoft, who will handle a certain volume of data if you have an account with them. For example, the Box service (https://apps.box.com/), which provides cloud content management, has proven to be useful within the scope of project solutions.

Compared to regular e-learning, problems arise in the design of visual and multimedia outputs. For example, about 10 years ago, a designer wanted to include Text to Speech (TTS) and Speech Recognition technologies in e-learning for undergraduates. While TTS was partially used for the semester papers of undergraduates, Speech Recognition did not prove itself even when the software was purchased for hundreds of dollars. In addition, after switching from 32-bit to 64-bit computers, the Dragon software did not work on them, and a new version had to be bought. While TTS is at an excellent level today (e.g. is used for multilingual e-learning outputs), so Speech Recognition is already at a fairly good level in English, but it is not available in the native language of the authors (Fortunately, Windows 1 already makes this possible).

Anyone who has been designing e-learning content for several years will find that a huge number of problems arise from the incompatibility between software and computer file formats and their shorter lifespans. When teaching undergraduates ten years ago, the Opera browser was excellent from a pedagogical point of view, and version 9.27 was implemented when teaching undergraduates. This version enabled collaborative internet searches, so that students could create sessions from search outputs and could simply send the corresponding files to the teacher. However, higher versions of Opera switched to Google Chromium code, and this feature no longer worked.

8 Conclusion

Teaching, learning, self-study, and all other teaching and educational activities are knowledge-based processes. Information and knowledge form the essence of educational content that can be created, processed and transmitted with the support of computers, which is the underlying principle of e-learning design. This research was initiated by the authors as part of their research on TEL, and aimed to address the automation of knowledge-based processes as human mental processes. It was assumed that if a knowledge process is to be automated, knowledge is logically its parameter and the slowest points of the process should be sought. However, questions arose as to what knowledge is and how to get it into the computer as quickly as possible, and how to make it work in bulk in larger batches.

It is surprising that an analysis of the scientific literature on e-learning design or the integration of IT into teaching indicates that the issue of what knowledge is has not been addressed. There is no universal definition of knowledge. It should also be mentioned that the field of computer science relates to information, and does not involve the concept of knowledge. After a few years, a foothold was established in the cybernetics literature, where learning was described as a cybernetic feedback system, and human-machine synergy could be achieved by saying that there must be some isomorphic switch between the mental processes of the human and the physical processes of the machine. This led to the formulation of a knowledge representation in a specific tabular structure, into which the teacher or user can insert knowledge in the form of natural language; the machine can process this as an information structure and deliver the result in a form that is comprehensible to a human. How this works has been explained and illustrated with screenshots: the in-house educational software WPad produces WPad tables, and since it also works as an HTML editor, the teacher can work in program mode or can produce HTML tables that open the default web browser in Windows (IE, Edge, Firefox, Opera, Google Chrome, Safari). Since the standard format for e-learning educational content is HTML, even users with only basic IT skills can design e-learning content on their laptop and transfer it to the web or share it on a virtual machine with Windows 2022 as needed. Collaborative e-learning design is currently being modelled on the principle of hybrid hypertext as part of the international V4 EDUPORT project. In future research, we also aim to design a 'primitive knowledge'; in other words, to allow the WPad table to function as an intelligent multi-media structure that can also manage itself. A further specific goal is the modification of WPad for e-learning design for visually impaired users, the real-world testing of which has been described in a section of this paper.

Regarding the main findings of the research, it can be mentioned that the presented internal software WPad (desktop application) and PIKS software (internet application) work as personal e-learning creation tools that allow the selection of tacit and educational knowledge (content) into 'virtual knowledge' (WPad table, PIKS table). These contain only selected, useful content and can be transferred between offline and online university environments as normal files. Since these tables can be read by both humans and computers, it also enables the automatic creation of learning packages from common computer files used by teachers (texts, audio, video), which can also be used to supplement 'Learning Management Systems' such as Moodle. From a didactic point of view, it is important to note that the mentioned IT tools enable interactive collaborative creation of educational content if there are appropriate didactic algorithms compiled by teachers for teaching.

Acknowledgments. This research work was supported by financial resources from the International Visegrad Fund under Strategic Grant No. 22210032, 'V4 Educational Academic Portal for Integrating IT into Education (V4-EDUPORT)'.

References

1. Littlejohn, A.: Key Issues in the Design and Delivery of Technology-Enhanced Learning. Facet Publishing, London (2005)

2. Svetský, Š.: Modelling of multilingual e-Learning and virtual learning space for R&D staff. In: Proceedings of the ICETA - 6th International Conference on Emerging E-Learning Technologies and its Applications, Stará Lesná. Elfa, Košice (2007)
3. Svetsky, S., Moravcik,O., Tanuska, P., Rehakova, A., Ruskova, D. The implementation of technology enhanced learning at dislocated university workplace. In: Proceedings of the ICETA - 6th International Conference on Emerging E-Learning Technologies and its Applications, Stará Lesná. Elfa, Košice (2008)
4. Laurillard, D.: Rethinking University Teaching: A Conversational Framework for the Effective Use of Learning Technologies. Routledge, London (2002)
5. Mayes, J.T., Fowler, C.J.H.: Learning technology and usability: a framework for understanding courseware. Interact. Comput. **11**, 485–497 (1999)
6. Salmon, G.: E-moderating: The Key to Teaching and Learning Online, 3rd edn. Routledge, New York (2011)
7. International Classification for Standards. ISO edition 7. https://www.iso.org/files/live/sites/isoorg/files/archive/pdf/en/international_classification_for_standards.pdf. Accessed 21 Jan 2023
8. EN ISO/IEC 19788–1:2012. Information technology - Learning, education and training - Metadata for learning resources - Part 1: Framework (ISO/IEC 19788–1:2011). https://standards.iteh.ai/catalog/standards/cen/e007380c-ba79-41cb-beef-d87c476631da/en-iso-iec-19788-1-2012. Accessed 21 Jan 2023
9. EN ISO/IEC 19788–2:2012 Information technology - Learning, education and training - Metadata for learning resources - Part 2: Dublin Core elements (ISO/IEC 19788–2:2011). https://standards.iteh.ai/catalog/standards/cen/02c3fbfb-a2e6-4e5b-aa52-dfb262f2e596/en-iso-iec-19788-2-2012. Accessed 21 Jan 2023
10. Haixia, H.E.: E-Learning theory in theoretical models for teaching and research by Joy Egbert and Mary Roe. https://opentext.wsu.edu/theoreticalmodelsforteachingandresearch/chapter/e-Learning-theory/. Accessed 21 Jan 2023
11. Mayer, R.E., Moreno, R., Sweller, J.: E-Learning theory. https://www.learning-theories.com/e-Learning-theory-mayer-sweller-moreno.html. Accessed 21 Jan 2023
12. Haythornthwaite, C., Andrews, R.N.L.: E-Learning Theory and Practice, 1st edn. Sage Publications Ltd., Thousand Oaks (2011)
13. Theriault, N.: Three models of e-Learning to improve pedagogy (2015). https://www.linkedin.com/pulse/three-models-e-Learning-improve-pedagogy-normand-theriault/. Accessed 21 Jan 2023
14. Svetsky, S.: The practical aspect of knowledge construction and automation of teaching processes within technology-enhanced learning and eLearning. Habilitation thesis, Slovak University of Technology (2012)
15. Mishra, P., Koehler, M.J.: Technological pedagogical content knowledge: a framework for integrating technology in teachers' knowledge. Teach. Coll. Rec. **108**(6), 1017–1054 (2006)
16. Zhang, W., Tang, J.: Teachers' TPACK development: a review of literature. Open J. Soc. Sci. **9**, 367–380 (2021)
17. Saljö, R.: Digital tools and challenges to institutional traditions of learning: technologies, social memory and the performative nature of learning. J. Comput. Assist. Learn. **26**, 53–64 (2010)
18. Goodman, P.S., et al.: Technology Enhanced Learning: Opportunities for Change. Laurence Erlbaum Associates, Mahwah (2002)

New Minds, New Methods: Pre-service Teachers Envision the Educational Future

Nina Bergdahl[1,2](✉) and Jeanette Sjöberg[1]

[1] Halmstad University, Halmstad, Sweden
nina.bergdahl@hh.se
[2] Stockholm University, Stockholm, Sweden

Abstract. What do people who study to become teachers dread and hope with regard to technology? Pre-service teachers (n = 21) were encouraged to consider the future of education by incorporating elements of, for example, augmented and virtual reality and Chat-GPT during three workshops that introduced them to the subject of sociotechnical imaginaries. Pre-service teachers were requested to write reflections and narratives for their logbooks. Individual thought expression can encourage group discussion and make instructors more confident in controlling future narratives. Such empowerment can mitigate risks and potential, inform elective courses of action, and improve understanding of the complexity of the future of digital education. Reflections were therefore done both individually and in groups. Data were analysed using the lens of Diffusion of Innovations theory, using thematic analysis. Three themes were identified, and key findings reveal opposing and ambivalent responses to the digitalisation of teaching, with plurality and flexibility on the one hand and restriction and isolation on the other. The results convey both passivity and hope as digital innovation was thought to solve problems by bridging the classroom to the outside world, leading to increased loneliness. The paper would interest stakeholders looking to leverage responsible digitalisation, the teaching community, and educational researchers.

Keywords: Socio-Technical Imaginaries · Pre-Service Teachers · Chat-GPT · Speculative Futures

1 Introduction

Technology and its potential and risks are central elements in the societal debate [1], not least concerning education. In education, the discussion on digitalisation of education has hardly changed, even though the technologies in focus have [2]. It has been argued that future teachers must ensure their plans are closer to the students' reality and the use of technology [3]. Following [4], this article posits that pre-service teachers' socio-technical imaginaries may offer insights into tomorrow's normative practices and plural ontologies. Rather than exploring existing practices, conceptualising future education or planning for protection against emergent fears [5], this article frames and theorises the nature of pre-service teacher articulation of future education and educational practices. Pre-service teacher socio-technical imaginaries are a relatively new concept, and there

K. Arai (Ed.): FTC 2023, LNNS 816, pp. 314–322, 2023.
https://doi.org/10.1007/978-3-031-47448-4_22

is still much that needs to be explored and understood about pre-service teacher anticipation. Exploring respondent narratives allows for raising awareness and could support emerging alternative paths. The study empowers pre-teachers to verbalise the unknown by sharing their views and positioning themselves and future education in context and with each other [6]. Pre-service teachers entering education are expected to teach for at least a few decades. They need to stay relevant and engage in professional development and educational practices. School development needs to reflect the digitalisation of society at large to prepare learners for an engaged life in an even more digital society. Indeed, the research found that the main difference between less and more experienced teachers lies not in the type of learning activities they undertake but rather in their attitudes toward learning, their learning outcomes, and how they are influenced by their context [7]. Thus, exploring pre-service teachers' attitudes and anticipation of educational futures will likely offer other insights into how they anticipate future schooling.

Research at the forefront has already described the potential implications of data-driven and automated assessment [8], learning analytics and artificial intelligence in the K-12 classroom [9]. However, when pre-service teachers engage in socio-technical imaginaries, they may not be updated with "the latest" in the tech industry. Instead, they may express their reflections concerning teaching and learning and technologies that do not yet exist. However, they are to become teachers; as such, they will need to use their prior knowledge and experience to prepare for challenges and scenarios in the classroom. This article explores the anticipation of future education through pre-service teacher socio-technical imaginaries. It discusses how these can help to improve the quality of teaching and learning, create better learning experiences for students, provide teachers with the necessary skills and knowledge to be better prepared for future challenges and scenarios and inform further development of pre-service teacher education through the lens of Diffusion of Innovations theory [10]. In the context of teacher education, [11] addressed how student teachers are being prepared to use and integrate technology into their programs. Teacher education programs must work to infuse technology better throughout the entire teacher education program and across different subject areas. Thus, while socio-technical imaginaries and speculative futures approaches have been conducted before, these studies are scarce in Sweden, and we did not identify any studies to date that include the stimulus of generative artificial intelligence. This study contributes insights in this matter by exposing pre-service teachers to emerging technologies and then exploring their socio-technical imaginaries on the future of education. The structure of this paper is as follows: the study continues with a theoretical background which frames socio-technical imaginaries, and then we present the Diffusion of Innovations theory, which guides our analysis. We then detail our methodology, present our findings, and discuss the results. Finally, we propose directions for future research and conclude the paper.

2 Theoretical Background

2.1 Socio-Technical Imaginaries

Pre-service teacher socio-technical imaginaries refer to the potential outcomes of students considering a career in teaching. Jasanoff defines socio-technical imaginaries as: "collectively held, institutionally stabilised, and publicly performed visions of desirable futures, animated by shared understandings of forms of social life and social order attainable through, and supportive of, advances in science and technology" [12]. Building on Jasanoff, the position here is that with socio-technical imaginaries, we move beyond the language of imagination and recognise the diverse pluralistic ontologies of cultures to develop inclusive frameworks for socio-technological system transitions. The introduction of standards-based instruction, the use of technology in the classroom, and the increased emphasis on teacher accountability have significantly impacted the teaching profession. All changes require teachers to be reflective and critical of their practice, necessitating a more in-depth exploration of the different elements that shape the teaching experience. At the same time, the rapidly changing landscape of education has created an increased need for prospective teachers to consider their future as educators. "How a person understands his or her relationship to the world, how that relationship is structured across time and space, and how the person understands possibilities for the future" [13].

Teachers will shape the future of learning, and their practices will be relevant for decades. Verbalising how and what pre-teachers believe, hope for, fear, and visualise the future is therefore of interest. Teachers' rights to verbalise priorities, learning studies, and practice development remain relevant to professional development. Visualising the expanded ideas can contribute to better preparation. Previous studies exploring imaginaries of the future roles of teachers lifted aspects like "AI taking over" and questions of who would be right: the teacher or the AI system?[14] Here we explore ontologies expressed by pre-teachers. An ontology formally represents knowledge or concepts within a domain (entities, attributes, processes). It reflects an understanding, worldview, and how it is conceived as a set of concepts [15]. Acknowledging that individuals often harbour diverse and incommensurable ontologies is essential. We need to accept the reality of pre-service teachers to achieve insights into their anticipation. Their stories can offer valuable insights into teachers' current positions on the role of educational technologies in the future.

2.2 Diffusion of Innovation

The Diffusion of Innovations theory was developed by sociologist Everett Rogers in 1962 and is based on social networks and how information and ideas spread within them. This theory explains how, why, and at what rate new ideas and technologies spread through a given population. The theory consists of five main components: the innovation itself, the communication channels used to spread the innovation, the social system in which the innovation is being spread, the time it takes for the innovation to spread, and the decision-making process of whether to adopt the innovation [10, 16]. According to the theory, the rate of adoption of an innovation is determined by five main factors: relative advantage, compatibility, complexity, trialability, and observability:

- Relative Advantage: refers to the extent to which an innovation is perceived as better than existing alternatives,
- Compatibility: refers to the extent to which an innovation is perceived as being consistent with existing values, needs, and experiences,
- Complexity: refers to the extent to which an innovation is perceived as being difficult to understand and use,

- Trialability: refers to the extent to which an innovation can be tried out on a limited basis,
- Observability: refers to the extent to which others can observe innovation results.

The Diffusion of Innovations theory can be applied to the use of technology in the classroom to understand the rate at which new technologies are adopted, how they are used, and how they may impact educational equity.

3 Method

The study was conducted within the teacher program at a Swedish university during November and December 2022. Pre-service teachers (n = 21) participated in a synchronous online course on Assessment in digital learning environments, during which times were designated to the socio-technical imaginaries. The students were encouraged to reflect in a logbook and asked to co-author a narrative following intervention scaffolding that supports teacher imaginaries [17]. While lesson time was limited, the pre-service teachers were encouraged to continue writing in their logbooks throughout the course. The data comprises 35 text entities: logbook entries (n = 30) and narratives (n = 5). In exploring narratives of the future and education or educational practices, the actions (interactions and communication) become essential in understanding the critique of eth current and the expectations ahead. The discovery becomes central because discovery means learning about one's surroundings, exploring potential goals, and directing intention towards specific affordances [17]. As such, these processes reflect how pre-service teachers think about future education and educational practices via the buildings, materials, objects, and characters interacting with them and each other.

The course was designed to introduce digital technologies in education, stimulate creative thinking and exploration, and highlight the risks and potential of artificial intelligence following the A-C sequence. A) Lectures 1–2, including orientation on "The changes of the written word" and "Future education" were conducted. Adjacent to these lectures, the instruction for reflections 1–2 was communicated. B) Lecture 1–2 was followed by a seminar in which students would present a situation with artificial intelligence, augmented reality, virtual reality, green screen, telepresence robots, or other digital technologies which they could see themselves using and describe how they engaged learners and executed their summative and formative teaching practices. C) Then, students participated in Lectures 3–4, which included assessment theories, digital examination, diagnostics, technologies to assess live practices, data-driven educational perspective, trying Chat-GPT and automatic assessment. Then, the students began writing their narratives in groups. The narratives were handed in the following week. The data was analysed using thematic analysis [18]. Notes were gathered in the familiarisation stage. Tentative themes and sub-themes were formulated and grouped. The themes were un-grouped and re-grouped when all data was coded to explore patterns and explore sub-grouping levels.

4 Results

Three themes were identified: 1. Joint development of educational and societal digitalisation, 2. Multiple self-identities, and 3. EdTech companies bridge isolation and redeem loneliness.

4.1 Theme 1: Joint Development of Educational and Societal Digitalisation

Teachers discussed changes in emerging curricula and modalities for knowledge acquisition and production and how the development was aligned with the needs of society, the world, and the labour market, with sustained influences from Big Tech and capitalist ventures.

"On the other hand, working life has changed, as robots take care of manufacturing and food deliveries, for example. One consequence is that there are new subjects in the curriculum. We learn to meet the needs of the labour market, which focuses on streamlining digitisation. A new school subject is health (which focuses on mental health), living in a digitised world. This new lifestyle demands a lot from us." (Students 4, 18, 17: collaborative reflection).

"The education system will look completely different but retain basic elements. /…/ First, you need to calculate the other changes in society, and this mainly centres around technological development. It is worth mentioning here something about the extremely exponential tendencies of technological development to suddenly take large steps in progress. Otherwise, we would have seen the development increase gradually. But I think it [digitalisation] will increase more and more intensively in the direction that we are developing toward." (Pre-service teacher 8: log 1).

The pre-service teachers discuss educational changes due to technological advancements, including introducing new subjects such as mental health and living in a digitalised world.

4.2 Theme 2: Multiple Self-identities

Pre-service teachers discussed how identity formation would develop into multiple choices of digital self-representations of self and of the group of learners, where the class composition and educational inclusion could stimulate specific choices.

"Some students choose to be represented by a scanned version of themselves in hologram form, others want to be present in virtual reality as an animated figure, and a third group of students choose to be present with surrogate robots." (Pre-service teachers 4, 18, 17: collaborative reflection).

"I need to be online at 9.30. I will be working with Colin from Scotland and Sally from South Africa; it was the computer that put the groups together. It is very fun to get to know many different people during my schooling. Together we will travel to two different landscapes via Google Earth and our digital glasses and examine these using a set of questions. The report must be submitted as a digital presentation in English. We are part of global digital classes and a home class. I am equally good at English as at Swedish." (Pre-service teacher 7: log 3).

The teachers talk about a virtual catchment area, with several class and group compositions and expanded ways to interact with a global society as digital selves. Yet, there is little recollection of physical (embodied) activity and interaction - instead of bridging the human-human distance – another human is replaced by robot interaction.

4.3 Theme 3: EdTech Companies Bridge Isolation and Redeem Loneliness

Somewhat paradoxically, the pre-service teachers discuss increased loneliness and isolation and how tech companies respond with digital support solutions.

"It is not only in the classrooms that we meet via robots, but in our free time we socialise via robots. In some seasons we can meet physically." (Pre-service teachers 4, 18, 17: collaborative reflection).

"… about school refusers [students who stay home and refuse to go to the physical school], will many people in the future will be tutored at home? Will that then create an A and B team? But who would be the beneficiaries, the A and B team, the stay-at-home-learners, or the ones at school? Who knows?" (Pre-service teacher 6: log 2).

"…interesting with thoughts about AI robots, because my son does not go to school [due to] autism, ADHD, depression, and has been without school for a long time. An AI robot could help, at least for a certain time. The same with assisting technologies that provide continuous feedback during the student's own work." (Pre-service teacher 10: log 2).

The pre-service suggests significant changes to the educational system, support functions and joint development of educational and societal digitalisation. The exponential tendencies of technological development are also mentioned, and it is predicted that digitalisation will continue to increase intensively in the direction society is developing.

5 Discussion

Exploring pre-service teachers' socio-technical imaginaries on the future of education, this study found that the narratives reflect complexities that may appear contradictory. One way to see the narratives is in dichotomies as a promising or terrifying future. However, the imaginaries are linguistically, socially, culturally, and historically constructed more or less clearly in the three identified themes: 1) Joint development of educational and societal digitalisation, 2) Multiple self-identities, and 3) EdTech companies bridge isolation and redeem loneliness. These themes are here discussed through the lens of the Diffusion of Innovations theory [10, 16] more precisely: the relative advantage, compatibility, complexity, trialability and observability.

The results highlighted the existence of the paradox of increased loneliness and machines facilitating and replacing human connection and enabling a multitude of short-term and context-dependent individual and group identities, which signals anonymity and distance. At the same time, the results highlight the interdependence of digital development as a response to societal, environmental, and psychological problems. So, digital technologies are expected to bring a solution to real-life problems. Addressing compatibility [10, 16], the narratives included emerging curricula, educational distributions, technologies, and modalities for knowledge acquisition. The teachers incorporated perspectives of students' reality [3] but also include the pressing societal demands that were reported to become more severe in parallel with the development of digitalisation. The results reflected the potential of advanced technologies to solve problems at a global level, including hunger and inequalities were incorporated into the curriculum alongside more global citizenship and network. These global-level imaginaries were more ideal than imaginaries at the individual level. These can be understood as more self-serving (such as having a robot, even though not ideal, that can redeem some of the personal angst). Thus, 'the relative advantage' was seen in relation to the individual, which can be contrasted to 'the compatibility' at the global level. The global-level events were expressed with a more idealistic framing that does not openly oppose, provoke, or problematise existing values.

As mentioned in earlier research, especially at the individual level, there is a lack of problematising continued or increased inequality, such as shift roles, where remote and home-schooling may become the beneficiary form [19]. In essence, those who could use it had access to it, could afford it, and where such use would benefit their advancement and social inclusion, viewed the innovation as superior to existing alternatives. The teachers then discuss weighing the benefits and drawbacks of the changes to decide whether they are worth attempting, which relates to trialability. These instructors want to use artificial intelligence and automated assessment in the classroom, but they don't have the resources. However, despite expressing fear and caution, teachers also claim to wish to put their ideas into practice and see how they work. Finally, at the level of observability, human networking, already international, due to social media, will become even wider: with fluent class compositions, global schools, interplanetary accommodation and fluency between languages and physical and non-physical interaction. Interestingly, the pre-service teachers' imaginaries of the teacher role differed from the previous [14].

The pre-service teachers' reflections may support verbalisation and action. Importantly, teachers will continue to shape the educational landscape, so they need to consider the potential implications of changes and how these may affect their educational practices. Contrary to Rogers' idea [10], the teachers did not see themselves as having a choice. Instead, digitalisation was happening to them. Considering the expressed lack of control, there were few dystopian visualisations. The author in [6] suggested that visualising dystopia by no means is predictive. Instead, they can be interpreted as visualisations and strategies to encounter fears, vulnerability, despondency, and danger. Further, she proposes that these could empower the individuals' sense of attending to their full range of unverbalised notions regarding the future and raise awareness and orientation levels. Such orientations could then guide actions and responses to avoid such a future. The author [6] also alerted that while open with ideas, adolescent socio-technological imaginaries shall not be mistaken for cutting–edge techno centrism or state-of-the-art digitalisation, as their desires and fears may be conservative. As described, the pre-service teachers' talk on the societal impact on education was much related to measures of control and inequality, as well as self-expression, pluralism, and flexibility. Future research could contribute to the field by offering cross-cultural or longitudinal comparative studies to capture how pre-service teachers develop their ideas and anticipation.

6 Conclusion

In conclusion, this paper has accomplished its aim and objectives to explore pre-service teachers' socio-technical imaginaries on the future of education, shedding light on diverse ontologies and the need for inclusive frameworks and by doing so contributing valuable insights to the field. Through meticulous data analysis and rigorous methodology, the study has provided a comprehensive understanding of pre-service teachers' anticipation for their profession in the future. The data signals that "imaginaries" can sometimes reflect incommensurable plural ontologies of pre-service teachers. With socio-technical imaginaries, we move beyond the language of imagination and recognise the diverse pluralistic ontologies of cultures to develop inclusive frameworks of anticipation. Key findings show that pre-service teachers anticipate that digitalisation will likely continue, primarily as a response to capitalist companies. The future is also seen as one with food shortages, increased isolation, and sustained mental health problems. To the escalation of these arenas, the pre-service teachers report that both education and digitalisation will come together to find solutions. Identity formation is viewed in an increasingly liberal way, where students can choose their preference for self-representation. When empowered, pre-service teachers could raise awareness through open discussions on the potential risks and benefits associated with emerging technologies while guiding actions and responses to steering clear of potential future issues while allowing for increased flexibility and pluralism. The findings have been consistent with previous research, reinforcing the existing knowledge base. Moreover, the study has addressed the limitations identified in earlier studies, filling important literature gaps. The achieved results not only deepen our understanding of the topic but also offer practical implications for stakeholders looking to leverage responsible digitalisation, the teaching community, and educational researchers, and in doing so, paving the way for future investigations and advancements in this field.

References

1. Kerschner, C., Wächter, P., Nierling, L., Ehlers, M.H.: Degrowth and technology: towards feasible, viable, appropriate and convivial imaginaries. J. Cleaner Prod. **197**, 1619–1636 (2018). https://doi.org/10.1016/j.jclepro.2018.07.147
2. Kerres, M.: Bildung in a Digital World: The social construction of future in education. In: David, K., Michael, P., Jesper, G., Birte, H.-K., (eds.) Bildung in the Digital Age: Exploring Bildung through Digital Media in Education, 1st ed., Routledge, pp. 29–46 (2022). https://doi.org/10.4324/9781003158851
3. Becuwe, H., Roblin, N.P., Tondeur, J., Thys, J., Castelein, E., Voogt, J.: Conditions for the successful implementation of teacher educator design teams for ICT integration: a Delphi study. Australas. J. Educ. Technol. **33**(2), 159–172 (2017). https://doi.org/10.14742/AJET.2789
4. Schelly, C., Gagnon, V., Arola, K., Fiss, A., Schaefer, M., Halvorsen, K.E.: Cultural imaginaries or incommensurable ontologies? Relationality and sovereignty as worldviews in socio-technological system transitions. Energy Res. Soc. Sci. **80**, 102242 (2021). https://doi.org/10.1016/j.erss.2021.102242
5. Facer,K., Enright, B.: Creating Living Knowledge. University of Bristol AHRC, Connected Communities Programme (2016)
6. Priyadharshini, E.: Anticipating the apocalypse: Monstrous educational futures. Futures **113**, 102453 (2019). https://doi.org/10.1016/j.futures.2019.102453
7. Kyndt, E., Gijbels, D., Grosemans, I., Donche, V.: Teachers' everyday professional development: mapping informal learning activities, antecedents, and learning outcomes. Rev. Educ. Res. **86**(4), 1111–1150 (2016). https://doi.org/10.3102/0034654315627864
8. Paiva, J.C., Leal, J.P., Figueira, Á.: Automated assessment in computer science education: a state-of-the-art review. ACM Trans. Comput. Educ. **22**(3), 1–40 (2022). https://doi.org/10.1145/3513140

9. Tedre, M., et al.: Teaching machine learning in K-12 Classroom: pedagogical and technological trajectories for artificial intelligence education. IEEE Access **9**, 110558–110572 (2021). https://doi.org/10.1109/ACCESS.2021.3097962
10. Rogers, E., Singhal, A., Quinlan, M.: Diffusion of Innovations. In: Stacks, D.W., Salwen, M.B. (eds.) An Integrated Approach to Communication Theory and Research, Routledge, pp. 432–448 (2014). https://doi.org/10.4324/9780203887011-36
11. Tondeur, J., Roblin, N.P., van Braak, J., Fisser, P., Voogt, J.: Technological pedagogical content knowledge in teacher education: in search of a new curriculum. Educ. Stud. **39**(2), 239–243 (2013). https://doi.org/10.1080/03055698.2012.713548
12. Jasanoff, S.: Future imperfect: science, technology, and the imaginations of modernity. In: Jasanoff, S., Kim, S.-H. (eds.) Dreamscapes of Modernity: Socio-Technical Imaginaries and the Fabrication of Power, Chicago, University of Chicago Press, pp. 1–33 (2015). https://doi.org/10.7208/chicago/9780226276663.001.0001
13. Norton, B.: Identity, and Language Learning: Gender, Ethnicity, and Educational Change. Longman Publishing Group, Harlow Essex (2000)
14. Hrastinski, S., et al.: Critical imaginaries and reflections on artificial intelligence and robots in Postdigital K-12 education. Postdigit. Sci. Educ. **1**(2), 427–445 (2019). https://doi.org/10.1007/s42438-019-00046-x
15. Uschold, M., Gruninger, M.: Ontologies: Principles, methods and applications. Knowl. Eng. Rev. **11**(2), 93–136 (1996). https://doi.org/10.1017/s0269888900007797
16. Rogers, E.M.: Diffusion of innovations. Free Press (1983)
17. Ross, J.: Speculative methods and digital futures research. In: Digital Futures for Learning, Routledge, pp. 169–186 (2022). https://doi.org/10.4324/9781003202134-12
18. Braun, V., Clarke, V.: Thematic analysis. In: Lyons, E., Coyle, A. (eds.) APA Handbook of Research Methods in Psychology: Volume 2. American Psychology Association, Washington DC, pp. 57–71 (2012). https://doi.org/10.1037/13620-004
19. Selwyn, N.: Less work for teacher? The ironies of automated decision-making in schools. In: Everyday Automation: Experiencing and Anticipating Emerging Technologies, pp. 73–86 (2022). https://doi.org/10.4324/9781003170884-6

Online Collaborative Learning: A Study of Communicative Interactions in a Latin American University Environment

Sandra Isabel Arango-Vásquez[1], Bell Manrique-Losada[1](✉), and Fernando Moreira[2,3]

[1] University of Medellin, Medellin, Antioquia, Colombia
fmoreira@upt.pt
[2] REMIT, IJP, Portucalense University, Porto, Portugal
[3] IEETA, University of Aveiro, Aveiro, Portugal

Abstract. The Collaborative Learning has been a reference learning theory in higher education, nevertheless, learners don't recognize and take advantage of its potential, or how to reach approaches and coregulate with peers. The current post-pandemic scenario has generated new opportunities for interaction in virtual environments. However, the low or inconstant students' participation in them has given way to research seeking to know the causes and whole scenario of process and suggest alternatives to enhance communicative interactions and collaboration on online platforms. In this article we study how can communicative interactions and collaboration mediated by information and communication technologies be fostered, enhanced, or increased in university e-learning environments. This research was developed using the qualitative perspective with a hermeneutic methodological approach. For the study, 124 primary studies of the literature and 42 e-learning environments of the Academic U-Virtual platform of the 'Universidad de Medellín' -Colombia were analyzed. The application of the research method allowed us to find four major components as a result: motivation and feedback, mediation/interaction, didactic, and technological. Such components can integrate and promote communicative interactions and online collaboration in university environments. As main conclusions, we found that the proposed components will allow: (i) to distinguish the particularities of communicative interactions; (ii) to facilitate tasks oriented to promote skills centered on collaboration, self-learning, and flexibility; and (iii) to design improved virtual environments by using synchronous and asynchronous tools that encourage the communicative interactions and the permanent collaboration of professors and students on e-learning platforms.

Keywords: Collaborative Learning · e-Learning Environments · Communicative Interactions · University Environments

K. Arai (Ed.): FTC 2023, LNNS 816, pp. 323–338, 2023.
https://doi.org/10.1007/978-3-031-47448-4_23

1 Introduction

The Information and Communication Technologies -ICT, in addition to its value as a mechanism or tool, are mediations that allow to improve learning. These have advantages for education, such as: students have large volumes of content; they allow flexible participation of students and professors regardless of space and time [1–6].

On the other hand, the COVID-19 pandemic streamlined the massive usage of ICT in the educational context and positioned the leading role of virtual learning environments -VLEs (classrooms, platforms, or environments on virtual world) in the 'new wave of normality' of teaching-learning processes. Thus, we have seen a very rapid transition from face-to-face to technology-mediated learning. This latest learning has favored a paradigm shift in all education levels; in the case of higher education, institutions have had to assume several challenges, mainly in the implementation of strategies, techniques, and methodologies to promote the interaction and participation of students and professors in the educational processes.

Undoubtedly, such digital or technology-mediated learning has brought with it great opportunities for communicative interaction that require, consequently, a capacity for adaptation capacities of individuals (students, professors, and educational administrators) and institutions [7]. Faced with this panorama, higher education institutions have undergone some changes to adapt academic processes to these contexts. One of the biggest challenges is to promote the permanent participation of students and professors under pedagogical mediation in e-learning environments.

The Universidad de Medellín -UM, a Colombian Higher Education Institution accredited with the highest quality standards by the National Accreditation Council, recognize the relevance of technology-mediated training and starting the 2020 defined policies to the ICT incorporation. Thus, at starting each academic semester is institutionalized the provision of virtual learning environments for each course, subject, or module to be offered in the academic programs. In addition, it declares that professors should promote educational strategies oriented to develop autonomous learning, teamwork, and knowledge skills in students, through: (i) approaching content by means of digital educational resources -DER (*e.g.* textual, audiovisual, or multimedia); (ii) synchronous and asynchronous communication tools that allow student participation in the learning activities; and (iii) learning evaluations by using the resources arranged on the UVirtual platform.

The Virtual Education and ICT department at UM carried out a study in the 2019–1 semester, which to show that of 2,520 VLEs arranged on the UVirtual platform, only 102 (4%) of them were published as DER and in 58 (2%) were used communication tools to generate several kinds of interaction among professors, students, and content.

With this background, coupled with the institutional interest and the E-Virtual and ARKADIUS research groups for increasing the publication of DER in VVLEs and generate new opportunities for interaction, the study "Communicative Interaction in VLEs of the UM UVirtual platform" was developed. This research studied how can communicative interactions be fostered, strengthened, or increased in university VLEs mediated by ICT. The objectives of the study were the following: (i) characterize the components of communicative interaction in VLEs; (ii) analyze the technological resources and tools used by UM professors to generate student participation in the e-learning environment, and (iii) formalize a strategy to increase the communicative interactions students-professors in VLEs.

Based on the research problems the importance of this initiative is grounded in the integration and promotion of the communicative interactions, and how they consolidate the online collaboration in the current e-learning university environments. Such collaboration arises from the permanent interaction students-professors under pedagogical mediation, adapting the capacities of both: individuals and institutions.

This article presents the results of the research, with the following structure: Sect. 2 describes the referential framework, highlighting online collaborative learning, communicative interactions theory, VLEs, and pedagogical mediation. Section 3 details the methodological development of the research, which includes the techniques and instruments, as well as the general analysis of the results from the documentary review and the contents analysis of the platform. Section 4 presents the results discussion. Finally, Sect. 5 summarizes the conclusions.

2 Referential Framework and Related Work

2.1 Collaborative Learning

The collaborative learning mediated by technology, integrates at least three aspects: cognitive, interpersonal, and social [8]. The *cognitive* is presented in the learning of the contents raised in the virtual environment and its application in solving the problems raised; the *interpersonal* is reflected in the motivation of students to develop teamwork; and the social is evidenced in the communicative interactions between the members of the group. This kind of mediated-learning has created the Online Collaborative Learning -OCL- theory. According to Harasim [9] OCL theory defines a learning model where students are stimulated to work collaboratively for creating knowledge through the following steps: idea generating, idea organizing, intellectual convergence, and knowledge application. With OCL, the aim is using the technology mainly to increase and promote communicative interactions among professor, students, and platforms, but not to replace the professor, in order to develop learning based on knowledge construction [10].

There are different strategies and tools to promote collaboration online or with the help of technology. For example, synchronous and asynchronous communication tools can be used to create discussion groups, monitor activities, and exchange feedback; while visual representation tools that allow you to create concept maps, diagrams, matrices, and lists, help to build representations, and share ideas. Likewise, using digital games or simulators generate an interactive and immersive learning environment [11]. Subsequent paragraphs, however, are indented.

2.2 Communicative Interactions

Communicative interaction is the process of reciprocal communication among participants that can induce behaviors, thoughts, or actions of others, through a dialogue or finite conversation [12]. Communicative interactions imply more than the information transfer because they are a "social process of creation, preservation, maintenance and transformation of social realities" [13].

Communicative interactions, in this research, are discourses that promote learning in VLEs, foster the creation of interpersonal relationships, and favor group relationships [14, 15]. In learning processes, student participation is done through the socialization of prior knowledge, ideas, experiences, abilities, and attitudes to build learning communities [16]. In e-learning environments, participation or interaction is done using communication tools, digital resources and oral or written language [17].

Technological advances and the ICT implementation in educational processes have expanded and have overcome space-time barriers, changing the way knowledge is built, how work is done in the classroom, and how people interact through virtual environments [18, 19, 20].

2.3 Virtual Learning Environments -VLEs

A VLE is a digital space that interrelate pedagogical, communicational, social, and affective features. VLEs comprise tools to facilitate knowledge' appropriation to the students. Following Maraza [21], a VLE intends to develop approaches for searching and identifying the best learning style of students. In the same lines, Silva [2] points out that VLE are computer applications implemented to enable pedagogical communication among educational participants. Thus, a VLE must be created with a pedagogical conception that integrates knowledge, resources, and communication skills into its mediating interface to promote learning [14].

2.4 Mediation

To understand mediation, it is important to refer to the concept of *mediating interface* by Serrano [22], who researched and developed a theory in the field of mediations and communication in two areas: 1) social and technological mediations of communication and 2) mediations of the uses and contents of communication. Lately, he radically transforms the study of the media because from the analysis of mediation it is shown that there are forms of social control that are not explicit in the narratives.

Barbero [23] states that the mediation process occurs in those places where the communication is given and is directly related to the cultural identity and production of culture of the subjects. In this way, the fact of elaborating meaning requires mediation. Following Vygotsky [24] the students don't learn or formalize concepts spontaneously and, therefore, intentional mediation is required. Therefore, the interaction professor-student changes when a mediation function is assigned for the development of the student' own structures, and for the professor be a guide in the VLE learning process.

Besides, pedagogical mediations comprise actions, resources, and didactic materials articulated to intervene and to facilitate the educational processes. The main purpose of mediation is facilitating communication student-professor to allow a broad approach to the ideas through the senses [25].

In the ICT context, Lima [26], Escalona [27], and Bustos *et al.* [28] state pedagogical mediation as an element that fosters interaction among professors, students, and contents. Professors implement methodologies and activities for students integrate theory and practice and they develop communication and technological abilities.

Therefore, the communicative interactions pedagogically mediated by VLEs represent one of the most creative and revolutionary training spaces to energize the teaching and learning processes. To García [29] VLEs are the implementations of spaces to make educational use of the Internet and must be hosted on a platform that integrates virtual communication tools, hardware, and software.

According to Barberá *et al.* [30], Hirumi [31] and Huss *et al.* [6], the communicative interactions are produced in VLEs in the following cases: 1) student-student, 2) student-content, 3) student-professor, and 4) student-interface interaction.

In this context, consolidating and promoting communicative interactions under pedagogical mediation in VLEs, allows us to considerate technological advances in educational processes, the resources and tools that are implicit, and change the ways of building knowledge and society from the classrooms.

3 Methodology and Research Results

3.1 Methodological Design

This research was developed with a qualitative perspective and a hermeneutic approach. Gadamer and Olasagasti [32] define hermeneutics as "all conceptualization work that pursues possible consensus, possible agreement, and must even be based on consensus if people are to understand each other" (p.116). This consensus was reflected in the interdisciplinary work of the researchers who contributed from the fore-knows, the documentary analysis, and the analysis of the VLEs contents.

In the research, a circular hermeneutic structure was used that allowed understanding the communicative interactions and collaboration as an object of study, which correspond to the *whole*, and the *parts* refer to: the features of the communicative interactions and collaboration in the VLE, the technological resources and tools used by UM professors to promote students participation and collaboration, and didactic strategies to increase communicative interaction and collaborative work among students. The circular hermeneutic structure is based on prejudices, analysis, comparison, reflection, understanding, interpretation, and synthesis.

Techniques and Instruments. To answer the research question, a set of "texts" was selected to apply the circular hermeneutic structure, referring to the primary studies selected in the literature review process. These texts are understood as "any action susceptible to translation and in a restricted sense, any fixation that generates a structure of meaning" [33]. In this way, the "text" is constituted in a meaning structure that contains the data or codes that allow passing from pre-knowledge to new knowledge.

The techniques used were documentary review and content analysis of VLEs. The literature review process followed the next protocol: (i) creation of search strings according to the specific objectives and researchers' pre-knowledge; (ii) definition of inclusion and exclusion criteria, including: observation period 2018–2021, and English and Spanish language; (iii) definition of the Google Scholar, Redalyc, SCOPUS, and ERIC databases. The instruments used to collect the information, are as follows:

- Basic documentary review protocol, comprising the results synthesis of the application of the search strings in the selected databases and the description of the inclusion criteria, defined in the selection of articles or primary studies.
- Document review file, for recording the information obtained from each of the 124 documents with the researcher's analysis notes.
- Characterization file, which allowed systematizing, analyzing, comparing, and synthesizing the information in aspects or subcategories.

For content analysis, a file was designed as an instrument for collecting data from VLE, related to four categories defined from the documentary review: (i) didactic and cognitive, (ii) communicative interactions and collaboration, (iii) technological aspects, and (iv) social-affective. This instrument was validated with a pilot test applied to five environments randomly selected on the UM Virtual platform. From this test, adjustments were made and then the refined instrument was applied to 42 VLEs used in the 2021–1 academic semester. For the selection of these environments, the following information was triangulated: 1) Database of 2560 VLEs published in the 2021–1 semester on the UVirtual platform; 2) List of 65 environments that presented a high communicative interaction; and 3) List of 25 professors participating in the Conference on "Good Pedagogical and Curricular Practices" carried out by the Academic Direction of the UM, between 2019 and 2021. The selection criteria were as follows: equitable representation of six VLEs for each Faculty (Law, Engineering, Administrative Economics, Communication, Social Sciences, and Basic Sciences) and six from the Language Center.

3.2 Results Analysis

The documentary review allowed to identify important findings regarding the importance of communicative interactions and collaboration in VLEs. Among these findings, the following can be prioritized. Motivation and feedback are a transversal and indispensable component in communicative interactions and collaboration in virtual environments mediated by technology. Barreto [34] found that motivation is related to the technological context, social interactions, and interactions with resources and content. Furthermore, Duarte-Herrera *et al.* [35] highlight that permanent follow-up and feedback to students allows them to motivate them to participate, to resolve or contribute to the doubts raised. On the other hand, Bylieva and others [36] reveal that the spaces for exchange and dialogue of ideas about the topics influence the organization and academic performance of students. Salam and Farooq [37] found that interaction and collaboration improve the educational experience and the sense of belonging when common activities or goals are proposed. In addition, Fernández [38] determined that collaborative learning supported by ICTs is a methodological strategy to promote participation.

As defined Pérez [39], to establish communicative interactions in VLEs, three dimensions must be kept in mind: cognitive, social, and affective. In the cognitive dimension, when VLEs are used, the dialogic relationship between the student and the published resources is evidenced, and teamwork is developed through collective strategies or techniques [36]. In the social dimension, students exchange knowledge and develop skills or competencies through interaction with students or professors [40]. The affective dimension is related to motivation and emotions, in the virtual environment, making it possible to consult the information, its analysis and interpretation for decision-making and subsequent responses. Despite differentiating cognition, sociability, and affectivity as independent processes, these occur simultaneously in educational settings [13, 35].

According to Aparicio and Ostos [41] the pedagogical and didactic aspects present in virtual learning environments can be established with emerging teaching and learning methodologies, which allow integrating pedagogy as a central element to establish the intention of the interactions between students, professors, environments, and contents. The authors recommend implementing active learning pedagogies based on e-learning, b-learning, m-learning, u-learning, and c-learning methodologies.

The characterization of the documents was carried out using a data processing tool, based on the reading and analysis of the 124 texts in relation to the objectives. Frequencies and relationships were established to group the concepts into categories. In Fig. 1 we show the four categories into which the concepts are grouped.

3.3 Content Analysis

We analyzed the 42 VLEs hosted on the UVirtual platform of the UM in the 2021–1 semester, based on a file created for grouping various components associated with the categories defined in the literature review.

In the *didactic and cognitive category*, aspects related to the following: general indications of the subject, clear instructions in the evaluation activities, evidence of accommodation of digital resources, logical order of the contents and activities for group reflection were considered. In Fig. 2 we present the findings in the 42 environments.

Figure 2 shows that 37 of the 42 environments present evidence of general indications, among which we can highlight: syllabus, activity guide, and professor information. As an important alert, finding the five environments that did not present this information, even though they are the basic elements available to the student to know the agenda of the contents and the activities to be carried out during the 2021–1 semester (modality at home for pandemic). On the other hand, it was possible to verify that in 12 VLEs there were no clear instructions for the development of the evaluation activities, furthermore the students asked for clarification in the "Ask the professor" forum; however, in 10 of these timely responses were not obtained. Regarding educational resources, it was possible to demonstrate their use and/or appearance in 27 settings. Regarding the organization of the environment, it was identified that 29 met the minimum guidelines established by the University, and only in 23 of them did the professors invite group or collaborative work.

In the *communicative interactions* category, evidence was sought on the participation and collaboration of students and professors in different spaces created for academic discussions, the development of activities that encourage collaborative work. The types of communicative interactions of the students with their peers, the professor, the contents, and the environment were also identified. In Fig. 3 we show the data obtained around seven aspects valued in the 42 environments analyzed.

Several aspects can be highlighted in this category, among them: in 34 VLEs there is evidence of academic discussions, and similarly the interest of the professor in solving concerns is appreciated. In 33 VLEs it was possible to verify that students use the published content. The interaction among students is a little less, it could only be verified in 30 VLEs. Only in 27 of them were activities that favor group reflection found. Regarding

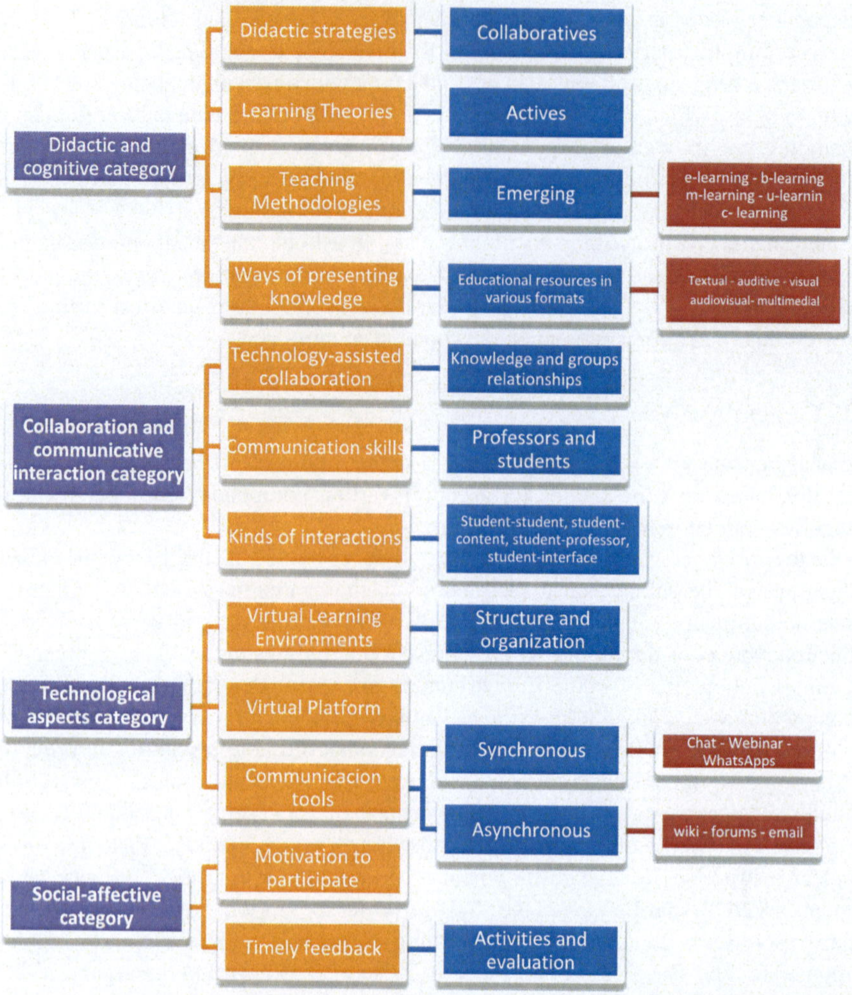

Fig. 1. Concept grouping categories based on documentary review.

the interactions among students to promote teamwork, it could only be verified in 23 environments. In general terms, it is only recognized in 20 of 42 activities that allow collaborative work.

Regarding technological category, the analysis was carried out by using the synchronous communication tools of the Uvirtual platform. The data obtained are presented in Fig. 4.

In the analysis of Fig. 4, it can be identified that the most used communication tools are the Forum (40 times), the WhatsApp messaging application (35), and the Homework/Task (27). The Blogs and Wikis (2 times respectively), the Workshop (6), and the counseling and grade book (13 times each) are identified as less used tools.

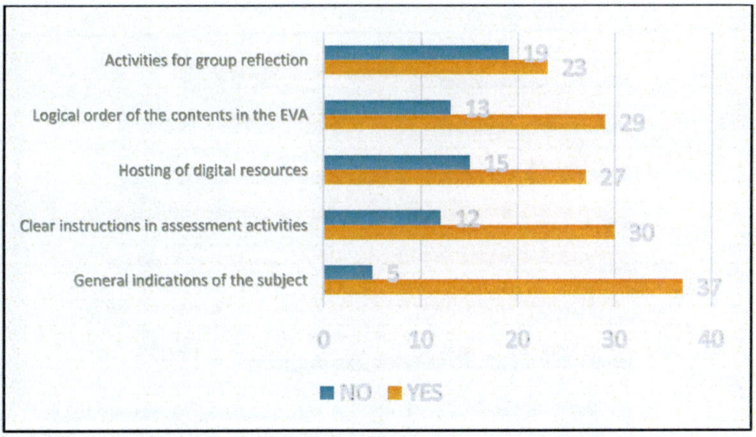

Fig. 2. Data analysis didactic and cognitive category.

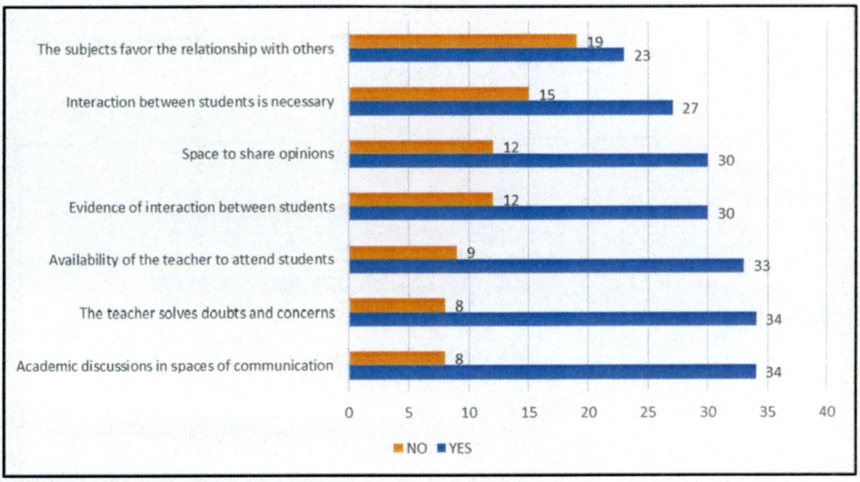

Fig. 3. Data category communication interactions and collaboration.

Five aspects were defined in the *social-affective* category, associated with the timely feedback, guidelines and motivation for participation, and the opinions in the environments. The specific data are in Fig. 5.

In Fig. 5, it can identify that in 38 VLEs the participation of students is encouraged. In 32 of them, the solution of doubts and the timely feedback of the professor, to the concerns of the students, are evident. Likewise, in 29 VLEs the possibility of expressing opinions by the actors is recognized. On the other hand, it is important to highlight that only in 18 of the 42 settings had guidelines for students' participation.

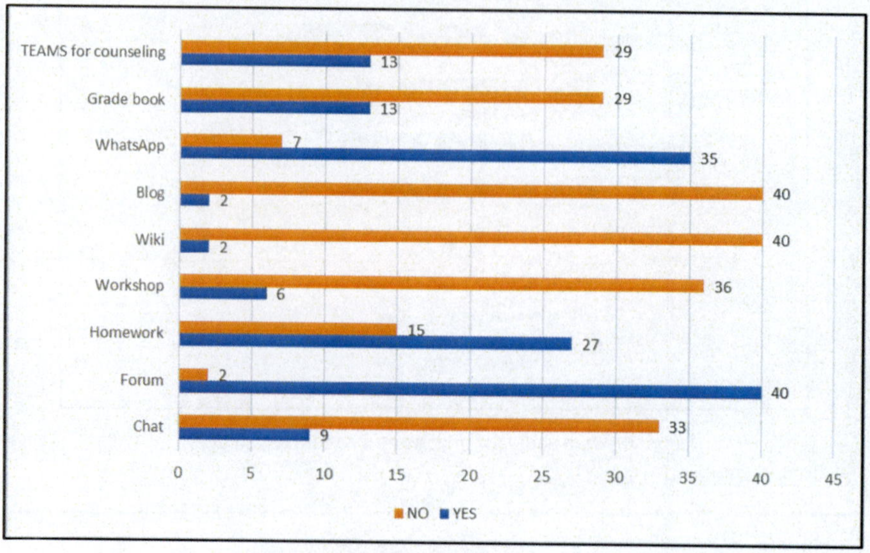

Fig. 4. Category technological aspects.

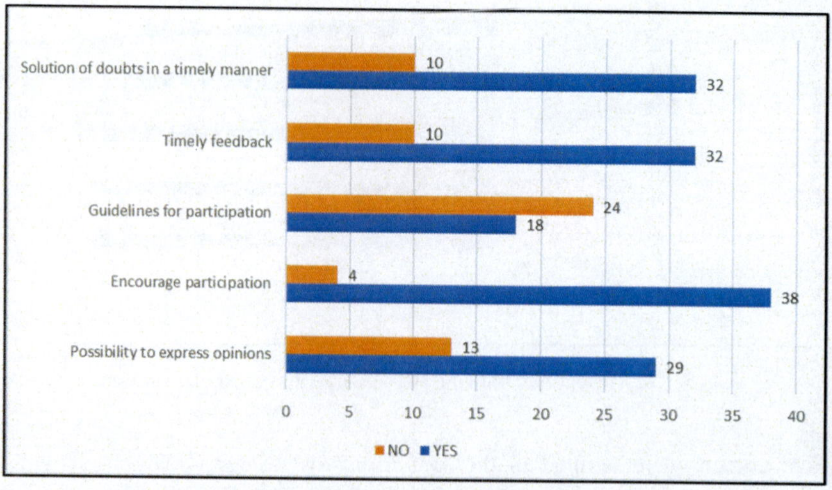

Fig. 5. Social-affective category data.

4 Findings Discussion

Based on the analysis, comparison, comprehension, and synthesis of the findings of the documentary review, it was possible to establish four components that allow to promote, enhance, or increase communicative interactions and the online collaborative learning. These components are the basis to the factors model proposed in Arango-Vasquez & Manrique-Losada [42], and they are integrated for: 1) didactic and cognitive, 2) mediation, 3) technological, and 4) motivation and feedback. Likewise, for each component a set of relevant characteristics was established. In Fig. 6 we present the four components that synthesize the results of the research.

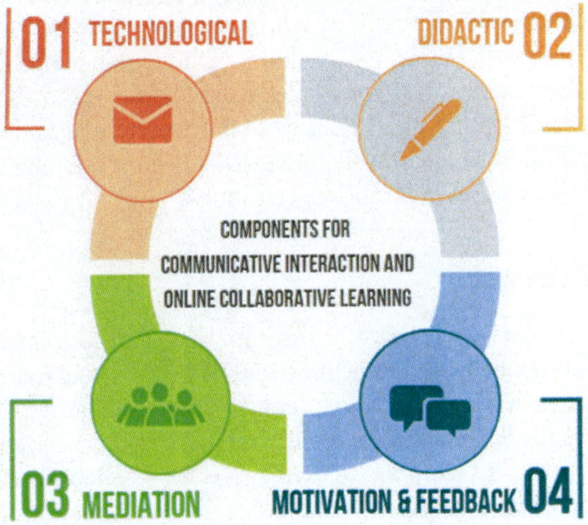

Fig. 6. Components of communicative interaction and collaboration in VLEs.

These components present an aggrupation of elements and tools for the professors to plan, organize, and facilitate communicative interactions and collaboration in technology mediated VLEs, as described below.

4.1 Technological Component

This component is referring to the learning platform, the VLEs and the communication tools used to foster interactions.

Learning platforms are computer applications that bring together a set of technological tools facilitating the execution of a pedagogical act and the development of teaching-learning processes. Currently, teaching-learning platforms must prioritize a socio-formative approach, so that knowledge is promoted more easily, and the user is given a clear vision of their training process, articulating technological resources such as content management, networks social, collaborative work and videoconferences, among others [43].

Communication tools in VLEs are means of interpersonal communication that through the Internet materialize in two ways: synchronous communication, in which users through a telematics network coincide in time/virtual space and communicate with each other via text, audio and/or video. In asynchronous communication, on the other hand, the participants use said communication system at different times. These communicational tools offer the possibility of overcoming space-time barriers in virtual training processes, but at the same time they also imply an initial consideration to be analyzed by the professor, since it differs from the face-to-face space that is identical for each of the users, since in virtual spaces, each actor has a different one for each one, opening the possibilities for cultural values and perceptions to be different in each case.

In this sense, in the teaching-learning processes, communicative interaction gains meaning and practical value through synchronous and asynchronous communication tools (such as chat, email, wikis, and forums). Thus, interaction, access, and use of information, and the generation of knowledge from the use of these tools are enhanced. Consequently, its assertive use in VLEs is something more than a simple change in the use of the means through which didactic interaction occurs, creating a new paradigm in e-learning to achieve maximum performance of these kinds of tools.

4.2 Didactic Component

Based on its educational purposes, the institutions decide a transformation process. Professors in this component can define the necessary conceptual foundations in relation to: (1) Learning theories such as active, constructivist, cognitive, collaborative, and cooperative learning; (2) Didactic strategies, such as problem-based learning, flipped classroom, and service-plus learning; (3) The professor's communication skills focused on the organization of DER in the VLE, guidance in the instructions for evaluation and learning activities, dynamization in feedback, assertive usage of web netiquette rules and permanent motivation to students to participate in activities arranged in the VLE.

4.3 Motivation and Feedback Component

This component defines the professor under the role of 'mediator' who promotes communicative interactions and collaboration among the different actors: professor, student, content, and interface; looking for the motivation and guidance to the students in their learning processes. According to Pérez [39] the motivation refers to the attitudes supporting and giving meaning to the teaching and learning processes, and it specific to everyone based on their goals and environment relationship. The student's motivation is promoted with the content design and interface design. Besides, timely feedback is fostered with clear evaluation criteria and the delivery to the student of information associated with the achievements of their learning process.

4.4 Mediation Component

Groups a set of synchronous and asynchronous student' participations related to DER, evaluative or learning activities, and questions or discussions deliberately organized by professors in the VLEs. The students develop their own social and cognitive structures in contexts of permanent collaboration.

In such components, are identified the interactions between student and:

Professor. Bidirectional interaction allowing the rebuilding of significances or senses. Professors can create opportunities for the student in the VLE, starting from its context, to go through the contents, make context-content relationships, build their own knowledge, and foster thinking competences, abilities, and attitudes. This mediation allows students to reach feedback, motivation, and dialogue with professors.

Student. Through comments in the forums and group activities students create interactions student-student. Like this, learning is developed collaboratively, and students can make their judgments or perceptions about specific topics to each subject.

Contents. Professors must be clear about the educational purpose and expected learning outcome for the students, and thus design DER in several formats: text, audio, images, video, and simulators, among others. Such a DERs favors for develop possibilities of reading and comprehension and, as well as the student motivation. For the content's development, professors must guide in this interaction the exploration, reading, and understanding of the student's relationship with the contents.

Interface. A VLE that promotes communicative interactions must be intuitive, navigable, and usable. Besides, needs to define visible and clear objectives, in an appropriate context, with an easy-to-understand structure, and permanent accompaniment and motivation from professors. An appropriate interface promotes the management of knowledge apprehension scenarios, experimentation, exploration, and self-learning, including broader contexts of merely instructional content.

5 Conclusions

As outputs of this study, we define four components to characterize communicative interactions within the framework of online collaborative learning: didactic, technological, motivation and feedback, and mediation,

Such components allow educational institutions to keep in mind different elements for promoting, encouraging, and spreading interactions and collaboration between professors and students in VLEs from Latin American university environment.

The professors' training needs to be permanent during all the teaching-learning processes which implies they have communication, digital, and methodological skills very solid to implement communicative interactions in VLEs and to apply didactic strategies focused on students.

Collaborative learning is one of the didactic strategies especially used in technology-mediated education, which, in addition to improve academic performance and student participation, enhances the development of critical thinking, decision-making, leadership, respect for differences, conflict resolution and assertive communication, as necessary skills to integrate into the work world and communities in general.

S*tudent-professor interaction*, beyond the masterful and professor leading role, should focus on the student. Thus, students based on techno-pedagogical training, is capable to suggest, drive, orientate, moderate, and direct approaches towards autonomous forms in the construction of knowledge. Professors can create opportunities for students in the VLE, from its context to contents, interface, and context-content relationships.

Student-student interaction needs to be motivated based on autonomy learning, by reducing the dependency to the professor and promoting knowledge self-management.

Student-content interaction demands student responsibility and discipline, to read, interpret, and learn in several ways (scenarios, formats, contents, resources).

The communicative interactions are produced from a student's trust approach, highlighting their attention, timely feedback, self-confidence, and interest in their own progress. The interaction goes beyond the academic aspects, influenced by student attitudes, potentials, and possibilities, to become it in a learning provider in interaction scenarios.

Before thinking about content and learning activities, the motivational and significant factors for the student should be examined, in order to stimulate participation and productivity in the learning process, after having the content, activities and participation in the VLEs on professor must be in permanent feedback.

Acknowledgments. This work was supported by the FCT – Fundação para a Ciência e a Tecnologia, I.P. [Project UIDB/05105/2020].

References

1. Castillo Salcedo, M.L.: La Interacción Virtual y su relación con el Aprendizaje en los Estudiantes del II Ciclo de la Facultad de Tecnología de la Universidad Nacional de Educación Enrique Guzmán y Valle, Durante el Ciclo Académico 2017. Universidad Nacional de Educación, Lima (2020)
2. Silva, J.: Un modelo pedagógico virtual centrado en las e-actividades(RED). Revista de Educación a Distancia (RED) **10**(53), 31–34 (2017)
3. Marmanillo, J. L.: Aplicación de las herramientas digitales y recursos Web 2.0 para el logro de competencias digitales de los docentes del Instituto de Educación Superior Privado de Formación Bancaria. Universidad Privada Norbert Wiener, Lima (2019)
4. Espinosa, R., Rodríguez, R., Olvera, M.G.: El uso de las TIC, TAC, TEP, para desarrollar competencias empresariales y comunicativas en los estudiantes universitarios. Revista TECSISTECATL 21(1) (2017)
5. Montenegro, J.: Interacción comunicativa con Blackboard Collaborate y el rendimiento académico en estudiantes de educación a distancia. HAMUT'AY **3**(2), 68–82 (2017)
6. Huss, J., Sela, O., Eastep, S.: A case study of online instructors and their quest for greater interactivity in their courses: overcoming the distance in distance education. Aust. J. Professor Educ. **40**, 72–86 (2015)
7. Salinas, J.: Modelos emergentes en entornos virtuales de aprendizaje. In: 1er Foro Nacional Virtual en Didáctica, Medios y TIC. Eductec, Colombia (2011)
8. García-Chitiva, M.: Aprendizaje colaborativo, mediado por internet, en procesos de educación. Revista Electrónica Educare **25**(2), 1–19 (2021)
9. Harasim, L.: Learning Theory and Online Technologies. Routledge, New York (2012)
10. Bates, A.W.: Teaching in a Digital Age, 2nd edn. Pressbooks, UK (2019)
11. Chen, J., Wang, M., Kirschner, P.A., Tsai, C.: The role of collaboration, computer use, learning environments, and supporting strategies in CSCL: a meta-analysis. Rev. Educ. Res. **88**(6), 799–843 (2018)
12. Vásquez, C., Ricaurte, A., Arango, S.: Capítulo 7: Interacciones comunicativas en un Entorno Virtual de Aprendizaje (EVA). In: Pensar la comunicación – Reflexiones y resultados de investigación (Vol. 2). Sello Editorial Universidad de Medellín, Medellín (2010)
13. Ogannisyan, L., Akopyan, M., Misirov, D., Borzilov, A., Semergey, S.: Study of interpersonal interaction of educational process subjects in social and pedagogical space of the university. In: Inclusive Development of Society, pp. 277–284. CRC Press, Taylor & Francis Group (2020)
14. Suárez, C.: Los entornos virtuales de aprendizaje como instrumento de mediación. Gredos (2003)

15. Hoppler, S., Segerer, R., Nikitin, J.: The six components of social interactions: actor, partner, relation, activities, context, and evaluation. Front. Psychol. Sec. Pers. Soc. Psychol. **10**(1), 743074 (2022)
16. Lehman, R., Conceição, S.: Creating a sense of presence in online teaching: how to "be there" for distance learners. Christ. Educ. J.: Res. Educ. Ministry **10**(1), (2010)
17. Kehrwald, B.: Understanding social presence in text-based online learning environments. Distance Educ. **29**(1), 89–106 (2008)
18. Bustos, A., et al.: Desarrollo, aprendizaje y enseñanza en la educación secundaria 1(1), Grao, España (2010)
19. Dieser, P., Sanz, C., Zangara. A.: Capítulo 2. Autorregulación del aprendizaje en contextos educativos mediados por tecnologías digitales. Aprendizaje y tecnologías. Habilidades del presente, proyecciones de futuro. Noveduc Gestión Editorial, Buenos Aires (2020)
20. Fajardo, J., Santana A., Caldas., C.: Capítulo 6. La construcción ecointeractiva de conocimiento en los ambientes virtuales de aprendizaje. Psicología Ambiental Volumen I. Experiencias, diálogos y perspectivas académicas, 1st ed. Ascofapsi, Bogotá (2022)
21. Maraza, B.: Hacia un aprendizaje personalizado en ambientes virtuales. Revista Campus Virtuales **5**(1) (2016)
22. Serrano, M.: La producción social de la comunicación. Signo y Pensamiento **5**(1), 47–57 (1986)
23. Barbero, J.M.: De los medios a las mediaciones: comunicación, cultura y hegemonía.2nd ed. Gustavo Gili, Barcelona (1991)
24. Vygotsky, L.S.: Pensamiento y lenguaje, 1st edn. Editorial Pueblo y Educación, Barcelona (1960)
25. Eisner, E., Castillo, O.: Cognición y curriculum: una visión nueva. Amorrortu Editores, México (1998)
26. Lima, S.: La mediación pedagógica con uso de las Tecnologías de la Información y las Comunicaciones. Editorial Pueblo y Educación, México (2005)
27. Escalona, M.: El Uso de recursos informáticos para favorecer la integración de contenidos en el área de Ciencias Exactas del preuniversitario. Instituto Superior Pedagógico José de la Luz y Caballero, Departamento de Ciencias Exactas (2007)
28. Bustos, A., et al.: Desarrollo, aprendizaje y enseñanza en la educación secundaria, 1st ed. Grao, Barcelona (2010)
29. García, L.: ¿Por qué va ganando la educación a distancia? UNED, Madrid (2009)
30. Barberá, E., Badia, A., Mominó, J.M.: La incógnita de la educación a distancia. Revista de Docencia Universitaria **1**(3) (2001)
31. Hirumi, A.: The design and sequencing of elearning interactions: a grounded approach. Int. J. E-Learning **1**(1), 19–27 (2002)
32. Gadamer, H.G., Olasagasti, M.: Verdad y método. 1st edn. Sígueme, Salamanca (1992)
33. González, E.M.: Sobre la hermenéutica o acerca de las múltiples lecturas de lo real. Discusiones Filosóficas **12**(18), 125–143 (2011)
34. Barreto, J.: Estudio de los mecanismos que inciden en la autorregulación durante el proceso de aprendizaje colaborativo en entornos personales de aprendizaje. 1st edn. Universitat Oberta De Catalunya, Barcelona (2020)
35. Duarte-Herrera, M., Valdes Lozano, D.E., Montalvo Apolín, D.E.: Estrategias disposicionales y aprendizajes significativos en el aula virtual. Revista Educación **43**(2), 30–37 (2019)
36. Bylieva, D., Bekirogullari, Z., Kuznetsov, D., Almazova, N., Lobatyuk, V., Rubtsova, A.: Online group student peer-communication as an element of open education. Future Internet **12**(9), 1–13 (2020)
37. Salam, M., Farooq, M.: Does sociability quality of web-based collaborative learning information system influence students' satisfaction and system usage? Int. J. Educ. Technol. High. Educ. **17**(26), 1–39 (2020)

38. Fernández, E.: Análisis de estrategias metodológicas docentes apoyadas en el uso de TIC para fomentar el Aprendizaje Cooperativo del alumnado universitario del Grado de Pedagogía. Revista Interuniversitaria de Formación Del Profesorado **34**(2), 79–100 (2020)
39. Pérez, M.: La comunicación y la interacción en contextos virtuales de aprendizaje. Apertura, Revista de Innovación Educativa **1**(1), 34–47 (2009)
40. Africano, B., Anzola, M.: El acto educativo virtual. Una visión desde la confianza. Educere **22**(73), 521–531 (2018)
41. Aparicio Gómez, O.Y., Ostos-Ortiz, O.L.: Pedagogías emergentes en ambientes virtuales de aprendizaje. Revista Internacional de Pedagogía e Innovación Educativa **1**(1), 11–36 (2021)
42. Arango-Vásquez, S.I., Manrique-Losada, B.: Interacciones comunicativas y colaboración mediada por entornos virtuales de aprendizaje universitarios. Revista de Educación a Distancia (RED) **23**(76), 6 (2023)
43. Tobón, B., Tobón, S., Veytia-Bucheli, M.G., Escudero, A.: Hacia un nuevo concepto: Plataformas Virtuales Socioformativas (PVS). Revista Espacios **39**(53), 27 (2018)

Value Co-creation in Education Services: A Flipped Classroom Approach in EIU University, Vietnam

T. D. Dang[1,2](✉) and M. T. Nguyen[2]

[1] Eastern International University, Thủ Dầu Một, Binh Duong, Vietnam
doan.dang@eiu.edu.vn, dtdoan.sdh221@hcmut.edu.vn
[2] HCMC University of Technology - VNU-HCM, Ho Chi Minh City, Vietnam
n.m.tuan@hcmut.edu.vn

Abstract. In recent years, the flipped classroom approach has emerged as a promising pedagogical strategy emphasizing active learning and student-centered teaching. This study presents a framework for implementing the flipped classroom approach in the Machine Learning for Economics course within the economics program at Eastern International University (EIU), Vietnam, by incorporating AWS Academy and Moodle platform. The framework comprises four phases: course selection and instructor training, development of pre-class and in-class materials, pilot testing, and complete implementation and evaluation. Our findings suggest that using AWS Academy for online video lectures with labs and Moodle platform for class management can improve students' engagement and learning outcomes, as evidenced by positive feedback from students and instructors. We discuss the implications of our findings for academic institutions, instructors, students, and recruiters, focusing on value co-creation through collaborative learning. The study framework provides a practical guide for instructors and academic institutions adopting the flipped classroom approach in economics education. Limitations and future research directions are also discussed.

Keywords: Flipped Classroom · AWS Academy Program · Cloud Computing skills · Economics Education · Value Co-creation

1 Introduction

The demand for digital skills has been increasing rapidly, and universities must prepare their students for the digital workplace [1]. However, students in economics may lack strong background knowledge in programming, data analytics, and other IT-related subjects, and universities must find effective ways to equip them with these critical skills [2]. According to the AWS Academy webpage, the program offers "cloud computing base courses designed to help students develop technical expertise" (AWS Academy, n.d.), making them a promising solution [3]. The COVID-19 pandemic has also accelerated the need for digital skills, with businesses and institutions shifting to remote operations [4]. Therefore, educational institutions are exploring new ways of incorporating digital skills

K. Arai (Ed.): FTC 2023, LNNS 816, pp. 339–348, 2023.
https://doi.org/10.1007/978-3-031-47448-4_24

into their curricula to meet the demands of the digital era [5, 6]. One approach that has gained attention is the flipped classroom approach [7]. In a flipped classroom, students engage with course content outside of class, typically through videos or online lectures, and then use class time to engage in hands-on activities, such as problem-solving, collaboration, and discussions [8]. The flipped classroom approach has improved student engagement, learning outcomes, and employability [9]. In addition to adopting the flipped classroom approach, this paper also explores the idea of value co-creation in education. Value co-creation is a collaborative process between teachers and students that jointly creates and delivers educational value [10]. By engaging students in the process of co-creation, teachers can foster a more student-centered approach to teaching and learning, which can lead to better learning outcomes and increased student satisfaction [11]. However, there is a lack of research on the effectiveness of using big tech academy programs and LMS in a flipped classroom model, particularly in economics education and even more so in non-native English language learning environments.

To fill this gap in the existing research, this study presents a case study of how the AWS Academy Program and Moodle will be incorporated into the economics curricula at EIU using a flipped classroom approach. Moodle is a Learning Management System (LMS) that allows instructors to create and manage online courses and resources [12]. The AWS Academy Program provides a comprehensive set of cloud computing courses covering topics from the fundamentals of cloud computing to advanced machine learning and security [13]. The program is designed to improve student engagement, learning outcomes, and employability through a collaborative, hands-on environment. By offering new and original research, this paper analyzes the potential impact of the AWS Academy Program and Moodle in a flipped classroom model in economics education. This research will benefit educators and industry partners who aim to create a skilled workforce that can meet the demands of the digital era.

The structure of the paper comprises four parts. Firstly, a literature review of relevant topics used in this study is presented. Secondly, the framework utilized to integrate the AWS Academy Program and Moodle into the economics curricula at EIU is presented. Thirdly, the program's potential benefits are analyzed, including its impact on student engagement, learning outcomes, and employability. Finally, the study's implications for educators and industry partners are discussed, and the need for future research is emphasized.

2 Literature Review

2.1 Academy-Industry Partnership

Academy-industry partnerships have become increasingly important for academic institutions and industry organizations to collaborate and provide students with relevant and up-to-date training to prepare them for careers in specific industries or technology areas [14]. These programs often offer academic and practical training and may include opportunities for students to gain real-world experience through internships, cooperative programs, or other industry-related activities [1]. Big tech academy programs, such as AWS Academy, Microsoft Imagine Academy, Google Cloud Training, IBM Skills

Academy, and Red Hat Academy, are examples of such partnerships that provide students with the skills and knowledge necessary to succeed in the technology industry [15]. These programs are often developed in collaboration with industry organizations to ensure the content and skills taught are relevant and up-to-date [1]. They may offer students opportunities to earn industry certifications or other credentials to enhance their employability [13]. In this study, we will focus specifically on the AWS Academy Program, which provides educators with resources and curriculum to teach cloud computing to students and aims to equip them with the practical skills and knowledge required for the digital workplace. The study of the AWS Academy Program is essential as it can provide insight into the effectiveness of academy-industry partnerships in addressing skills gaps and preparing students for high-growth industries like technology [16]. While these programs are designed to offer students the skills and knowledge required to succeed in the technology industry, there are some limitations to consider. For example, while the resources from Big Tech companies like AWS Academy are often free for universities when they join the partnership program, the availability and accessibility of these resources may still challenge some institutions. Additionally, the focus on industrial skills and knowledge may not necessarily align with the broader educational goals of some institutions or may not fully address the skills gaps or needs of other industries [6].

2.2 Flipped Classroom Model

The flipped classroom model is a pedagogical approach that has gained significant attention recently [7]. In this approach, students engage with course content outside of class, typically through videos or online lectures, and then use class time to engage in hands-on activities, such as problem-solving, collaboration, and discussions [8, 17]. The flipped classroom model has improved student engagement, learning outcomes, and employability (Baepler et al., 2016; Lo and Hew, 2017). According to a study by Ahmad Uzir et al. (2020), this model can improve learning outcomes, as students have more time to work on activities and receive personalized feedback during class time. In a study by Day (2018), students in a flipped classroom model scored higher on exams and demonstrated greater retention of course content than in traditional classroom instruction.

Additionally, a meta-analysis by Bredow et al. (2021) found that the flipped classroom model can improve student engagement and promote higher-order thinking skills. The effectiveness of the flipped classroom model in improving student learning outcomes, especially in IT-related subjects, has been supported by research [18, 19]. For economics students studying IT-related subjects, the flipped classroom model may be precious in helping students to develop the technical skills and knowledge necessary for careers in the technology industry [1, 20]. In addition, the flipped classroom model is a promising approach to teaching and learning [18]. As such, the flipped classroom model has the potential to transform traditional classroom instruction and improve student success in a variety of academic settings [21].

However, there are also some limitations to consider. For non-native English students like Vietnamese students, language barriers and study habits may challenge engaging with course content outside of class [22]. Additionally, the flipped classroom model may not be suitable for all subjects or learning objectives, and some students may struggle to adapt to the increased responsibility for their learning [23]. Finally, while the flipped classroom model can promote critical thinking and problem-solving skills, it may not be the best approach for all students, and some may prefer more traditional classroom instruction [24].

2.3 Value Co-creation in Education

Value co-creation has become essential in education, emphasizing the importance of collaborative learning and student engagement [11, 25]. Recent studies have shown that value co-creation can improve student learning outcomes and promote the development of critical thinking and problem-solving skills [26]. In the context of higher education, value co-creation has been applied to various disciplines, including business, economics, and engineering, with promising results. In particular, the flipped classroom approach has been recognized as a valuable strategy for promoting value co-creation, as it emphasizes the active participation of students in the learning process [11, 27]. This paper builds on this literature by presenting a framework for implementing the flipped classroom approach in the Machine Learning for Economics course, incorporating AWS Academy resources and Moodle to promote value co-creation in business and economic education.

2.4 Building the Flipped Classroom Use Resources from Big Tech Academy and LMS

Building a flipped classroom model with resources from Big Tech companies could be a good strategy for several reasons. Firstly, the flipped classroom model effectively improves student learning outcomes, particularly in IT-related subjects [28]. Secondly, resources from Big Tech companies are often free for universities when they join partnership programs, making it a cost-effective way to provide students with the latest tools and technologies [14]. Thirdly, economics students often focus on developing industrial skills and knowledge and using resources from industry leaders can be a shortcut to acquire the necessary technical skills for careers in the tech industry [20, 29, 30]. Finally, partnering with big tech companies can allow students to network with industry professionals and gain valuable work experience through internships or other industry-based programs [4, 16]. In addition to resources from Big Tech companies, the flipped classroom model can be enhanced by using learning management systems (LMS) such as Moodle. Moodle provides a platform for students to access course materials and engage with interactive content outside of the classroom [31]. Furthermore, Moodle's interactive features, such as discussion forums and quizzes, can promote collaboration and student engagement. By combining resources from Big Tech companies with an LMS like Moodle, the flipped classroom model can provide an effective and efficient way for students to develop technical skills and knowledge [15].

However, as mentioned earlier, the availability and accessibility of resources may pose a challenge for some institutions [16]. In addition, language barriers for non-native English students may make it difficult for them to engage with the course content using English [32]. Finally, while the flipped classroom model can promote critical thinking and problem-solving skills, some students may struggle to adapt to this approach and prefer more traditional classroom instruction [33]. Despite these limitations, the flipped classroom model with resources from Big Tech companies offers a valuable opportunity for students to gain the technical skills and knowledge necessary for careers in the technology industry [20]. At EIU, we plan to incorporate Moodle as part of our flipped classroom model, in conjunction with the AWS Academy program, to provide a comprehensive and engaging learning experience for our students in IT-related economics subjects. In this study, by combining this approach with the AWS Academy program and Moodle for IT-related business and economic curriculum subjects at EIU, we believe that students can better understand the subject matter and apply their knowledge in a real-world context.

3 Methodology

To introduce the flipped classroom approach in the Machine Learning for Economics course at Becamex Business School (BBS) at EIU in Binh Duong province, Vietnam, this study follows and modify the iterative design thinking process proposed by Plattner et al. (2009), we developed a four-phase framework that comprised of (1) course selection and instructor training, (2) development of pre-class and in-class materials, (3) pilot testing of the flipped classroom approach, and (4) complete implementation and evaluation.

Phase 1: Course selection and instructor training involved selecting the appropriate course for the flipped classroom approach, training the instructor on the approach, including best practices, course design, and online tools, and partnering with AWS Academy to access their IT-related resources. The instructor was trained to use online tools such as Moodle, AWS Academy resources, and other relevant tools. The AWS Academy resources, which include course materials and hands-on experiences, were integrated into the course design to enhance students' technical skills and knowledge in IT-related subjects. Moodle was the primary platform for delivering pre-class materials, including videos, readings, quizzes, and tracking student progress and engagement.

Phase 2: Development of pre-class and in-class materials involved creating and organizing online pre-class materials such as videos, readings, and quizzes using Moodle, as well as in-class materials such as worksheets, group activities, and discussions. The AWS Academy resources were also incorporated into the materials development to ensure students were exposed to real-world examples and hands-on experiences.

Phase 3: Pilot testing of the flipped classroom approach was conducted to evaluate the effectiveness of the pre-class and in-class materials, identify any issues with the online tools, and gather feedback from instructors and students. A small group of students participated in the pilot test and were interviewed after completing a sample lesson using the flipped classroom approach.

Phase 4: Complete implementation and evaluation of the flipped classroom approach is ongoing and involves the full implementation of the approach in the course and collecting and analyzing data on its effectiveness. The four-phase framework is illustrated in Fig. 1.

Overall, this framework provides a structured approach for introducing the flipped classroom approach in a course, intending to improve student engagement, participation, and learning outcomes. By implementing and evaluating the framework, we can gain valuable insights into the approach's effectiveness and identify areas for improvement and best practices for future implementation.

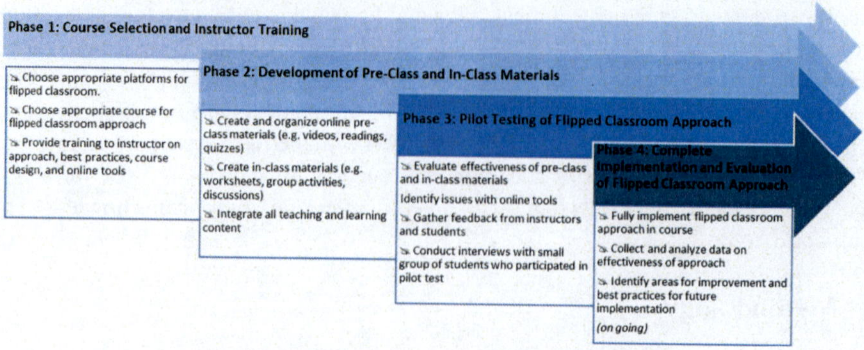

Fig. 1. Four phases framework for the flipped classroom.

4 Results and Discussion

Utilizing a four-phase framework, this research successfully introduced the flipped classroom approach in the Machine Learning for Economics course at EIU—some results and findings described in detail for four phases.

In the first phase, the course selection process involved identifying the BBS's Machine Learning for Economics course as an appropriate course for the flipped classroom approach. The instructor was provided with training on the flipped classroom approach, including best practices, course design, and online tools such as Moodle and AWS Academy. The training helped the instructor better understand the flipped classroom approach and how to design and implement it effectively.

In the second phase, developing pre-class and in-class materials involved creating and organizing online pre-class materials such as videos, readings, and quizzes using Moodle and in-class materials such as worksheets, group activities, and discussions. The use of AWS Academy resources helped to provide up-to-date and relevant content for the course, and the integration of Moodle provided a convenient platform for students to access the materials and engage with the course content outside of class.

In the third phase, the pilot testing of the flipped classroom approach in the BBS Machine Learning for Economics course resulted in positive feedback from the instructors and students. The pre-class materials, made available on Moodle, were well-received by students, who appreciated the flexibility of completing the online activities at their own pace. The in-class activities, which included group discussions and problem-solving exercises, were more engaging and interactive than traditional lectures. Instructors noted that the flipped classroom approach allowed for more individualized attention to students' learning needs and provided opportunities for deeper discussions and critical thinking. During the interviews, students expressed satisfaction with the flipped classroom approach, with many stating that it helped them to understand the course material better and improved their confidence in their ability to apply their knowledge. They also appreciated the opportunity to collaborate with peers and engage in discussions, which they felt helped to reinforce their learning. Instructors similarly found that the flipped classroom approach led to higher student engagement and participation levels, with students taking a more active role in their learning. Overall, the results of the pilot testing suggest that the flipped classroom approach with resources from Big Tech companies, such as AWS Academy, can be an effective method for improving student engagement, participation, and learning outcomes. The structured framework developed in this study provides a valuable guide for instructors looking to implement the approach in their courses. However, further evaluation and refinement of the approach are needed to ensure its continued effectiveness and address any challenges or limitations.

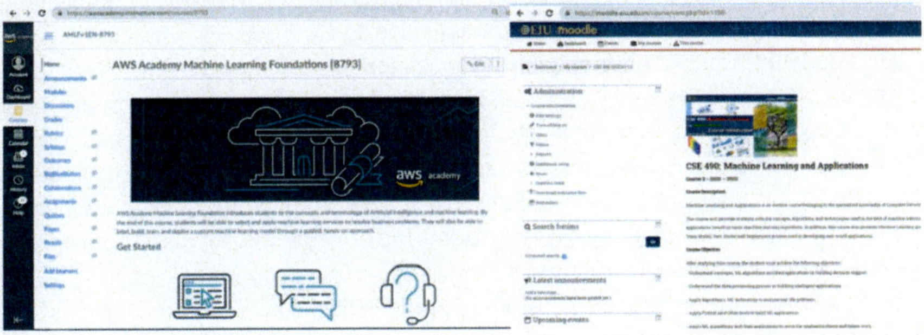

Fig. 2. AWS Academy LMS and EIU moodle from the pilot phase.

The ongoing fourth phase involves implementing and evaluating the flipped classroom approach with a larger group of students [34]. Based on the success of the first three phases, we are confident that the complete implementation phase will yield positive results.

In terms of value co-creation in education, the framework presented in this study incorporates value co-creation principles throughout its four phases. In the course selection and instructor training phase, instructors are encouraged to collaborate with AWS Academy and Moodle platform to identify content that aligns with industry needs and fosters student engagement (see Fig. 2). In developing the pre-class and in-class materials phase; instructors design learning activities promoting active student participation and collaboration. In the pilot testing phase, instructors gather student feedback and use it to refine the course materials and teaching methods. Finally, in the complete implementation and evaluation phase, instructors assess student learning outcomes and continuously improve the course. By incorporating value co-creation principles into the flipped classroom framework, this study provides a practical guide for instructors and academic institutions to deliver high-quality business education that aligns with industry needs and fosters student engagement and success.

Overall, the results of the four-phase framework showed that the flipped classroom approach using resources from Big Tech companies such as AWS Academy and online learning platforms such as Moodle could effectively improve student learning outcomes in IT-related subjects such as Machine Learning for Economics. The pilot testing and evaluation of the approach provided valuable insights into the effectiveness of the approach and identified areas for improvement and best practices for future implementation. The use of online resources and platforms can help to provide convenient and accessible course materials for students, while in-class activities and discussions can help to promote active learning and critical thinking skills.

5 Conclusion

In conclusion, the flipped classroom approach was successfully introduced in the Machine Learning for Economics course at EIU using a four-phase framework consisting of course selection and instructor training, pre-class and in-class materials development, pilot testing, and complete implementation and evaluation. The results of the pilot testing phase indicate that the flipped classroom approach improved student engagement, participation, achievement, and instructor satisfaction.

This study has important implications for academia and management. For academia, the flipped classroom approach can be a useful pedagogical tool for improving student outcomes and engagement, particularly in technical and data-driven courses. Additionally, the four-phase framework can serve as a helpful guide for instructors and institutions looking to adopt the flipped classroom approach. This study contributes to the literature on value co-creation in education by demonstrating how the flipped classroom approach can create value for students in business and economics education.

For management, the flipped classroom approach has benefits beyond the classroom. The approach can help produce more job-ready graduates with practical skills, a crucial need for many companies, particularly those in the technical and data-driven industries. By recruiting graduates from institutions that adopt the flipped classroom approach, companies can access employees better equipped to handle job responsibilities and contribute to the organization's growth.

However, this study has limitations, including a small sample size during the pilot testing phase and the restriction to a single technical course. Future research can explore the flipped classroom approach in different courses and disciplines, using larger sample sizes and diverse student populations. Additionally, future studies can evaluate the long-term impact of the flipped classroom approach on student outcomes and achievement and the benefits for employers beyond the classroom.

Acknowledgment. The authors extend their appreciation to Eastern International University (EIU), Vietnam, for funding this research.

References

1. Brahma, M., Tripathi, S.S., Sahay, A.: Developing curriculum for industry 4.0: digital workplaces. High. Educ., Skills Work-Based Learn. **11**(1), 144–163 (2021)
2. Goulart, V.G., Liboni, L.B., Cezarino, L.O.J.I., Education, H.: Balancing skills in the digital transformation era: The future of jobs and the role of higher education. Ind. High. Educ. **36**(2), 118–127 (2022)
3. Ozalp, H., Ozcan, P., Dinckol, D., Zachariadis, M., Gawer, A.: "Digital Colonization" of highly regulated industries: an analysis of big tech platforms' entry into health care and education. Calif. Manag. Rev. **64**(4), 78–107 (2022)
4. Webb, A., McQuaid, R.W., Webster, C.W.R.: Moving learning online and the COVID-19 pandemic: a university response. Technol. Sustain. Dev. **18**(1), 1–19 (2021)
5. Amhag, L., Hellström, L., Stigmar, M.: Teacher educators' use of digital tools and needs for digital competence in higher education. J. Digit. Learn. Teach. Educ. **35**(4), 203–220 (2019)

6. Voogt, J., Erstad, O., Dede, C., Mishra, P.: Challenges to learning and schooling in the digital networked world of the 21st century. J. Comput. Assist. Learn. **29**(5), 403–413 (2013)
7. Chen, F., Lui, A.M., Martinelli, S.M.: A systematic review of the effectiveness of flipped classrooms in medical education. Med. Educ. **51**(6), 585–597 (2017)
8. Mok, H.N.: Teaching tip: the flipped classroom. J. Inf. Syst. Educ. **25**(1), 7 (2014)
9. Murillo-Zamorano, L.R., Sánchez, J.Á.L., Godoy-Caballero, A.L.: How the flipped classroom affects knowledge, skills, and engagement in higher education: effects on students' satisfaction. Comput. Educ. **141**, 103608 (2019)
10. Dollinger, M., Lodge, J., Coates, H.: Co-creation in higher education: Towards a conceptual model. J. Market. High. Educ. **28**(2), 210–231 (2018)
11. Usković, V.: Flipping the flipped: the co-creational classroom. Learning **13**(1), 1–24 (2018)
12. Cavus, N., Alhih, M.S.: Learning management systems use in science education. Procedia-Soc. Behav. Sci. **143**, 517–520 (2014)
13. Goteng, G.L., Shohel, M.M.C., Tariq, F.: Enhancing student employability in collaboration with the industry: case study of a partnership with amazon web services academy. Educ. Sci. **12**(6), 366 (2022)
14. Ankrah, S., Omar, A.T.: Universities–industry collaboration: a systematic review. Scandinavian J. Manag. **31**(3), 387–408 (2015)
15. Khmelevsky, Y., Voytenko, V.: Cloud computing infrastructure prototype for university education and research. In: Proceedings of the 15th Western Canadian Conference on Computing Education, pp. 1–5 (2010)
16. Meyer Jr, L., Billionniere, E.: AWS academy vs Microsoft Learn for educators vs IBM Skills Academy: the educators choice. In: Society for Information Technology & Teacher Education International Conference, pp. 528–534. Association for the Advancement of Computing in Education (AACE) (2021)
17. Ahmad Uzir, N.A., Gašević, D., Matcha, W., Jovanović, J., Pardo, A.: Analytics of time management strategies in a flipped classroom **36**(1), 70–88 (2020)
18. Bredow, C.A., Roehling, P.V., Knorp, A.J., Sweet, A.M.: To flip or not to flip? A meta-analysis of the efficacy of flipped learning in higher education **91**(6), 78–918 (2021)
19. Toto, P.: Flipped classrooms and translation technology teaching: a case study. In: Empirical Studies of Translation and Interpreting, Routledge, pp. 240–258 (2021)
20. Morin, J., Tamberelli, F., Buhagiar, T.: Educating business integrators with a computer-based simulation game in the flipped classroom. J. Educ. Bus. **95**(2), 121–128 (2020)
21. Brewer, R., Movahedazarhouligh, S.: Successful stories and conflicts: A literature review on the effectiveness of flipped learning in higher education. J. Comput. Assist. Learn. **34**(4), 409–416 (2018)
22. Mehring, J.G.:An exploratory study of the lived experiences of Japanese undergraduate EFL students in the flipped classroom, Pepperdine University (2015)
23. Rotellar, C., Cain, J.: Research, perspectives, and recommendations on implementing the flipped classroom. Am. J. Pharm. Educ. **80**(2) (2016)
24. Burke, A.S., Fedorek, B.: Does "flipping" promote engagement?: A comparison of a traditional, online, and flipped class. Act. Learn. High. Educ. **18**(1), 11–24 (2017)
25. Könings, K.D., Mordang, S., Smeenk, F., Stassen, L., Ramani, S.: Learner involvement in the co-creation of teaching and learning: AMEE Guide No. 138. Med. Teach. **43**(8), 924–936 (2021)
26. Temple Clothier, A., Matheson, D.: Using co-creation as a pedagogic method for the professional development of students undertaking a BA (Hons) in education studies. J. Further High. Educ. **43**(6), 826–838 (2019)
27. Lubicz-Nawrocka, T.M.: Students as partners in learning and teaching: the benefits of co-creation of the curriculum. Int. J. Stud. Partners **2**(1), 47–63 (2018)

28. Shaw, G., Northedge, A.: Learning and teaching in the online era. Successful Univ. Teach. Times Divers., 83–109 (2018)
29. Kamp, A., Delft, T.U.: Science & technology education for 21st century Europe (2019)
30. Oke, A., Fernandes, F.A.P.: Innovations in teaching and learning: exploring the perceptions of the education sector on the 4th industrial revolution (4IR). J. Open Innovat.: Technol., Mark. Complex. **6**(2), 31 (2020)
31. Artz, B., Johnson, M., Robson, D., Siemers, S.: Live or lecture capture: Evidence from a classroom random control trial. Int. Rev. Econ. Educ. **40**, 100240 (2022)
32. Alaofi, S.: The impact of English Language on non-native English speaking students' performance in programming class. In: Proceedings of the 2020 ACM Conference on Innovation and Technology in Computer Science Education, pp. 585–586 (2020)
33. Lapitan, L.D., Jr., Chan, A.L.A., Sabarillo, N.S., Sumalinog, D.A.G., Diaz, J.M.S.: Design, implementation, and evaluation of an online flipped classroom with collaborative learning model in an undergraduate chemical engineering course. Educ. Chem. Eng. **43**, 58–72 (2023)
34. Plattner, H., Meinel, C., Weinberg, U.: Design-Thinking. Springer, Cham (2009)

Classification of Software Tools Available for Download from Servers Hosted at Educational Institutions

Sylvia Encheva(✉)

Western Norway University of Applied Sciences, P. O. Box 7030, 5020 Bergen, Norway

sbe@hvl.no

Abstract. Faculty members are expected to carry out an increasing number of teaching and administrative tasks, as well as high level research. Their work is likely to be supported by a variety of computer systems, tools and programs. This immediately raises the question of how to select those tools that are genuinely useful and do not have a steep learning curve. This work is an attempt to create a structure for hosting computer-based tools at a university, rather than presenting them by title alone. The approach is based on an eight-value logic and aims to classify these tools according to their frequency of use.

Keywords: Classification · Eight Valued Logics · Educational Software

1 Introduction

Staff in higher education institutions are typically required to use a variety of IT systems and software tools for their daily interactions with students, colleagues, and administrative units. Since most of these systems do not communicate with each other, the task of feeding them with data is left to faculty members. The number of such systems and tools varies from one institution to another, but within each institution, it is constantly increasing. Systems tend to be updated much more frequently than the corresponding manuals. When the latter do exist, they are voluminous and, therefore, require even more time for users to find what they are looking for.

We are not aware of any statistics on the amount of time teachers spend selecting and learning to use different tools. However, growing frustration is a fact, and it definitely affects productivity.

In order to facilitate the processes of selection of systems and tools that an educational institution offers to its employees, as well as to reduce the burden on each faculty member to find out which software tool to download and try, we propose the application of eight valued logics. This particular logic is chosen for its ability to "allow sources not to hide the lack of knowledge but to express it explicitly", [10].

K. Arai (Ed.): FTC 2023, LNNS 816, pp. 349–355, 2023.
https://doi.org/10.1007/978-3-031-47448-4_25

The rest of this paper is organized as follows. Section 2 gives some background on many valued logics. Section 3 illustrates how eight valued logics can be used to classify software tools according to their frequency of use in an educational institution. The benefits of such a classification for staff members and IT departments are also presented. The paper ends with a conclusion in Sect. 4.

2 Background

While Kleene's logic, [4], accommodates partial uncertainty by introducing a third truth value called 'unknown' in addition to the classical 'true', 'false', Belnap's logic, [1], operates with four truth values, 'true', 'false', 'both', and 'neither'. The latter was intended to help computers draw conclusions in the presence of inconsistent data.

Zaitsev, [10] argues that four-valued logic, although very useful, does not accommodate cases where a computer does not know the answer. To facilitate computations involving such cases he introduced eight-valued logic. In his work he provides a model for a computer to make decisions based on information from different sources, which may send incomplete and/or inconsistent data. To do this, the author uses the following truth values:

N $= \emptyset$, **T** - told 'true', **U** - a 'source does not know', **F** - told 'false', **TU** - asserted and uncertain, **FU** - denied and uncertain, **B** - contradiction, **BU** - asserted, denied and uncertain information.

A paraconsistent eight valued logis was later on introduced in [9].

For definitions and statements concerning lattices see [3]. Bilattices were first presented in [8] and studied later on by a number of authors [2,4–7], etc.

A tetralattice $EIGHT_4$ is an algebraic structure introduced in [10] and illustrates the following ordering relations - information, truth, falsity, and uncertainty.

3 Classification Based on Eight Valued Logic

Below, we first go through some challenges faced by university staff when selecting educational software tools for teaching and research purposes, and provide a solution to them afterwards.

The ever-growing number of suggestions and recommendations concerning use of software is overwhelming. Many rankings are often based on reviews or the experiences of a single individual. While they provide general information about advantages and disadvantages, they may also be biased and heavily influenced by the authors' skills and experience. In addition, such rankings do not reflecting on types and levels of studies and students.

Most staff prefer software tools that require minimal time to become proficient and are often interested in hearing about their colleagues' experiences.

Let's focus on educational institutions which, for security reasons, restrict downloading of software tools on computers provided to their employees. Such tools can be those the institutions either do not have a licence for or are not on the list of those available for download from their servers. In practices, a user would be able to choose among a number of software tools listed on the institution's website. Well, the degree of usefulness can vary considerably depending on the content of the subject, the type of students, the skills and interests of the lecturers, etc.

It's essential to remember that gathering and maintaining up-to-date information on other users' experiences can be time-consuming for ordinary users. IT departments however can track all downloads, installations, update requests, and uninstalls of software tools available on the institution's servers for staff PCs where administrative rights are granted. The aforementioned information can then be used to help employees choose which tools to use.

This can be implemented by establishing a mapping between the eight truth values and the actions of employees. Our suggestion for mapping is presented in Table 1.

Table 1. Truth values for software tools' status.

Symbol	Status
N	no downloads registered so far
T	installations followed by requests for updates
U	downloads are registered but no installations
F	all downloads and installations are followed by uninstall
TU	some of the installations are followed by requests for updates
FU	downloads only some of which are followed by uninstall
B	both downloads only and some of installations
BU	downloads only, installations, and some requests for updates

Summarized outcomes are graphically represented in Fig. 1, which is adopted from [10] in a way that the nodes in the tetralattice in [10] are described with some elements.

Notations used in Fig. 1: The axises **a**, **d**, **u**, **i** in Fig. 1 represent four independent orders: **a** - order (truth order), **d** - order (falsity order), **u** - order (uncertainty order), and **i** - order (information order). Arrows and explanations notes are collared in blue to make them easier to distinguish them from the edges of the tetralattice.

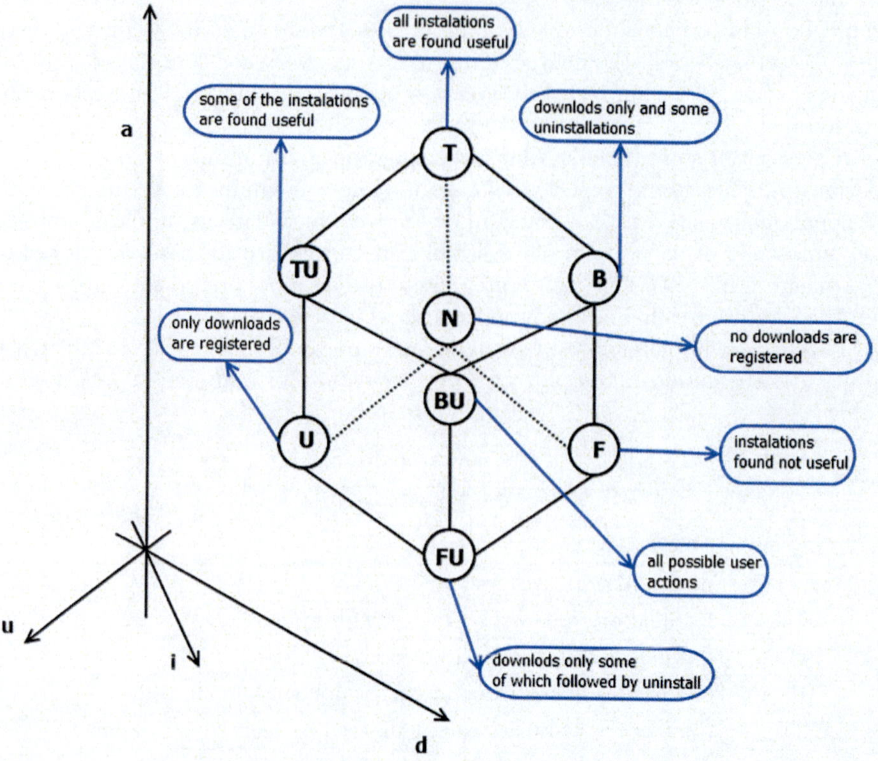

Fig. 1. The tetralattice $EIGHT_4$ with explanations related to usability of suggested software tools.

3.1 Advantages for an IT Department

IT departments can take the following recommended actions for software tools placed in each node of the tetralattice in Fig. 1:

It is important to note that these recommendations are not absolute and should be adapted to the specific context of each institution and its staff members.

IT departments can also use the data collected from user actions to make informed decisions about which software tools to acquire and support. For example, if a tool has a high frequency of use, the IT department may prioritize acquiring and supporting that tool. On the other hand, if a tool has little to no user response, it may be a sign that it is not meeting the needs of the users.

Overall, the use of the eight-valued logic provides a systematic approach for IT departments to manage and recommend software tools to their users based on the collective experiences of the institution's staff.

N - delete from the list of programs suggested to employees or placed in archive if they are not downloaded by anybody after a certain period of time decided by a corresponding IT department

T - place in a highly recommended category. This category can eventually be split in subcategories according to f. ex number users from social and natural sciences, undergraduate or graduate students, etc.

F - discourage the use of the software tool and remove it from the institution's servers

TU - placed a recommended category

FU - placed in a temporary category and removed after a certain period of time decided by a corresponding IT department

B - place in a recommended category

BU - place in a category without particular recommendation.

3.2 Advantages for Stuff Members

Typically, users are presented with a list of software tools to download, sorted alphabetically or 'latest first'. However, if an IT department has placed software tools into different nodes according to user actions or lack of actions, it can present users with a list of software tools classified into the following categories 'Highly recommended', 'Recommended', 'Of some interest', 'Of hardly any interest' and 'Without user's response', Fig. 2.

To provide additional information to stuff members about other users' actions related to software tools placed in the category - Highly recommended, and Recommended, we propose two coefficients:

$$\alpha = \frac{RP}{D} \text{ and } \beta = \frac{I}{D}$$

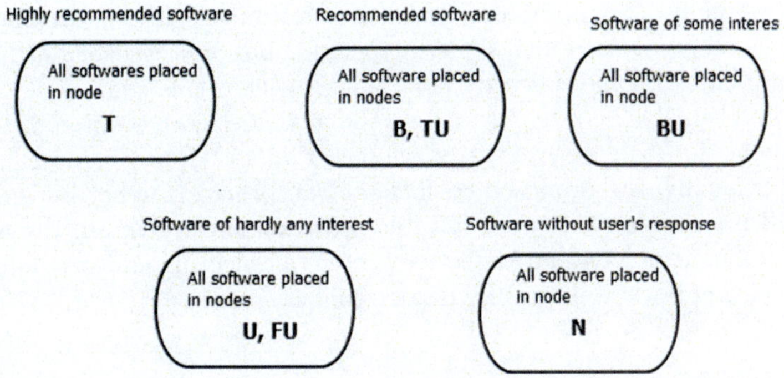

Fig. 2. User recommendations.

where RP denotes number of requests for updates, I denotes number of installations, and D denotes number of downloads. Table 2 and Table 3 show examples of how these coefficients can be presented to users.

Table 2. Information about highly recommended tools.

Highly recommended tools	Requests for updates α	Installations β
TeXstudio, [12]	0.5	0.8
Orange, [13]	0.3	0.7
Dia, [16]	0.1	0.4

Table 3. Information about recommended tools.

Recommended tools	Requests for updates α	Installations β
Texmaker, [11]		0.05
GIMP, [15]		0.06
Weka, [14]	0.09	0.02

To avoid expressing personal preferences, we do not provide examples of commercial software tools that could be placed in any of the categories in Fig. 2.

4 Conclusion

This approach would help users quickly identify the most useful software tools for their needs and reduce the time and effort required for them to find the right tool. Additionally, this approach would also encourage users to explore tools that they might not have considered otherwise but are categorized as potentially useful based on the collective experience of their colleagues. The presented classification can also help to ensure that the tools are more tailored to the needs and preferences of the users, as they are based on actual user actions and feedback. Additionally, the proposed coefficients can provide further information to assist staff in selecting the most suitable software tools. Overall, the approach can lead to a more efficient and effective process of selecting educational software tools, benefiting both staff and IT departments.

References

1. Anderson, A.R., Belnap, N.D., Dunn, J.M.: Entailment: The Logic of Relevance and Necessity, vol. II. Princeton University Press, Princeton, NJ (1992)
2. Arieli, O., Avron, A.: Reasoning with logical bilattices. J. Logic Lang. Inform. **5**, 25–63 (1996)
3. Davey, B.A., Priestley, H.A.: Introduction to Lattices and Order. Cambridge University Press, Cambridge (2005)
4. Fitting, M.C.: Bilattices in logic programming. In: Epstein, G. (ed.) The Twentieth International Symposium on Multiple-Valued Logic, pp. 238–246. IEEE (1990)
5. Fitting, M.: Bilattices and the theory of truth. J. Philos. Log. **18**, 225–256 (1989)
6. Fitting, M.: Bilattices and the semantics of logic programming. J. Log. Program. **11**, 91–116 (1991)
7. Fitting, M.: Bilattices are nice things. In: Hendricks, V.F., Pedersen, S.A., Bolander, T. (eds.) Self-Reference. CSLI Publications, Cambridge University Press (2004)
8. Ginsberg, M.: Multi-valued logics. In: Proceedings of AAAI-86, Fifth National Conference on Artificial Intelligence, pp. 243– 247. Morgan Kaufman Publishers (1986)
9. Kamide, N.: An eight-valued paraconsistent logic. Rep. Math. Logic **49**, 3–21 (2014)
10. Zaitsev, D.: A few more useful 8-valued logics for reasoning with tetra lattice EIGHT_4. Stud. Logica. **92**(2), 265–280 (2009)
11. https://www.xm1math.net/texmaker/
12. https://www.texstudio.org/
13. https://orangedatamining.com/
14. https://www.cs.waikato.ac.nz/ml/weka/
15. https://www.gimp.org/
16. http://dia-installer.de/

PySkuwela: An Online Python Learning Tutorial and Mobile Application for Senior High School S.T.E.M. Students Specializing in Information Technology

Irish C. Juanatas[1](✉), Roben A. Juanatas[2], Jeremiah B. Sto. Domingo[1], and Rene Christopher R. Tio[1]

[1] FEU Diliman, 1101 Quezon City, Philippines
{icjuanatas,rrtio}@feudiliman.edu.ph
[2] National University, 1008 Manila, Philippines

Abstract. Learning programming languages like Python can be difficult, especially for those who have no background in the subject. This study intends to design a mobile application and online Python learning lesson to solve the challenges experienced by senior high school S.T.E.M. students. S.T.E.M. students in senior high school who participated in a survey found that academic pressure, a lack of engagement, and a lack of practice chances were the main issues affecting their learning experience. For beginner to advanced levels, the PySkuwela provides a user-friendly interface with video tutorials, hands-on activities, and assessments. The PySkuwela forum promotes student cooperation and knowledge exchange. Students may study Python conveniently and without added academic pressure thanks to the self-paced learning strategy. The FURPS evaluation gave the system an overall acceptability rating of 4.21, which is a high level of agreement. According to the study's findings, the created system is advantageous and appropriate for senior high school S.T.E.M. pupils, offering a thorough platform for learning Python.

Keywords: Python · Web-based Learning · Mobile Application

1 Introduction

Learning programming languages can be a challenging endeavor, particularly for individuals without prior knowledge in the field [1, 2]. This struggle is especially evident among S.T.E.M. students at FEU Diliman who are grappling with the complexities of Python programming language [3]. To gain insights into their difficulties, the researchers conducted a survey involving 200 Grade 12 students from the S.T.E.M. program. Through this survey, three key factors that significantly impact the students' learning experience were identified: lack of engagement, limited practice opportunities, and academic pressure.

Some students report feeling less engaged when learning Python, especially when doing so online. Their involvement is hampered by the lack of depth in the learning

K. Arai (Ed.): FTC 2023, LNNS 816, pp. 356–361, 2023.
https://doi.org/10.1007/978-3-031-47448-4_26

material's explanation of how the code works, which they perceive to be inadequately captivating. Additionally, it is believed that current practice options are insufficient to provide a more immersive and exploratory Python learning experience. The students also emphasize the significant impact that academic pressure has had on their learning process. They are under pressure to continue studying Python, but the language's complexity and the abundance of resources make it difficult for them, leaving them feeling confused and intimidated. The use of technical language can be difficult for beginners. The students also indicate difficulty understanding some lessons, which makes it more difficult for them to keep up with the learning process. As a result, their levels of anxiety and despair are increased.

In order to help senior high S.T.E.M. students become proficient in a highly preferred and extensively used programming language, this study provides thorough tutoring and support. The user-friendly dashboard of the online Python learning lesson includes a variety of video courses, hands-on activities, and evaluations ranging from beginner to intermediate levels. To foster a collaborative learning environment, the tutorial includes a dedicated section called the PySkuwela forum. Students can engage in friendly interactions by answering questions from their peers, posting their own queries, and comments, and even showcasing customized Python projects or outputs to inspire and guide fellow students using the system. This platform also encourages students to provide feedback, suggestions, and constructive comments on each other's ideas, projects, or outputs [4]. One significant advantage of this study is its self-paced learning approach, which allows students to study Python at their own convenience without feeling overwhelmed by the pressures of other subjects they are concurrently undertaking [5]. The subsequent sections outline the Methodology, which includes the study design, data collection methods, and instruments utilized in gathering the necessary data. The Results and Discussion section presents the results and provides a summary of the findings. Finally, the paper ends with the Conclusion and Recommendations section.

2 Methodology

System architecture refers to the structural design of computer software that operates all the tasks inside the PySkuwela [6, 7], it serves as a guide for the developer teams to efficiently collaborate, communicate, develop, maintain, and upgrade the system. In contrast to Sequence diagrams, which represent the progression of events over time, use case diagrams are simply another diagram type that illustrates a user's interaction with the system. Figure 1 illustrates the system architecture, which provides access to both web and mobile applications for administrators and students. Both the administrator and student can utilize various modules within PySkuwela's website and mobile applications, including the landing page, dashboard, learning materials, PySkuwela forum, and learning material customization. However, the ability to customize specific courses is limited to the administrator.

The presented use case diagram in Fig. 2 illustrates the interactions and exchanges that occur between the students and the administrator [8]. The student has the ability to log in to the application and access various features such as learning materials, the PySkuwela forum, and messages, which are also associated with the administrator. The

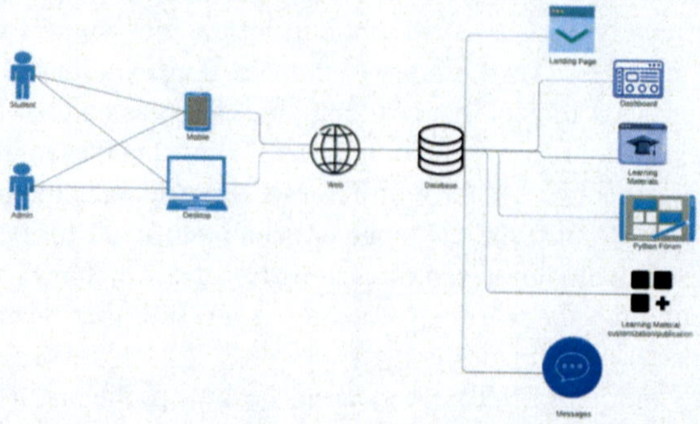

Fig. 1. System architecture of the study.

login use case serves as the base case, encompassing one included use case, which is the "verify password" use case, and one extended use case, which is the "display login error" use case. The "verify password" use case is executed every time a student logs in to verify their password. On the other hand, the extended use case, "display error", is executed only when the student's login credentials are invalid. Furthermore, the "learning materials" use case is a generalized or parent use case, as it encompasses three specialized or child use cases: "beginner", "intermediate", and "advanced". These specialized use cases cater to different levels of learning materials within the system.

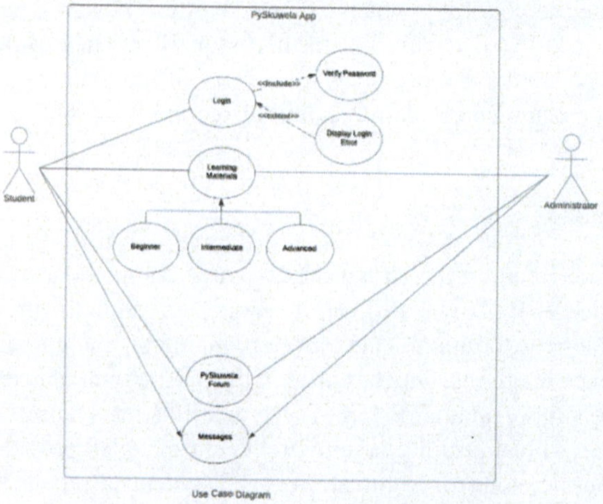

Fig. 2. Use case diagram of the study.

The sequence diagram of the study presented in Fig. 3 illustrates the sequence of messages exchanged between objects within the system during an interaction [9]. The sequence begins with the student attempting to log in to the PySkuwela website or mobile app. The PySkuwela server verifies the login credentials provided by the student and sends a "return message" indicating whether the login was successful or unsuccessful. Upon successful login, the student gains access to the learning materials. The server sends another "return message" containing the published learning materials, which are then made available to the student through the website or mobile app. The student may also choose to participate in the PySkuwela forum, where the server sends a "return message" providing the PySkuwela forum data, which can be accessed by the student through the website or mobile app. Additionally, the student has the option to send messages to the administrator for any concerns. The administrator's reply, originating from the PySkuwela server, is directed back to the user to address their inquiries.

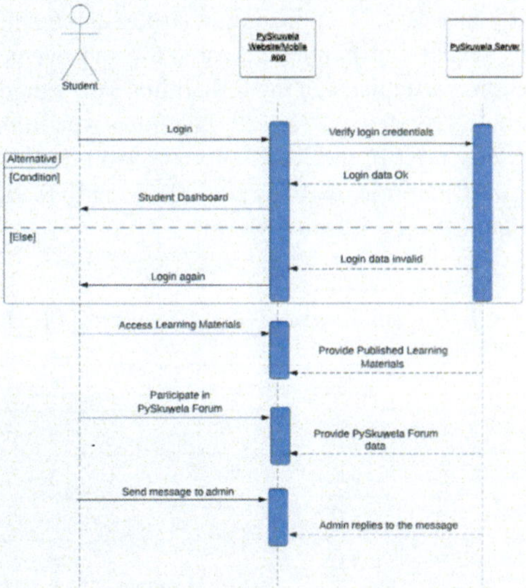

Fig. 3. Sequence diagram of the study.

Responses from the gathered data are subjected to statistical analysis, utilizing the data required by the system. The researchers use a Likert Scale to assess the system, as shown in Table 1. A rating scale from 1 to 5, with 1 denoting the lowest rating and 5 denoting the highest, is used in the Likert Scale [10]. This rating scale enables the respondents to express their opinions, preferences, and needs regarding the proposed system.

Table 1. Likert scale.

Point	Scale Range	Equivalent
5	4.20–5.00	Strongly Agree
4	3.40–4.19	Agree
3	2.60–3.39	Neutral
2	1.80–2.59	Disagree
1	1.00–1.79	Strongly Disagree

It also assists the researchers in assessing whether the system meets the requirements of the students and fulfills their expectations.

3 Results and Discussion

Table 2 presents the summary of the weighted mean from the FURPS assessment conducted by the researchers. Based on the evaluation of the respondents, the system received an overall acceptability rate of 4.21, indicating a strong agreement. This suggests that the system is generally acceptable, but there is room for improvement in certain areas to enhance user satisfaction, particularly in the Reliability, Performance, and Supportability criteria, which were interpreted as "Agree". In comparison to the related literatures, the lowest rating was observed in the Reliability criterion, indicating the presence of several issues that require attention, particularly in relation to the learning materials, as they are a critical component of a learning website.

Table 2. Overall Weighted Mean Summary of FURPS.

Criteria	Weighted Mean	Interpretation
Functionality	4.40	Strongly Agree
Usability	4.32	Strongly Agree
Reliability	4.06	Agree
Performance	4.09	Agree
Supportability	4.18	Agree
Overall Weighted Mean	4.21	Strongly Agree

For students ranging from beginners to experts, the system was able to handle activities including offering video training, hands-on exercises, and assessments. It contains an online platform that will assist students in becoming familiar with Python while also streamlining the learning process to better meet their needs. Through the effectiveness of both the website and the mobile application, this design enhanced customer satisfaction and information acquisition.

4 Conclusion

In conclusion, this study proved to be a suitable and beneficial learning tool for the senior high school S.T.E.M. students specializing in Information Technology, offering a comprehensive platform for learning Python programming and promoting engagement, collaboration, and ease of use. According to the results, students gave the system a favorable overall acceptance rating of 4.21, demonstrating their pleasure with its usability and functionality. To improve learning and encourage a deeper grasp of Python, this contains video courses and demos. By enabling users to ask for assistance and exchange expertise about Python programming, the addition of the PySkuwela forum further encouraged user collaboration. Additionally, the development of an Android mobile application provided students with the flexibility to learn Python on the go, catering to their preferences and convenience. The researchers also tested the system using FURPS (Functionality, Usability, Reliability, Performance, and Supportability) to assess its overall capability and came to the conclusion that both the website and mobile application are adequate for users. The positive responses from the survey of the students confirmed the system's efficiency and user satisfaction.

5 Recommendations

Based on the development of the online Python learning tutorial and mobile application for Senior High School S.T.E.M. Students specializing in Information Technology, several recommendations can be made to further enhance the system. Add an online Python sandbox compiler module, so that users that have devices that cannot handle Python applications can test and formulate their Python programs on the website and mobile application, without opening any external applications. Improve the framework of both the website and mobile application by utilizing frameworks like Codeigniter or Laravel in developing the system, which helps both the website and mobile application to improve code reusability, eliminating the need to write much repetitive code. Extend the learning capabilities of the system, by implementing several programming language lessons (C++, JavaScript, PHP, etc.), so that users can also learn several programming languages other than Python. Add more activities to add interactivity towards lessons, as well as an article creation module, in order to create articles containing the summary of the lessons that the users can read and examine.

References

1. Matthes, E.: Python Crash Course: A Hands-On, Project-Based Introduction to Programming, 3rd edn. No Starch Press, San Francisco (2023)
2. AMZ Press: Python Programming for Beginners (2022)
3. Ahmed, Z., Kinjol, F., Ananya, I.: Comparative analysis of six programming languages based on readability, writability, and reliability. In: International Conference on Computer And Information Technology (ICCIT), pp. 1–6 (2021)
4. Millington, R.: Buzzing Communities: How to Build Bigger, Better, and More Active Online Communities. FeverBee Limited, London (2019)
5. Sweigart, A.: Automate the Boring Stuff with Python: Practical Programming for Total Beginners, 2nd edn. No Starch Press, San Francisco (2019)
6. Bernus, P., Fox, M., Gabrysiak, G.: Enterprise Architecture for Business Success. Springer, Heidelberg (2020). https://doi.org/10.1007/978-3-642-01310-2
7. Rozanski, N., Woods, E.: Software Systems Architecture: Working with Stakeholders Using Viewpoints and Perspectives. Addison-Wesley Professional, Boston (2020)
8. Weilkiens, T.: UML Use Case Diagrams: A Practical Guide. Rocky Nook, Kingston (2020)
9. Sowa, J.: UML Sequence Diagrams: A Practical Guide. Morgan & Claypool Publisher, San Rafael (2020)
10. Simmons W.: Likert Scale Questionnaires: Unleash the Power of Data Collection (2021)

Introducing Coding to Children with Scratch: A Pilot Study

Check-Yee Law(✉), Kah-Ong Goh, Shih-Yin Ooi, Lee-Ying Chong, Wen-Kang Siow, and Boon-Jin Wong

Multimedia University, Melaka, Malaysia
{cylaw,michael.goh,syooi,lychong}@mmu.edu.my

Abstract. As digital literacy is necessary in the modern workplace, programming or coding skills have been deemed as an important competence in 21st century education. However, learning to code or program is generally perceived as a difficult task. To this end, various game-based platforms have been established to encourage and engage novice programmers in learning coding and programming. Scratch, a game-based platform was used in this pilot study to introduce coding to children. Two Scratch teaching and learning modules were prepared to help young children learn programming concepts and acquire coding skills. Participant quiz results and mini project scores were collected and analyzed to examine to what extent the participants had learnt Scratch action scripts and coding blocks from the Scratch teaching and learning modules as planned. The results show that students had acquired basic coding knowledge and skills from Scratch. In the future, more Scratch modules will be created to introduce more Scratch functions with different level of difficulty for young children to progressively learn programming concepts and acquire coding skills.

Keywords: Coding Skills · Computational Thinking · Digital Literacy · Programming · Scratch · Young Children

1 Introduction

In this digital era, the world is hyperconnected. Computing devices have become indispensable tools in our daily lives. As digital literacy is necessary in the modern workplace, programming or coding skills have been deemed as an important competence in 21st century education [1, 2]. Coding skills include problem solving, logical thinking, computational thinking, and design skills [3]. Computational thinking is the mental skill of designing systems, solving problems, and finding solutions derived from computer science principles and concepts [4, 5]. Coding skills are important for students to possess computational and critical thinking skills so that they become active creators instead of passive consumers in this digital era [5]. However, learning to code or program is generally perceived as a difficult task [6]. Many novice programmers find it hard to apply programming concepts in code writing [7, 8]. Introductory programming courses that are introduced to students in secondary school or university level in enterprise-grade

K. Arai (Ed.): FTC 2023, LNNS 816, pp. 362–371, 2023.
https://doi.org/10.1007/978-3-031-47448-4_27

programming language such as Java and C++ cause many challenges in teaching programming to novice programmers. To this end, various game-based platforms have been established to encourage and engage novice programmers in learning coding and programming [9]. In an attempt to expose young children to basic computer coding skills, Scratch has been developed. Scratch, which was launched in May of 2007, has now become the largest coding community for kids worldwide [10].

This paper reports a pilot study of introducing coding to children with Scratch. Section 2 describes the Scratch software. The methodology for conducting the pilot study is presented in Sect. 3. This is followed by data analysis results and discussion in Sect. 4. Section 5 presents the threats to validity in this research study. The paper is then concluded with some future work in Sect. 6.

2 Scratch

The Scratch software is built with visual programming language which makes it simple for kids to create their original interactive entertainment (music, games, art, and interactive stories). Children can post their creations to the Scratch website and share them with everyone, which promotes teamwork [11]. The Lifelong Kindergarten Group of the MIT Media Lab originally developed Scratch [12]. Scratch builds on the constructivist tenets of Logo and Etoys. Scratch is designed to assist young people (ages 8 and older) in the development of 21st-century learning skills, according to its website. Through the usage of Scratch, individuals can improve their ability to "reason logically, collaborate cooperatively, and think creatively." Furthermore, young people who use Scratch have the opportunity to deal with significant computational and mathematical issues including variables, conditions, parallelism, and other [13]. One of the features that make Scratch a popular programming environment for the integration of programming in the primary school curriculum is its capacity to let young programmers build videos, multimedia stories, and games. The Scratch programming environment was designed with a low floor that enables the development of really simple programmes with very little formal knowledge and a high ceiling that enables the development of quite complicated applications [11]. Figure 1 shows the project editor of Scratch.

- A–Code Tab (Blocks Palette): Let a user to drag blocks into the field to form scripts
- B–Costumes/Backdrop Tab (Paint Editor): For a user to edit the sprite or backdrop's costume.
- C–Sound Tab (Sound Editor): A user can add, edit, and delete sound
- D–Menu Bar: Enable a user to change language, save Scratch file, load Scratch file, rename file, edit, and show tutorials
- E–Script: The set of instructions that are connected with one another and used to program in Scratch
- F–Sprite: The image on the Scratch computer program screen
- G–Stage Size: For a user to change the size of the stage
- H–Code Area: For a user to insert codes
- I–Extensions Library: A user can add extension and the extension added will appear in the block categories
- J–Scroll Bar: For a user to scroll and view the blocks

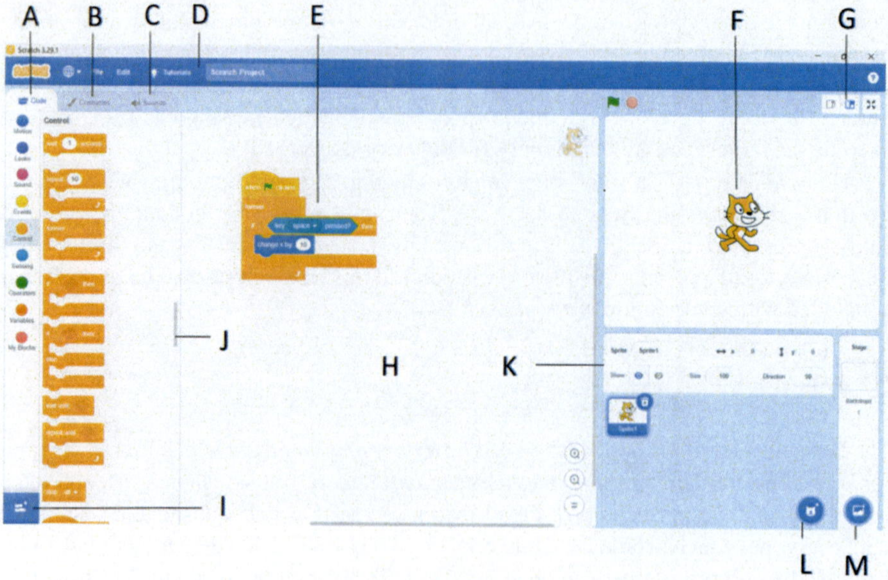

Fig. 1. The project editor of Scratch.

- K–Sprite Header: A user can rename, change x and y axis, change size, and direction of the sprite, hide and show the sprite
- L–Choose a Sprite: Let a user to choose, paint, upload a sprite, or pick a random sprite that is recommended by the software
- M–Choose a Backdrop: Enable a user to choose, paint, upload a backdrop, or pick a random backdrop that is recommended by the software

3 Methodology

Two Scratch teaching modules were prepared. A pilot study was carried out to expose participants to Scratch coding class by using the planned teaching modules. The study was conducted in May to June 2022 in a local kindergarten–Tadika Voyage. Five participants who were new to Scratch attended, including two 7-year-old girls and three 9-year-old boys. All the five participants were Chinese. This course was implemented in five sessions. Module 1 was conducted in the first session followed by Module 2 in the next three sessions. The last session was a revision session in which the participants were required to complete a mini game. Each session was conducted by an instructor together with a facilitator for two hours with a short break of 15 min.

The Scratch teaching modules consist of Module 1 and Module 2 that introduce participants to the Code blocks in Scratch. Module 1 is Make it Fly that we adopt from the Scratch official website (https://resources.scratch.mit.edu/www/cards/en/fly-cards.pdf) whereas Module 2, named Under the Sea is adopted from one of the educational games available at Tadika Voyage. There are seven activities in Module 1–Make it Fly. These activities enable the participants to create the game for a game player to move the Cat up, down, left, and right by pressing the up, down, left, and right keys on the keyboard. At the same time, the player (Cat) will need to touch as many Hearts as possible to increase the Score in the game. As for Module 2–Under the Sea, there are eight activities that guide the participants to create a game for a shark to eat fish, gain score and to level up. In this game, a player will use the mouse (not keyboard) to play as a shark, eat fish to gain score and to level up. The player (shark) will get 1 score when eating a fish. When the score reaches a multiple of 10, there will be a level up. The spawn rate of the fish decreases as the level increases. This increases the difficulty level of the game when level is up. There is a timer in the game, when the timer drops to 0, the game will end. Table 1 presents the content summary of the two teaching modules.

Table 1. Content summary of the two teaching modules.

No	Activity	Content Summary
	Module 1: Make it Fly	
1	Choose a Character	Participants were introduced to Choose a Backdrop and Choose a Sprite. Cat Flying to start a Scratch project. They then made the Sprite to greet audience by using the Say script and Go To script in the Looks code
2	Start Flying	Participants chose Buildings from the Choose a Sprite function and added the script to Buildings. In this activity, participants also were introduced to the x and y-axis. The scripts include: when green flag clicked, forever, set x to, repeat, and change x to
3	Switch Looks	Participants changed the look of Buildings by adding the Next Costume block into the script of Buildings
4	Make it Interactive	Participants were taught to make the game interactive by adding scripts to Cat Flying to move it accordingly when a player pressed up, down, left, and right key on the keyboard. Script blocks include: when __ key pressed, change x/y by blocks
5	Floating Clouds	Participants were guided to make the Clouds float in this activity. They chose Clouds in the Choose a Sprite function and added scripts to the Clouds. The scripts include: when green flag clicked, forever, set y to, pick random_ to_, set x to, next costume, repeat, change x by
6	Flying Hearts	Participants chose Heart from the Choose a Sprite function and added Code to it. The codes include: when green flag clicked, forever, go to random position, set x to, repeat, change x by
7	Collect Points	To increase the point when the Cat touches the Heart, participants were taught to make a Variable called Point and added Code to Cat Flying. The codes include: when green flag clicked, set Points to, forever, if…then, touching heart, change Points by, wait 1 s
	Module 2: Under the Sea	
1	Choose a Character	Participants chose two sprites from the Choose a Sprite function. The sprites were Shark 2 and Fish. They were introduced to Paint in the Choose a Sprite function to write the word "Time's up" for the Timer sprite

(*continued*)

Table 1. (*continued*)

No	Activity	Content Summary
2	Start Swimming	Participants added some Codes to make Shark 2 swim. The codes include: go to mouse-pointer, point in direction, forever, if…then
3	Switch Looks	Participants were taught to add Code to Shark 2 to make it switch Looks when eating fish such that it would open and close its mouth when eating. The codes include: when I receive eat, switch costume to shark2-b (open mouth), play sound Bite until done, wait 0.1 s, and switch costume to shark2-a (close mouth)
4	Swimming Fish	Some codes were added to Fish for it to swim around. The codes include: create clone of myself, when I start as a clone, move 10 steps, if on edge, bounce, show, switch costume to, pick random, point in direction, repeat until, touching Shark 2, and delete this clone
5	Add sound and music	Participants added the background music and Sound when Shark 2 was eating and when it was Time's up. The codes include: when green flag clicked, hide, when I receive time's up, start sound Lose, and show
6	Collect Score	Participants added the Score variable to the game and add the Collect Score code to the Fish sprite. The codes include: set Score to, change Score by
7	Add Game Level	Participants added the Level variable to the game and added the Level code to the Fish and the Shark 2 sprites. The codes include: set Level to, change Level by, if, =, score, >, when green flag clicked, and forever
8	Add timer	Participants added the Timer variable to the game and add the Timer code to the Fish and Shark 2 sprite. The codes include: set timer to and change timer by

For each teaching and learning session, PowerPoint slides were used to display and explain the Scratch functions to the participants. This was followed by demonstration of Scratch functions through a touch screen television that run on Windows operating system. Figure 2 shows the classroom setting. It was a hands-on learning session in which the participants would then follow the demonstration and practise the Scratch functions by using their own laptops. The participants were required to download the Scratch application via https://scratch.mit.edu/download and saved their work in their laptops.

For each teaching module, we evaluated the participants' knowledge and skills through pretest and posttest quizzes via the Quizizz platform. Pretest quiz was conducted prior to the commencement of the teaching and learning session whereas posttest quiz was held at the end of the teaching and learning session. The last session of the course required the participants to demonstrate their knowledge and skills learnt through a hands-on project that was to create a mini game. Finished game illustration and game requirements were given to participants as reference as in Table 2. Project evaluation was carried out to assess to what extent the participants had grasped the Scratch functions to create the mini game with specific game requirements.

Fig. 2. Classroom setting.

Table 2. Mini game illustration and game requirements.

<table>
<tr><td>Finished game:</td><td>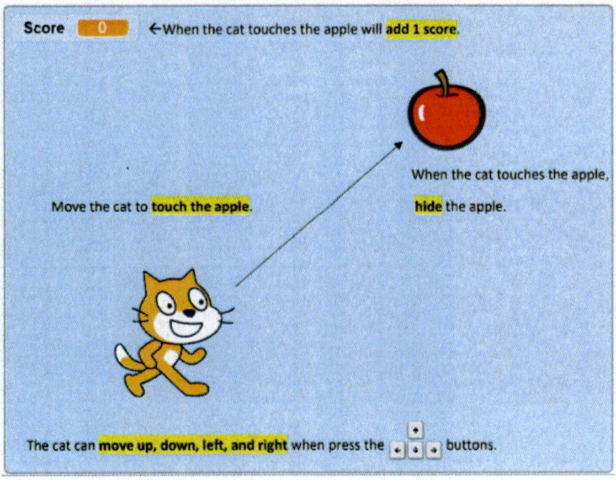

- Remember to change background, add sound, and switch costume.</td></tr>
<tr><td>Requirements:</td><td>1. Move Cat up, down, left, and right.
2. Display an Apple sprite.
3. When the Cat touches the Apple, hide the Apple.
4. Create Score variable.
5. When Cat touches Apple, add 1 point to Score.
6. Change Backdrop.
7. Add any sound to the game, for example, make the cat meow or add the sound of eating apple.
8. Switch any Costume of a sprite in the game, for example, change the costume of the Cat from Cat-a to Cat-b.</td></tr>
</table>

Participants' quizzes and project achievement scores were recorded to track their learning progress and performance. Data collected were analysed to answer the research questions as follows:

Research Question 1 (RQ1): To what extent participants have learnt Scratch functions from the Scratch teaching and learning modules?

Research Question 2 (RQ2): To what extent participants have acquired the knowledge and skills to create a mini game with the Scratch functions?

4 Data Analysis and Discussion

To answer RQ1, the quiz results collected from Module 1 and Module 2 were organised in a tabular form as in Table 3 and Table 4. Mean scores were calculated to examine the average performance of the participants in both the pre-test and post-test for Module 1 and 2. For Module1, the mean score of the pre-test is 48 as compared to the mean score of 92 in the post-test. For Module 2, the mean score of the pre-test is 30 whereas the mean score of the post-test is 70. The results indicate that there is an improvement in the performance score after the participants have attended the Scratch course. This means the participants have learnt Scratch functions from the teaching and learning modules as planned. Figure 3 presents the bar charts showing the comparison of participants' pre-test and post-test performance scores.

Table 3. Pre-test and post-test results of Module 1.

Participant	Pre-test scores	Post-test scores
1	60	100
2	50	90
3	50	100
4	50	90
5	30	80
Mean score	48	92

Table 4. Pre-test and post-test results of Module 2.

Participant	Pre-test scores	Post-test scores
1	30	70
2	10	70
3	20	60
4	50	80
5	40	70
Mean score	30	70

To address RQ2, participants' projects were evaluated based on the Scratch functions applied to create the mini game. One (1) mark was given for the correct Scratch

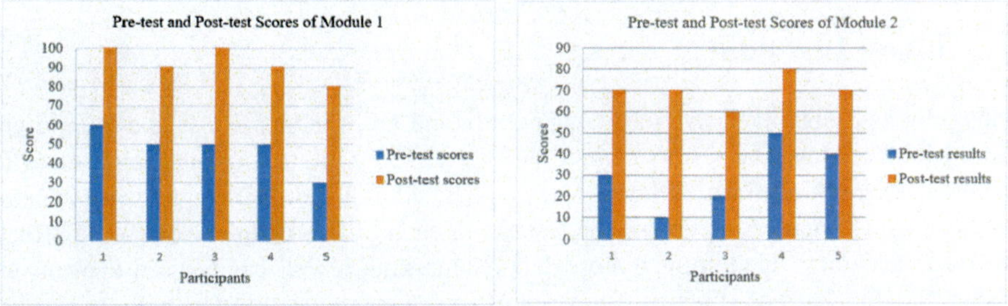

Fig. 3. Comparison of the pre-test and post-test performance scores in Module 1 and Module 2.

function applied whereas zero (0) mark for incorrect/no Scratch function applied to the mini game. Performance score in percentage was calculated based on the total correct Scratch functions applied to create the mini game. Table 5 presents the evaluation results. The mean performance score of 92.5% shows that the participants had acquired the knowledge and skills to create a mini game with the Scratch functions.

Table 5. Mini game performance scores.

Game requirements	Participant 1	Participant 2	Participant 3	Participant 4	Participant 5
1	1	1	1	1	1
2	1	1	1	1	1
3	1	1	1	1	0
4	1	1	1	1	1
5	1	1	0	1	1
6	1	1	1	1	0
7	1	1	1	1	1
8	1	0	1	1	1
Total correct Scratch functions applied	8	7	7	8	6
Performance (%)	100	87.5	87.5	100	75
Mean performance (%)	92.5				

From the participants' mini game performance scores, we can see all participants were able to create a simple mini game. They had a better understanding of using the Scratch software after attending the planned lessons. The average mean performance score of 92.5% indicates that they were able to apply most of the Scratch functions learnt to create the mini game.

5 Threats to Validity

This pilot study involved five participants only and all of them were Chinese children from Melaka. Thus, the results cannot be generalized without more participants with different backgrounds from different locations. Besides that, only one quiz was given to evaluate student understanding of the lessons in each of the module. This could affect the result accuracy. In addition, a more detailed marking rubrics can be used to evaluate each action block used to create the mini game so that the mini game performance results can better reflect student performance.

6 Conclusion

Two Scratch teaching and learning modules have been prepared to help young children learn programming concepts and acquire coding skills. Through the pilot study, the participants were able to progressively learn basic computer coding skills through Scratch action scripts and coding blocks. From the participants' performance results, they had learnt Scratch functions from the Scratch teaching and learning modules as planned. They also had acquired the knowledge and skills to create a mini game with the Scratch functions. In the future, we will create more Scratch modules that introduce more Scratch functions with different level of difficulty for the participants to progressively learn programming concepts and acquire coding skills.

References

1. Ng, W.S.: Coding education for kids: what to learn? How to prepare teachers. In: Proceedings of ICICTE (2017)
2. Papadakis, S.: Apps to promote computational thinking and coding skills to young age children: a pedagogical challenge for the 21st century learners. Educ. Process Int. J. (EDUPIJ) **11**(1), 7–13 (2022)
3. Tuomi, P., Multisilta, J., Saarikoski, P., Suominen, J.: Coding skills as a success factor for a society. Educ. Inf. Technol. **23**, 419–434 (2018)
4. Wing, J.: Computational thinking. Commun. ACM **49**(3), 33–35 (2006)
5. Papadakis, S., Kalogiannakis, M., Zaranis, N., Orfanakis, V.: Using Scratch and App Inventor for teaching introductory programming in secondary education: a case study. Int. J. Technol. Enhanc. Learn. **8**(3–4), 217–233 (2016)
6. Malan, D.J., Leitner, H.H.: Scratch for budding computer scientists. ACM Sigcse Bull. **39**(1), 223–227 (2007)
7. Derus, S.R.M., Ali, A.M.: Difficulties in learning programming: views of students. In: 1st International Conference on Current Issues in Education (2012)
8. Roy, K., Rousse, W., DeMeritt, D.: Comparing the mobile novice programming environments: app inventor for Android vs. GameSalad. In: IEEE Frontiers in Education Conference (FIE), New York (2012)
9. Combéfis, S., Beresnevičius, G., Dagienė, V.: Learning programming through games and contests: overview, characterisation and discussion. Olympiads Inform. **10**(1), 39–60 (2016)
10. Scratch_Foundation: About: Our Story (2022). https://www.scratchfoundation.org/our-story

11. Olabe, J.C., Olabe, M.A., Basogain, X., Castaño, C.: Programming and robotics with Scratch in primary education. In: Education in a Technological World: Communicating Current and Emerging Research and Technological Efforts, pp. 356–363 (2011)
12. Scratch: Scratch (2023). https://scratch.mit.edu/
13. Papadakis, S., Kalogiannakis, M., Orfanakis, V., Zaranis, N.: The appropriateness of scratch and app inventor as educational environments for teaching introductory programming in primary and secondary education. In: Early Childhood Development: Concepts, Methodologies, Tools, and Applications, pp. 797–819 (2019)

Online Exam Proctoring: Challenges, Timeline, and a Proposed Custom Rule-Based Online Proctoring Exam Framework

Ahmad Rawashdeh[1](✉) and Mohammad Rawashdeh[2]

[1] Yarmouk University, Irbid, Jordan
ahmad.rawashdeh@yu.edu.jo
[2] University of Central Missouri, Warrensburg, MO, USA
rawashdeh@ucmo.edu

Abstract. Online Exam Proctoring has become a major obstacle in education at universities around the globe since the beginning of the COVID-19 pandemic; which also has imposed a challenge on every university's administration and instructors as well, and has made it more difficult to grade students successfully and fairly. Additionally, one important challenge is detecting student's cheating in the exams. Current research literature lacks a description of an online proctoring exam which has as an input a set of rules represented as numerals that determine how many times a cheating attempt needs to be repeated until it is identified/classified as cheating. So, this paper begins by providing information about online exam proctoring including definition, various ways students cheat in online exams, challenges and limitations of online exam proctoring and some of the current Online Proctoring Systems, timeline of research on this field, and proposed a framework for online proctoring system. This framework, the input of which is the rules for cheating detection, is described in this paper.

Keywords: Online · Proctoring · Exams · Automatic Proctoring · Online Proctoring · Universities · Test Taking · Examinee

1 Introduction

During the COVID-19 period, many universities have switched to online course delivery which requires the students to take online exams. However, the need for using Automatic Exam Proctoring has continued and it has never ceased to exist even after the end of COVID-19. On the other hand, hiring and employing human proctors comes with a cost of time, money, human management, and administration, in addition to the cost for administering the test centers. Utilizing online proctoring will contribute to reducing these costs. This paper provides information about online exam proctoring, which is the type of exam conducted using software or service implemented using machine learning approaches or various other techniques or algorithms. It also describes a rule-based Exam Proctoring Framework.

K. Arai (Ed.): FTC 2023, LNNS 816, pp. 372–377, 2023.
https://doi.org/10.1007/978-3-031-47448-4_28

The following subsection provides a definition of Online Exam Proctoring. After that, the answer to the question of how students cheat is provided. Then, the Literature Review including Challenges and Limitations and some of the current Exam Proctoring systems is discussed. Next, is the Proposed Framework and then the timeline. Finally, the conclusion and future work is presented.

1.1 Definition

Online Exam proctoring or Automatic Exam Proctoring is monitoring exams taken by test takers, i.e., examinees, to detect any attempt of non-permissible cheating. In a broader sense, it is ensuring that the exam is done according to the rules of the examiner. For example, if the exam is closed book, then it must prevent any attempt to open any web document or local material using technology such as lockdown browsers which uses active windows detections and locking of exam screen to prevent using any other window (web browser, or file browser). If it is an open book, then it must capture any attempt of cheating from other students as well. Also, it should have one way of alerting the admin of any attempt of cheating or violating the rules. This requires student verification as well as student id recognition.

The tools and technologies used by existing online proctoring systems are either completely software based (no use of hardware cameras), or rely more on using hardware such as webcam, wear-Cam, and mic. In [1], a Multimedia exam proctoring was described. In the first place, this raises the question of "How do students cheat in exams?". The next section provides details about that.

2 How do Students Cheat

There are various ways on which students cheat in exams. For example, superiors and seniors may take the test first, then they record the questions and the answers and share it with fresh/first-year students. Students also may prewrite the answer to the questions before the test on a small piece of paper and refer to it during the exam without getting caught. For online exams, students may not abide by the rules and look for answers on the web or in the textbook stored on local computer. Students also may text their colleagues asking for answers during exams. For multiple choice exams, it is easier to cheat, especially if the answers and the questions are not being shuffled and randomized. For multiple choice exam, students also could have prior agreement that a one nudge indicates that the answer is the first option, two nudges indicates the answer is second choice etc. students also can obtain test questions from universities libraries or from seniors [2]. Nowadays, students are more digital literate and familiar with technologies than older generations, and also, they are more likely to utilize and use technologies such as smart watches, mobiles, smart glasses, and other technologies/smart-devices to share the exam or to know the answers for the questions. About eight Online exam control procedures (OECPS) and online testing plan were described in [2]. The described online testing plan consists of: Limiting exam time, limiting student access, changing test characteristics, and verifying student ID.

3 Literature Review

This section provides an overview of the research on this field. It is comprised of two subsections: the first describes the challenges and limitations of existing online proctoring systems, and the second lists two existing online proctoring systems.

3.1 Challenges and Limitations

How to make a decision/judgement on a student, who has taken the test and made enough attempts of cheating, as cheater and thus deserves to fail the test (what is the threshold of cheating)? How to accurately out rule any other suspicious attempt that was determined to be a cheating and distinguish it from non-cheating attempts (false positive)? Is Id verification good enough? That and more technical problems are discussed.

There are multiple objects that need to be detected: text, speech, phone, active window detection in addition to user verification which is required to authenticate the test taker. Also, gaze estimation is needed. Detecting text is challenging, since the test taker could cheat by using a text with small font size or place the text far away from the camera. Also, the text on paper needs to be differentiated from text on computer screen or keyboard. So, the work in [1] only focuses on printed since detecting handwritten text is challenging. Since detecting handwritten text requires estimating the eye gaze of the user. For detecting text, a training dataset consisting of 186 positive training images that contain text in a typical office environment and 193 negative training images were used to training.

The work in [3] discusses the negative impacts of using online exam proctoring on the educational outcome, pedagogical relationships, and student subjectifies of students in Korea. Online exams with online proctoring are no different from traditional exams in which students have exhibits psychological reactions including stress and anxiety which in turn influenced their performance and well-being even after the test. The research in [4] has addressed that issue. Moreover, the work in [5] conducted a comparative study on the impact of proctoring environment, offline and online, and whether it influences the student performance. It was found out that, after considering students sample test of 1,762 students from 2009 to 2016, the environment has no impact on student performance. Although online proctoring seems more profound and robust, there are known vulnerabilities to using it. For more details the reader may refer to [6].

3.2 Current Exam Proctoring Systems

Kryterion [7] and ProctorU [8] are one step ahead of traditional exam proctoring which requires the examinee to be present in physical location with proctors. Kryterion still relies on human proctor. However, the proctor uses a cam to monitor the examinee. The proctors are trained to watch and listen for any unusual activity. The proctor watches the examinee through a webcam from a remote location.

The current systems lack the support for writing custom rules used for cheating detection and lack the support for running the language used for writing the rules, in the first place, which needs parsing and compilation. This framework will be very flexible when used by admins since it allows them to write their own rules for cheating detection.

4 The Proposed Framework (Rule-Based Exam Proctoring)

The way the current exam proctoring system works is by flagging the suspicious actions of the examinee. For example, if the examiner (test taker) looks away from the screen, then that might be a cheating attempt. To implement the decision on the actions of the examinee one needs to decide on what threshold of repetition of suspicious behavior is going to be followed. So, a general framework for implementing the custom rules which are used to decide whether a student has cheated or not is described conceptually in this paper (no empirical experiment yet). In this paper, the author is proposing a Custom Rule Based Online Exam Proctoring Framework the architecture for which is described in Fig. 1. In this proposed rule-based online exam proctoring, the examiner input the custom rules for detecting cheating which will be used to mark a student attempt of the exam as cheating or not. Each rule has two parts: the condition/action, and the decision with confidence. Examples of such rules are as follows:

- Look away from the screen 5 times to different directions -> 50% cheating
- Look away from the screen to the same direction 5 times -> 90% cheating
- Text identified within the view of the camera -> 90% cheating
- If test takers look at the identified text once-> 100% cheating

So, for example, the first rule says that if the examinee looks away from the screen 5 times, then he is cheating with probability of 50%. Since he could look at the wall while thinking of solutions to some difficult question which require his retention and remembering of the answer. Some actions could be classified as cheating immediately with even only one occurrence. Example if the examinee looks for a recognized text once.

The admin defines the rules: for example, looking away from the monitor three times should be flagged as cheating. Speaking more than twice should be flagged as cheating as well. Opening a book or looking at notes even only for once will be flagged as well. The Input is a text file which defines the rules and the threshold (frequency).

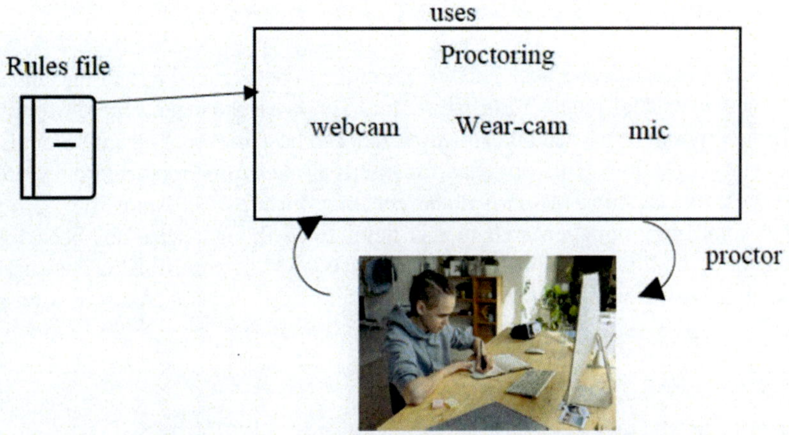

Fig. 1. The architecture of the rule based exam proctoring.

In its advanced form and as a future work, the framework could also be made to support its own language for writing the rules which have been described in previous paragraphs as examples. However, implementation details of this framework are outside the scope of this current research paper.

The next section provides a timeline of the last major research effort in the last 20 years.

5 Timeline (Major Three Research in Recent Era) and RoadMap

Figure 2 illustrates the timeline of three of the major research on Online Exam Proctoring.

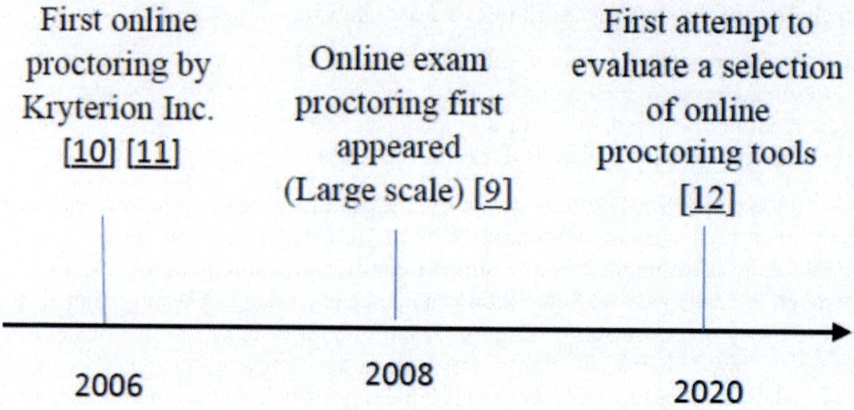

Fig. 2. Timeline of research on online exam proctoring.

6 Conclusion

Online Exam Proctoring has been and will continue to be a crucial part of the education systems. However, additional efforts in research to improve and make it more widespread and effective should be carried out. However, that can be achieved, first and foremost, by addressing its limitations. This work has discussed Online Exam Proctoring and proposed a framework for detecting plagiarism and cheating during such exams. In addition, it surveyed some important research in this field. Even though there has been lots of research about the topic, it seems clear that there is no well-prescribed framework that is based on using customized rules for cheating detection and here comes this research.

7 Future Work

As a future work, more highlight of research in this field needs to be added to the timeline. Also, a road map and implementation of the described framework needs to be clearly provided in detail.

References

1. Atoum, Y., et al.: Automated online exam proctoring. IEEE Trans. Multimedia **19**(7), 1609–1624 (2017)
2. Cluskey, G.R., Jr., Craig, R.E., Mitchell, H.: Thwarting online exam cheating without proctor supervision. J. Acad. Bus. Ethics **4**(1), 1–7 (2011)
3. Lee, K., Mik, F.: Online exam proctoring technologies: educational innovation or deterioration? Br. J. Edu. Technol. **53**(3), 475–490 (2022)
4. Kolski, T., Jennifer, W.: Examining the relationship between student test anxiety and webcam based exam proctoring. Online J. Distance Learning Admin. **21**(3), 1–15 (2018)
5. Lee, J.W.: Impact of proctoring environments on student performance: online vs offline proctored exams. J. Asian Finance Econ. Bus. **7**(8), 653–660 (2020)
6. Li, H., et al.: A visual analytics approach to facilitate the proctoring of online exams. In: Yoshifumi, K., et al. (eds.) Proceedings of the 2021 CHI Conference on Human Factors in Computing Systems, Yokohama, Japan, pp. 1–17 (2021). https://arxiv.org/pdf/2101.07990.pdf
7. kryterion. kryterion (2022). https://www.kryterion.com/. Accessed 21 Nov 2022
8. proctoru (2022). https://www.proctoru.com/. Accessed 21 Nov 2022
9. Lee, K., Fanguy, M.: Online exam proctoring technologies: educational innovation or deterioration? Br. J. Educ. Technol. **53**(3), 475–490 (2022). https://doi.org/10.1111/bjet.13182
10. Jiyou, J., Yunfan, H.: The design, implementation and pilot application of an intelligent online proctoring system for online exams. Interact. Technol. Smart Educ. **19**(1), 112–120 (2022)
11. Foaster, D., Bonsall, R., Caddell, J., Dorman, W., Perryman, L., Peeke-Vout, J.: Inventors. System for the Administration of a Secure. Proctored Examination (2011). 12/723,663. https://patents.justia.com/inventor/russ-bonsall
12. Hussein, M.J., Yusuf, J., Deb, A.S., Fong, L., Naidu, S.: An evaluation of online proctoring tools. Informit. **12**(4), 509–525 (2020)

Relating Network Behavior to Demand-Response During dDOS Attack in the Smart Grid

Rajesh Manicavasagam(✉), Mike Rogers, Satish Mahajan, Robert Craven, and Chikezie Emeghara

Tennessee Technological University, Cookeville, TN 38501, USA
{rmanicava42,mrogers,SMahajan,rcraven,cmemeghara42}@tntech.edu

Abstract. Smart Grid data networks that support Demand Response programs are unlike commodity data networks because they make extreme use of heterogeneous physical, data link, network, session, and application layer protocols. Often these protocols are vulnerable to various types of attacks because they are implemented on devices that lack hardware powerful enough to implement sophisticated security techniques. The attacks on the vulnerabilities of the data networks of smart grids can not only cause disruptions of service but also can be used to skew the reporting of consumer behavior. This skew can thus fool the provider into giving the consumer either more or fewer credits than what they deserve as participants in the demand-response program. The impacts can be both increased provider costs and decreased consumer satisfaction. Our goal is to investigate a real-world Demand Response program by simulating various attacks on the various protocol layers commonly found in Smart Grid data networks, and determine how these real-world attacks can affect the efficacy in terms of consumer credits earned for the demand response program. Additionally, we evaluate if correlations of network statistics and consumer consumption indicate that further investigation of the network is advisable to see if tampering was done that caused deviation in consumer consumption.

Keywords: Smart Grid · Protocols · Network Vulnerabilities · Cyber-Attacks · Demand Response Program · Consumer Behavior · Credits · Throughput · Correlation

1 Introduction

Smart grids are next-generation power grids. Smart grids control the processes that generate electricity in power plants, transmit the electricity across distances, distribute the electricity to consumers, and monitor the power consumed by the consumers in residences, institutions, and industries. *Demand-response* (DR) programs are gaining momentum to shed load during periods when the load is high. These programs integrate consumers into the Smart Grid information

K. Arai (Ed.): FTC 2023, LNNS 816, pp. 378–393, 2023.
https://doi.org/10.1007/978-3-031-47448-4_29

loop. Smart Grids provide real-time communication to consumers for important DR events and accurate real-time energy prices and billing updates. Such programs empower consumers to have more control over energy decisions so that they can reduce their consumption during peak times and save money during the next billing cycle. Additionally, Utilities can monitor consumers' real-time information about the amount of consumed energy, and credit the consumer for load shedding. Thus, the Smart Grid facilitates real-time supply and demand stability on the electrical grid [1].

A critical part of smart grids that is necessary to support any demand response program is its communication network. The communication network transfers important information to and from devices, as well as to energy providers and consumers. The Smart Grid is extremely heterogeneous and uses many protocols for communication. It consists of four types of communication networks, namely Home Area Networks, Field Area Networks, Local Area Networks, and Wide Area Networks, and thus is composed of protocols from all three of these types of networks. Well-known protocols include TCP/IP, Ethernet, Modbus, DNP3, MQTT, OpenADR, OPC, Wi-Max, and others. The protocols differ in their standards, implementations, intended applications, and the amount of security they provide. Unfortunately, many of these network standards do not include network security, and even if they do, often these features are disabled due to the inability of the computationally underpowered devices to implement them. Therefore, network communication and remote access capability have made smart grids more vulnerable to various cyber-attacks. Critical control systems such as power grids that can be compromised are a threat to the security of the country as well as a severe threat to human life [3].

Attacks on the data communications of a Smart Grid can skew the reporting of consumer behavior, and thus thwart the purpose of a DR program. This skew can fool the provider into giving the consumer either more or fewer credits than what they deserve as participants in the DR program. The results can be both increased provider costs and decreased consumer satisfaction. Determining how an attack affects the DR program can be very complex for the following reasons: 1) Various attacks exist, and each different type of attack can affect the network and, in turn, the demand response program in different ways, and 2) Many protocol layers exist, and determining how an attack at a particular layer affects the network, and thus the demand response program, can be difficult.

Attacks that occur in lower-layer protocols may exhibit no noticeable abnormal behavior in an upper-layer protocol or may exhibit abnormal behavior in an upper-level protocol that is not immediately attributed to an attack on a lower-layer protocol. For example, a TCP reset, to someone examining the TCP traffic logs, may simply look like an overloaded router at the IP layer, and may not be immediately recognizable as the result of a Denial of Service attack at the MODBUS layer. Proper forensics must be done to determine if the application or transport level behavior is due to an attack, network failure, device failure, or actual consumer behavior during a Demand Response event. The process of performing forensics to identify the patterns of network behavior is a time-consuming but important step to identifying an attack and taking necessary steps to mitigate the results of that attack and future attacks.

If the perceived consumer behavior indicates that further forensics investigation is necessary, then that investigation needs not only to be done thoroughly but also quickly. A quick but thorough investigation will allow a quick response to mitigate the possible attack. So any information, particularly information that is readily available or easily collected, that can be used to shorten the forensics investigation without compromising correctness is valuable. Therefore, we want to answer the following research questions:

- How does a network attack impact perceived consumer behavior, and therefore impact the DR program?
- What readily available network statistics can indicate whether a change in perceived consumer behavior is due to a network attack?
- How can readily available network statistics indicate the root cause? In particular, what network statistics can help focus the investigation by indicating where the attack may have occurred?

The goals of the work presented in this paper are to investigate an end-to-end application of DR to identify the impact on credits received by consumers who participate in the demand response program when the network is under attack and determine the relationships between the attack, the protocol affected, and the impact on the credits received. Additionally, we evaluate if correlations between network behavior and consumer power consumption indicate that further investigation of the network is advisable in order to see if tampering was done that caused deviation in consumer consumption. Furthermore, we then investigate whether network behavior can provide clues that indicate which communication layer was attacked. For this initial investigation, we chose the popular protocols Modbus over RS485, TCP, and HTTP.

We have found that correlations between network behavior and perceived consumer behavior exist and can be measured, that certain correlations may indicate tampering, and that these correlations differ according to the network layer attacked.

The rest of this paper is organized as follows. Section 2 describes the related works. Section 3 describes our threat model. Section 4 explains the research methodologies and experimental setup. Section 5 gives the experimental results. Section 6 presents our analysis of those results. Section 7 presents our future work, and Sect. 8 describes our conclusions.

2 Related Works

Our research is closely related to work in the fields that includes *smart grid communication*, *protocols*, *vulnerabilities and attacks*, *network traffic analysis*, *demand response and consumer behavior*, *attacks on DR programs*, and *correlation statistics*.

A smart grid like any other commodity network contains multiple layers. These layers typically implement one or more of the layers defined in the OSI (Open Systems Interconnection) model that includes a physical layer, data link layer, network layer, transport layer, session layer, presentation layer, and application layer. The communication between each of these layers requires a set of rules and standards within a network that is known as protocols [11]. Various works have been done in the area of Smart Grid protocols and vulnerabilities. In [5], the authors discuss industrial control system protocols, standards, and vulnerability and countermeasures for protocols that include DNP3, Modbus, IEC 60870-5-104, IEC 61850, IEC 61400-25 as well as IEEE C37.118. In [8], the authors provide a comprehensive list of smart grid protocols that are related to different areas of the environment that includes wide area monitoring, distributed generation, substations, demand response programs, metering, electric vehicles, and cyber security. In [6], the authors focus mainly on IoT-specific protocols that include the Manufacturing Message Specification (MMS) protocol and the eXtensible Messaging and Presence Protocol (XMPP), and their standards and vulnerabilities in a smart grid communication network. The authors give suggestions on what protocols should be used for LAN (Local Area Network) and WAN (Wide Area Network).

Communication network protocols have vulnerabilities that make them a target for various attacks. The vulnerabilities exist due to the simplicity of protocol standards that lack encryption, authentication, congestion control, etc. Various works have been done in the area of Smart Grid attacks. In [14], the authors discuss in detail the impact of dDoS attacks on AMI (Advanced Metering Infrastructure) networks and categorize attacks into hardware-based attacks, software-based attacks, and network topology-based attacks, and depict them using a simulation. In [3], the authors focus on different attacks that included flooding and spoofing attacks and use the round trip time to identify the attacks. In [9], the authors provide a survey of various DoS attacks and solutions to address the attack in a Smart Grid. The authors also categorize the DoS attacks into four categories, namely Layers, Targets, Techniques, and Sources, and provide defense methodologies for each category. In [10], authors discuss solutions to detect distributed DoS attacks in AMI networks using honeypots to provide an infrastructure to gather, detect and, prevent the attacks. In [13], the authors evaluate the impact of Dos/dDos attacks on AMI networks using packet queue length and received TCP packets. The authors discuss Demand Response but do not provide any detail on the impact on perceived consumer behavior due to an attack. In [12], the authors introduce a framework for cyber-physical systems and testbeds that will detect and mitigate DoS and MITM attacks. In [15], authors classify DoS attacks into power grid segments, networking layers, communication protocols, and power grid applications, and provide a remedial approach for counterbalancing the attacks.

Network traffic has to be thoroughly analyzed in order to detect an attack. Network packets have to be examined carefully for packet injection and modification attacks Various tools and methods have been used to detect an attack.

Various works have been done in the area of network analysis. In [2], the authors describe a Linux-based tool that detects basic packet sniffing for analysis of abnormality in network activity. In [3], the authors survey various network traffic analysis tools that include Wireshark, NTOP, Microsoft Message Analyzer, and Paessler Router Traffic Grapher (PRTG), their advantages and disadvantages, and the cases where they should be used and avoided. In [4], the authors provide an extensive comparison of widely used network analyzing tools that include Wireshark and tcpdump, and the authors describe the advantages of Wireshark over tcpdump. The authors present a strong case for using Wireshark.

Demand Response programs implemented in a smart grid network provide opportunities to reduce both costs and load. In [17], the author surveys Demand Response and smart grids and provides a detailed account of Demand Response that includes different pricing models, classification of consumers based on power consumption, examples of advanced metering infrastructures, and benefits of Demand Response programs. In [18], authors provide an optimization model that supports RTP (Real Time Pricing) which consumers can utilize to adjust their consumption and get the maximum benefits. In [16], authors provide a literary review of the benefits and challenges when incorporating Demand Response on a power system and provide insight into modeling Demand Response programs based on fixed linear price-demand relationship for price-responsive demand. Other works that discuss on challenges of implementing an effective Demand Response program included but are not limited to [27–29].

In [23], the author discusses the importance to fulfill consumer needs by examining the role of AMI and perform a thorough analysis of data to cluster by consumer behavior and load pattern disaggregation. In [19], the authors discuss the hurdles in implementing an effective Demand Response program for residential customers and propose a hybrid engineering approach that will make use of social psychology and economic behavior model to address those hurdles. In [7], the authors propose a method to model physical system parameters, external influences, and user behaviour and preferences. They then use the model to determine if the consumer behavior is normal or abnormal. In [22], authors focus on time series data that incorporate the Auto Correlation Function (ACF) and the Auto Regression Integrated Moving Average (ARIMA) to evaluate consumer behavior.

In [24], the authors focus on the attack on Demand Response programs by dynamically altering the load and using trajectories to simulate the attacks in a testbed. In [25], the authors propose a method to apply the Stackelberg leadership model that aims at increasing productivity for energy providers and customers. Furthermore, they simulate a price modification attack that impacts the Demand Response program and provide a solution for the attack. In [26], the authors propose machine learning algorithms to detect false data injection attacks that will impact pricing in the residential Demand Response program.

Correlation is a statistical way of measuring the dependence between two variables and identifying anomalous behavior in the network. Various work has been done in employing correlation for anomaly detection. In [20], the authors discuss correlation-based anomaly detection for ICS (Industrial Control Systems). The anomaly detection method employs recurrent neural networks along with correlation analysis methods. The correlation is analyzed between the fields in the dataset. In [21], authors propose the use of the information theory method Kolmogorov Complexity for detecting DoS attacks that uses Deep Packet Inspection (DPI).

At the time of the writing of this paper, little work has been done on the effect of perceived consumer behavior and network attacks, and these works do not consider the complex relationship between network layers. These complex relationships could lead to new ways of early detection and mitigation of the attacks. The challenges for determining these relationships are 1) integrating DR programs and the needed network infrastructure into smart grid simulations, which requires co-simulation and/or hardware-in-the-loop as discussed in Sect. 4.1, 2) statistically analyzing the correlation between consumer behavior and network behavior to identify anomalous behavior for an attack, and 3) statistically analyzing the relationships between one network layer and another during an attack.

3 Threat Model

The threat model which pertains to our work is shown in Fig. 1. Three types of actors exist in the threat model: a provider, consumers, and attackers. The provider is a utility company that is responsible for the generation, transmission, and distribution of power to consumers. Consumers receive power from the utility. They also receive updates via the communication network and notifications from the provider about their power consumption. An important notification that the consumer receives from the provider is a notification of a DR event, which is a near future period when the load on the smart grid will be high. The notification gives the consumer the opportunity to respond by reducing their consumption. Likewise, the network provides information and notifications to the provider. However, the provider information and notifications are data about the operations of the smart grid and the power consumption of the consumers. The provider collects information about the consumers' power use through an Advance Metering Infrastructure that consists of smart meters that measure consumer consumption. The information/notifications may travel from one device to another using multiple protocols before it reaches the provider.

Attackers attack the data network of the provider. The intent of the attacker is to fool the provider into believing that consumers are behaving differently from the consumers' actual behavior. The end result is that the consumers are rewarded either too much or too little for their participation in the DR program. The attacker has the ability to monitor and modify network traffic at any protocol layer. In other words, attackers can inject frames into the data link layer, inject packets into the network or transportation layer, fake field devices, and spoof IP addresses.

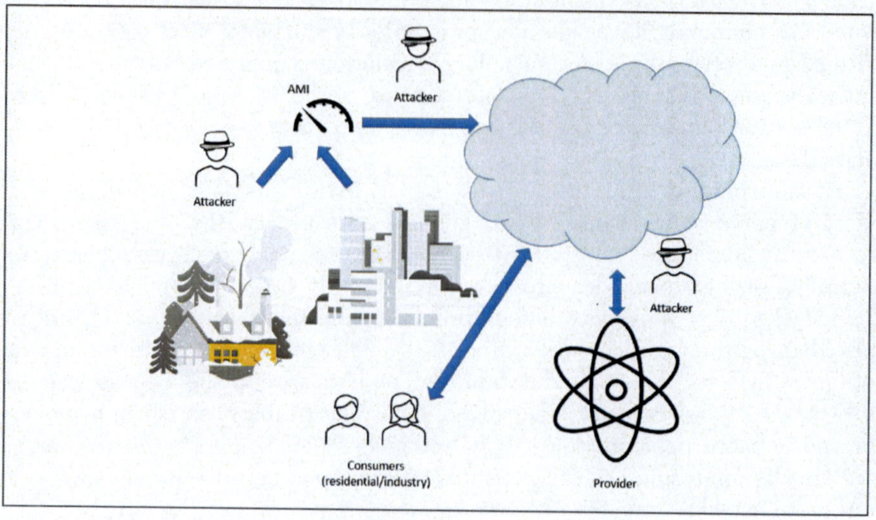

Fig. 1. Threat model: attackers can attack any protocol layer.

4 Research Methodology and Experimental Setup

Our research methodology used collected, real-time consumer data to drive a simulation of a simplified smart grid that used a common data link layer protocol and protocols used in the OpenADR specifications (MODUS, TCP, HTTP). Details of our experimental setup and research methodology follow.

4.1 Experimental Setup

So that we could properly investigate network behavior, we conducted our experiments using a smart grid testbed that simulates consumer behavior as well as smart grid devices and network behavior in software. It also allows the connection of hardware for Hardware-In-the-Loop (HIL) experiments. The testbed consists of physical smart meters (Schneider PM870), a simulated grid in Typhoon HIL [32], and OpenADR-compliant Virtual Top Nodes (VTN) and Virtual End Nodes (VEN) [30] that run OpenLeadr [31]. The VEN runs on a Raspberry PI and reports to the VTN. The VTN calculates DR credits and generates billing information and other Statistical reports. The communication between the VEN and the VTN uses HTTPS and the TCP protocol. The VEN uses MODBUS as the data link layer on top of RS485 as the physical layer to pull data from the smart meters. Typhoon HIL was used to design, implement and simulate smart grid SCADA models and will directly interact with smart meters as well as software-based simulated meters. The data for consumer behavior is based on anonymized data obtained during an actual DR program that was provided by a partnering utility.

4.2 Research Methodology

We ran our simulations on the data and established a baseline behavior for consumer power consumption.

Our choice of attack was a dDoS attack because it is widely used in many network-based attack simulations, and it is a common one indicated in the literature as discussed in [3,9,10,12–15,21].

We ran the simulation and performed DoS (Denial of Service) attacks for each of the protocols (MODUS and TCP), and collected network statistics for analysis of the correlations to perceived consumer behavior. Our main parameters included the frequency of polls from VENs to VTN, frequency of power consumption report generation from VENs, how often attacks occur for a given time period, duration of attack that included short and long time periods, and the number of packets injected.

For this paper, we focus on throughput analysis. We used a statistical method to help determine if a relationship exists between the behavior of the network and the perceived behavior of the consumer. Correlation is a statistical way of measuring the dependence between two variables. Correlation coefficient values are in the ranges from −1 to +1. The value of +1 denotes that the compared variables are a perfect positive correlation, 0 denotes no correlation between the variables, and −1 denotes a perfect negative or inverse correlation between the variables. We chose throughput in this study because it is an easily obtainable network metric that can be computed by using common tools. Throughput is the total number of network packets that were successfully transmitted without any errors for a specific time period. The correlation between consumer consumption and throughput was computed to evaluate the network behavior based on whether they are positively correlated, negatively correlated, or had no correlation. We evaluated the consumer behavior during the Demand Response event when an attack occurred and compared the correlation statistic to the baseline correlation when there was no attack, which was a zero correlation.

5 Experimental Results

Our experiments were conducted based on the research methodology given in Sect. 4.2 and experimental setup in Sect. 4.1. Our simulation performed DoS attacks at the MODBUS and TCP/IP layers, but all the network statistics were computed at TCP/IP layers. Collecting TCP/IP layer statistics is more likely to occur in real-world scenarios as TCP/IP layers are easier to measure and collect than retrieving network measures from the data link layers of remote field devices, which may number in the thousands.

First, we ran the simulation to determine the baseline usage during normal hours and during Demand Response event days. The statistical model was used to predict consumer behavior and was compared to *perceived consumer behavior*. Perceived consumer behavior is the behavior of the consumers as reported by the VTN. Wireshark was used to capture the network traffic and in the generation of throughput. Figure 2 shows perceived consumer behavior to statistical model predicted behavior during a Demand Response event.

5.1 Consumer Behaviour

Figure 2 shows the perceived consumer behavior compared to model prediction during the Demand Response event day.

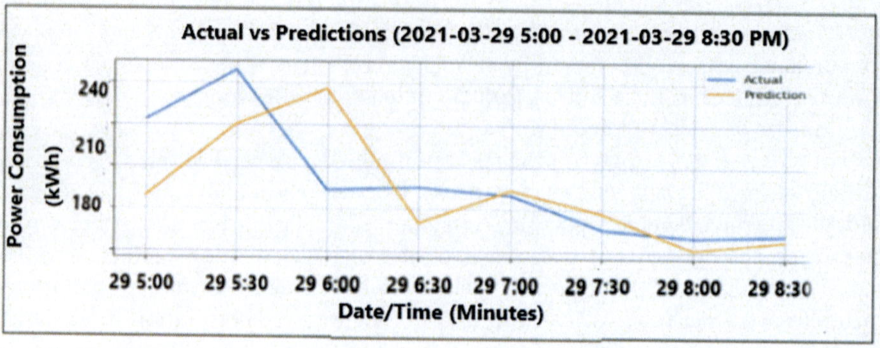

Fig. 2. Perceived behavior versus normal (predicted) behavior.

For the first experiment, DoS attacks were performed on the MODBUS protocol. We used ModPoll an open-source tool that will continuously poll the meter. The DoS attacks were performed at various time intervals. In the experiment, the VEN is polling the meters to obtain the data. Due to the increased number of MODBUS frames, the meter will begin to fall behind when responding to the VEN's polling. In other words, responses from the meter will slow, and the inter-arrival times between responses will increase. Likewise, because the VEN sends reports to the VTN once it has data from all the meters, those reports will also slow and the inter-arrival times between responses will increase. Eventually, the VEN will receive no response from the meter, and, in turn, will send no reports of the meter readings to the VTN.

We computed throughput measurements and compared them to power consumption during the time intervals. The time intervals included the time of the attack. As shown in Fig. 3, the throughput dropped during the attack. Likewise, power consumption also dropped. Power consumption when the attack occurred is marked as a red 'x'. The lack of reports from the VEN eventually results in the VTN computing an average consumption of zero.

Correlation statistics were calculated on throughput and power consumption. First, we ran the simulation to show the throughput and power consumption during normal network activity when no attacks occurred. *No Correlation* graph shown in Fig. 4 shows no fluctuation (straight red line) in the throughput denoting normal network behavior and no significant changes to normal activity. Therefore, No correlation exists between network activity and consumer

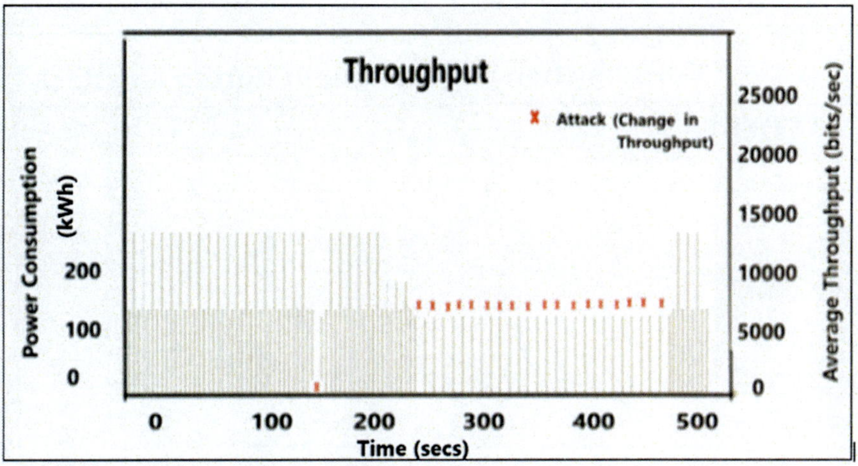

Fig. 3. Throughput.

consumption when the network is not being attacked. The lack of correlation is due to the regular activity of the VEN polling the meters, and that activity does not change regardless of the meter readings.

Then we performed DoS attacks during various time intervals that included a Demand Response event and evaluated the correlation between throughput and power consumption. Again, the throughput and power consumption go down as the attack time interval increases. Eventually, both of them become zero. This behavior causes a *positive correlation* between throughput and power consumption during the time interval when an attack occurs. Positive correlation denotes that an increase or decrease in the values of one variable will have the same impact on the second variable as shown in Fig. 5.

Consumers receive credit when they participate in the Demand Response program and reduce their consumption during a DR event. Even though the consumer did *not*, in this case, positively respond during the DR event, they received a *$220 credit*. Figure 6 shows actual consumer behavior to statistical model predicted behavior during the DR event. The reason for the miscalculation of consumer credits is due to the DoS attack impacting the perceived power consumption decrease during the demand response period, thus giving higher credits to consumers than they actually would have received.

Our next DoS attack was performed on TCP/IP layer by using a packet-injecting tool that generates packets and floods the network layer with requests to the VEN, causing an increase in the throughput in TCP/IP layer. The increase in throughput is due to the flooding of packets into the network and causing an increase in network activity and throughput. As before, we performed the DoS attacks during various time intervals that included demand response events and evaluated the correlation between throughput and power consumption. The DoS attacks caused increased throughput in the TCP/IP layer and made the VEN

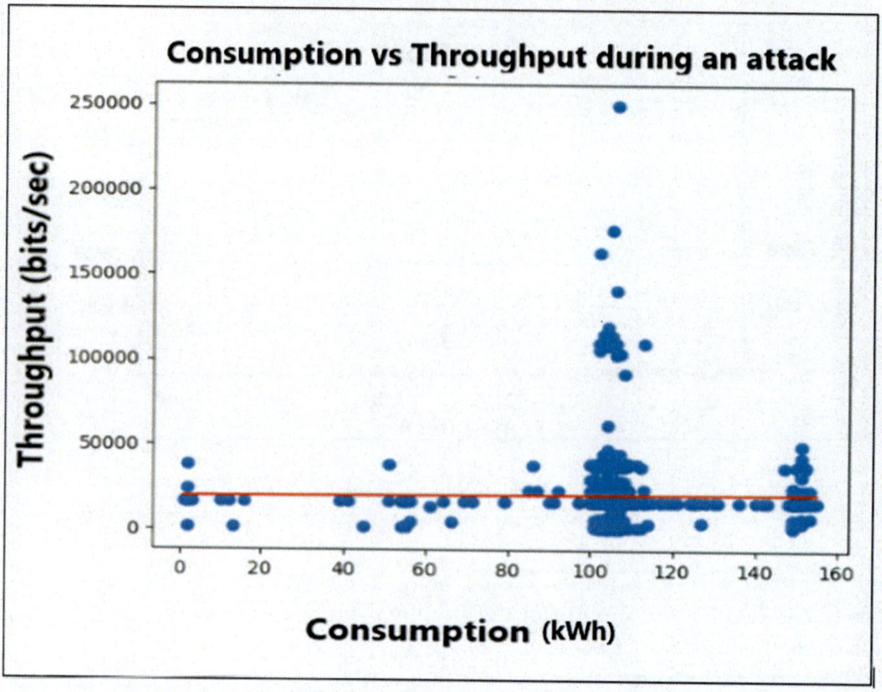

Fig. 4. Normal network behavior - no correlation. (Color figure online)

unable to send meter readings to VTN, and power consumption decreased to zero. Figure 7 shows the negative correlation between throughput and consumer power consumption. A negative correlation denotes that an increase or decrease in the values of one variable will have an inverse impact on the second variable.

Again, even though the consumer did *not* positively respond during the DR event, they received a *\$220 credit*. Figure 8 shows actual consumer behavior to statistical model predicted behavior during the DR event. Similar to the DoS attack at the MODBUS layer, the reason for this perceived consumer behavior is that the VTN calculated consumption based on the data obtained in reports from the VEN and, as those reports became less frequent, the average consumption decreased. Thus, the consumer was given more credits than they actually should have received.

6 Analysis

To accomplish our goals discussed in Sect. 1, we performed Dos attacks on both the MODBUS and TCP/IP layers during the time interval that also included the Demand Response event. Our metrics included the evaluation of throughput, the correlation between throughput and power consumption, and consumer behavior during the Demand Response event. The experimental results discussed in Sect. 5

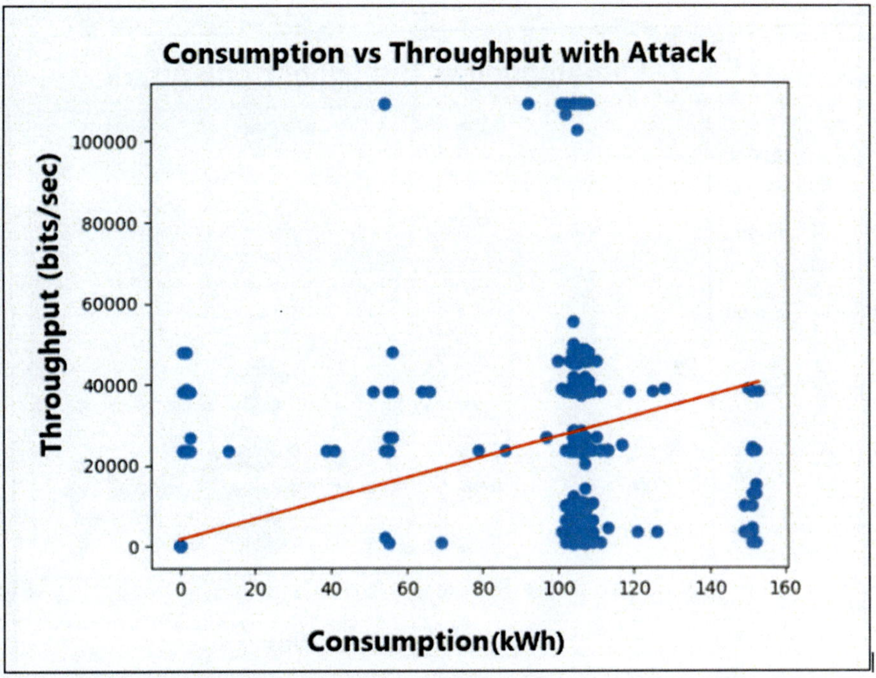

Fig. 5. MODUS DoS attack - positive correlation.

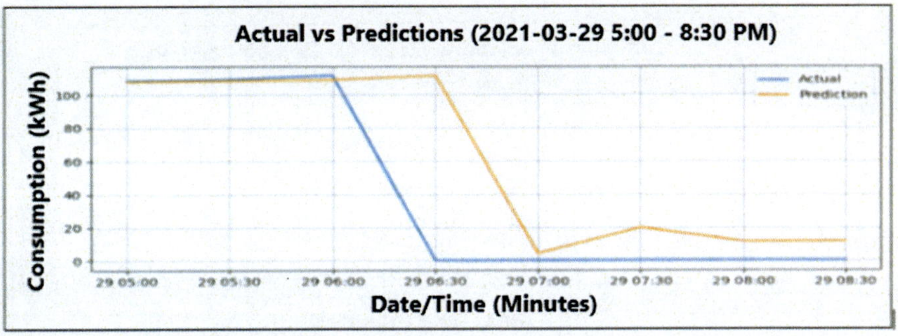

Fig. 6. Demand response - high consumer credits.

show that the attack on both protocols caused power consumption to go down to zero during the Demand Response period causing the increase in credits given to the consumer more than they actually would have received. Our second goal was to analyze if the correlation between throughput and power consumption can provide insight into network behavior. Our experiment results showed that there was a positive correlation when DoS attacks were performed in MODBUS layer because of the lack of meter readings sent to the VEN which in turn, caused the VEN to fail to send reports to the VTN, and both throughput and

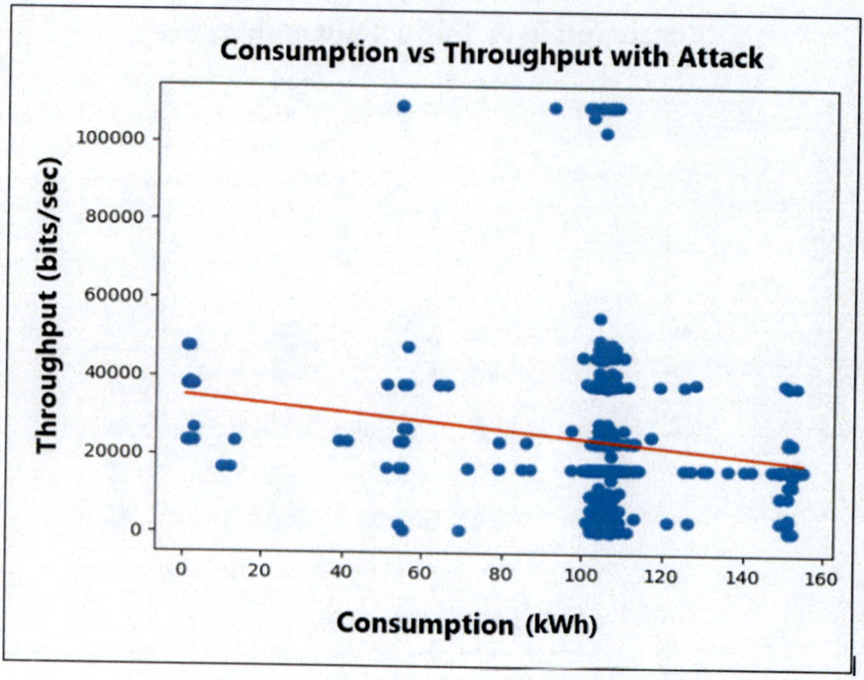

Fig. 7. TCP/IP layer DoS attack - negative correlation.

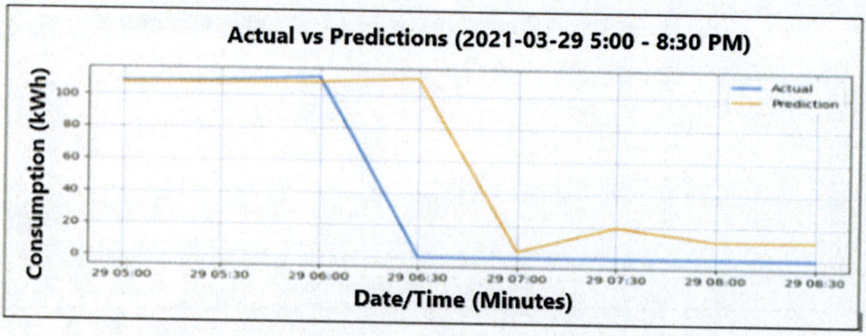

Fig. 8. Demand response - high consumer credits.

perceived power consumption dropped (and eventually became zero). The inverse correlation occurs when DoS attacks were performed in TCP/IP layers due to packet flooding causing the increase in throughput in network traffic, but power consumption decreased because the VEN was unable to send meter readings to VTN.

No real correlation between throughput and power consumption exists during normal behavior. Our results showing a positive correlation in the MODBUS layer indicates that forensic analysis should begin at the MODBUS layer. Likewise, a negative correlation indicates that forensic analysis should begin at the TCP/IP layers.

In summary, we found that there was a correlation between network throughput and perceived consumer behavior, and the type of correlation, either positive or negative, indicated where the DoS attack originated. This information can be used to narrow the search for which layer to look for that behavior and is valuable to reduce the time taken for the forensics investigation.

7 Future Works

Our future work will investigate various attacks that are specific to various smart grid protocols at different communication layers and network behavior. Our analysis will include not only network statistics but other metrics that can guide us in evaluating perceived consumer behavior and behavior during an attack. We will also evaluate Demand Response scenarios that will incorporate automated Demand Response to determine the exact layer where the attack originated and where the forensics has to start.

8 Conclusion

Creating effective DR programs requires identifying any anomalous network and consumer behavior quickly to prevent huge financial losses for providers and improve consumer satisfaction and trust. We were able to perform DoS attacks on MODBUS and TCP/IP layers protocols at different communication layers and experiment results showed that both attacks caused the perceived power consumption to decrease during the Demand Response period which in turn, resulted in giving consumers higher credits than they deserved. Experimental results also showed how positive and negative correlations between throughput and consumption can be applied to identify abnormal network activity during the attack and guide us to the origin of the attack and where the forensics analysis should focus instead of brute force analysis of every communication layer.

References

1. El Mrabet, Z., et al.: Cyber-security in Smart Grid: survey and challenges. Comput. Electr. Eng. **67**, 469–482 (2018)
2. Qadeer, M.A., Iqbal, A., Zahid, M., Siddiqui, M.R.: Network traffic analysis and intrusion detection using packet sniffer. In: 2010 Second International Conference on Communication Software and Networks, pp. 313–317 (2010). https://doi.org/10.1109/ICCSN.2010.104
3. Mistry, D., Modi, P., Deokule, K., Patel, A., Patki, H., Abuzaghleh, O.: Network traffic measurement and analysis. In: 2016 IEEE Long Island Systems, Applications and Technology Conference (LISAT), pp. 1–7 (2016). https://doi.org/10.1109/LISAT.2016.7494141

4. Goyal, P., Goyal, A.: Comparative study of two most popular packet sniffing tools-Tcpdump and Wireshark. In: 2017 9th International Conference on Computational Intelligence and Communication Networks (CICN), pp. 77–81 (2017). https://doi.org/10.1109/CICN.2017.8319360
5. Xu, Y., Yang, Y., Li, T., Ju, J., Wang, Q.: Review on cyber vulnerabilities of communication protocols in industrial control systems. In: 2017 IEEE Conference on Energy Internet and Energy System Integration (EI2), pp. 1–6 (2017). https://doi.org/10.1109/EI2.2017.8245509
6. Tightiz, L., Yang, H.: A comprehensive review on IoT protocols' features in smart grid communication. Energies **13**(11), 2762 (2020)
7. Heussen, K., Tyge, E., Kosek, A.M.: Residential demand response behaviour modeling applied to cyber-physical intrusion detection. In: 2017 IEEE Manchester PowerTech, pp. 1–6 (2017). https://doi.org/10.1109/PTC.2017.7981209
8. Kuzlu, M., Pipattanasompom, M., Rahman, S.: A comprehensive review of smart grid related standards and protocols. In: 2017 5th International Istanbul Smart Grid and Cities Congress and Fair (ICSG), pp. 12–16 (2017). https://doi.org/10.1109/SGCF.2017.7947600
9. Huseinović, A., Mrdović, S., Bicakci, K., Uludag, S.: A survey of denial-of-service attacks and solutions in the smart grid. IEEE Access **8**, 177447–177470 (2020)
10. Wang, K., Du, M., Maharjan, S., Sun, Y.: Strategic honeypot game model for distributed denial of service attacks in the smart grid. IEEE Trans. Smart Grid **8**(5), 2474–2482 (2017)
11. Tanenbaum, A.S.: Network protocols. ACM Comput. Surv. **13**(4), 453–489 (1981). https://doi.org/10.1145/356859.356864
12. Chen, B., et al.: Implementing attacks for modbus/TCP protocol in a real-time cyber physical system testbed. In: 2015 IEEE International Workshop Technical Committee on Communications Quality and Reliability (CQR). IEEE (2015)
13. Sgouras, K.I., Birda, A.D., Labridis, D.P.: Cyber attack impact on critical smart grid infrastructures. In: ISGT 2014. IEEE (2014)
14. Asri, S., Pranggono, B.: Impact of distributed denial-of-service attack on advanced metering infrastructure. Wireless Pers. Commun. **83**, 2211–2223 (2015)
15. Huseinovic, A., Mrdovic, S., Bicakci, K., Uludag, S.: A taxonomy of the emerging denial-of-service attacks in the smart grid and countermeasures. In: 2018 26th Telecommunications Forum (TELFOR), Belgrade, Serbia, pp. 1–4 (2018). https://doi.org/10.1109/TELFOR.2018.8611847
16. O'Connell, N., Pinson, P., Madsen, H., O'Malley, M.: Benefits and challenges of electrical demand response: a critical review. Renew. Sustain. Energy Rev. **39**, 686–699 (2014). https://doi.org/10.1016/j.rser.2014.07.098. ISSN 1364–0321
17. Siano, P.: Demand response and smart grids-a survey. Renew. Sustain. Energy Rev. **30**, 461–478 (2014)
18. Conejo, A.J., Morales, J.M., Baringo, L.: Real-time demand response model. IEEE Trans. Smart Grid **1**(3), 236–242 (2010). https://doi.org/10.1109/TSG.2010.2078843
19. Gyamfi, S., Krumdieck, S., Urmee, T.: Residential peak electricity demand response-Highlights of some behavioural issues. Renew. Sustain. Energy Rev. **25**, 71–77 (2013)
20. Jadidi, Z., Pal, S., Hussain, M., Nguyen Thanh, K.: Correlation-based anomaly detection in industrial control systems. Sensors **23**(3), 1561 (2023)
21. Kim, S., Kim, B.C., Lee, J.Y.: DDoS analysis using correlation coefficient based on Kolmogorov complexity. In: Park, J.J.J.H., Arabnia, H.R., Kim, C., Shi, W.,

Gil, J.-M. (eds.) GPC 2013. LNCS, vol. 7861, pp. 443–452. Springer, Heidelberg (2013). https://doi.org/10.1007/978-3-642-38027-3_47
22. Hansen, P.R., Lunde, A., Lunde, A.: Estimating the Persistence and the Autocorrelation Function of a Time Series that is Measured with Error. CREATES Research Paper No. 2010–8, 4 February 2010. Available at SSRN: https://ssrn.com/abstract=1550269 or https://doi.org/10.2139/ssrn.1550269
23. Chicco, G.: Consumer behavior and data analytics. In: 2016 International Conference and Exposition on Electrical and Power Engineering (EPE), Iasi, Romania, pp. 771–779 (2016). https://doi.org/10.1109/ICEPE.2016.7781443
24. Amini, S., Mohsenian-Rad, H., Pasqualetti, F.: Dynamic load altering attacks in smart grid. In: 2015 IEEE Power & Energy Society Innovative Smart Grid Technologies Conference (ISGT). IEEE (2015)
25. Maharjan, S., Zhu, Q., Zhang, Y., Gjessing, S., Basar, T.: Dependable demand response management in the smart grid: a Stackelberg game approach. IEEE Trans. Smart Grid **4**(1), 120–132 (2013). https://doi.org/10.1109/TSG.2012.2223766
26. Dayaratne, T., et al.: False data injection attack detection for secure distributed demand response in smart grids. In: 2022 52nd Annual IEEE/IFIP International Conference on Dependable Systems and Networks (DSN). IEEE (2022)
27. Samad, T., Koch, E., Stluka, P.: Automated demand response for smart buildings and microgrids: the state of the practice and research challenges. Proc. IEEE **104**(4), 726–744 (2016)
28. Nolan, S., O'Malley, M.: Challenges and barriers to demand response deployment and evaluation. Appl. Energy **152**, 1–10 (2015)
29. Medina, J., Muller, N., Roytelman, I.: Demand response and distribution grid operations: opportunities and challenges. IEEE Trans. Smart Grid **1**(2), 193–198 (2010)
30. Open Automated Demand Response (OpenADR). https://www.openadr.org
31. OpenADR implementation for Python. https://openleadr.org/docs/
32. Typhoon HIL. https://www.typhoon-hil.com/

Computational Approaches to Persuasion Detection and Potential of Use in Social Engineering

Subia Ansari[1](✉) and Mohammad Zaiyan Alam[2]

[1] Purdue University, West Lafayette, IN 47906, USA
ansari0@purdue.edu

[2] Information Sciences Institute, University of Southern California, Los Angeles, CA 90007, USA
mzalam@isi.edu

Abstract. Social Engineering (SE) is a major threat to cybersecurity and seeks to manipulate a target to divulge information or take an action that may or may not be in their best interest. One of the ways social engineers manipulate their targets to perform an action is by utilizing persuasion techniques that prey on qualities of human nature. In this paper, a comprehensive literature review was conducted to analyze current computational approaches to persuasion detection and identify what approaches may address their limitations in the SE context. We found that persuasion detection has so far been studied in different contexts: argumentative discourse, recommendation systems, propaganda, and SE. We review a total of thirteen approaches, discuss their advantages and disadvantages and surmise approaches that may be feasible to address the gaps identified based on current advancements in the field. We also observe that currently, only a few approaches to persuasion detection in SE exist and typically rely on basic feature engineering. The reviewed approaches in other domains, however, demonstrate the potential of deep learning to capture the essence of persuasion.

Keywords: Persuasion Detection · Social Engineering · Phishing · Principles of Influence

1 Introduction

The study of persuasion, a process of social influence, investigates different ways in which one can influence an individuals' attitudes and behaviors [1]. Recent technological advancements show that computer systems can influence an individual to change their attitudes and behavior. Persuasion has thus found its application in commerce through technology – mainly emphasizing on leveraging it for social good and encouraging positive behavior such as healthy lifestyle choices, reduce speeding, and in advertising sector [2–4]. In argumentative discourse, the goal of computational persuasion is capturing human ability to persuade by considering the arguments themselves, beliefs and concerns of the individual being persuaded, and strategies to maximize likelihood of

K. Arai (Ed.): FTC 2023, LNNS 816, pp. 394–409, 2023.
https://doi.org/10.1007/978-3-031-47448-4_30

being persuaded [4]. Despite positive applications, computational persuasion also has negative applications such as propaganda in political campaigns, social media, and in phishing through social engineering (SE). Propaganda utilizes persuasive messages to advance a specific agenda and often exploits social networks to reach users sensitive to specific topics. In this context, persuasion techniques are often a subset of a larger group of propaganda techniques used to leverage human emotions that are modeled using deep learning approaches [5]. Computational approaches to persuasion have found success in different domains as discussed above, and their potential of application in SE seems promising, however, current approaches in this domain are limited.

In information security, SE exploits weakness in human nature by utilizing persuasion techniques to obtain unauthorized information [6, 7]. SE attacks are typically carried out through phishing, where persuasion techniques are used to not only trick users by appearing legitimate, but also avoid detection by spam filters [8, 9]. It can be argued that detection of SE phishing attacks is not a trivial problem as only those SE attempts with malicious intent must be detected. Research on SE detection is limited and typically relies on approaches that require natural language processing (NLP) techniques such as question extraction, topic identification [10], and predefined blacklists of words [11, 12]. Some approaches were able to successfully determine requests or 'asks' in SE attacks [13] with the aim to countering such attempts. Automated detection of phishing emails has also been studied by researchers and these techniques utilize email structure such as script-based features which look for pop-up window codes, URL-based characteristics that look for IP address used instead of domain names and check for spelling errors [14, 15]. Deep learning methods have also been used to tackle this problem [16, 17], however, they do not explain why an email was classified as phishing – which is a potential gap that persuasion detection could address. Researchers have argued that identification of phishing emails must incorporate persuasive content scrutiny rather than mere evaluation of structure of emails as it may provide explainability about their classification [18–20]. Deeper understanding and detection of the persuasion techniques may help achieve effectiveness in detecting phishing and in training users by increasing their awareness. In this research, we try to address challenges to computational persuasion in social engineering by considering the following research questions:

RQ1: What are the advantages and disadvantages of existing methodologies in computational approaches to persuasion?

RQ2: How much, if any, do these approaches draw from pertinent information from non-computational fields?

RQ3: What research methods/approaches would be feasible to address some of the gaps identified in RQ1 in the context of social engineering?

Social Engineering is a major threat to cybersecurity that has impacted not only individuals but also corporations and governments. Since these attacks target vulnerabilities in human behavior, approaches to SE detection have been focused on training humans to recognize such attacks. However, SE continues to be successful in phishing and misinformation campaigns with very little research in how advancements in NLP and deep learning can be used to address the problem. Based on the analysis of existing literature on persuasion in SE, this paper will summarize current approaches to persuasion and postulate what approaches might be feasible to address the identified gaps. This paper tries to address how the problem of SE based phishing can be combated using the rapid advancements in NLP. The significance of this paper is that it helps map out a niche field in technology to help advance research community's efforts towards solving the problem of SE by using persuasion detection to address the explainability gap.

The methods discussed in this study are not to be considered exhaustive as both SE attack techniques and NLP techniques are evolving rapidly. This work will help provide the research community with an overview of current landscape in persuasion detection and possible future avenues of research. In Sect. 2, the paper discusses foundations of persuasion and lays out the theoretical basis for pursuing persuasion detection. Section 3 describes the research methodology. In Sect. 4, an analysis of current approaches is presented, and gaps are identified through a systematic literature review. Section 5 discusses potential methods to address these gaps and finally, in Sect. 6, conclusions of this research are discussed.

2 Overview of Persuasion Literature

2.1 Theoretical Foundations of Persuasion

Persuasion principles were studied in depth by Cialdini [1] who primarily focused on major factors that target compliance with a request, which are useful in detecting phishing attempts that employ SE. Cialdini [1] describes six factors that influence behavior as follows - *reciprocation, consistency/commitment, social proof/social validation, liking/similarity, authority*, and *scarcity* (see Table 1). Cialdini model was further extended by adding *distraction* as a principle of influence pertinent to phishing [6].

Table 1. Cialdini's principles of persuasion.

Principle of Persuasion	Description
Reciprocation	Repaying favors out of obligation
Consistency	Being consistent with past behaviors
Social Proof	Being guided by the behavior of others
Liking	Complying with individuals' someone likes
Authority	Complying with authoritative requests
Scarcity	Attributing more value to limited availability

Marwell and Schmitt [21] also studied persuasion in terms of gaining compliance and identified the following sixteen techniques:- *promise of a reward* if the person being persuaded complies, *threat of punishment* for non-compliance, *positive/negative expertise* where the individual is awarded/punished due to the nature of things, *liking, pre-giving* where reward is given before compliance, *aversive stimulation* which involves punishing the target until they comply, *debt* where compliance is owed due to past favors, *moral appeal* where non-compliance is equated to immorality, *self-feeling (positive/negative)* where the target is made to feel better/worse for compliance/non-compliance, *altercasting (positive/negative)* is implying that a "good"/"bad" person would comply/not comply, *altruism* is seeking compliance for someone else's sake, *esteem (positive/negative)* where others would think better/worse if they complied/didn't comply. These techniques were grouped into five major clusters of: *rewarding activity, punishing activity, activation of impersonal commitments, and activation of personal commitments.* Overall, these factors were grouped into socially acceptable and unacceptable factors of persuasion.

Specifically for the SE context, Gragg [22] defined psychological triggers that "exhibit some kind of power to influence or persuade people" which are include: creating a *strong affect* in the beginning of the conversation, *overloading* the target with information, *reciprocation, deceptive relationships* formed with the intent to deceive, *diffusion of responsibility and moral duty* to others when compliance is gained, *authority and integrity,* and *consistency.*

Other persuasion techniques have been discussed by researchers in social and cognitive psychology; however, the three models discussed above are widely accepted models in the research community and have overlapping common factors. Cialdini's model is the most widely accepted and researched model in this domain and represents the six core persuasion principles represented in Table 1. As we will later see, most computational approaches to persuasion draw information from the Cialdini model and a combination of the models discussed above.

2.2 Persuasion in Information Technology

Several studies in the domain of human-computer interaction have attempted to understand the relationship between persuasion, SE, and phishing. They have focused on phishing susceptibility amongst victims and their personalities [23] and evaluating persuasion techniques utilized in phishing emails. A phishing simulation conducted in German municipalities found that the most effective persuasion techniques in their domain were authority, urgency, benefit, and danger, and that people succumb to spear phishing much faster when it uses persuasion [24].

A temporal analysis of the trends in usage of persuasion principles in phishing emails retrieved from three educational institutes over a period of five years found that usage of authority, liking, commitment, and scarcity had increased, while social validation and reciprocation had decreased over time [25]. They also found that an increasing number of emails now use logos to appear reliable and that request for replies over emails decreased as did the number of emails from financial institutions.

Another study showed that principles of persuasion are effective for phishing and identified authority and distraction principles are the main drivers for the success of an attack [7]. Researchers also found that certain persuasion techniques were frequently used together by studying different types of scams [20]. The combinations were commitment, reciprocation, consistency, and distraction in cases of Business Email Compromise (BEC), money transfer and travel bookings followed by liking, similarity, and deception principles in the case of romance scam settings. The study also indicates that word frequencies as a feature in phishing detection approaches may be useful [19]. Akbar [26] found that different persuasion techniques were used for different targets; likability was used to target social media users, consistency and social proof was used to target governments, and scarcity was used to target financial, and e-commerce sectors.

These studies reveal that a relationship may exist between the target's personality and persuasion used in phishing. Lawson et. al. [27] confirmed that such a relationship exists, for example, extraversion in individuals was predictive of their susceptibility to persuasion. Another study used signal detection theory to understand differences between persuasion principles and individuals' susceptibility to them and found that participants were most skeptical of authority and scarcity techniques – generating more false alarms for non-phishing attempts, and less skeptical to liking, with high miss rates for phishing attempts [28]. These experiments were conducted in a lab setting and may not represent real-world behavior. However, all studies conducted in this domain have found relationships between persuasion techniques and phishing/SE, hence providing theoretical grounds to explore automated detection of persuasion.

3 Research Methodology

In this paper, a systematic literature review was conducted by performing a comprehensive search on several databases to identify literature relevant to the domain. The following steps were followed:

1. Defining the research questions and search keywords.
2. Developing search strategy and inclusion and exclusion criteria.
3. Extracting, summarizing, and analyzing information from identified relevant research papers.

3.1 Search Strategy and Inclusion/Exclusion Criteria

To locate relevant research papers in persuasion detection, we identified our data sources as the following databases: ACM Digital Library, Google Scholar, Science Direct, and Web of Science. The keywords for search were defined based on research questions: persuasion detection, computational persuasion, influence detection, persuasion in social engineering, and social engineering detection. The search resulted in a total of 72 research articles that were filtered based on relevance to the research questions (see Fig. 1 and Table 2). We included some non-peer reviewed articles in persuasion detection from arXiv, as they are still moderated to gain diverse perspectives and additional context not covered in peer-reviewed articles, as this is an emerging area of research. However, non-peer reviewed articles have their limitations, including a lack of rigorous scrutiny and high bias. Despite this, including non-peer reviewed articles helps provide a comprehensive overview of approaches and identify new research avenues. In this paper, we acknowledge non-peer reviewed papers and discuss their limitations.

3.2 Evaluation and Analysis

The selected articles that met the inclusion criteria were reviewed for their content thoroughly; their methodology was summarized and then scrutinized to identify their limitations and advantages. Non-peer reviewed articles are acknowledged in the discussion of their limitations.

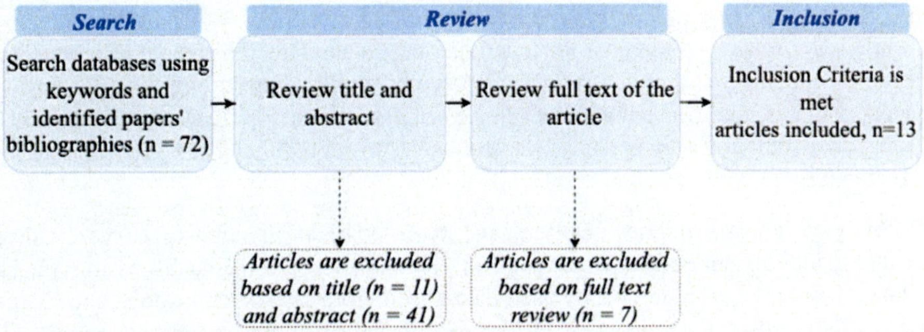

Fig. 1. Representation of search process and inclusion criteria.

Table 2. Inclusion/exclusion criteria.

Phase	Inclusion/Exclusion Criteria
Keyword Search	Include peer-reviewed articles, and only non-peer reviewed articles from arXiv
Title Review	Include articles that discuss persuasion
Abstract Review	Include technical articles that utilize persuasion to classify text
Full-text Review	Include technical articles that perform persuasion detection. For SE, include articles that consider persuasion in any capacity

4 Computational Approaches to Persuasion: A Review

4.1 Persuasion Detection in Argumentative Discourse

Statistical Approaches. Iyer et al. [29] proposed an unsupervised and domain-independent approach to detect persuasion principles in text using inherent sentence structure. Their process involves extracting arguments from text, classifying them into different persuasion tactics by using prototype parse strings for classification. During training, prototype parse strings were calculated to represent each persuasion principle using parse trees and removing leaf nodes that contain words. To determine if a given test argument was persuasive, its parse string was computed and compared with the prototype parse strings using edit distance and assigned to the persuasion category prototype string with least edit distance.

The advantages of this approach are that it does not require labeled data, obtaining which is a major challenge in SE due to privacy concerns. Its robustness and stability across datasets of different sizes, using four different datasets was demonstrated through a sensitivity analysis. The approach provides versatility by relying on parse strings and not lexical terms, i.e., it is not vocabulary dependent. The method was also able to avoid heavy computational costs and maintain good performance on a small subset of data and was faster than Doc2Vec by a factor of 1.5.

However, the model has limitations, such as only being tested on argumentative discourse domain dataset and not considering certain lexical terms like "because", "if", "while", etc., that may be persuasion technique-dependent rather than domain dependent. Another limitation is that it assumes that one argument contains only one persuasion tactic, which is not always the case even for sentence-level classification in SE context. Additionally, the fourteen persuasion techniques considered may not be appropriate for SE context.

Deep Learning Approaches. Dutta et. al. [30] proposed an attention-based hierarchical LSTM to identify successful persuasive conversations in social media discussions on the ChangeMyView sub-reddit. They used data-specific features, such as delta and karma scores, to formulate the task of persuasion modeling as a sequence classification problem. The model constructs a vector representation for each sentence in a chain of comments and computes weight distribution over all sentences. The comment space is mapped to a binary classification of being awarded a delta score or not, i.e., successful persuasion or not. While this approach demonstrated that addition of attention mechanism to an LSTM improves its performance, it uses dataset specific delta and karma scores that do not apply to SE persuasion and cannot identify the persuasion technique used as it only performs binary classification (persuasion or no persuasion). It also requires a large sequence of sentences to draw conclusions from data, which not only poses a challenge in the SE context but can also be computationally intensive and domain dependent.

Hidey et. al. [31] researched persuasion taking into consideration its subjective nature. They also used the 'ChangeMyView' dataset and developed an LSTM model to predict the persuasiveness of an argument as a binary classification problem. The model used a weighted average of word embeddings and semantic frames from FrameNet [32] to represent sentences and showed promise in capturing long-term dependencies and implicit social interactions that may not be explicitly expressed in words. However, the model is limited to binary classification and does not indicate which persuasion technique was used. It is also heavily dependent on the domain and the dataset (using karma scores) and may not be generalizable to the SE context.

Feature Engineering and Machine Learning Approaches. Shi et. al. [33] proposed Lumen, a machine learning framework, which aims to detection persuasion cues used as well as framing and objectivity or subjectivity of the argument. They leveraged linguistic inquiry word count (LIWC), sentiment analysis, and topic modeling to extract features that would ultimately be fed to a random forest classifier. This approach is simple, but it does not take advantage of recent advancements in word embedding representation that the other papers discussed above do.

4.2 Persuasion Detection in Propaganda

Deep Learning Approaches. Yu et. al. [5] proposed a supervised approach for detecting specific persuasion techniques used in news articles by using a multi-granularity neural network. The model has separate classification layers for each granularity level, i.e., sentence, word, and character. Each layer generates an output that generates a weight for the next layer. The approach can identify specific propaganda techniques in the sentence, which can help users understand why a sentence was classified as propagandist. The authors adjusted the loss function to address the imbalance in data, which decreased accuracy but improved the detection of propagandist techniques. However, direct applicability of this approach in SE is unknown; customization may be required due to the unknown effectiveness of propaganda techniques in SE. Another limitation we would like to acknowledge is that this approach was published on arXiv, which is not peer-reviewed, but it is moderated.

Gupta et. al. [34] proposed a method to detect persuasion techniques in propagandist memes as part of SemEval contest [35] which was also published on arXiv. They fine-tuned BERT and RoBERTa models with a feed-forward network on [CLS] token embedding, with two linear layers and default activation and dropout layers. RoBERTa was found to be a promising model for persuasion detection in text. This approach also addressed class imbalance by customizing the loss function. However, the paper lacks details on how text was extracted from memes and how labels were retained. Additionally, it is not peer-reviewed.

4.3 Persuasion Detection in SE

Statistical Approaches. Heijden et al. [36] built on information from econometrics and natural language processing and proposed a model that predicts the success of a phishing attack based on quantitative measures of vulnerability triggers used in emails. They formulated this problem as the likelihood that an individual would comply with requests in the phishing email. Their approach derives a triaging model that predicts how many clicks a phishing email can generate based on its cognitive features. They modeled phishing success as a Poisson process and evaluated the explanatory power of each persuasion principle, followed by conducting ANOVA tests to identify factors that can predict phishing success. The study found that generated clicks are positively associated with consistency, scarcity, and authority, and negatively with reciprocity. This approach uses simple models and less data to make estimations for triaging of phishing attacks and could be effective in quickly identifying threats. However, the work used small dataset affecting generalizability of this approach – raising the question if the results will be consistent on a diverse dataset. Additionally, it does not detect persuasion, but uses similar "cognitive vulnerabilities" to predict phishing success.

Feature Engineering and Machine Learning Approaches. Derakhshan et. al. [37] detected telephone-based SE attacks using scam signatures. They defined scam signatures as "unique utterances to a class of SE attacks" and were identified manually, embedded using Google's Universal Sentence Encoder and then classified using K-means clustering. A conversation from the dataset is then compared to all scam signatures using a similarity score, and the highest scoring signature class, if above a set threshold, was assigned to the input conversation. This simple approach contains persuasion techniques within the scam signatures, even if not explicitly mentioned, for example, in the signature: "If you don't pay money immediately, you would be arrested", authority is used. However, their dataset was limited to 140 non-scam and 75 scam emails, which could affect generalizability. The challenge with this approach is that scam signatures are pre-defined sentences that can quickly evolve hence, false negative rate would be high, making it less viable in a real-world context.

Li et. al. [38] proposed a method to detect phishing emails by detecting the presence of persuasion principles in them. They used a phishing email dataset from monkey.org with a sample of 1000 phishing emails and their approach has three steps: basic feature extraction, persuasion-related feature extraction, and machine learning-based binary classification into phishing or non-phishing emails. They considered authority, reciprocation, and scarcity as the most prevalent persuasion principles and used TF-IDF algorithm to obtain a list of high-importance words for each principle, which are used as features too. K-Nearest Neighbors, Decision Tree, and Naïve Bayes classifiers were used for training and testing, outperforming two models proposed in [14, 15]. However, their approach only considers three persuasion principles, and the authors do not specify how persuasion principles are coded once the corresponding words are extracted.

Edwards et. al. [39] studied persuasion in advance fee frauds in a scamming and scam-baiting context. They used SVM, Naïve Bayes and Binomial Logistic Regression to classify scam-baiting exchanges from normal email traffic and scammer messages from scam-baiting messages. They performed feature engineering to retrieve features using basic techniques such as Bag-Of-Words and were able to identify scammer communications and label involved parties accurately. This is a simple model that did not detect persuasion techniques used, but the authors were able to identify different linguistic markers in the scammers' persuasive process which may help future work in persuasion detection.

Deep Learning Approaches. Valecha et. al. [18] conducted a study to evaluate if persuasion cues would be effective in detecting phishing emails and considered three cases: gain persuasion cues, loss persuasion cues and both gain and loss persuasion cues. They performed manual annotation of a small subset of phishing emails which was validated and then used to train a Bi-LSTM model to generate loss and gain persuasion labels for the entire dataset. Word2Vec model was used to predictively learn word embeddings and understand context – they trained four Word2Vec models – one for testing effectiveness of loss cues, gain cues, both gain and loss cues and no cues. These Word2Vec models were then fed to classical machine learning models: random forest, logistic regression, support vector machine and naïve bayes classifiers. Their results indicated that models with persuasion cues outperformed models without persuasion cues. This paper was able to show that persuasion techniques are effective predictors of phishing emails and may be able to add explainability to classifier decisions. However, the reduction of persuasion techniques to gain and loss cues may have resulted in information loss that the model could have learned.

Tsinganos et. al. [40] formulate the problem of recognizing persuasion methods in chat-based SE as a sentence level classification problem. They train their model on data obtained from SE dark websites, forums, tutorials, books, and logs, and compiled 3800 sentences. They proposed a convolutional neural network as an encoder and feature extractor in order to retain a sentences' spatial information to classify if a given sentence contains persuasion principles or not. The model uses FastText which, compared to Word2Vec, can provide better embeddings for morphologically rich languages which may be useful to look at. Additionally, it provides a fresh perspective towards detection of persuasion by using spatial information and demonstrated that it may be useful. However, there are several limitations, primarily, efficient training is of concern due to the model using grid search to find optimal hyperparameter settings. Secondly, the intent of capturing spatial information over sequential information by using a CNN is not clear, what specific advantage is gained? The authors state that spaces between words capture hidden meaning, intent, and emotion but this claim is not accompanied by supporting literature. Finally, the size of the data the model was tested on is very small, the annotation process is not described, and its validity is not discussed.

4.4 Miscellaneous Domains: Deep Learning Approaches

Graph Neural Network Approach. Liu et. al. [28] studied the problem of modeling persuasion factor for user decision in recommendation systems by attempting to learn persuasion factors via representation learning. To accomplish this, they build a graph with users and recommended items as nodes and persuasion factors as edges, which are only added if a user clicked on a recommended item under a persuasion factor. They argue that learning effective embedding of persuasion factors is not trivial in this context and perform information propagation using graph convolutional neural networks to learn the corresponding representations of persuasion factors. This approach shows potential for application of a similar representation learning approach to model SE behavior and relevant persuasion techniques. However, it assumes that persuasion factors are independent and disentangled, i.e., each latent variable is responsible for only one generative factor in the data – indicating the need for factor analysis of persuasion techniques observed in the data – which makes the result less generalizable to different contexts. Another limitation is that persuasion factors in this paper are unique to recommendation systems and not associated with the Cialdini model mentioned earlier.

Fine-tuning Approach. Pöyhönen et. al. [41] presented a multilingual approach to persuasion detection using dialogues form role playing video games. They fine-tuned a variation of language-specific BERT model for each language in the dataset – English, French, Italian, German and Spanish and also fine-tuned a single multilingual BERT using the entire dataset with all languages The models intake a sentence and predicts its label as persuasive or non-persuasive. They showed satisfactory results from simply fine-tuning a BERT model, however, they found that the performance of multilingual BERT was lower than monolingual BERT models. They attributed this to a decrease in out-of-vocabulary words which facilitates fine-tuning for monolingual BERT. This approach is fairly simple in nature in the sense that it only performs fine-tuning of an established model on new data. However, the dataset was annotated by video game developers and even though it may be the largest, multilingual dataset as claimed, its validity has not been tested and the paper does not discuss any inter-rater agreements. Additionally, the dialogues do not indicate the specific persuasion techniques used and the model only performs high level binary persuasion classification. This paper is also not peer-reviewed.

5 Prospective Approaches in SE

Computational approaches to persuasion within the context of social engineering (SE) represents a nascent area of inquiry, which has garnered heightened attention in recent years, particularly from 2020 onwards as seen in the previous section. At present, extant research employs basic feature engineering processes and classical machine learning algorithms to address the task at hand. However, multi-label persuasion detection in SE poses numerous challenges, including the acquisition of a suitably extensive corpus of phishing emails and messages, apt for deep learning applications, as well as the annotation of said corpus. Thus, prospective research avenues may delve into potential strategies to tackle these obstacles. In this section, we will surmise certain research possibilities that may help address some of these concerns.

5.1 Applicability of Current Approaches in SE Domain

The computational approaches to persuasion described in the previous section demonstrate the potential of benefiting from deep learning techniques. In [30], researchers successfully applied an attention mechanism that focused on argumentative sentences while learning persuasion, it may be interesting to experiment with a similar model to see if there exists a possibility to focus on sentences with malicious/SE intent while learning persuasion. Their approach uses sub-reddit specific scores for classification task, however, could these scores be replaced by variables related to SE? One potential avenue of research could be utilizing quantitative measures of cognitive triggers and likelihood of click generation demonstrated by Heijden [36] as variables for SE persuasion detection task.

Another approach towards persuasion detection in SE would be to focus on sentence structures of persuasion techniques, for example, by evaluating the applicability of domain-independent, unsupervised approach presented in [29] in the context of SE, as all the data used in the study were a part of argumentative discourse. It would be interesting to observe if the argument extractor model can capture SE/phishing requests for compliance.

5.2 Data Augmentation and Imbalance

In order to address the difficulty of obtaining large datasets, one question that arises is if there are any data augmentation techniques that can be used? Data augmentation can also strengthen decision boundaries of a classification problem. Typically, data augmentation results in samples that are slightly different from the original text, which familiarizes the model with the local space around the samples. One such technique is rule-based augmentation which inserts and re-arranges existing data which includes random swaps, random insertion, random deletion, and random synonym replacement [42]. Another interesting and relatively new approach is graph-structured augmentation wherein text data is represented as graph structures. This type of augmentation could capture structural information by looking at knowledge graphs containing relation and entity encoding and grammatical structures of syntax trees [42]. A graph convolutional layer was successfully used for modeling persuasion as demonstrated by Liu et. al. [28]

in the domain of recommendation systems. It would be an interesting avenue of research to model persuasion in SE by using this technique in conjunction with a deep learning model. In [18], researchers showed that there is potential model persuasion automatically given a smaller set of manually annotated data.

5.3 Intent Classification

Non-phishing messages may contain elements of persuasion as well, i.e., persuasive messages are not always sent with malicious intent. We know that phishing messages are usually sent to the receiver with intent to mislead them [43], so in order to ensure that persuasion techniques are only detected for phishing messages, it would be interesting to see if such an approach benefit from intent classification. In [10], an approach to detect malicious intent by extracting the command and question topics by exploiting linguistic structure of such requests. In an alternative approach, Dorr et. al. [13] leveraged computational sociolinguistics, and media-specific structural clues to extract asks, i.e., requests and framing, i.e., consequences. A literature review of intent classification methods was not conducted as it would be out of scope for this paper but can be considered as future work.

5.4 Leveraging Large Language Models (LLMs)

Exploring the possibility of using more sophisticated techniques based on recent advancements is also a suggested avenue of research. Recent research has showed that large language models (LLMs) such as BERT, GPT-2 and GPT-3 have been able to capture phishing techniques and have been used in simulated phishing campaigns [44]. Exploring if large language models such as BERT and GPT can capture persuasion techniques used in SE/phishing or assist in their classification is also a possible avenue of research.

In this section, possible approaches were described to address the limitations and concerns observed in the methods surveyed in the paper. Since persuasion detection in SE is an emerging field, one possible avenue we identified was to study and transform existing approaches in other domains (argumentative discourse and propaganda) to SE domain and SE specific data. In this case, the context of the attack must be addressed too – phishing, misinformation campaign or chat-based SE. Secondly, there is a problem of data imbalance in SE – some techniques are used more often than others, which may result in some persuasion classes being learned better and some worse. Potential approaches to address these issues were discussed and the community would benefit from a comparative analysis of which approach works better. We also suggest intent classification as an added step in the detection process to distinguish genuine persuasive messages from those sent with malicious intent. Finally, we suggest leveraging LLMs to perform multi-label classification of persuasion and evaluate their effectiveness in capturing persuasion.

6 Conclusion

In this paper, we identified existing computational approaches to persuasion by surveying thirteen different methods in different domains – propaganda, argumentative discourse, and SE. We found that deep learning approaches were used successfully in the domains of propaganda and argumentative discourse, each with their own sets of advantages and disadvantages. Since the models were all heavily domain dependent, in order to test their applicability in the context of SE, these models would need to be customized according to the dataset collected. Yet, certain aspects of these approaches render them promising in the context of SE. We also observe that there is potential for using social psychology literature theories to provide explainability and interpret deep learning models that act like a black box. Current state of research shows that thus far approaches to persuasion detection in SE/phishing rely on feature engineering to obtain words or sentences characteristic of a persuasion principle by drawing information from the field of linguistics and social psychology. While the review showed some promising techniques such as topic modeling, leveraging sentence structures and using semantic frames to capture additional meaning and context, there are some undesirable assumptions made. Some research we looked at formulate the problem as a binary classification task wherein an instance is either persuasive or non-persuasive. Another common undesirable assumption observed was that one instance contains one persuasive technique, although in reality, a combination of techniques may be in use rather frequently. Current weakness in the reviewed articles also arises from the fact that recent advancements in NLP such as stronger word or document embedding techniques were not used to represent complex SE messages. Particularly for certain SE detection approaches discussed in Sect. 4.3, sentence representations of SE attacks were used to identify new instances that would result in high miss rates as attackers evolve their techniques. Future research could focus on addressing some of these concerns and limitations and identify ways to improve these approaches. For future avenues of feasible research, we suggest studying data augmentation techniques, automated data annotation techniques, intent classification and leveraging large language models to evaluate their effectiveness in capturing and detecting persuasion. Looking at modification of approaches used in different domains discussed in the paper and analyzing their applicability to SE along with prior SE research could be beneficial as well. Finally, we acknowledge the discussion of certain non-peer reviewed articles that may contain high bias; however, this discussion was done to critically assess the proposed approaches and present diverse ideas in this emerging field.

References

1. Cialdini, R.B.: The science of persuasion. Sci. Am. **284**, 76–81 (2001)
2. Fogg, B.: Persuasive computers: perspectives and research directions. In: Proceedings of the SIGCHI Conference On Human Factors In Computing Systems - CHI '98, pp. 225–232. ACM Press, Los Angeles, California, United States (1998). https://doi.org/10.1145/274644.274677
3. Stock, O.: A (Persuasive?) speech on automated persuasion. In: Proceedings of the 9th ACM Conference on Recommender Systems, pp. 1–2. Association for Computing Machinery, New York, NY, USA (2015). https://doi.org/10.1145/2792838.2799503

4. Kadowaki, K., Kobayashi, K., Kitamura, Y.: Influence of social relationships on multiagent persuasion. In: Proceedings of the 7th International Joint Conference on Autonomous Agents and Multiagent Systems - Volume 3, pp. 1221–1224. International Foundation for Autonomous Agents and Multiagent Systems, Richland, SC (2008)
5. Yu, S., Martino, G.D.S., Nakov, P.: Experiments in Detecting Persuasion Techniques in the News, http://arxiv.org/abs/1911.06815. (2019). https://doi.org/10.48550/arXiv.1911.06815
6. Ferreira, A., Coventry, L., Lenzini, G.: Principles of Persuasion in Social Engineering and Their Use in Phishing. In: Tryfonas, T., Askoxylakis, I. (eds.) Human Aspects of Information Security, Privacy, and Trust: Third International Conference, HAS 2015, Held as Part of HCI International 2015, Los Angeles, CA, USA, August 2-7, 2015. Proceedings, pp. 36–47. Springer International Publishing, Cham (2015). https://doi.org/10.1007/978-3-319-20376-8_4
7. Ferreira, A., Lenzini, G.: An analysis of social engineering principles in effective phishing. In: 2015 Workshop on Socio-Technical Aspects in Security and Trust, pp. 9–16 (2015). https://doi.org/10.1109/STAST.2015.10
8. Dhamija, R., Tygar, J.D., Hearst, M.: Why phishing works. In: Proceedings of the SIGCHI Conference on Human Factors in Computing Systems, pp. 581–590. ACM, Montréal Québec Canada (2006). https://doi.org/10.1145/1124772.1124861
9. Dong, X., Clark, J.A., Jacob, J.L.: Defending the weakest link: phishing websites detection by analysing user behaviours. Telecommun. Syst.. Syst. **45**, 215–226 (2010). https://doi.org/10.1007/s11235-009-9247-9
10. Sawa, Y., Bhakta, R., Harris, I.G., Hadnagy, C.: Detection of social engineering attacks through natural language processing of conversations. In: 2016 IEEE Tenth International Conference on Semantic Computing (ICSC), pp. 262–265 (2016). https://doi.org/10.1109/ICSC.2016.95
11. Lan, Y.: Chat-oriented social engineering attack detection using attention-based Bi-LSTM and CNN. In: 2021 2nd International Conference on Computing and Data Science (CDS), pp. 483–487 (2021). https://doi.org/10.1109/CDS52072.2021.00089
12. Lopez, J.C., Camarago, J. E.: Social engineering detection using natural language processing and machine learning. In: 2022 5th International Conference on Information and Computer Technologies (ICICT) (2022). https://doi.org/10.1109/ICICT55905.2022.00038
13. Dorr, B.J., et al.: Detecting Asks in SE attacks: Impact of Linguistic and Structural Knowledge, http://arxiv.org/abs/2002.10931, (2020). https://doi.org/10.48550/arXiv.2002.10931
14. Fette, I., Sadeh, N., Tomasic, A.: Learning to detect phishing emails. In: Proceedings of the 16th International Conference on World Wide Web, pp. 649–656. Association for Computing Machinery, New York, NY, USA (2007). https://doi.org/10.1145/1242572.1242660
15. Zhang, Y., Hong, J.I., Cranor, L.F.: Cantina: a content-based approach to detecting phishing web sites. In: Proceedings of the 16th International Conference on World Wide Web, pp. 639–648. Association for Computing Machinery, New York, NY, USA (2007). https://doi.org/10.1145/1242572.1242659
16. Dawabsheh, A., Jazzar, M., Eleyan, A., Bejaoui, T., Popoola, S.: An enhanced phishing detection tool using deep learning from URL. In: 2022 International Conference on Smart Applications, Communications and Networking (SmartNets), pp. 1–6 (2022). https://doi.org/10.1109/SmartNets55823.2022.9993984
17. Shalini, L., Manvi, S.S., Gowda, N.C., Manasa, K.N.: Detection of phishing emails using machine learning and deep learning. In: 2022 7th International Conference on Communication and Electronics Systems (ICCES), pp. 1237–1243 (2022). https://doi.org/10.1109/ICCES54183.2022.9835846
18. Valecha, R., Mandaokar, P., Rao, H.R.: Phishing email detection using persuasion cues. IEEE Trans. Depend. Secure Comput. **19**, 747–756 (2022). https://doi.org/10.1109/TDSC.2021.3118931

19. Ferreira, A., Jakobsson, M.: Persuasion in Scams. In: Jakobsson, M. (ed.) Understanding Social Engineering Based Scams, pp. 29–47. Springer New York, New York, NY (2016). https://doi.org/10.1007/978-1-4939-6457-4_4
20. Ferreira, A., Teles, S.: Persuasion: how phishing emails can influence users and bypass security measures. Int. J. Hum. Comput. Stud.Comput. Stud. **125**, 19–31 (2019). https://doi.org/10.1016/j.ijhcs.2018.12.004
21. Marwell, G., Schmitt, D.R.: Dimensions of compliance-gaining behavior: an empirical analysis. Sociometry. **30**, 350–364 (1967). https://doi.org/10.2307/2786181
22. Gragg, D.: A Multi-Level Defense Against Social Engineering. Presented at the (2003)
23. Sheng, S., Holbrook, M., Kumaraguru, P., Cranor, L.F., Downs, J.: Who falls for phish? a demographic analysis of phishing susceptibility and effectiveness of interventions. In: Proceedings of the SIGCHI Conference on Human Factors in Computing Systems, pp. 373–382. Association for Computing Machinery, New York, NY, USA (2010). https://doi.org/10.1145/1753326.1753383
24. Koddebusch, M.: Exposing the Phish: the effect of persuasion techniques in phishing E-Mails. In: DG.O 2022: The 23rd Annual International Conference on Digital Government Research, pp. 78–87. Association for Computing Machinery, New York, NY, USA (2022). https://doi.org/10.1145/3543434.3543476
25. Zielinska, O.A., Welk, A.K., Mayhorn, C.B., Murphy-Hill, E.: A temporal analysis of persuasion principles in phishing emails. Proc. Human Factors Ergonom. Society Annual Meeting. **60**, 765–769 (2016). https://doi.org/10.1177/1541931213601175
26. Akbar, N.: Analysing Persuasion Principles in Phishing Emails, http://essay.utwente.nl/66177/. Accessed 22 Feb 2023
27. Lawson, P.A., Crowson, A.D., Mayhorn, C.B.: Baiting the Hook: Exploring the Interaction of Personality and Persuasion Tactics in Email Phishing Attacks. In: Bagnara, S., Tartaglia, R., Albolino, S., Alexander, T., Fujita, Y. (eds.) Proceedings of the 20th Congress of the International Ergonomics Association (IEA 2018): Volume V: Human Simulation and Virtual Environments, Work With Computing Systems (WWCS), Process Control, pp. 401–406. Springer International Publishing, Cham (2019). https://doi.org/10.1007/978-3-319-96077-7_42
28. Liu, C., et al.: Modeling persuasion factor of user decision for recommendation. In: Proceedings of the 28th ACM SIGKDD Conference on Knowledge Discovery and Data Mining, pp. 3366–3376. Association for Computing Machinery, New York, NY, USA (2022). https://doi.org/10.1145/3534678.3539114
29. Iyer, R.R., Sycara, K., Li, Y.: Detecting type of Persuasion: is there structure in persuasion tactics?. In: Proceedings of the 16th edition of the International Conference on Artificial Intelligence and Law, Washington, DC, USA (2017)
30. Dutta, S., Das, D., Chakraborty, T.: Changing views: persuasion modeling and argument extraction from online discussions. Inf. Process. Manage. **57**, 102085 (2020). https://doi.org/10.1016/j.ipm.2019.102085
31. Hidey, C., McKeown, K.: Persuasive influence detection: the role of argument sequencing. Proc. AAAI Conf. Artif. Intell. **32**, 1 (2018). https://doi.org/10.1609/aaai.v32i1.12003
32. Johnson, C.R., Petruck, M.R.L., Baker, C.F., Ellsworth, M., Ruppenhofer, J., Fillmore, C.: FrameNet: Theory and Practice. Presented at the (2003)
33. Shi, H., Silva, M., Giovanini, L., Capecci, D., Czech, L., Fernandes, J., Oliveira, D.: Lumen: a machine learning framework to expose influence cues in texts. Front. Comput. Sci. **4**, 929515 (2022)
34. Gupta, K., Gautam, D., Mamidi, R.: Volta at SemEval-2021 Task 6: Towards Detecting Persuasive Texts and Images using Textual and Multimodal Ensemble, http://arxiv.org/abs/2106.00240 (2021)

35. Dimitrov, D., et al.: SemEval-2021 Task 6: Detection of Persuasion Techniques in Texts and Images, http://arxiv.org/abs/2105.09284 (2021). https://doi.org/10.48550/arXiv.2105.09284
36. Heijden, A. van der, Allodi, L.: Cognitive Triaging of Phishing Attacks. Presented at the 28th USENIX Security Symposium (USENIX Security 19) (2019)
37. Derakhshan, A., Harris, I.G., Behzadi, M.: Detecting telephone-based social engineering attacks using scam signatures. In: Proceedings of the 2021 ACM Workshop on Security and Privacy Analytics, pp. 67–73. Association for Computing Machinery, New York, NY, USA (2021). https://doi.org/10.1145/3445970.3451152
38. Li, X., Zhang, D., Wu, B.: Detection method of phishing email based on persuasion principle. In: 2020 IEEE 4th Information Technology, Networking, Electronic and Automation Control Conference (ITNEC), pp. 571–574 (2020). https://doi.org/10.1109/ITNEC48623.2020.9084766
39. Edwards, M., Peersman, C., Rashid, A.: Scamming the scammers: towards automatic detection of persuasion in advance fee frauds. In: Proceedings of the 26th International Conference on World Wide Web Companion, pp. 1291–1299. International World Wide Web Conferences Steering Committee, Republic and Canton of Geneva, CHE (2017). https://doi.org/10.1145/3041021.3053889
40. Tsinganos, N., Mavridis, I., Gritzalis, D.: Utilizing convolutional neural networks and word embeddings for early-stage recognition of persuasion in chat-based social engineering attacks. IEEE Access. **10**, 108517–108529 (2022). https://doi.org/10.1109/ACCESS.2022.3213681
41. Pöyhönen, T., Hämäläinen, M., Alnajjar, K.: Multilingual persuasion detection: Video Games as an Invaluable Data Source for NLP, http://arxiv.org/abs/2207.04453 (2022). https://doi.org/10.48550/arXiv.2207.04453
42. Shorten, C., Khoshgoftaar, T.M., Furht, B.: Text data augmentation for deep learning. J. Big Data. **8**, 101 (2021). https://doi.org/10.1186/s40537-021-00492-0
43. Jagatic, T.N., Johnson, N.A., Jakobsson, M., Menczer, F.: Social phishing. Commun. ACM. ACM **50**, 94–100 (2007). https://doi.org/10.1145/1290958.1290968
44. Karanjai, R.: Targeted Phishing Campaigns using Large Scale Language Models, http://arxiv.org/abs/2301.00665 (2022). https://doi.org/10.48550/arXiv.2301.00665

Investigate the Suitability of Adversarial Perturbation in Preserving Privacy in the Context of Photos

Srinivas Madhisetty(✉) and Vaishvi Patel

University of Technology Sydney, Ultimo, Australia
Srinivas.madhisetty@uts.edu.au

Abstract. Post-publication of Photos and Videos privacy settings may not be as its original intention for publication was. Sometimes inadvertently sensitive information may be disclosed to others, or a more alarming aspect is the advent of Deepfake technology which can manipulate images and videos to make them available for public viewership distorting its privacy settings partially or completely. A photo or video can be manipulated in such a way that it can remove its original context and purpose. This ability has serious consequences for the public regarding their privacy concerns. This paper showcased the role of Adversarial Perturbation (AP) is becoming more critical, as it may be seen as one of the ways to mitigate problems of unintended image or video manipulation. This paper studies numerous articles and uses a methodological approach to demonstrate the advantages of AP attacks in this study while discovering that pre-trained data classification models might be fooled by instilling a greater noise gradient into a picture in our proposed trials.

Keywords: Privacy in Photos · Videos · Deepfakes

1 Introduction

Privacy is not well-defined or articulated. Perhaps contextualizing privacy may be considered a challenge due to the subjective nature of the concept. When Deepfake technology is used, it has the potential to make sharing photos and videos unsafe for individuals.

Although many social media applications are adopting timely deletion of data in the form of text, images, and videos, Snapchat has implemented a solution where media such as photos or videos are deleted after a period of a specified time, making it a popular social media application [13].

The rapid growth of AI (Artificial Intelligence) programs has created new challenges in preserving privacy. This paper explores the role of preserving privacy by taking advantage of adversarial perturbation by incorporating defects that are not visible to the human eye. However, this approach can create hurdles when machine learning tools are used to analyze a photo or video.

When images are uploaded to platforms such as social media platforms users may sometimes voluntarily or involuntarily reveal information about themselves, which may lead to a loss of their privacy.

K. Arai (Ed.): FTC 2023, LNNS 816, pp. 410–418, 2023.
https://doi.org/10.1007/978-3-031-47448-4_31

Management of privacy, when photographs or videos are shared, is that proper grounding in terms of context may not be available. A document is contextually rich and observations of loss of privacy by reading a document may be more precise. This is because photos may contain rich tacit knowledge.

The main artifact identified through the research is that a photo or video may contain insufficient information to establish its context. With the availability of Deepfake technology, such content can be easily manipulated, potentially causing harm to the publisher.

The proposed solution in this paper is to utilize Adversarial Perturbation (AP) by adding a destructive noise pattern to an image, making it difficult for an image classifier algorithm to operate. This could limit the fidelity of such algorithms and prevent published image and video content from being transformed for other purposes using Deepfake technology.

As the image or video is distorted by adding such noise, data detection systems may not function as intended. This process may reduce the efficiency of the programs that use algorithms or a convolution of neural networks, leading to significant misrepresentation due to the distortion of the original image or video.

The significance of using such technology to counter the power of machine learning algorithms may give more confidence to the general public but may also hinder big corporations from mining privacy-relevant data from an image or video.

Perhaps limit the fidelity of such an algorithm to transform a published image and video content for other purposes using Deepfake technology.

As the image or video is distorted by adding such noise, data-detecting systems may not work as intended. This process may reduce the efficiency of a program that uses data mining algorithms or a convolution of neural networks as this allows significant misinterpretation due to distortion of the original image or video. "However, we would like to state a ground truth about adversarial examples, which is the fact that the model should first classify the original example well so that the small perturbation will misclassify the resulting adversarial example with high confidence." [1].

The significance of using such technology to counter the power of machine learning algorithms may afford more confidence to the general public but also may hinder big corporations from mining privacy-relevant data from an image or a video. An agile neural network may be required to combat the unintended consequence of the loss of privacy of an individual who publishes content. A deeper study is required to understand and fine-tune the variables that are required to cause effective distortion but preserve the image or video aesthetics for the viewer of the image.

The association between the imposed noise gradient to a visual as well as the misinterpretation efficiency when examined by an image classifier will be the focus of our research.

Therefore, a highly adaptable architecture must be designed to enable DNNs architecture to comply with AP attacks, reducing the possibility of leaking personal data to an intruder. To do so, we must first comprehend the variables that cause adversarial examples to harm deep learning models. The association between the imposed noise gradient to a visual as well as the misinterpretation efficiency when examined by an image classifier will be the focus of our research.

This paper intends to float the idea to further study the feasibility to derive the so-called sweet spot between the applied random gradient and the loss in classification accuracy as distortion is applied through the noise. According to [11] Xue and others in 2020, there is some evidence of such misclassification as different levels of noise that distorts the image when applied.

Therefore, the assumption made in this paper assumes that there is a correlation -the higher amount of noise applied may result in the inaccuracy of classification models to administer image privacy.

What variables allow AP attacks to trick neural network modules and algorithms, such that a stronger AP method can be investigated further? According to the hypothesis, there is a substantial favourable relationship between the imposed random gradient to the visual and the drop in classification accuracy when more disturbance is provided to an image. This assumption has been supported by the investigation of [11] which showcases different possible misclassifications as a varied amount of destructive noise is added to the images.

As a result, it can be comprehended the more noise added to an image, the greater the possibility of an object being misclassified when it passes through the architecture of a neural network or image classifier dataset library.

The assessment of existing scenarios among organizations and individuals who use neural networks, as well as the usage of such Adversarial disturbance to enhance individual's privacy.

2 Background

Individuals share photos and videos via social media with others sometimes selecting an appropriate audience. According to an article by [11], data analytics companies may be using such information to provide customized solutions to their users.

Liu [7] and [11] Xue in 2020 have showcased an adversarial artifact that can mask personal data from it being revealed to an AI-based program or a neural network.

Liu's [7] 2019 work entails using distortion of image data with noise in the whole image whereas [11] is working on portions of the image which is more effective in preserving privacy.

Adversarial perturbation can also be applied to videos that can distinguish between Deepfake-generated videos according to [4] Hussain and others in 2021. However, such detection may be circumvented by modifying video material using Deepfake synthesis in an adversarial manner.

Therefore, the proposal in this paper is to use different adversarial techniques applied to combat the loss of individual privacy. The assumption is that the Adversarial Perturbance technique may also be used to preserve privacy but can also be used as a possible threat to deceive pre-trained classifiers that use a neural network to categorize distorted visual data.

The above assumption is made due to a lack of fundamental understanding of factors that influence the interactions within the framework of Adversarial Perturbance models. Future work in this area may bring more information about the relationship between the algorithmic evaluations to assess the absolute effectiveness to detect sensitive information.

3 AP Advantages and Shortcomings

When an image is encrypted using Adversarial models based on artificial intelligence, can be used to test to challenge image classifiers and detectors that employ machine learning to preserve privacy in an image. However, when this AP technique is used as a method to hack into a data set with limited scope may lead to unintended consequences.

Litan [6] looked at how the ordinary sharing of images may lead to breaches of privacy. Particularly how they can be turned into pornographic content. These may be used to extort money from unsuspecting people whose images have been misused. These Deepfake images may be used by the abuser as a means of revenge porn.

Cybercriminals get images of individuals that may belong from social networking servers and change these by appending sexually explicit content over the identity of the victim. However, according to the author, this pattern is evolving, and complex forgeries are constantly being used to gain private or financial information (Litan, 2020).

One of the most known and exceedingly used strategies for producing adversary instances is indeed the FGSM, short for Fast Gradient Sign Method, [5] Khalid and others (2019) use the variance in generated pre-processing image noise filters on perturbed depictions using the stated methods, along with their implication on the image classification model, Very Deep Convolutional Networks, VGGNet.

Thus, according to their research design and observations, all the approaches studied pose a serious risk to neural network models and automated systems in their capacity to generate convincing Deepfakes. When compared to other attacking strategies, FGSM has the least impact on the throughput of pattern recognition frameworks, but it initiates the adversarial assault the fastest, empowering it to be one of the most prominent attacking techniques.

This study primarily identifies potential shortcomings in the current (Dynamic Neural Network) DNN models and methodologies to assist future academics and researchers in developing an acceptable technique for defending neural networks from adversarial threats [5].

This could pose a significant threat to businesses and users who rely on AI models that use network architecture and pre-trained data libraries to collect or produce intended outcomes for future personal or professional tasks.

As a result, many enterprises may be forced to compromise their privacy practices, placing their personal and sensitive information in danger. Although, with the changing needs of the cybersecurity industry, AP can also preserve personal and sensitive data while supporting big organizations and mainly individuals to secure their visual data when they share them on social platforms or elsewhere.

The objective is to protect users' privacy from AP assaults on deep learning models so that it can protect one's confidentiality on social networking sites, as investigated in numerous articles. The goal of AP assaults is to deceive neural networks and drive Machine learning (ML) algorithms to give false results.

Adversary models may be used for image encryption and manipulation, or other appropriate strategy stated by AI techniques. When challenged with criminal misuse, this process could be used to trick image classifiers and detectors that employ machine learning by conducting an ant attack strategy.

AI can also be used for society's good such as to also increase efficiency in providing health care during the Covid pandemic. The successes in chest radiologic image recognition and assisting healthcare providers enabled us to meet expectations for efficient performance in the outbreak area of covid cases. [8] Pal analyzes the execution of complex learning models and AI to trace the symptoms of the virus, COVID-19.

Neural network-based methods focus on readily available, pre-trained learning or classification models; this may limit variation in the radiological image-detailing setting. The paper written by [8] Pal in 2021 demonstrated the successful deployment of the Fast Gradient Sign Attack (FGSM).

This permitted the readers to understand how different frameworks can make the detection and diagnosis sensitive scale toward adversary models. Thereby heightening the danger to the organizational ability to promptly deliver appropriate remedies to COVID sufferers.

4 Methodology

The methods to effectively address the challenge of investigating and analyzing the relationship between AP gradient noise and classification precision, as well as evaluating the level of threat is the objective of this study.

To accomplish this, the primary goal of this study is to create an adversarial example utilizing the FGSM, the far most dangerous AP procedure, with varying amounts of gradient noise applied to the image.

A future goal is to employ the adversarial examples produced that integrate with deep learning models for object categorization, testing with the distinct gradient thresholds, and observing the key variables getting altered in output. Finally, conduct a correlation analysis of all these factors and parameters to investigate the effects of AP assaults on classifiers at various degrees and concentrations of gradient noise.

The noise concentration interval between each trial can be versatile and modified to achieve more relevant and consistent outcomes while testing and investigating the relationship between the gradient noise intensity and misclassification accuracy.

This means that the less the interval gap, the more fundamental information can be accessed and analyzed, allowing for legitimate and trustworthy results as well as appropriate interpretation. Furthermore, the quantitative information generated as a relationship among the variables analyzed will be established.

This methodology, as previously said, is heavily evaluated experimentally to acquire core data and analysis for future review. To complete the investigation and analyze the results, this methodology needs a computer with at least four gigabytes of memory as well as an Intel 1.6 gigahertz engine that's interoperable with Anaconda, an accessible Python distribution, and other languages for programming and developing automated systems. This will enable the use of TensorFlow, a neural network framework, to determine the differential rate of loss and make the implementation of the algorithm easier.

5 Discussion

Investigating and analyzing the aspects involved in adversarial perturbation attacks on misleading image classifiers, as well as producing in-depth analyses of the factors that are prompted or changed during the assault, would pave the way for the upcoming generation of reliable AP methods and algorithms.

Such robust techniques could potentially protect secrecy by using data mining algorithms against themselves and concentrating on managing the elements that are being provoked or modified. The target audience for the published content is to be relevant and engaging while preserving privacy. Microservices could be utilized as Adversarial Perturbance agents which can generate gateways to accessing photos and videos. As a result, the powerful APs would be able to deter intruders from destroying or obtaining one's accurate personal information. Many advantages can be gained by adequately identifying and safeguarding private information with adversary noise deployment.

This primarily benefits large corporations by allowing them to communicate their visual data in ordinary work processes using a strong encryption method that machine learning systems cannot decrypt. Such businesses can defend themselves from today's increasing cybercrime and reduce the risk of losing all their sensitive material.

Deepfake images may be used to generate pornographic content without the knowledge or consent of the person in the photo or video. Cybercriminals exploit images of individuals that may be belonging to social networking websites or other professional work-related websites to change these by appending sexually explicit content over the identity of the victim.

Many victims claimed that they revealed that photos and videos were shared by their ex-partners without their consent. Other victims claimed that their cellular devices were hacked, and explicit content was stolen [12].

However, according to the author, this pattern is evolving, and complex forgeries are constantly being used to gain private or financial information [6] according to Litan, in 2020.

Post-privacy management would greatly inhibit and advance the individual scope of retaining privacy by discovering their visual data and maintaining privacy on social media platforms.

As a result, this investigation is critical to establish a strong encryption method for visual data while utilizing Adversarial Perturbation (AP) attacks and DNNs to provide various industries, enterprises, and individuals with enhanced privacy measures as technology advances.

6 Conclusion

The ability of Deepfake technology to manipulate available photos and videos is deeply concerning. The availability of cloud services to remotely process image and video data has resulted in the technology of developing Deepfake videos in common reach. There are sophisticated and high-fidelity photos and videos being generated using technology.

This study will feed a neural network with a pre-trained data library and classifiers with several photos that differ in the intensity of noise to expand the sample group and emphasize any anomalies that could be evaluated from the data, according to the technique. Many AP algorithms could be used to attack convolutional neural networks and misidentify the entities in an image. One of the most known and exceedingly used strategies for producing adversary instances is indeed the FGSM, short for fast gradient sign method, L-BFGS, short for Limited-Memory Broyden-Fletcher-Goldfarbshanno and BIM, short for Basic Iterative Method, approaches.

All of the approaches studied may pose a serious risk to neural network models and automated systems. When compared to other attacking strategies, FGSM had the least impact on the throughput of pattern recognition frameworks, but it initiates the adversarial assault the fastest, empowering it to be one of the most prominent attacking techniques.

Increase the introduction of mandatory artifacts, such as encryption of pixels, etc., Artefacts that can create an audit trail on how and where the content such as photos and video were to be published may be vital to trace the perpetrators of such crimes.

[9] Xiao and Pun in 2021 introduced a revolutionary Constrained Iterative–Fast Gradient sign method (CI-FGSM) to address the issues of image content deformation since the enforced intervention can be explicitly acknowledged in perturbed depictions.

The authors chose to develop further into CI-FGSM because FGSM and BIM are the most extensively utilized attack methods on DNNs. The scholars claim that, in comparison to many other gradient-based adversarial attack algorithms, their proposed approach encompasses minimizing the excess modifications on immaculate input data while upholding the optimal effect of adversarial attacks in a misleading classification algorithm.

This is achieved with distinguishable behavior constraint approaches after evaluating the impact of the threat proposed on several different neural networks dataset library resources [9]. This research allows us to understand the process of the gradient level assigned to a graphic affects the categorization accuracy of a disrupted image.

This broadens the readers' ability to comprehend and enables them to assess for any possible quantitative or qualitative association between the variables of the gradient rate of the AP algorithm and image classifiers' outcomes.

We assume a high positive association between the misclassification precision variable and the distortion gradient component that was added to the objects based on the examination of the data. [3] Fawzi in 2018 explore further the interesting instability of classifiers when perturbed by adversaries. With the exception of [2] Choi and others in 2021, indicated article provides a detailed framework for evaluating the strength and resilience of the prediction model to adversarial perturbation instances by obtaining an arbitrary limit on sequential and non-linear classification methods, where the upper estimate varies depending on a distinguishability way of measuring that encapsulates the assumption of classification task complexity. The "iterative method calculates the current gradient at each step in the iterative process and adds the gradient result to the adversarial sample." [10].

According to their findings, because of the clear limited utility of distinguishability, any prediction with a low classifier performance is vulnerable to adverse events (Fawzi et al., 2018) [3].

This consents researchers to assess a reduced percentage of misinterpretation as an element that is linked to the perturbation noise concentration applied to the tampered image data, as well as undertake additional research to determine what factors in the transformation of adversarial examples affect the nonlinear classification algorithm to maintain high detectability and thus be more robust than that of a linear classification model.

Identifying the process of AP attacks is critical for identifying and improving the consequences of these changes. As a result, studying the process in which attacks can be carried out may be a useful approach for designing better algorithms.

Furthermore, AP can improve a company's performance which requires strong image encryption or security measures to protect sensitive data when shared publicly, which can benefit such businesses through secure business operations. It can also help protect one's personal information from being abused.

Litan's 2020 [6] study, in which AI and image detectors are used to recognize an individual's identity and advertise it further for fake pornography films, is a true previously examined example of this.

As a result, building a robust AP algorithm that cannot be detected by human eyes is critical for protecting susceptible data gathered from social media platforms or pre-trained data models.

References

1. Ai, S., Voundi Koe, A.S., Huang, T.: Adversarial perturbation in remote sensing image recognition. Appl. Soft Comput. **105**, 1–13 (2021). https://doi.org/10.1016/j.asoc.2021.107252
2. Choi, S.-H., Shin, J., Liu, P., Choi, Y.-H.: EEJE: two-step input transformation for robust dnn against adversarial examples. IEEE Trans. Netw. Sci. Eng. **8**(2), 908–920 (2021). https://doi.org/10.1109/TNSE.2020.3008394
3. Fawzi, A., Fawzi, O., Frossard, P.: Analysis of classifiers' robustness to adversarial perturbations. Mach. Learn. **107**(3), 481–508 (2018). https://doi.org/10.1007/s10994-017-5663-3
4. Hussain, S., Neekhara, P., Jere, M., Koushanfar, F., McAuley, J.: Adversarial deepfakes: evaluating vulnerability of deepfake detectors to adversarial examples. In: Proceedings of the IEEE/CVF Winter Conference on Applications of Computer Vision (WACV), pp. 3348–3357 (2021)
5. Khalid, F., Hanif, M.A., Rehman, S., Qadir, J., Shafique, M.: FAdeML: Understanding the impact of pre-processing noise filtering on adversarial machine learning. In: 2019 Design, Automation & Test in Europe Conference & Exhibition (DATE), pp. 902–907 (2019). https://doi.org/10.23919/DATE.2019.8715141
6. Litan, A.: Deep Fakes are coming to your Enterprise – Are You Ready? https://blogs.gartner.com/avivah-litan/2020/12/02/deep-fakes-are-coming-toyour-enterprise-are-you-ready/?_ga=2.65835883.433126625.1630835860-1119423799.1630835860 (2020)
7. Liu, B., Ding, M., Zhu, T., Xiang, Y., Zhou, W.: Adversaries or allies? privacy and deep learning in the big data era. Concurr. Comput. **31**(19), e5102 (2019). https://doi.org/10.1002/cpe.5102

8. Pal, B., Gupta, D., Rashed-Al-Mahfuz, M., Alyami, S.A., Moni, M.A.: Vulnerability in deep transfer learning models to adversarial fast gradient sign attack for COVID-19 prediction from chest radiography images. Appl. Sci. **11**(9), 1–17 (2021). https://doi.org/10.3390/app11094233
9. Xiao, Y., Pun, C.-M.: Improving adversarial attacks on deep neural networks via constricted gradient-based perturbations. Inf. Sci. **571**, 104–132 (2021). https://doi.org/10.1016/j.ins.2021.04.033
10. Xu, J.: Generate Adversarial Examples by Nesterov-momentum iterative fast gradient sign method. In: Proceedings of the IEEE International Conference on Software Engineering and Service Sciences, ICSESS, 2020, pp. 244–249 (2020). https://doi.org/10.1109/ICSESS49938.2020.9237700
11. Xue, H., Liu, B., Din, M., Song, L., Zhu, T.: Hiding private information in images from AI. In: ICC 2020 - 2020 IEEE International Conference on Communications (ICC), 2020-, pp. 1–6 (2020). https://doi.org/10.1109/ICC40277.2020.9148656
12. Kamal, M., Newman, W.J.: Revenge pornography: mental health implications and related legislation. J. Am. Acad. Psych. Law Online **44**(3), 359–367 (2016)
13. https://help.snapchat.com/hc/en-us/articles/7012334940948-When-does-Snapchat-delete-Snaps-and-Chats-Viewed (2023)

Requirements for Work Products for ASPICE and Cybersecurity

Christian Schlager[1(✉)], Georg Macher[1], Richard Messnarz[2], Damjan Ekert[2], and Eugen Brenner[1]

[1] Technical University of Graz, Graz, Austria
christian.schlager@student.tugraz.at, {georg.macher,brenner}@tugraz.at
[2] ISCN, Graz, Austria
{rmess,dekert}@iscn.com
https://www.tugraz.at/institute/iti/home/, https://www.iscn.com

Abstract. ASPICE has been designed to help automotive suppliers build embedded systems and assess the effectiveness of workflows. Processes for software, systems, quality assurance, configuration management, problem solving, change management, project management, and supplier monitoring are included in the ASPICE framework as defined by the VDA (Verband der Automobilindustrie). It outlines a technique for determining the maturity of the process by placing the PAM at its core. In addition, HW Spice is considered since, in contrast to mechanical Spice, it is crucial for cybersecurity. Based on ASPICE, VDA has developed the procedures for cybersecurity engineering, cybersecurity risk management, and supplier request and selection. The technical team and assessors have more work to do when the ASPICE assessment is expanded to include cybersecurity processes in order to comply with standard ISO/SAE 21434. The VDA also includes Guidelines and Recommendations for rating techniques in accordance with the NPLF system of ASPICE for processes listed in VDA Scope and for Cybersecurity. This study examines the questions that ought to be covered throughout the assessments with a concentrate on work products in order to raise the caliber of the output.

Keywords: ASPICE · Cybersecurity · Assessment · Maturity · Rating

1 Introduction

For instance, in order to comply with UN 155 [31] and UN 156 [32], European automakers must be able to show that they developed a thorough cybersecurity case [2,6,25,28,30]. In order to give the Cybersecurity Management Systems (CSMS) the clear guidance required to comply with the UN 155 recommendation, the ISO 21434 [12] norm was developed and issued in August 2021. Also, the UN 156 regulation will use the standard for a SUMS (Software Update Management System) as a guidance starting in February 2023. Following this, the

K. Arai (Ed.): FTC 2023, LNNS 816, pp. 419–432, 2023.
https://doi.org/10.1007/978-3-031-47448-4_32

German Automotive Association developed an Automotive for Cybersecurity assessment methodology [36] and an ACSMS audit [33] book based on UN 155 and ISO 21434. Each project having external interfaces that could be intentionally attacked by hackers is subject to the Automotive for Cybersecurity review, which is carried out every three years as part of the organization's cybersecurity management system audit [1]. INTACS's cybersecurity training materials explain three different types of examinations that have been established (see Fig. 1):

1. Automotive Cybersecurity Management System (ACSMS) Audit [6,17–19,33]
2. Automotive SPICE for Cybersecurity [3,9,10,13–16,21,24,27,34–36]
3. ISO 21434 Product Assessment [12]

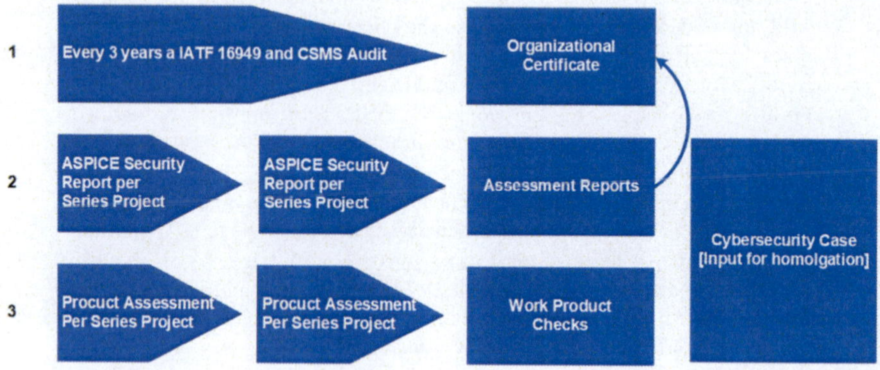

Fig. 1. Overview of audits.

The question answered in Sect. 4 is:

How can one type of cybersecurity assessment enhance the other?

By answering this research question a time reduction performing an assessment can be shown. This paper is organized the following way: Sect. 2 shows the work, which has already been done in this area, Sect. 3 lists the required assessments for cybersecurity, Sect. 4 an example for mutual enhancement to reduce time performing assessment, Sect. 5 the validation of the concept and finally Sect. 6 the conclusion.

2 Related Work

2.1 ASPICE

In 2001, the AUTOSIG (Automotive Special Interest Group) initially developed ASPICE [34] to assess the evolution of embedded systems. AUTOSIG's members include non-German Manufacturers like Fiat, Ford, Land Rover, and Volvo as well as Audi, BMW, Daimler, Porsche, and VW. The ASPICE model shown in Fig. 2 encompasses a number of process types. They include management processes (like MAN.3), acquisition processes (like ACQ.4), support processes (like SUP.1, SUP.8 to SUP.10), and software processes (like SWE.1 to SWE.6). HW Spice [9] (HWE.1-HWE.6) is a component of the V model as described by SPICE, while not yet being a part of ASPICE. At that time, the mechanical aspect (MEE.1-MEE.6) of a system, as described in [10], was not included in ASPICE. Moreover, ASPICE provides recommendations and guidelines [35] that highlight the connections between various practices both within and between processes. The directive may be applied to any sort of development [35] (agile and non-agile development).

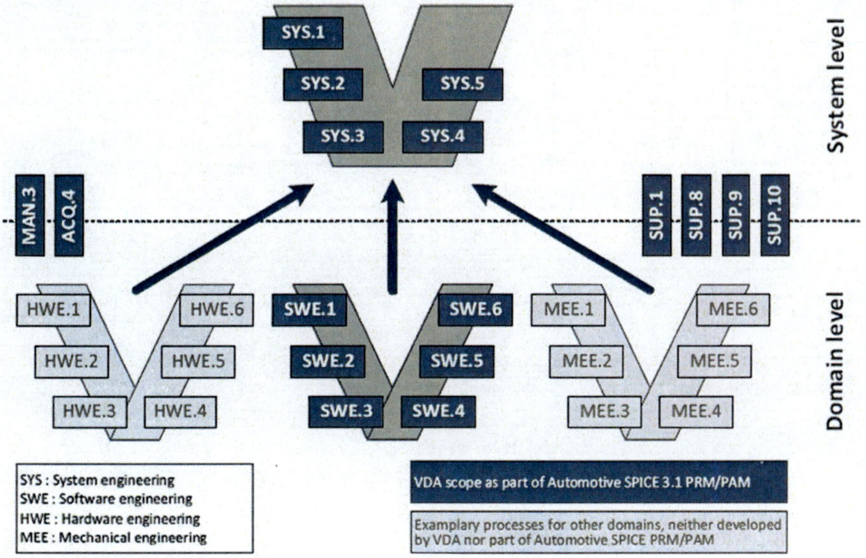

Fig. 2. Process assessment model (PAM) of ASPICE. [34]

2.2 Cybersecurity

In addition to specifying that cybersecurity must be taken into account [6,14,15,25,28], ISO/SAE 21434 [12] outlines the standards for developing systems for the automotive industry. This standard incorporates cybersecurity engineering during project development, manufacturing, use, and decommissioning. Additionally, it considers project-specific cybersecurity management in organizations. It considers the following cybersecurity issues:

- Key Exchange [7]
- Spoofing of messages [1,3,26]

– Treatment of Cybersecurity Requirements [16]

In addition to ASPICE, VDA [36] defines processes for cybersecurity engineering, supplier selection, and cybersecurity risk management in order to comply with the requirements of ISO/SAE 21434 [12].

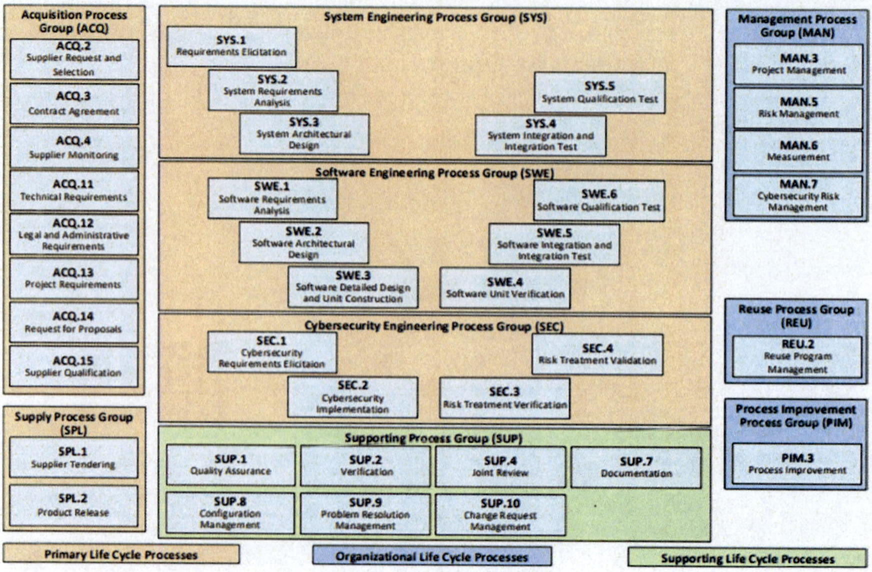

Fig. 3. ASPICE and cybersecurity. [36]

The additional processes are [36]:

1. SEC.1 - Requirement Elicitation
2. SEC.2 - Cybersecurity Implementation
3. SEC.3 - Risk Treatment Verification
4. SEC.4 - Risk Treatment Validation
5. ACQ.2 - Supplier Request and Selection
6. MAN.7 - Cybersecurity Risk Management

Figure 3 shows how the ASPICE model which is extended with the additional cybersecurity processes. The processes SEC.1-SEC.4 are defined by the *Cybersecurity Engineering Process Group*. The processes MAN.7 and ACQ.2 are a part of existing process groups that ASPICE has created. The process *ACQ.4 - Supplier Monitoring*, which has already been defined by ASPICE [34], has been modified to meet the requirements of cybersecurity.

2.3 Assessment with Automotive SPICE for Cybersecurity PAM

In order to analyze the processes described in Sect. 2.2 in accordance with the NPLF scheme defined in [34], the extra procedures outlined for cybersecurity have already been implemented in the tool Capability Advisor [23,24]. The following characteristics have been found:

- It is very helpful to take into account ASPICE, safety, and security elements [17] regarding the system to be evaluated
- Assessors must have a thorough understanding of the work product [4]
- Management and support processes were found to be lacking [24]
- The system being evaluated is not stand-alone, thus the system's interfaces must be taken into account [21]

3 Types of Audits

3.1 Automotive SPICE for CyberSecurity

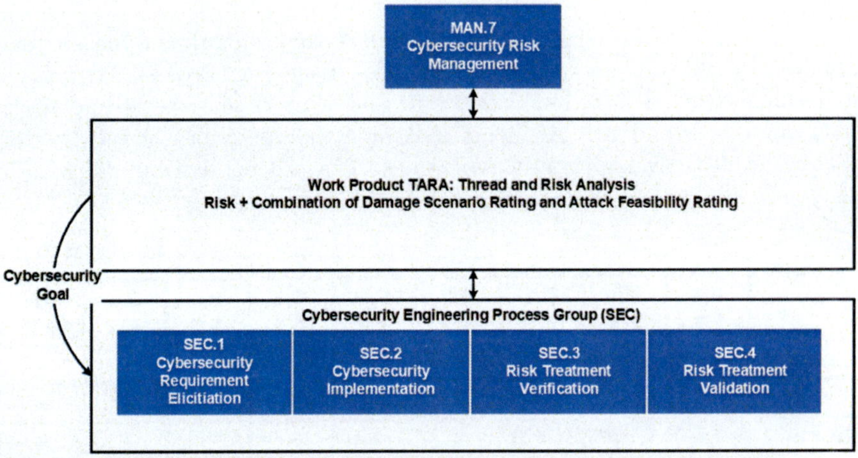

Fig. 4. TARA

New cybersecurity-related processes [36] to be examined as described in Sect. 2.2 - Cybersecurity were included to the Automotive SPICE 3.1 assessment model's 2021 extension [36], which was published. The connections between the processes MAN.7 and SEC.1 to SEC.4 are highlighted in Fig. 4. MAN.7 evaluates TARA (Threat and Risk Analysis) management and produces cybersecurity objectives and risk management decisions [22]. SEC.1 takes the cybersecurity goals as top-level requirements and further develops them into cybersecurity system, software, and hardware requirements [5,25] while also taking threat modeling at various tiers into account. SEC.2 improves the design, incorporates cybersecurity measures [4,20–22,26], and enriches them with controls. The cybersecurity standards are assessed as part of SEC.3's planning and implementation of the cybersecurity verification. SEC.4 organizes and carries out a validation of the product using fictional hackers (e.g., by penetration testing by an independent expert team). Further vulnerabilities and risks could be found in each of the processes SEC.1 to SEC.4, which would need updating the TARA (Threat and Risk Analysis). In an ASPICE for cybersecurity assessment the processes are rated at capability levels 1 to 3:

1. Performed
   - Process Attribute (PA) 1.1: Process Performance
2. Managed
   - PA 2.1 Performance Management
   - PA 2.2 Work Product Management
3. Established
   - PA 3.1 Process Definition
   - PA 3.2 Process Deployment

The rating scale is defined in ISO 33002 [11] and is used as a measurement framework in the well-known ASPICE model. Each PA (Process Attribute) is rated using N(ot), P(artially), L(argley), and F(ully) scale according to ISO 33002, and the standard also includes a table that describes how to create a capacity level by averaging process ratings. The process attribute(s) of levels below must be F in order to reach level ≥ 2. (See Fig. 5)

	PA 1.1	PA 2.1	PA 2.2	PA 3.1	PA 3.2	PA 4.1	PA 4.2	PA 5.1	PA 5.2
Level 0	-	-	-	-	-	-	-	-	-
Level 1	L / F	-	-	-	-	-	-	-	-
Level 2	F	L / F	L / F	-	-	-	-	-	-
Level 3	F	F	F	L / F	L / F	-	-	-	-
Level 4	F	F	F	F	F	L / F	L / F	-	-
Level 5	F	F	F	F	F	F	F	L / F	L / F

Fig. 5. Levels for ASPICE.[34]

Assessors use the information from the NPLF to evaluate processes in accordance with the predetermined scope. Before the assessment begins, the assessor and the project's responsible agree on the scope of the processes to be evaluated. The scope of the evaluation relies on the scope of the project and may incorporate the processes of VDA (Verband der Automobilindustrie) scope. The

Table 1. Capability profile.

Process	PA 1.1	PA 2.1	PA 2.2	PA 3.1	PA 3.2	Level
MAN.7	F	L	F	X	X	2
SEC.1	F	L	F	X	X	2
SEC.2	L	L	F	X	X	1
SEC.3	L	L	F	X	X	1
SEC.4	L	L	F	X	X	1
ACQ.2	F	L	F	X	X	2

capacity profile is one of the evaluation's outcomes. Table 1 provides an illustration of a competency profile for security-related processes.

All automakers currently demand their suppliers to have at least level 1 capability in each cybersecurity process and no level 0 capability in any cybersecurity process. Premium automakers already checking for level 2 capability in all cybersecurity processes. Table 1 shows a possible result of an assessment which has Cybersecurity in scope.

3.2 Automotive Cybersecurity Management System (ACSMS) Audit

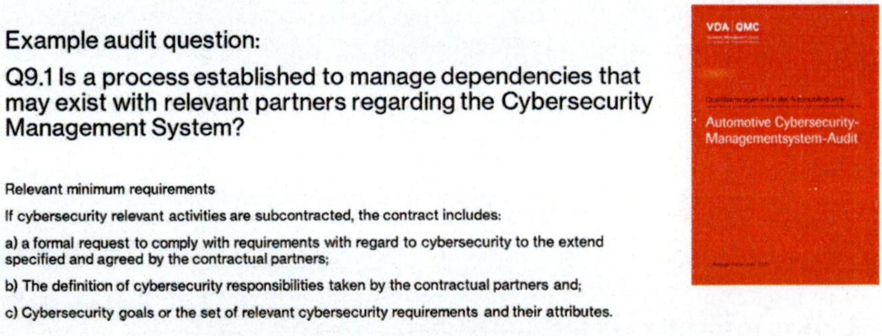

Fig. 6. Example ACSMS audit question.

The IATF 16949 auditor performs an organizational audit known as the Automotive Cybersecurity Management System audit every three years utilizing the VDA Red Book checklist. The IATF 16949 [8] auditor either possesses cybersecurity expertise or hires a cybersecurity specialist to join his audit team. The ACSMS audit book [33], which VDA published in December 2020, has a list of the checkpoints that must be used. The questions in that book (see example question in Fig. 6) determine whether a cybersecurity management system has been built by the organization supporting all projects (and including the complete life cycle). The outcome is an audit protocol that either has no deviations or does. If there are no deviations, a CSMS certificate will be given to the organization; if there are deviations, an improvement action plan will be developed to improve CSMS of organizations.

3.3 ISO 21434 Product Assessment

The work products are examined by independent cybersecurity specialists as part of the ISO 21434 product assessment. The work products that need to be examined in the context of automotive pertain to embedded systems [25]. The state-of-the-art is used as an argument because the ISO standards and UN regulations do not specify the expertise of these specialists. Professionals with demonstrated experience in cybersecurity, for instance, will compare the work items to industry standards. Working parties currently discuss work products' cutting-edge content, but this phrase will change with time. State-of-the-art, for instance, can alter throughout time. Checklists for specialized reviews of each work product are used. Deviations are recorded and need to be fixed before manufacturing can begin. If they are unsolvable, a report with residual risks must be delivered to the client.

Work Product Review Checklist. SEC.1) Work Product: 17-11 Software requirements specification expected content based on state of the art (example review checklist from SOQRATES [29] working group):

- Is Key Management defined?
- What kind of encryption is used?
- Is HSM used?
- Random generator with seed key used?
- Etc.

4 Commonalities Between Automotive SPICE for Cybersecurity and ISO 21434 Product Assessment

The evaluation for ISO 21434 Product Assessment includes technological considerations and work product related review checklists, whereas the assessment for Automotive SPICE for Cybersecurity follows a formal approach. Results from a recent PhD study are published in this part, demonstrating the relationship between the work product assessments of ISO 21434 [12] and ASPICE for Cybersecurity [36], as well as how to monitor their consistency. Also, it is assumed that by connecting them, time can be saved by combining the information learned from the two different types of assessments. The methods used in the two types of assessments (Automotive SPICE for Cybersecurity and ISO

21434 Product Assessment) vary. Discovering the foundation for their assessment result is helpful for improving one form of assessment by the other. When process assessment and product assessment are performed, it becomes clear that the majority of the queries concern established work products, such as the *17-11 Software requirements specification*. The questions for the two distinct assessment kinds vary. Hence, the entities Attribute and Question will be added to the process assessment model (PAM) as defined by [34]. In Fig. 7, these recently introduced entities are highlighted in color. The entities in Fig. 7 with white backgrounds and highlighted text do not distinguish between the two forms of evaluation. The additional entity categories have been developed to address the many issues raised by the two methods of assessment. The example in Table 2 below demonstrates how Automotive SPICE for Cybersecurity (Assessment 1) and ISO 21434 Product Assessment (Assessment 2) may both enhance another.

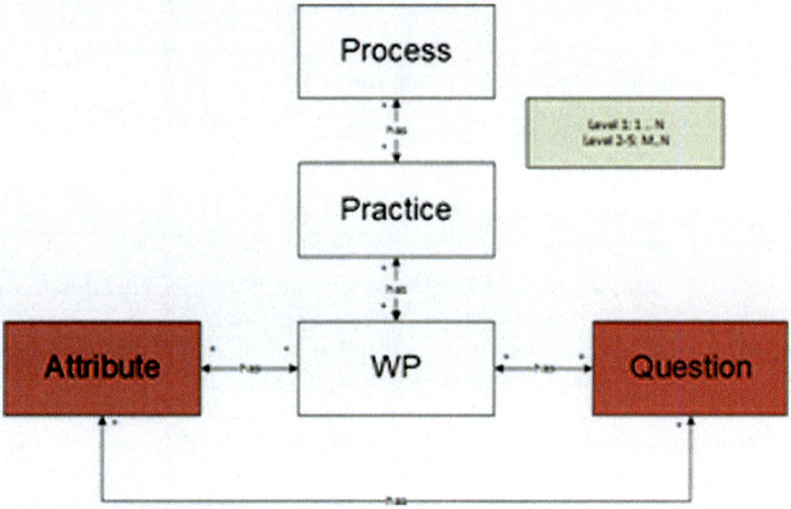

Fig. 7. Extended relationship model.

As shown in Table 2 the assessment Automotive SPICE for Cybersecurity (Assessment 1) supports the assessment ISO 21434 Product Assessment Assessment 2). For example, the question *Are cybersecurity specific questions part of the checklist?* answers one question for attribute *Review* of assessment Automotive SPICE for Cybersecurity for work product *17-11 Software requirements specification*. The response to this question may also be used for assessment ISO 21434 Product Assessment. The questions asked during assessment ISO 21434 Product Assessment are not used by assessment Automotive SPICE for Cybersecurity because assessment ISO 21434 Product Assessment focusses on the technical content of the work product while the focus of assessment Automotive SPICE for Cybersecurity is on the underlying process and therefore follows a formal procedure.

Table 2. Attributes and questions for 17-11 software requirements specification.

Attr.	Question	Ass.
Text	Is Key Management defined?	2
Text	What kind of encryption is used?	2
Text	Is HSM used?	2
Text	Random generator with seed key used?	2
Text	Is One Time Password (OTP) used?	2
User	Is the responsible defined?	1
Text	Is the requirement testable?	1
State	Do requirements have a state?	1
State	How many requirements are released?	1,2
Review	Do all released requirements have a review?	1,2
Review	Are cybersecurity specific questions part of the checklist?	1,2
Link	Are the requirements linked correctly?	1
...	...	...

5 Validation by Review

Major automobile suppliers validated the procedure after reviewing it with future combined security evaluations in mind. The concept covered in Sect. 4 was introduced to the SOQRATES participants. Tier 1 suppliers like Bosch, Hella, Conti, and ZF are a part of the working group SOQRATES [29] directed by Richard Messnarz from the company ISCN. The suggestions made by the working group have been implemented to improve the existing concept. For instance, the first plan called for adding attributes and questions to the PAM of ASPICE. This extension is recommended since it will make it easier for assessors to do their work and cut down on the amount of time needed. Although this is a novel strategy that expands the PAM, the working group came to the conclusion that it needs to be verified in multiple conducted assessments. The current format was presented and discussed three times at SOQRATES over a nine-month period. The goal of the working group is to reduce the time that both types of assessments now take to finish an evaluation. The answers to separate questions about the aspects of time reduction introduced in Sect. 4 are described in Sects. 5.1 and 5.2.

5.1 Attributes

Concentrating on work products and, consequently, work product attributes, is a novel way to shorten the time needed for an evaluation. The ASPICE and cybersecurity processes call for a wide variety of work products. If the characteristics of both types of evaluations were taken into consideration, the length of the assessment might be decreased.

5.2 Questions

The questions and traits are strongly related, according to Sect. 4. When the traits of typical work products that fulfill requirements for ASPICE and cybersecurity are taken into account, questions will automatically be taken into account.

5.3 Summary

The general reaction of the working group SOQRATES is that the recommendations presented in Sect. 4 may lead to a reduction in the time required to complete an evaluation. To confirm that the proposals lead to a time savings, several assessments encompassing both ASPICE and cybersecurity processes must be performed.

6 Conclusion and Further Research

Processes for evaluating the competence level in the range 0–5 are defined by ASPICE. To evaluate the process attributes of one process, the NPLF Scheme is used. System development, software development, support, management, and acquisition are all topics covered by the ASPICE processes. Moreover, ASPICE for Cybersecurity defines acquisition (ACQ.2), cybersecurity risk management (MAN.7), and security engineering processes (SEC.1-SEC.4). The evaluation will take a lot of time if both sorts of assessments are included in its scope. Concentrating on work products can help shorten the assessment process. It enables assessors who are carrying out evaluations within the given scope to shorten assessment times. Because the used work products are the same or one output work product may at least be a component of the other, this approach for shortening assessment execution time is feasible. This study outlines the discovery of shared characteristics and assessment-related questions (see Sect. 4). The research question:

How can one type of cybersecurity assessment enhance the other?

can be answered the following way: Assessment Automotive SPICE for Cybersecurity follows a very formal procedure as defined by ASPICE and the answers of this assessment can be used for assessment ISO 21434 Product Assessment. The assessment ISO 21434 Product Assessment focusses on technical details of work product like for *17-11 Software requirements specification* and does not give any input to assessment Automotive SPICE for Cybersecurity. This means, in order to save time performing the assessments mentioned above it is recommended to execute assessment Automotive SPICE for Cybersecurity before assessment ISO 21434 Product Assessment. In future the time reduction performing an assessment, using the proposal is investigated. Finally, the time reduction will be measured, caused by the proposals described in Sect. 4.

Acknowledgments. We are grateful to a working party of Automotive suppliers SOQRATES (www.\discretionary-soqrates.de) who exchanged knowledge about such assessment strategies. This includes: Boehner Martin (Elektrobit), Brasse Michael

(HELLA), Bressau Ernst (BBraun), Dallinger Martin (ZF), Dorociak Rafal (HELLA), Dreves Rainer (Continental Automotive), Ekert Damjan (ISCN), Forster Martin (ZKW), Geipel Thomas (BOSCH), Grave Rudolf (Elektrobit), Griessnig Gerhard (AVL), Gruber Andreas (ZKW), Habel Stephan (Continental Automotive), Haellmayer Frank (Software Factory), Haunert Lutz (Giesecke & Devrient), Karner Christoph (KTM), Kinalzyk Dietmar (AVL), Koenig Frank (ZF), Lichtenberger Christoph (MAGNA ECS), Lindermuth Peter (Magna Powertrain), Macher Georg (TU Graz & ISCN), Mandic Irenka (Magna Powertrain), Maric Dijas (Lorit Consultancy), Mayer Ralf (BOSCH Engineering), Mergen Silvana (TDK/EPCOS), Messnarz Richard (ISCN), Much Alexander (Elektrobit), Nikolov Borislav (msg Plaut), Oehler Couso Daniel (Magna Powertrain), Riel Andreas (Grenoble INP & ISCN), Rieß Armin (BBraun), Santer Christian (AVL), Schlager Christian (Magna ECS), Schmittner Christoph (Austrian Institute of Technology AIT), Schubert Marion (ZKW), Sechser Bernhard (Process Fellows), Sokic Ivan (Continental Automotive), Sporer Harald (Infineon), Stahl Florian (AVL), Wachter Stefan (msg Plaut), Walker Alastair (Lorit Consultancy), Wegner Thomas (ZF).

References

1. Ahmad, F., Adnane, A., Franqueira, V., Kurugollu, F., Liu, L.: Man-in-the-middle attacks in vehicular ad-hoc networks: evaluating the impact of attackers strategies. Sensors **18** (2018). https://doi.org/10.3390/s18114040
2. Brennich T., Moser M.: Automotive Security auf dem Pruefstand. In: ATZelectronics, Month 1+2, pp. 48–53 (2020)
3. Cheng, B., Doherty, B., Polanco, N., Pasco, M.: Security patterns for connected and automated automotive systems. Automot. Softw. Eng. **1**(1), 51–77 (2021). https://doi.org/10.2991/jase.d.200826.001
4. Dobaj, J., Ekert, D., Stolfa, J., Stolfa, S., Macher, G., Messnarz, R.: Cybersecurity threat analysis and risk assessment and design patterns for automotive networked embedded systems: a case study. JUCS - Univ. Comput. Sci. **27**(8), 830–849 (2021). https://lib.jucs.org/article/72367/
5. Dobaj, J., Macher, G., Ekert, D., Riel, A., Messnarz, R.: Towards a security-driven automotive development lifecycle. J. Softw. Evolut. Process. (2021). https://doi.org/10.1002/smr.2407. WILEY Online Library
6. Ebert, C.: Efficient implementation of standards for security, safety and UNECE. In: ATZelectronics Worldwide, pp. 40–43 (2020)
7. Groza, B., Murvay, P.: Identity-based key exchange on in-vehicle networks: CAN-FD and FlexRay, Sensors **19**(22) (2019). https://doi.org/10.3390/s19224919
8. IATF. IATF 16949 Anforderungen an Qualitaetsmanagementsysteme fuer die Serien- und Ersatzteilproduktion in der Automobilindustrie (2016)
9. intacs. HW Spice, intacs Working Group HW Engineering Processes (2019)
10. intacs. Process Assessment Model SPICE for Mechanical Engineering, intacs Working Group MECH Engineering Processes (2020)
11. ISO. ISO 33002 Information technology - Process assessment - Process measurement framework for assessment of process capability (2019)
12. ISO, SAE. ISO/SAE DIS 21434. Strassenfahrzeuge, Cybersecurity Engineering (2020)
13. Ivančič, J., Riel, A., Ekert, D.: An interpretation and implementation of automotive hardware SPICE. In: Yilmaz, M., Niemann, J., Clarke, P., Messnarz, R. (eds.)

EuroSPI 2020. CCIS, vol. 1251, pp. 684–695. Springer, Cham (2020). https://doi.org/10.1007/978-3-030-56441-4_51
14. Jadhav, A.: Automotive cybersecurity. In: Kathiresh, M., Neelaveni, R. (eds.) Automotive Embedded Systems. EICC, pp. 101–114. Springer, Cham (2021). https://doi.org/10.1007/978-3-030-59897-6_6
15. Kim, S., Shrestha, R.: Introduction to automotive cybersecurity. In: Automotive Cyber Security, pp. 1–13. Springer, Singapore (2020). https://doi.org/10.1007/978-981-15-8053-6_1
16. Laborde, R., Bulusu, S., Wazan, A., Oglaza, A., Benzekri, A.: A methodological approach to evaluate security requirements engineering methodologies: application to the IREHDO2 project context. Cybersecur. Privacy **1**(3), 422–452 (2021). https://doi.org/10.3390/jcp1030022
17. Macher, G., Schmittner, C., Dobaj, J., Armengaud, E.: An Integrated View on Atomotive SPICE and Functional Safety and Cyber-Security (SAE Technical Paper) (2020). https://doi.org/10.4271/2020-01-0145
18. Macher, G., Schmittner, C., Veledar, O., Brenner, E.: ISO/SAE DIS 21434 automotive cybersecurity standard - in a nutshell. In: Casimiro, A., Ortmeier, F., Schoitsch, E., Bitsch, F., Ferreira, P. (eds.) SAFECOMP 2020. LNCS, vol. 12235, pp. 123–135. Springer, Cham (2020). https://doi.org/10.1007/978-3-030-55583-2_9
19. Macher, G., Armengaud, E., Messnarz, R., Brenner, E., Kreiner, C., Riel, A.: Integrated Safety and Security Development in the Automotive Domain (2017). https://doi.org/10.4271/2017-01-1661
20. Macher, G., Much, A., Riel, A., Messnarz, R., Kreiner, C.: Automotive SPICE, Safety and Cybersecurity Integration. In: Tonetta, S., Schoitsch, E., Bitsch, F. (eds.) SAFECOMP 2017. LNCS, vol. 10489, pp. 273–285. Springer, Cham (2017). https://doi.org/10.1007/978-3-319-66284-8_23
21. MacGregor, J., Burton, S.: Challenges in assuring highly complex, high volume safety-critical software. In: Gallina, B., Skavhaug, A., Schoitsch, E., Bitsch, F. (eds.) SAFECOMP 2018. LNCS, vol. 11094, pp. 252–264. Springer, Cham (2018). https://doi.org/10.1007/978-3-319-99229-7_22
22. Messnarz, R., Ekert, D., Macher, G., Stolfa, S., Stolfa, J., Much, A.: Automotive SPICE for cybersecurity MAN.7 cybersecurity risk management and TARA. In: Yilmaz, M., Clarke, P., Messnarz, R., Woeran, B. (eds) Systems, Software and Services Process Improvement. EuroSPI 2022. Communications in Computer and Information Science, vol 1646. Springer, Cham (2022). https://doi.org/10.1007/978-3-031-15559-8_23
23. Messnarz, R., Ekert, D., Zehetner, T., Aschbacher, L.: Experiences with ASPICE 3.1 and the VDA automotive SPICE guidelines – using advanced assessment systems. In: Walker, A., O'Connor, R.V., Messnarz, R. (eds.) EuroSPI 2019. CCIS, vol. 1060, pp. 549–562. Springer, Cham (2019). https://doi.org/10.1007/978-3-030-28005-5_42
24. Messnarz, R., et al.: First experiences with the automotive SPICE for cybersecurity assessment model. In: Yilmaz, M., Clarke, P., Messnarz, R., Reiner, M. (eds.) EuroSPI 2021. CCIS, vol. 1442, pp. 531–547. Springer, Cham (2021). https://doi.org/10.1007/978-3-030-85521-5_35
25. Moselhy, N., Ali, Y.: Impact of the new A-SPICE appendix for cybersecurity on the implementation of ISO26262 for functional safety. In: Yilmaz, M., Clarke, P., Messnarz, R., Reiner, M. (eds.) EuroSPI 2021. CCIS, vol. 1442, pp. 122–136. Springer, Cham (2021). https://doi.org/10.1007/978-3-030-85521-5_9

26. Petho, Z., Khan, I., Torok, Á.: Analysis of security vulnerability levels of in-vehicle network topologies applying graph representations. J. Electron. Test. 1–9 (2022). https://doi.org/10.1007/s10836-021-05973-x
27. Schlager, C. Messnarz, R., Sporer, H., Riess, A., Mayer, R., Bernhardt, S.: Hardware SPICE extension for automotive SPICE 3.1. In: Larrucea, X., Santamaria, I., O'Connor, R.V., Messnarz, R. (eds.) EuroSPI 2018. CCIS, vol. 896, pp. 480–491. Springer, Cham (2018). https://doi.org/10.1007/978-3-319-97925-0_41
28. Singh, M.: Cybersecurity in automotive technology. In: Information Security of Intelligent Vehicles Communication. SCI, vol. 978, pp. 29–50. Springer, Singapore (2021). https://doi.org/10.1007/978-981-16-2217-5_3
29. SOQRATES, Task Forces Developing Integration of Automotive SPICE, ISO 26262, ISO21434 and SAE J3061. http://soqrates.eurospi.net/
30. Stolfa, J., et al.: DRIVES EU blueprint project for the automotive sector a literature review of drivers of change in automotive industry. J. Softw. Evolut. Process **32**(3) (2020). Special Issue: Addressing Evolving Requirements Faced by the Software Industry
31. UN. UN Regulation No. 155 - Cyber security and cyber security management system (2021)
32. UN. UN Regulation No. 156 - Software update and software update management system (2021)
33. VDA QMC. Automotive Cybersecurity Managementsystem Audit (2020)
34. VDA QMC. Automotive SPICE Process Reference Model/Process Assessment Model (2015)
35. VDA QMC. Automotive Spice Guidelines, 2nd edn (2017)
36. VDA QMC. Automotive SPICE for Cybersecurity Process Reference and Assessment Model (2021)

How Zero Trust as a Service (ZTaaS) Reduces the Cost of a Breach

A Conceptual Approach to Reduce the Cost of a Data Breach

Yuri Bobbert(✉) and Tim Timmermans

ON2IT, Zaltbommel, Netherlands
{yuri.bobbert,tim.timmermans}@on2it.net

Abstract. Nowadays, the cost of data breaches is immense for organizations and society. Organizations can benefit from understanding the core components influencing a data breach's likelihood and impact. Zero Trust (ZT) strategies have proven to be more effective in addressing cyber risk resilience than the more traditional 'implement-a-framework' approach because the ZT focus on segmenting the environment by ringfencing sensitive data sets, limiting lateral movements and reducing the potential blast radius in case of a breach. However, Zero Trust has traditionally focused on technical measures and associated verifications. In isolation, technological measures won't offer the optimal data-breach-cost-reduction. Company owners and boards that neglect their responsibilities are liable to hefty financial fines and individual liability. This paper describes how Zero Trust and Zero Trust as a Service (ZTaaS) decrease the economic impact of a data breach. It concludes with a conceptual approach that practitioners can immediately leverage to reduce the potential cost of a data breach and lower insurance premiums. Insurers can benefit from the simultaneous reduction of likelihood and impact.

Keywords: Cybersecurity · Zero Trust · Information Economics · Risk Management · Incident Response · Case analysis

1 Introduction

Every year Ponemon publishes its report on the cost of a data breach and the core cost categories that make the overall cost [1]. Some highlights from the 2021 Ponemon report:

- "Data breach costs rose from \$3.86 million to \$4.24 million, the highest average total cost in the history of this report. Costs were significantly lower for some organizations with a more mature security posture, and higher for organizations that lagged in areas such as security AI and automation, zero trust and cloud security.
- 287 is the average number of days to identify and contain a data breach. The longer it took to identify and contain, the more costly the breach.
- Costs varied widely across industries, and year over year. Healthcare data breach costs increased from an average total cost of \$7.13 million in 2020 to \$9.23 million in 2021, a 29.5% increase. Costs in the energy sector decreased from \$6.39 million in 2020 to an average \$4.65 million in 2021. Costs surged in the public sector, which saw a 78.7% increase in average total cost from \$1.08 million to \$1.93 million.

K. Arai (Ed.): FTC 2023, LNNS 816, pp. 433–454, 2023.
https://doi.org/10.1007/978-3-031-47448-4_33

- $1.76m Cost difference in breaches where mature zero trust was deployed vs. no zero trust.
- A zero-trust approach helped reduce the average cost of a data breach.
- The average cost of a breach was $5.04 million for those without zero trust deployed. Yet in the mature stage of zero trust deployment, the average cost of a breach was $3.28 million, $1.76 million less than organizations without zero trust, representing a 2.3% difference.
- Ransomware attacks cost an average of $4.62 million, more expensive than the average data breach ($4.24 million). These costs included escalation, notification, lost business and response costs but did not include the cost of the ransom. Malicious attacks that destroyed data in destructive wiper-style attacks cost an average of $4.69 million."

Ponemon identifies four cost centers regarding the costs of a data breach:

1) *Detection* – The costs of activities that enable a company to detect the breach reasonably.
2) *Notification* – The costs of activities that enable the company to notify data subjects, data protection regulators and other third parties.
3) *Lost business* – The costs of activities that attempt to minimize the loss of customers, business disruption and revenue losses.
4) *Post-breach response* – The costs of activities to help breach victims communicate with the company and redress activities to victims and regulators.

Ponemon refers to data breaches like ransomware, data leakage, theft and technical failures. Over the past, we have seen additional cost categories arise next to the one in Ponemons' research, such as missed revenue, key personnel leaving due to lack of trust, share prices drop or other intangible costs. These costs are not directly tackled by technology, but effective technology helps to get the fact-based response better and avoid speculation, and therefore, the perception of being 'in control' is better, resulting in minimal damages.

The Role of Regulatory Fines in Cost of Data Breaches. In the majority of the cases, legal fines for violating personal data regulations (GDPR (EU), CCPA (California), LGPD (Brazil), HIPAA (US), etc.) are by far the most significant cost category. Enforcementtracker.com keeps track of fines issued by European Data Protection Authorities because of GDPR violations. As of 23 December 2021, the highest fines per type of violation are listed in Table 1.

We observe most violations relate to GDPR articles on:

- The obligations and legal basis (art. 5, 6, 13).
- The required technical and organizational measures to ensure information security (art. 32).

According to Kuijper (2020), "the data privacy authorities (DPA's) mainly fine the visible symptoms of GDPR non-compliance (the materialized risks) rather than fining and describing the underlying root cause(s) of those non-compliance symptoms like, e.g., failing to analyze the data processing risks, a lack of governance or controls, etc."

Table 1. Biggest GDPR penalties taken from GDPR enforcement tracker, tracked by CMS. Retrieved on 23 December 2021.

Sum of Fines	Violation	Category	GDPR Article
€ 784,694,744	Non-compliance with general data processing principles	Compliance	Art. 5 Principles relating to the processing of personal data (proportionality)
€ 234,966,595	Insufficient fulfillment of information obligations	Compliance	Art. 13 Information to be provided where personal data are collected from the data subject (Data localization)
€ 192,414,588	Insufficient legal basis for data processing	Compliance	Art. 6 Lawfulness of processing (Legal ground)
€ 69,733,969	Insufficient technical and organizational measures to ensure information security	Technical, Organisational	Art. 32 Security of processing

Regulatory fines are the most significant cost of the breach; they sometimes can increase to 4% of the annual global revenue of the violating company. In this paper, we take GDPR-related fines after a breach as a tangible cost example since these are transparent to the public (see Table 1) regarding the exact amount. We then zoom into GDPR article 32 related causes, being technical measures. Ponemon, however, refers to generic costs hidden from the public and does not specify the cause per case. In line with the Data Protection regulations, announcing upcoming regulatory requirements for the EU, such as NIS2, will impose even higher costs.

Ponemon states that the implementation of Zero Trust, which focuses on protecting valuable data, applications, assets, and services (DAAS), can significantly reduce the cost of the breach. However, the report needs to be more precise on which Zero Trust principles were taken into account and needs a clear description of what Zero Trust comprises. Our intermediate finding is that organizations generally struggle with understanding and implementing sufficient technical security measures that prevent data breaches from a likely chance of occurrence or reduce the impact of a data breach. It requires expert knowledge to distinguish this, and that remains a problem for practitioners wanting to implement Zero Trust. "Despite the widespread usage of Zero Trust concepts in the practitioner's community, critics are on the rise, since Zero Trust remains covered in mystic and misconceptions" [2].

The sole *assumption* that Zero Trust can reduce the costs of a breach is quite a claim. Therefore, this asks for a more fine-grained analysis that underpins the different components that reduce the risk of a data breach. We will highlight the difference between Zero Trust as a technology-oriented strategy (i.e., design and implementation of technical security measures) versus a holistic 'as a Service' approach that also considers operational processes along with the suitable steering mechanism (i.e., dashboarding).

With the objective to provide a conceptual framework with clear guidelines of how Zero Trust can be implemented and how it contributes in lowering the financial impact of a breach.

2 Methodology

For this paper, we have used a mixed-method approach. We start by introducing the *why* of this topic via contemporary literature and benchmark data (Ponemon, CMSlaw). After introducing the basic concepts of Zero Trust, we elaborate on what Zero Trust as a Service (ZTaaS) is. Then we elaborate on the cost reduction components based upon previous executive master breach-scenario-studies conducted at Antwerp Management Schools Cybersecurity Management track. These components are used as a structured 'frame of reference' to detail *how* they reduce the *likelihood* and *impact* of a breach. We will conclude with a conceptual framework that practitioners can apply to minimize a breach's cost, which academics can use to further research.

3 Core Concepts

3.1 What Are the Zero Trust Concepts?

Zero Trust, in its densest form, is positioned as a security strategy, assuming that *inherent trust is removed from the network and all network traffic is explicitly verified.* The Zero Trust strategy is based on four core principles as defined by John Kindervag (2010) [3]:

1) *Design from the inside out* means critical assets are protected up close through Protect Surfaces.
2) *The principle of least privileges* means access is granted to systems and users only on a need-to-know basis.
3) *Inspect and log all traffic* to quickly detect and respond to a malicious intent to ensure complete visibility.
4) *Never trust, always verify*, which means inherent trust is removed from the network, and everything requires explicit verification.

Traditional Zero Trust focuses predominantly on technological measures in IT and less on the organizational capabilities and processes required to monitor, handle and respond to events and incidents. It focuses on describing the *why* and *what* and less on the *how* of cybersecurity people and processes. This is why many organizations struggle with implementing Zero Trust on a strategic level.

3.2 What Are the Zero Trust as a Service (ZTaaS) Concepts?

Zero Trust as a Service (ZTaaS) has incorporated additional cybersecurity services involving the design of Zero Trust environments, implementing relevant technical measures, monitoring and response capabilities, and dashboarding analytics presented via a single pane of glass. ZTaaS consists of the following elements:

Design of Protect Surfaces. A process according to the five steps of Zero Trust Creator John Kindervag to identify business-relevant data, applications, assets, and services to design Protect Surfaces and thereby segment the environment and assign measures.

Implementation of Technical Measures. An organization implements technological measures to Identify, Protect, Detect, respond to, and recover cybersecurity. The selection of technical measures is part of designing Protect Surfaces. These measures use a high level of automation and orchestrated event handling. Multiple vendors like Palo Alto Networks, Microsoft, or Fortinet offer these measures, regardless of whether these are on-premise or in the cloud. Consider measures such as Endpoint protection, Firewalls, Cloud Access Security Brokers (CASB), cloud controls, etc., to leverage security up to layer seven in the OSI reference model [4].

Dashboarding. After the implementation of technical security measures like firewalls, endpoints, etc. Also known as Policy Enforcement Points (PEPs), these must be managed and maintained in a Security Orchestration and Automation Response (SOAR) platform with dashboarding capabilities. That also displays the status of the effective working of these measures. This is where the use of the measures in the Business as Usual (BAU) is displayed, and during security events and incidents, the board and management can make informed decisions. These platforms with dashboards must comply with information such as risk scores on high-value assets and regulatory frameworks. Following the Zero Trust Framework's proposition, previous publications from Bobbert and Scheerder elaborate on building and engineering a portal technology [5] and how SOC and CSIRTs can benefit from this.

SOC. A Security Operations Center (SOC) represents the technical, human, and organizational capabilities to monitor and respond to security events. For this paper, we refer to the official capabilities of a SOC presented by Schinagls paper "A Framework for Designing a Security Operations Centre (SOC)" [6]. In addition, Appendix 2 (Table 6) refers to Security Operations Center capabilities.

CSIRT. A Computer Security Incident Response Team (CSIRT) represents the technical, human, and organizational capabilities to respond to cybersecurity incidents. For this paper, we refer to the official capabilities presented by ENISA [7] and Mooi & Botha [8] for CSIRTs.

3.3 What Are the Risk Management Concepts?

In recent decades IT security has become 'information security' [9]. The International Organization for Standardization (ISO) specifies information security as "*protecting information assets from a wide range of threats in order to ensure business continuity, minimize business risk and maximize return on investment and business opportunities*" [10–12]. The core concepts for information security are confidentiality, integrity, and availability, usually referred to by the acronym CIA [10, 13]:

- *Confidentiality* means preventing the disclosure of information to unauthorized parties.
- *Integrity* seeks to prevent unauthorized modification of information.

- *Availability* is the term for all kinds of ways to make the necessary information available to users within an enterprise and the 'extended enterprise'.

In addition, ISO added other properties such as authenticity, accountability, non-repudiation[1], reliability, and auditability due to audit and compliance regulations. This information security deals with the assurance of a certain level of system quality [14]. Information security is directly related to information risk management. Several risk & security methods have been developed over the last years, such as CRAMM[2], OCTAVE [15], National Institute of Standards and Technology (NIST) [16], and ISFs' IRAM [17], particularly into risk analysis and risk assessments to analyze threats, vulnerabilities and the impact on information systems and derive measures for mitigation. To determine the information security requirements, e.g., measures in the form of process measures, technical measures, or people measures is based on the risk and impact estimation on the critical *business* assets. Zero Trust looks at this from a more holistic approach, considering not only the asset but also associated applications and services that form a logical data segment. Also known as the 'Protect Surface'. This means that risk assessments need to be conducted on the Protect Surface rather than the siloed asset.

Gantz et al. (2013) [18] state "Risk determination assesses threats and vulnerabilities to consider the likelihood that known threat sources will be able to exploit identified vulnerabilities to cause one or more adverse events and the consequences if such events occur. Depending on the type of threat under analysis and the nature of the risk assessment, likelihood and impact may be determined using relative ratings or quantitative estimates. Many factors contribute to likelihood, including some that are difficult to measure accurately, such as ease of exploitation, skill level or sophistication of adversaries, visibility of the organization, and attractiveness of the organization or its assets to attack."

The relationship of Risk Management to Risk Assessment and information security control setting is visualized in Fig. 1 and is adapted from OCTAVE [15].

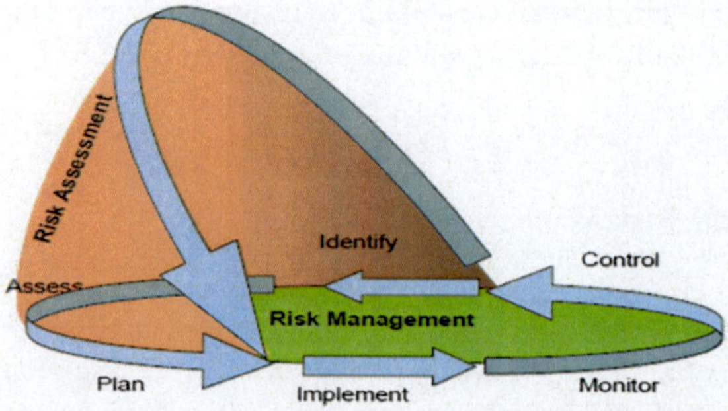

Fig. 1. The relationship between risk management and risk assessment. Taken from OCTAVE.

[1] ISO 2.54: The ability to prove the occurrence of a claimed event or action and its originating entities.

[2] CRAMM (CCTA Risk Analysis and Management Method) is a risk management methodology, currently in its fifth version, CRAMM Version 5.0.

Risk Determination provides a quantitative risk value representing the system's exposure to a threat exploiting a particular vulnerability after current controls have been considered. This quantitative value is in the form of a risk score. A risk score follows the formula:

$$Risk = Impact \times Likelihood$$

The computation for the risk value is a relatively straightforward multiplication of the Impact and Likelihood scores. This computation was performed for all threat and vulnerability pairs for all systems in scope and can be visualized as the heat map displayed in Table 2.

Table 2. Risk heat map: tools for developing a tailored approach to risk management. Scott McKay, AICPA, 2011.

Potential impact						
	Extreme	5	10	15	20	25
	High	4	8	12	16	20
	Medium	3	6	9	12	15
	Low	2	4	6	8	10
	Negligible	1	2	3	4	5
		Remote	**Unlikely**	**Possible**	**Likely**	**Probable**
		Likelihood				

In this case, we do not quantify the specific risks and measures but propose a simplified way of risk modeling of likelihood and impact while assessing each Zero Trust as a Service component's contribution to reducing the costs.

4 How ZTaaS Reduces the Risks of a Data Breach

If one can reduce the risks related to data breaches, this will impact the associated costs and fines. Since the components of risk are *likelihood* and *impact*, the approach to this risk reduction is twofold:

1) A *preventive function* to avoid materialization risks, i.e., reducing the *likelihood* of a data breach. This relates to the Identify, Protect, and Detect functions of the NIST Cybersecurity Framework.
2) A *reactive function* on already materialized risks, i.e., reducing the *impact* of a data breach ('assume breach'). This relates to the Respond and Recover functions of the NIST Cybersecurity Framework.

For each of the five ZTaaS components, we will discuss to what extent it has a preventive or reactive function in data breach risk mitigation. We also assess per ZTaaS component the likelihood and impact contribution via the following scales:

- No reduction
- Low reduction
- Moderate reduction
- High reduction
- Maximum reduction

4.1 Design of Protect Surfaces

The initial step of Zero Trust Creator John Kindervag's five steps involves identifying Protect Surfaces. Each Protect Surface is a segment containing data, applications, assets, or services critical for business processes. As part of the Protect Surface identification process, you will inventory where data resides and what types and locations are critical.

Assigning a relevance score is a fundamental step in identifying each Protect Surface. This score consists of the elements that combine information security and system quality assurance: Confidentiality, Integrity, Availability, Personal Data, Subject to Audit (covering Non-Repudiation) and Business Criticality.

Each element is assigned a value based on the strictest or most sensitive value identified within the Protect Surface. Therefore, the relevance score reflects the risks associated with a Protect Surface, enabling you to identify high-risk environments explicitly. This score is a crucial factor for determining what technical security measures must be implemented for the Protect Surface, as this score reflects the risk of a Protect Surface. Appendix 1 (Tables 4 and 5) describes the potential value scales applicable to each element.

Identifying these high-risk environments supports decision-making on selecting suitable preventative measures (e.g., firewall configuration, eXtended Detection and Response (XDR), Data Leakage Prevention (DLP)) and the appropriate post-breach measures (e.g., backups, disaster recovery, activity logging).

Assessment. Although the design of a Protect Surface itself does not bring any immediate security (and therefore cost reduction), it does have a moderate contribution on both the impact and likelihood as it can be considered a fundamental aspect of a Zero Trust architecture and strategy.

Reduction of impact: *moderate*

Reduction of likelihood: *moderate*

4.2 Technical Security Measures

To provide insights on potential threat exposure, MITRE (mitre.org), for example, provides an instrumental and globally accessible knowledge base of adversary tactics and techniques based on real-world observations to which organizations are potentially exposed. Analyzing the most notorious groups and their Techniques, Tactics, and Procedures (TTP) reveals a common path of most used TTPs in the Lockheed Kill Chain [19].

Looking into the commonly used techniques, MITRE also proposes mitigations that ON2IT mapped in her Zero Trust as a Service offering. These measures are listed earlier [20], and vary from segmenting the network to Role-Based Access Control (RBAC), inspecting and logging all traffic to encryption, etc. For this paper, the focus is on demonstrating that the majority of techniques used by hackers are addressed by the measures in the Zero Trust Framework of ON2IT and utilized via a SOAR-based platform called AUXO [5]. Each implemented technical security measure is mapped on the MITRE mitigations within this AUXO platform. This provides the SOC/CSIRT analyst with insights on how to respond to events and threats.

In 2019 MITRE reported over 48 new vulnerabilities a day, of which +60% are classified as HIGH and CRITICAL (CVSS3.0). MITRE proposes mitigating measures. As an example, the illustration below provides further insights on the protection level against Advanced Persistent Threat (APT29), classified by the US Federal Government as APT29, a Russian hacker group believed to be associated with one or more intelligence agencies of Russia (Cozy Bear) and the technology providers coverage to avert the TTP's used by Cozy Bear group. In this case, you see that Palo Alto Networks technology usage significantly reduces the risk of manifestation to 90%. Using intelligent security measures like XDR from Palo Alto Networks – sometimes consumed from the cloud- or other A-brand technology providers – lowers the likelihood of a breach and the impact, mainly expressed in financial damage. The effect depends on some additional factors such as:

- Type of industry (regulated, listed, etc.)
- The magnitude of the legal fines, e.g., as part of annual turnover
- Inadequate response to an incident, e.g., lack of explanation, communication, and transparency
- The perceived company value; where governments are mainly valued on trust and reliability, the value of privately listed companies is primarily based on their stock price

In earlier publications by Bobbert and Scheerder [20], distilled Zero Trust *Measures* as a comprehensive *baseline* of security measures[3] following the AFCEA [21] guidelines such as Access Management, Change management, Configuration management, Segmenting, Cryptography, etc. and additional measures to master more sophisticated TTP's from attack groups. On top of these standard security measures, we amend these with an extra set of likelihood- and impact-reducing measures:

- Behavior analysis to early detect abnormal behavior to reduce the likelihood of risk manifestation. As an example, Cortex XDR analytics on breach investigation and Root Cause Analysis to respond quickly and limit the impact of reputational damage.
- Automated response via SOAR technology and not manual SIEM event handling for quick response to reduce the likelihood and impact of risk manifestation.
- Proactive content inspection of traffic flows between Protect Surfaces, including containers and pipelines, to reduce the likelihood of malicious traffic and the impact of malicious content inside Protect Surfaces and infecting other Protect Surfaces, reducing blast radius and collateral damage in the entire value chain.

[3] The study refers to measures where some studies refer to controls. In terms of auditing and assurance "controls" are broader, e.g. soft controls, process controls etc. Measures represent *technological* functions.

- Data minimization principles and data retention schemes designed and effectuated per Protect Surface reduce the likelihood of leakage of large and depreciated data sets and minimize the impact of data leakage due to the limited amount of data processed.
- Backup and restore of data and images of machines that automatically restore clean after a breach to reduce the impact of long-lasting and troublesome data recovery.

Applying the Kipling Method to Specify the Effective Working of the Measure The Kipling method [22] is also known as the 5W + 1H method by Rudyard Kipling. Kipling is common practice in the auditing field [23]. Mills stated, "The auditor should always use Kipling's checklist… during this stage of the auditing process". [24] Mahopo et al. quote, "The Kipling method is always recommended for use because it helps to explore the problem by probing the thinking of the problem solver with the questions: what, why, how, who, when and where."

In Zero Trust, conditions for access are formulated in an abstract (i.e., technology-independent manner) via the Kipling method: *who* gets access, *what* exactly access is provided, *when*, (from) *where*, *why*, and *how* (i.e., under which conditions is access granted).

The access defined in this way is operationalized by implementing granular and precise (targeted) access measures with substantial restrictions and under-enforcement of 'security measures' via so-called 'policy enforcement points. User identity, membership of (functional) user groups, (geographical or organizational) origin, time, specific (and enforced) network application, the 'compliance' of the user and his workplace concerning the established frameworks and additional substantive validation/checking of network traffic are part of this.

Technical implementation of this could be, for example, in an environment in which Oracle database systems are hosted in a VMware environment, where a *policy* is enforced at the level of VMware networking (NSX), using Palo Alto Networks technology, and using Application ID, User ID, Content ID, and Host Information Profiling. The relevant policy is then as follows: users in the user group 'Oracle DBA' are given permission every Wednesday between 18:00 and 21:00 to access Oracle database hosts via the 'Oracle' network application from a 'compliant' configured workstation that is located either is located at an internally earmarked location or is connected within the Benelux via client VPN. This 'oracle' network traffic is inspected and alerted/mitigated when malicious traffic is detected.

More refined implementations are possible; for example, the relevant 'Oracle' traffic can be reserved for only some internal, highly guarded systems - which, under the conditions mentioned earlier, is the select administrator group. Implementing "need to know" throughout the organization is not only a technical matter but also requires proper role assignment (by Human Resources), including attribute assignment (by Risk Management and process owner). If and where this has not been done or has not been done sufficiently in the past or has been a settlement under pressure from everyday dynamics, this should be included in the implementation of Zero Trust. Advice is to do this small (per, say, five Protect Surfaces). Short iterations provide learning and improvement points for subsequent iterations. Process and asset owners should be closely involved in Zero Trust *use case* development, implementation, and evaluation. This collaborative process of small iterations of design and implementations is described in previous papers such as: "Zero Trust Validation: From Practical Approaches to Theory" and is called 'Zero Trust Readiness' [20] and is demonstrated in the paper "On the Design and Engineering of a Zero Trust Security Artefact" [5].

Assessment. Technical measures are a fundamental part of Zero Trust. Many security vendors provide technical measures that can provide great value. Most measures are oriented toward preventing and detecting security events or data breaches. It is worth mentioning that security vendors are commonly judged on the *number* of events they can see rather than their capability actually to *solve* events. Bruce Schneier once said: "*the medic who looks over your bleeding body, saying: "That looks like a sucking chest wound. I'd get that checked if I were you*" [25] makes the analogy with companies focusing on buying prevention and detection systems, without offering a remediation.

Reduction of impact: *high*

Reduction of likelihood: *moderate*

Assessment with Additional Kipling Method. Lily Hay Newman from Wired stated in her article about 'the cybersecurity world's favorite catchphrase': "*Vendors hear new buzzwords, and then they try to package a product they already have into that.* [26]" As stated, technical measures effectively reduce impact and likelihood. Still, the real finesse of utilizing Zero Trust design principles lies in using Kipling, which is equal to actually 'fastening your seat belt.'

Reduction of impact: *maximum*

Reduction of likelihood: *high*

4.3 Dashboarding

An effective Zero Trust cybersecurity dashboard covers all five steps of Zero Trust Creator John Kindervag and displays relevant information for all levels, i.e., strategy, management, and operational.

The localization and registration of Protect Surfaces are done in the AUXO portal. As the next step, flows between Protect Surfaces are identified, mapped, and registered in the portal. After this, the right architecture decisions and security measures are decided, implemented, and integrated with AUXO. The dashboard in AUXO provides current information and updates about the protection status of the defined Protect Surfaces and technical measures.

The portal embeds the principles of the ON2IT Zero Trust Framework detailed in the paper Zero Trust Validation: From Practical Approaches to Theory [20], approaching cybersecurity as a more holistic approach by encompassing the following levels:

- *Strategy* displays information needs during Business as Usual (BAU) and during an incident where boards can steer. They must comply with information such as risk scores on high-value assets and regulatory frameworks.
- *Management* on defining and managing critical business processes and architectural design.
- *Operations* contain the technological measures companies implement to Identify, Protect, Detect, Respond, and Recover regarding cybersecurity. These measures can be consumed by multiple vendors like Palo Alto Networks, Microsoft, or Fortinet, regardless of whether these are on-premise or in the cloud. Multiple technologies contain data feeds that can be connected to the portal. This enables real-time insights into security events, threats, and the organization's cybersecurity posture.

Assessment. Although dashboarding itself is not a direct measure that contributes to the cost reduction of a breach, it aligns the different organizational levels (strategy, management, operations), the five steps of Zero Trust and equips the SOC and CSIRT with fact-based information to adequately act, respond and report in case a breach has occurred. Every judging regulator will consider the 'level of control' when penalizing the offender.

Reduction of likelihood: *low*

Reduction of impact: *moderate*

4.4 SOC

SOC analysts perform triage, and additional research on threats and security events that automated mitigation measures could not mitigate, block, or resolve. Immediate response due to 24/7 availability contributes to quicker detection, notification, and treatment of an event, preventing an event from becoming an actual security incident in most cases.

SOC engineers can process configuration changes and system maintenance to ensure a whole level of protection. Maintenance can be scheduled to minimize disruption and maintain the current protection status. In a security incident, the SOC can directly initiate incident response via CSIRT and document evidence on what has happened and what steps have been taken. This extensive documentation can help lower legal fines as it can be demonstrated that there was no negligent behavior or bad practice.

Assessment. Technical measures generate events and logs that must be analyzed and handled. A common pitfall for organizations is that they invest heavily in technical measures and equipment but underestimate the organizational burden of event management and the associated required resources (i.e., technical experts). Organizations overlook that a SOC is fundamental for adequate incident response, e.g., CSIRT. The SOC provides 24/7 expertise to alert on a possible malfunction.

Reduction of impact: *moderate*

Reduction of likelihood: *moderate*

4.5 CSIRT

The Computer Security Incident Response Team (CSIRT) performs incident response activities after a data breach. It follows the PICERL cybersecurity incident response process as described by SANS [27]:

1) *Preparation*: Team members are trained to deal with the latest developments on cybersecurity threats. Communication plans, procedures, and handling checklists are periodically tested and updated. Contact with relevant authorities is maintained.
2) *Identification*: Advanced automated detection of threats and vulnerabilities to proactively alert and take action. Customers can report a security incident 24/7 to the SOC. The priority level, severity, and potential impact are determined. The incident is assessed and triaged by the SOC First Responder.
3) *Containment*: Primary damage control is initiated according to established protocols. Evidence is secured, and thorough analysis is performed to identify the impact and mitigation strategy. Frequent status updates are scheduled with relevant stakeholders to communicate progress and define the next steps.
4) *Eradication*: Clean-up and remove attack traces, infections, or backdoors with the customer. Additional investigation is performed to determine if other environments are potentially affected and require protective measures.
5) *Recovery*: Bringing systems back to operations with the customer and performing a final sanity check to validate mitigation. Systems might be subject to intensified monitoring to detect suspicious behavior.
6) *Lessons learned*: A post-incident root cause analysis (RCA) is performed to identify improvements and security improvement advisories (SIAs) that aim to avoid the reoccurrence of a similar incident. An evaluation meeting is scheduled to discuss post-mortem analysis and lessons learned and monitor follow-up on SIAs.

Assessment. A CSIRT acts solely in case of a data breach and therefore has a mitigating role. Organizations might find it challenging to identify which CSIRT capabilities are required for their specific situation, even though ENISA developed a CSIRT Capability Maturity Model[4] that can help. In addition, they find it cumbersome to create a Target Operating Model or SLA if outsourced, even knowing that legislation such as the NIS2 Directive and Digital Operational Resilience Act (DORA)[5] will be enforced on many organizations.

Reduction of likelihood: *none*

Reduction of impact: *high*

[4] ENISA CSIRT maturity assessment model: https://www.enisa.europa.eu/publications/study-on-csirt-maturity

[5] Regulation of the European Parliament and of the Council on digital operational resilience for the financial sector and amending Regulations.

5 Findings: How ZTaaS Lowers the Financial Impact of a Breach

Reducing the risks involved with data breaches increases how an organization is 'in control.' Fundamental in ZTaaS is the effective working of the technical measures that the rest of the "as a Service" relies on to master the biggest cost category, non-compliance to "insufficient technical and organizational measures to ensure information security."

Lowering the financial impact of a breach via Zero Trust makes sense, but Ponemon lacks accurate actuarial data in its breach modelling. It is unclear which Zero Trust principle is enforced at the Zero Trust Journey stage. Analyzing individual cases is a more suitable method than generalizing. Antwerp Management School (AMS) conducted several breach simulations, under supervision by first author prof. Yuri Bobbert, on different organizations in multiple industries. Based on this multi-case analysis, we distilled key causes and their contribution to the most significant cost categories. And presented solutions that can reduce the financial impact. We have categorized them per Zero Trust as a Service component:

1) *Finding*: Very few of the examined cases had a good asset inventory. CMDBs were old and needed to be adequately updated and maintained. Nor was there, in most cases, a list of High-Risk Applications (HRA). This means companies had no minimal insights into what to protect. In some regulated companies, an HRA list was maintained but not viewed holistically. Meaning connections to other systems or services were not present. Nor the owners' names (asset owners) or primary security contact in case of a breach.
   Solution: This can quickly be addressed in the ZT's first step of identifying the Protect Surfaces and DAAS elements.
2) *Finding*: In most cases, easy access to networks and the compromization of high privileged access rights to management systems were discovered. And in some cases, take complete control over security measures.
   Solution: This can be prevented by proper RBAC-based granular access to highly valuable assets and continuous inspection via App ID and User ID functions.
3) *Finding*: moving laterally was possible in most cases once in the network.
   Solution: This can be mitigated by defining and implementing Protect Surfaces and utilizing granular access for Data, Apps, and Services (including 3rd parties). The access is regulated, and the blast is in case of a breach.
4) *Finding*: the absence of endpoint monitoring on malicious behavior from the user and his device. An endpoint detection mechanism like XDR was not implemented, and in one case possible to turn it off due to compromised high privileged rights. Quickly identifying and analyzing a breach and determining the event's impact was, in all other cases, ad-hoc and the absence of an incident response plan. In one case, not having this led to massive vaporization of the share price. This was one case where AMS examined a real hack and its aftermath, Altran[6].
   Solution: This can be addressed via XDR, SOAR, and a professional SOC/CSIRT that can handle quick triage and event.

We only presented the main flaws in these cases because there were many more shortcomings that led to a breach. And there are also more mitigations in these cases,

[6] The Altran hack explained: https://mobile.reuters.com/article/amp/idUSKCN1PM0IJ

but we wanted to focus on those solutions that contribute the most in reducing the impact on the most significant cost categories:

- Legal costs: litigation costs and Public Relations (PR) cost to do -publicly- damage control and avoid and avert claims
- Recovery cost: all the costs for complete recovery and bringing the business out of the ropes and back in the game
- Penalties as a result of bad practices and unable to prove what was done to prevent the catastrophe.

The actual cost of the breach depends on several factors: size, responsiveness, revenue, communication, etc. Therefore, putting a precise reduction number to this potential reduction is cumbersome, but we estimate here based on our experience. We have learned from previous breaches that proper response positively affects the legal judgment of regulatory bodies preparing their claims. Experience also teaches that if you can explain quickly and to a certain precision what has happened that stakeholders perceive you more trustworthy and in control; resulting in lower fines and sometimes only a warning. This early detection, containment, and triage is something ZTaaS can bring due to leveraging sophisticated technology, strict processes, and professional staffing. Also, the quick event handling and documenting of every step is something ZTaaS does, regardless of whether it is a low-rated event or a significant incident.

6 Results: Cost Reduction Scheme

As discussed, each ZTaaS component can contribute to cost reduction regarding data breaches. To make this more tangible, we have defined a cost reduction scheme that illustrates how each element contributes to a cost reduction. We use a point system where 100 points is the maximum amount that can be achieved when all ZTaaS components are fully implemented. We use the following scales:

- 0 = No reduction
- 5 = Low reduction
- 10 = Moderate reduction
- 15 = High reduction
- 20 = Maximum reduction

The contribution to reducing the total cost of a breach is based on descriptions provided in the earlier section.

With the framework given in Table 3, we propose a prototype that can be applied in practice as a conversation starter for C-level people in a decade of Zero Trust. We do not claim this is an exhaustive and fully representative overview but an exercise in analyzing to analyze the cost-reducing ZTaaS components and assess the Return on Investments (ROI) scenarios in cybersecurity programs.

Table 3. ZTaaS cost reduction framework.

ZTaaS component	Short description See earlier sections for detailed description per service	Contribution to total breach cost reduction (in points, max. 100)		
		Max. Reduction of *likelihood*	Max. Reduction of *impact*	Total reduction (*likelihood* + *impact*)
1. Design of Protect Surfaces	Protect Surfaces with a relevance score	10	10	20
2. Technical measures	Implementation of technical measures (e.g., Firewalls, XDR, CASB, SASE, IAM, PAM, etc.) *Additional implementation of security measures using Kipling method (+5)*	15 *(20)*	10 *(15)*	25 *(35)*
3. Dashboarding	AUXO real-time dashboarding and reporting platform based on the Five Steps of Zero Trust	5	10	15
4. SOC	24/7 available analysts and engineers	10	5	15
5. CSIRT	Available for incident response, mitigation, and improvements	0	15	15
		Total reduction		**100 points**

7 Discussion

One could think that applying all these ZTaaS services the environment always reduces the *likelihood* and *impact*. Achieving Zero Trust from scratch in an existing environment is most likely cumbersome and potentially very expensive. The misconception is that you must buy new technologies to implement Zero Trust, but many of these features are already in technology. However, a dedicated SOC or CSIRT is often not feasible for many organizations due to talent scarcity, funding, or a capable standing organization with processes and structures that can deal with information. In addition, it is often challenging to select a comprehensive platform solution that can monitor, report and provide full coverage and assurance over the entire Zero Trust environment, with a strong (near) real-time dashboarding and reporting capability.

8 Conclusion

Considering the above and when utilizing all five Zero Trust steps on the majority >75% of the highly valuable assets represented in Protect Surfaces, we estimate that the cost of a data breach can be reduced to *approximately 50%*.

With the assumption of:

- a relevance score of >50 (on a 1–100 scale, as explained in earlier publications field [5]) with maturity level 3;
- which means Kipling-style policy enforcement;
- and a high level of automation of event handling and behavior analysis via Machine Learning.

On top of that, we postulate that for organizations that have implemented the ZTaaS components throughout their entire organization on all assets (100%) we estimate a 75% reduction of the breach costs. Assuming that they consume all ZTaaS functions as described in Table 3, such as technical measures, AUXO dashboard, and SOC/CSIRT.

A positive side-effect of ZTaaS can be the reduction of cyber insurance premiums due to additional assurance safety nets the organization put into place to reduce damages and lower the risk of hefty fines due to bad (governance) practices.

Appendix 1: Define a Protect Surface and Assign a Relevance Score

Table 4. Protect surface metadata sample of HR system.

Step	Field	Example value
1. Provide a name, description and owners of the Protect Surface	Name of Protect Surface	*HR Management System*
	Description of Protect Surface	*Manages human resources and related processes throughout employee lifecycle*
	Business Contact	*Harry Johnson*
	Security Contact	*John Doe*
2. Calculate the relevance score	Confidentiality	*Very high*
	Integrity	*Very high*
	Availability	*Medium*
	Personal data	*Very high*
	Business criticality	*Low*
	Subject to audit	*Yes*
	Relevance score	*75*

Table 5. Relevance score guidelines.

Relevance score component	Low	Medium	High	Very high
Confidentiality	1. No personal or financial data 2. No possible legal, contractual or regulatory impact 3. No reputational damage possible 4. Low financial impact < 5K	1. Limited amount of personal data, no sensitive data, limited number and amount of financial data 2. No possible legal, contractual or regulatory impact 3. Minor local reputational damage possible 4. Medium financial impact < 100K	1. Personal data (but no sensitive personal data) or financial data available 2. Legal, contractual, or regulatory impact possible 3. Reputational damage can be locally impacted 4. High financial impact < 250K	1. Sensitive personal data or sensitive financial data available 2. Serious legal, contractual, or regulatory impact (serious fines, suspension or loss of license) possible 3. Risk of sustained (inter)national reputational damage 4. Very high financial impact > 250K
Integrity	1. No personal or financial data 2. No possible legal, contractual or regulatory impact 3. No reputational damage possible 4. Low financial impact < 5K	1. Limited amount of personal data, no sensitive data, limited number and amount of financial data 2. No possible legal, contractual or regulatory impact 3. Minor local reputational damage possible 4. Medium financial impact < 100K	1. Personal data (but no sensitive personal data) or financial data available 2. Legal, contractual, or regulatory impact possible 3. Reputational damage can be locally impacted 4. High financial impact < 250K	1. Sensitive personal data or sensitive financial data available 2. Serious legal, contractual, or regulatory impact (serious fines, suspension or loss of license) possible 3. Risk of sustained (inter)national reputational damage 4. Very high financial impact > 250K
Availability	No RTO or RPO required or defined	RTO and RPO are defined, but implemented in organization	RTO and RPO are defined and effectively implemented and followed	RTO and RPO are defined, followed and time-critical (< 60 min)

(*continued*)

Table 5. (*continued*)

Relevance score component	Low	Medium	High	Very high
Personal data	No personal data	1. Limited amount of personal data (<4 different data types) 2. Limited amount of data subjects	1. Personal data (>4 data types) and significant amount of data subjects 2. No special personal data	1. Personal data and significant amounts of data subjects 2. Special personal data (photos, racial or ethnic origin, religion, political affinity, sex life, trade union membership, physical or mental health including an opinion thereof, or criminal offences photos, racial or ethnic origin, religion, political affinity, sex life, trade union membership, physical or mental health including an opinion thereof, or criminal offences)
Subject to audit	Not subject to audit	Subject to audit	N/A	N/A
Business criticality	None	Low disruption of supporting business process	Moderate disruption of core business process	Immediate significant impact on core business process (e.g. Industrial Controls System with high availability requirements)

Appendix 2: Assign Zero Trust Measures to a Protect Surface

Table 6. Zero trust measures assigned for an example protect surface.

Category	Measure	Example value
Encryption	SSL inspection	*Yes*
	Encryption at rest	*Yes*
	Encryption in transit	*Yes*
Identity	Centrally managed IAM	*Yes*
	Restrict control	*Yes*
	MFA	*No*
	Logging on deviations	*Yes*
DDoS	Volume attacks	*Yes*
	Targeted attacks	*Yes*
Endpoint protection	Exploit prevention	*Yes*
	Malware prevention	*Yes*
	Ransomware protection	*Yes*
Traffic flows	Segments	*Yes*
	Restricted traffic	*Yes*
	Application based	*Yes*
	Content inspection	*Yes*
	URL based	*Yes*
	Behavioral analytics	*Yes*
Data	Backup & restore	*Yes*
	Credential phishing prevention	*No*
	DLP controls	*Not applicable*
	Data classification	*Yes*
DevSecOps	Secure build programs	*Yes*
Threat management	Threat intelligence	*Yes*
	Vulnerability management	*Yes*
Secure systems	Hardened software	*Yes*
	Hardened hardware	*Yes*
	Updates	*Yes*
Orchestration	Rules of engagement	*Yes*
Reporting	Auditable/Evidenced	*Yes*
	Key Risk Indicators	*No*
	Key Performance Indicators	*Not applicable*
Special	Technical validation	*Yes*

References

1. Ponemon. Cost of Data Breach Study: Global Analysis. Ponemon Institute LLC (2016)
2. Bobbert, Y., Scheerder, J., Timmermans, T.: Perspectives from 50+ years practical zero trust experience and learnings on buyer expectations and industry promises. In: SAI Conference, London (2022)
3. Kindervag, J.: No more chewy centers: introducing the zero trust model of information security. Forrester Research (2010)
4. Kindervag, J.: All Layers Are Not Created Equal; How the Principles of Journalism Help Define Zero Trust Policy (2019). https://www.paloaltonetworks.com/blog/2019/05/network-layers-not-created-equal/. Accessed 19 Nov 2021
5. Bobbert, Y., Scheerder, J.: On the design and engineering of a zero trust security artefact. In: Future of Information and Communication Conference (FICC), Vancouver (2021)
6. Schinagl, S., Schoon, K., Paans, R.: A framework for designing a security operations centre (SOC). In: 2015 48th Hawaii International Conference on System Sciences, Hawaii (2015). https://doi.org/10.1109/HICSS.2015.270
7. ENISA. How to Setup Up CSIRT and SOC Good Practice Guide, Belgium (2020). ISBN 978-92-9204-410-7. https://doi.org/10.2824/056764
8. Mooi, R.D., Botha, R.: Prerequisites for building a computer security incident response capability. In: Proceedings of Information Security South African (ISSA), Johannesburg (2015)
9. Siponen, M.: An analysis of the traditional IS security approaches: implications for research and practice. Eur. J. Inf. Syst. **1**(14), 303–315 (2005)
10. ISO/IEC27001. Information Technology - Security Techniques - Code of Practice for Information Security Management. ISO/IEC, Geneva (2005)
11. ISO/IEC27001:2013. ISO/IEC 27001:2013 Information Technology – Security Techniques – Information Security Management Systems – Requirements. ISO/IEC, Geneva (2013)
12. Calder, A.: Implementing Information Security Based on ISO 27001/ISO 27002. Van Haren Publishing, Zaltbommel (2009)
13. Overbeek, P., Spruit, M., Lindgren, E.: Informatiebeveiliging Onder Controle. Pearson Education Uitgeverij, Netherlands (2004)
14. Cherdantseva, Y., Hilton, J.: A reference model of information assurance & security. In: IEEE Proceedings of ARES, SecOnt Workshop, Regensburg (2013)
15. Christopher, J., Alberts, Dorofee, A.: OCTAVE method implementation guide version 2.0. Carnegie Mellon University Software Engineering Institute, Pittsburgh (2001)
16. Stonenburner, G., Goguen, A., Feringa, A.: NIST Special publications 800-27 Risk Management Guide for Information Technology Systems. National Institute of Standards and Technology, Gaithersburg (2002)
17. ISF. IRAM: Information Risk Assessment Methodology 2. Information Security Forum (2016). https://www.securityforum.org/tool/information-risk-assessment-methodology-iram2/
18. Gantz, S., Philpott, D.: FISMA and the Risk Management Framework (2013)
19. MITRE. Attack Groups (2021). https://attack.mitre.org/groups/
20. Bobbert, Y., Scheerder, J.: Zero trust validation: from practical approaches to theory. Sci. J. Res. Rev. **2**(5) (2020). https://doi.org/10.33552/SJRR.2020.02.000546
21. AFCEA. The economics of cybersecurity: a practical framework for cybersecurity investment. In: AFCEA Cyber Committee (2013)
22. Kipling, R.: Just so stories. Double Day Page (1902)
23. Mills, D.: Working methods. In: Quality Auditing, pp. 122–142 (1993). https://doi.org/10.1007/978-94-011-0697-9_10

24. Mahopo, B., Abdullah, H., Mujinga, M.: A formal qualitative risk management approach for IT security. In: Information Security for South Africa (ISSA), pp. 1–8, (2015). https://doi.org/10.1109/ISSA.2015.7335053
25. Scheier, B.: Secrets and Lies: Digital Security in a Networked World. John Wiley & Sons Inc. (2000)
26. Newman, L.: Wired. https://www.wired.com/story/what-is-zero-trust/. Accessed 19 Nov 2021.
27. Kral, P.: The Incident Handlers Handbook. SANS Institute (2011)

Autonomous Vehicle Cybersecurity: A Pre-test

Warren King[1] and Leila Halawi[2(✉)]

[1] Capitol Technology University, Laurel, MD 20708, USA
[2] Embry Riddle Aeronautical University, Daytona Beach, FL 32114, USA
halawil@erau.edu

Abstract. The automotive industry understands that gaining consumer trust is vital to accepting autonomous vehicles; however, trust is complex and multi-faceted. This research introduced a new application of the TAM and provided the automotive industry with quantifiable metrics that link cybersecurity perceptions to increasing trust in autonomous vehicles.

Keywords: Autonomous Vehicles · Cybersecurity · Technology Acceptance Model (TAM)

1 Introduction

Automotive manufacturers and technology companies have been racing to be the first to bring autonomous automobiles to the market, ensuring they capture a sizable market share [1]. The Society of Automotive Engineers (SAE) trademarked an industry-adopted standard that defines the levels of autonomous driving. The SAE driving levels are divided into two categories, driver support features and automated driving features. Driver support features provide advanced driver assistance systems (ADAS) at lower levels that support the driver but do not provide autonomous capabilities. However, automated driving features offer higher-level vehicle driving capabilities without driver intervention [2].

In the past decade, vulnerability analysis and hacking studies have proven that automobiles are susceptible to remote cyber-attacks and pose a threat to drivers [3–5]. Hackers have demonstrated that vehicles can be remotely hacked while being driven, which in one instance resulted in over 1.4 million Chrysler vehicles being recalled addressing the vulnerability [3, 6]. Upstream Security [7] recently reported a 99% increase in vehicle cybersecurity incidents since 2018, with year-over-year growth since 2016. Top cybersecurity vulnerabilities and incidents in 2019: 90% of 237 vehicles assessed across 33 automotive brands all had exploitable cybersecurity vulnerabilities that enabled hackers to unlock and steal vehicles [7]; A hacker stole thousands of iTrack, and ProTrack user accounts, which provided access to monitor locations and shut off engines in tens of thousands of vehicles [7, 8]. Vulnerabilities in Tesla's keyless entry systems were exploited, allowing vehicles to be stolen in under 30 s [7, 9]. Cyber-attacks targeting the auto industry and connected vehicles pose risks to human life.

K. Arai (Ed.): FTC 2023, LNNS 816, pp. 455–467, 2023.
https://doi.org/10.1007/978-3-031-47448-4_34

Researchers continuously assess consumers' intentions toward trusting and adopting autonomous vehicles [10–14]. Since 2016, the American Automobile Association (AAA) [15] has been conducting an annual quantitative study to understand better consumer attitudes toward self-driving vehicles, driver attitudes toward sharing the road with fully self-driving vehicles, and their desire for advanced driver assistance systems in their next vehicle [16–18]. According to AAA [15], only 14% of drivers trust riding in an autonomous vehicle; 86% are afraid to ride in or unsure about trusting autonomous vehicles [17, 18].

This study presents the research for addressing a gap by measuring the magnitude of the correlation between the perceptions of autonomous vehicle cybersecurity and its relationship to consumer trust and acceptance of autonomous vehicle technology utilizing the technology acceptance model (TAM) among licensed US drivers. This study is the condensed version of the dissertation [19].

We begin by discussing the relevant literature focused on the central topics of this study. Additionally, we provide a historical overview of automated driving systems, current research on automated driving systems, and autonomous vehicle trust, describe the vehicle cyber landscape, and introduce and describe the theoretical model. A description of the survey that was the basis for our empirical investigation and the pre-test results follow.

2 Literature Review

Automotive manufacturers are competing to gain market share in the autonomous vehicle market; however, they must address the lack of trust by consumers toward autonomous vehicles [1, 20]. Furthermore, Baidu [21] argues that the mass adoption of autonomous vehicles can only become a reality with consumer trust. Despite the safety benefits of autonomous vehicles, the advanced technologies will undoubtedly grow, increasing vulnerabilities and the threat of being hacked [22–24]. Additionally, autonomous vehicle hacking threats, publicized accidents, and crashes have caused consumers to be apprehensive about trusting autonomous vehicles [25].

2.1 Vehicle Safety

Historically, reducing vehicle accidents and losing human life has been the catalyst for increasing vehicle safety. The US House of Representatives [26] reports that in 1965; there were over 49,000 deaths from motor vehicle crashes. With predictions of future increases in fatalities, President Lyndon B. Johnson compelled Congress to enact The National Traffic and Motor Vehicle Safety Act of 1966. The legislation subsequently led to the Department of Transportation (DOT), which laid the groundwork for establishing vehicle safety standards to reduce accidents and resulting deaths [26]. Shortly after, the National Highway Traffic Safety Administration was formed under the DOT, primarily responsible for reducing deaths, injuries, and economic losses resulting from motor vehicle crashes and setting safety performance standards for motor vehicles [27].

Automotive manufacturers incorporated basic safety systems to make automobiles safer to protect and prevent injuries to vehicle occupants [28]. In 1959 the three-point seatbelt, a passive safety system and significant improvement over the essential lap seatbelts, was first introduced by Volvo Corporation; however, it was not until 1968 that seatbelts were mandated to be installed in all motor vehicles [29, 30]. As additional safety systems became standard in vehicles, whether due to regulation or attempts to differentiate their brands, auto manufacturers also began to integrate active driver assistance systems in motor vehicles to make the vehicle safer and prevent crashes [28].

Fast forward more than fifty years after the formation of the DOT to 2019, when more than 36,000 people died in motor vehicle crashes, and more than 20,000 killed in motor vehicle crashes in the first six months of 2021 [31]. The NHTSA [22] attributes increased vehicle crashes to human error and argues that accident prevention is the primary catalyst for introducing vehicle driver assistance technologies. The NHTSA [22] further states that driver assistance technologies in motor vehicles keep drivers and passengers safe. However, drivers must understand that some technologies are designed to warn, and others are designed to take control to avoid a crash.

In 2012 the California Senate passed Bill 1298, which authorized the California Department of Motor Vehicles (CADMV) to formulate regulations to permit autonomous vehicles on public roadways [32]. Shortly after, in 2014, the first regulations geared toward testing autonomous vehicles on public roads with a driver went into effect. Finally, in 2018, the CADMV began approving applications for autonomous testing without a driver on public roads. Soriano [32] states that as part of the program, approved test permit holders must report accidents within ten days of occurrence and report accidental disengagements of the autonomous technology annually [32].

The Safely Ensuring Lives Future Development and Research in Vehicle Evolution (SELF DRIVE) Act was a congressional bill introduced in 2017 and passed in the US House of Representatives [33]. The Self-Drive Act [33] was developed to institute legislation establishing the federal government's role in autonomous vehicles and vehicles equipped with automated driving systems [33]. The Act contained four significant components: Prevent individual states from banning autonomous vehicles or automated driving systems, give authority to the NHTSA to regulate autonomous vehicles, grant safety standard exemptions for auto manufacturers' first 100,000 vehicles, mandate protections for consumer privacy, and prevent cyber-attacks on autonomous vehicles [33].

2.2 Drivers Assistance Systems

In 1995 Mitsubishi was one of the first auto manufacturers to install a safety-based driver assistance system. The Preview Distance Control System could slow the vehicle down as the distance between vehicles in front decreased. However, to reduce speed, the system could only control the accelerator pedal, not the brakes [34]. With braking ability, adaptive cruise control as we know it today was introduced by Mercedes-Benz as DISTRONIC in 1998. This was the first true ADAS capability introduced and could be considered the first technology for autonomous driving [34, 35].

Although the integration of driver assistance systems (DAS) and advanced driver assistance systems (ADAS) in consumer automobiles began decades ago, they are distinctively different [36]. DAS systems know vehicle conditions and rely upon simple sensors, such as the speedometer or wheel speed, to measure vehicle state. ADAS systems are also aware of vehicle conditions through sensors. However, ADAS is further aware of the vehicle's surroundings and environment. ADAS systems rely on inputs from others.

2.3 Current Research on Autonomous Vehicles Trust and Cybersecurity

The AAA Foundation for Traffic Safety (AAAFTSA) conducts annual research surveys to explore sentiment, trust, and intentions toward accepting automated driving systems [12]. The study concluded that trust in automated driving systems was notably higher with L2 and L3 than with L4 and L5 automated driving systems. Additionally, the study concluded that there is a preference for owning a vehicle with lower levels of automation [12]. Kaur and Rampersad [37] quantitatively investigated the key factors influencing the adoption of autonomous vehicles. The study concluded that performance expectations and reliability of autonomous vehicles are crucial key adoption determinants towards acceptance [37]. Jin et al. [38] experimented to quantitatively investigate how experience with an L3 autonomous vehicle influenced trust. After each of the four experimental sessions using a simulated L3 autonomous vehicle, participants completed surveys based on a five-point Likert scale to gauge their trust levels. The results concluded that increased driving experience and exposure to L3 autonomous driving significantly positively impacted trust [38]. Khan et al. [39] proposed implementing digital labels to increase awareness and trust in autonomous vehicles. The architecture, called Cybersecurity Box (CSBOX), supports the digital label that will display a cybersecurity safety rating on the dashboard of new and used autonomous vehicles [39]. The research collected two random probability survey samples from 1,004 participants to understand drivers' perceptions and attitudes toward sharing the road with self-driving vehicles [40]. The report concluded that the increase in trust was not attributable to a single reason; however, the researcher hypothesized that increased exposure and understanding of the technology played a role [40].

An analysis of the current literature indicates a solid reluctance to trust fully autonomous vehicles. However, there are purported safety benefits; and several significant stakeholders are investing significantly [41]. However, the literature indicates increased trust and acceptance of automated driving systems up to L2, which allow the driver to maintain control of the vehicle [42]. Furthermore, several studies quantify trust but do not take a holistic view of other factors influencing trust [43]. Furthermore, the most prominent research results on trust converge to conclude that trust is dynamic, complex, and multifaceted, making actual usage difficult to predict [43].

Current research on vehicle cybersecurity indicates that similar challenges with securing IT systems from cyber-attacks also are present in connected and autonomous vehicles [44]. Additionally, vehicles are not impervious to cyber-attacks [45]. Several studies confirm that vulnerabilities in vehicle technology are already being exploited, causing unsafe conditions [46–48]. The research examined vehicle attack surfaces and proposed solutions to protect the CAN by implementing similar traditional IT-based solutions such as firewalls and detection systems [49]. The most prominent study indicated that vehicle cyber-attacks have increased yearly since 2016 [50].

3 Methodology

This study expanded independent variables associated with the TAM: Perceived Usefulness (PU), and Perceived Ease of Use (PEOU), by incorporating Perceived Cyber Threat (PCT) to gauge the Level of Trust (LOT), to investigate acceptance [51] as presented in Fig. 1.

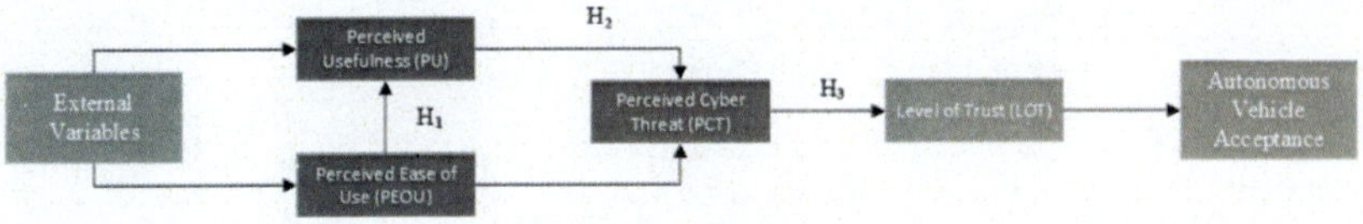

Fig. 1. Adapted technology acceptance model.

This study incorporated external factors such as age, gender, and ethnicity for demographic and categorization purposes within the scope of the sample. However, this study did not seek to investigate the underlying causes of attitudes, perceptions, how demographics influence perceptions or levels of influence. The instrument items used in this research are adapted to autonomous vehicles from the TAM determinants, Perceived Usefulness (PU), and Perceived Ease of Use (PEOU). Furthermore, researchers have assessed and confirmed a high correlation between PU and PEOU constructs using Cronbach's Alpha (α) [51–55]. Additional instrumentation items investigated perceived cyber threats and level of trust.

The perceived usefulness and perceived ease of use elements of the survey instrument in this study are based on the empirically tested TAM model. However, because this study investigates a different population and technology, the questions were modified, and a pilot study was conducted to ensure validity and reliability. The purpose of the pilot study in this study is to validate the data collection instrument and constructs to ensure it is within respected levels of validity [56].

Using convenience sampling, the researcher emailed 25 diverse individuals to participate in the pilot study. Of the 25 invitations emailed, 18 agreed to participate in the pilot study. However, one participant consented but did not fully complete the questionnaire. Therefore, this pilot study consisted of 17 participants who fully completed the questionnaire. This sample was sufficient to analyze the validity and reliability of the survey instrument.

4 Pre-test Results

The pilot study was administered online using the Survey Monkey platform, with a questionnaire comprised of 34 questions in four sections. Two of the four sections, Perceived Ease of Use (PEOU) and Perceived Usefulness (PU), consisted of closed-ended questions based on TAM constructs adapted to capture perceptions of autonomous vehicles. The remaining two survey sections, Perceived Cyber Threat (PCT) and Level of Trust (LOT), also centered on autonomous vehicle perceptions. Additionally, the researcher limited demographic information collected to the participant's age, state of residence, gender, and ethnicity. The distribution of pilot study participant demographics is shown in Table 1.

Table 1. The distribution of pilot study participants by state.

State of Residence	Frequency	Percent	Valid Percent	Cumulative Percent
Connecticut	1	5.6	5.6	5.6
Delaware	2	11.1	11.1	16.7
Florida	4	22.2	22.2	38.9
Georgia	1	5.6	5.6	44.4
Maryland	3	16.7	16.7	61.1
New Hampshire	1	5.6	5.6	66.7
North Carolina	1	5.6	5.6	72.2
Ohio	2	11.1	11.1	83.3
Pennsylvania	2	11.1	11.1	94.4
Texas	1	5.6	5.6	100
Total	18	100.0	100.0	

Table 2 presents the distribution of pilot study participants by age, followed by Table 3, which displays the distribution of participants by gender identity.

Table 2. The distribution of pilot study participants by age.

Age	Frequency	Percent	Valid Percent	Cumulative Percent
25–34	1	5.6	5.6	5.6
35–44	2	11.1	11.1	16.7
45–54	9	50.0	50.0	66.7
55–64	4	22.2	22.2	88.9
65+	2	11.1	11.1	100.0
Total	18	100.0	100.0	

The analysis of the data collected during the pilot study resulted in the rewriting of five PCT questions due to a low increase in internal coefficient and reliability. Construct

Table 3. The distribution of pilot study participants by gender identity.

Gender Identity	Frequency	Percent	Valid Percent	Cumulative Percent
Man	8	44.4	44.4	44.4
Woman	10	55.6	55.6	100.0
Total	18	100.0	100.0	

questions were dropped to optimize factor loading and validity. The participant ethnicity question was changed from multiple choice to single choice to allow for better analysis of participant demographic data.

The researcher iteratively examined factor analysis results to determine items that did not increase the factor loading of TAM constructs. Hence, items that did not meet acceptable levels (>0.50) were dropped from the [57]. The remaining items were reanalyzed until each construct was optimized with the highest factor loading.

Reliability is primarily concerned with the extent to which a measuring procedure yields the same results on repeated trials [56, 58]. Reliability is comprised of a true score while considering the measurement error. The higher the reliability coefficient, ranging from 0.00 to 1.00, the higher the level of reliability [59]. Furthermore, pilot testing allows the researcher to refine the scores by minimizing measurement errors [59]. One of the most widely used methods for measuring internal consistency reliability is Cronbach's alpha [58, 59].

The PU construct consists of eight items adapted from the TAM to assess participants' perceptions of autonomous vehicle usefulness. The items on this scale are loaded on a single component. As shown in Table 4, the items were above the threshold and were retained.

The PEOU construct consists of eight items adapted from the TAM to assess the participant's perceptions of autonomous vehicle ease of use. The items on this scale are loaded on a single component. As shown in Table 5, the items were above the threshold, and all items were retained.

The PCT construct consists of nine items to assess the participant's perceived cyber threats in autonomous vehicles. The items on this scale are loaded on three components. As shown in Table 6, six items below the threshold (PCT3, PCT4, PCT5, PCT7, PCT8, and PCT9) were eliminated, and the remaining items were reanalyzed.

The LOT construct consists of nine items to assess the participant's level of trust in autonomous vehicles. The items on this scale are loaded on two components. As shown in Table 7, a single item (LOT1) below the threshold was eliminated, and the remaining items were reanalyzed.

The PU construct consists of eight items adapted from the TAM to assess the participant's perceptions of autonomous vehicle usefulness. As shown in Table 8, removing one item (PU2) yielded an alpha of .931.

Table 4. Component matrix for perceived usefulness.

Component Matrix Component 1	
PU1	.841
PU2	.650
PU3	.858
PU4	.849
PU5	.866
PU6	.858
PU7	.726
PU8	.862
Extraction Method: Principal Component Analysis	

Table 5. Component matrix for perceived ease of use.

Component Matrix Component 1	
PEOU1	.785
PEOU2	.824
PEOU3	.784
PEOU4	.781
PEOU5	.922
PEOU6	.882
PEOU7	.943
PEOU8	.874
Extraction Method: Principal Component Analysis	

Table 6. Component matrix for a perceived cyber threat.

Component Matrix Component 1	
PCT1	.833
PCT2	.908
PCT6	.768
Extraction Method: Principal Component Analysis	

Table 7. Rotated component matrix for level of trust.

Component Matrix Component 1	
LOT2	.883
LOT3	.861
LOT4	.866
LOT5	.713
LOT6	.855
LOT7	.677
LOT8	.791
LOT9	.838
Extraction Method: Principal Component Analysis	

Table 8. Optimized internal reliability coefficient (α) for perceived usefulness.

Reliability Statistics		
Cronbach's Alpha	Cronbach's Alpha Based on Standardized Items	N of Items
0.931	0.931	7

The PEOU construct consists of eight items adapted from the TAM to assess the participant's perceptions of autonomous vehicle ease of use. As shown in Table 9, the alpha is .940, well above the acceptable threshold. This is the highest alpha in this case, so no items were removed.

Table 9. Internal reliability coefficient (α) for perceived ease of use.

Reliability Statistics		
Cronbach's Alpha	Cronbach's Alpha Based on Standardized Items	N of Items
0.940	0.945	8

The PCT construct consists of nine items to assess participants' perceptions of autonomous vehicle cyber threats. As shown in Table 10, removing five items (PCT4, PCT5, PCT7, PCT8, and PCT9) increased the alpha to .831.

The LOT construct consists of nine items to assess the participant's level of trust in autonomous vehicles. As shown in Table 11, removing one item (LOT1) increased the alpha to .919, which is well above the acceptable threshold.

Table 10. Optimized internal reliability coefficient (α) for perceived cyber threat.

Reliability Statistics		
Cronbach's Alpha	Cronbach's Alpha Based on Standardized Items	N of Items
0.831	0.839	4

Table 11. Optimized internal reliability coefficient (α) for level of trust.

Reliability Statistics		
Cronbach's Alpha	Cronbach's Alpha Based on Standardized Items	N of Items
0.919	0.926	8

5 Conclusion

Automated driving systems have been developed to improve safety; however, as we saw from the literature, consumers still lack complete trust in them. Furthermore, the issue of cybersecurity in autonomous vehicles hurts people's trust in the technology, ultimately hindering widespread adoption.

A pilot study is crucial as it enables the researchers to evaluate the survey instruments' validity and internal consistency and adjust survey questions and content. The pilot study aimed to validate the data collection instrument and constructs to ensure that they meet the accepted levels of validity. This study aims to address two gaps in the current research by expanding the TAM theoretical model to investigate how perceptions of cybersecurity influence the prediction of autonomous vehicle acceptance. This will enhance the understanding of TAM and contribute to the knowledge of autonomous vehicles. The authors have another article currently under review, which will present the model's findings and discuss this topic's theoretical and practical implications following the pre-test.

References

1. Masters, B.: Self-driving car companies find that going it alone is difficult. Financial Times (2019)
2. SAE International. https://www.sae.org/news/2019/01/sae-updates-j3016-automated-driving-graphic. Accessed 07 Jan 2019
3. Lee, Y., Woo, S., Song, Y., Lee, J., Lee, H.: Practical vulnerability-information-sharing architecture for automotive security risk analysis. IEEE **8**(1), 120009–120018 (2020)
4. Trend Micro. https://newsroom.trendmicro.com/2021-02-16-Connected-Cars-Technology-Vulnerable-to-Cyber-Attacks. Accessed 16 Feb 2021
5. Forbes. https://www.forbes.com/sites/forbestechcouncil/2021/12/10/cyberattacks-on-vehicles-pose-a-threat-to-drivers-and-manufacturers/?sh=7d2645346208. Accessed 10 Dec 2021
6. CNN. https://money.cnn.com/2015/07/24/technology/chrysler-hack-recall/index.html. Accessed 24 July 2015

7. UpstreamSociety. https://upstream.auto/upstream-security-global-automotive-cybersecurity-report-2020/. Accessed 2020
8. HackersOnlineClub. https://hackersonlineclub.com/gps-tracking-app-hack-allow-hacker-to-control-your-cars-remotely/. Accessed 27 Apr 2019
9. ThreatPost. https://threatpost.com/tesla-hacked-stolen-key-fob/161530/. Accessed 24 Nov 2020
10. Adnan, N., Nordin, S., Bahruddin, A., Ali, M.: How trust can drive forward the user acceptance to the technology? In-vehicle technology for autonomous vehicles. Transport. Res. Part A Policy Pract. **118**(1), 819–836 (2018)
11. Araujo, H.L.S., et al.: Trusted autonomous vehicles: an interactive exhibit. IEE, 386–393 (2019)
12. Kim, W., Kelley-Baker, T.: Users trust in and concerns about automated driving systems. In: AAA Foundation for Traffic Safety (2021)
13. Nastjuk, I., Herrenkind, B., Marrone, M., Brendel, A.B., Kolbe, L.M.: What drives the acceptance of autonomous driving? An investigation of acceptance factors from an end-user's perspective. Technol. Forecast. Soc. Change (2020)
14. Yuen, K.F., Chua, G., Wang, X., Ma, F., Li, K.X.: Understanding public acceptance of autonomous vehicles using the theory of planned behavior. Int. J. Environ. Res. Publ. Health **17**(12), 1–19 (2020)
15. American Automobile Association. http://media.acg.aaa.com/content/1205/files/E_1_%20Research%20Report%20-%20Evaluating%20ADA%20(FINAL%207-13-20)(1).pdf. Accessed Aug 2020
16. AAA Automotive Engineering. https://www.oregon.gov/ODOT/Get-Involved/Documents/AVTF-AV-Consumer-Survey-Fact-Sheet-FINAL-2-28-19.pdf. Accessed Mar 2019
17. AAA Newsroom. https://newsroom.aaa.com/2020/03/self-driving-cars-stuck-in-neutral-on-the-road-to-acceptance/. Accessed Mar 2019
18. AAA Newsroom. https://newsroom.aaa.com/2021/02/aaa-todays-vehicle-technology-must-walk-so-self-driving-cars-can-run/. Accessed Feb 2021
19. King, W.A.: A quantitative analysis of autonomous vehicle cybersecurity as a component of trust (Order No. 29395747). Available from ProQuest Dissertations & Theses Global (2728451800). http://ezproxy.libproxy.db.erau.edu/login?url=https://www.proquest.com/dissertations-theses/quantitative-analysis-autonomous-vehicle/docview/2728451800/se-2
20. Abraham, H., Autonomous vehicles and alternatives to driving: trust, preferences, and effects of age. Transport. Res. Board (2017)
21. MIT Technology Review. https://www.technologyreview.com/2020/12/16/1014672/building-a-self-driving-car-that-people-can-trust/. Accessed 16 Dec 2020
22. NHTSA. https://www.nhtsa.gov
23. IIHS. https://www.iihs.org/topics/advanced-driver-assistance
24. Watney, C., Draffin, C.: Addressing new challenges in automotive cybersecurity. RStreet Policy Study **118**, 1–16 (2017)
25. CPO Magazine. https://www.cpomagazine.com/cyber-security/autonomous-vehicles-and-the-threat-of-hacking/. Accessed 2018
26. U.S. House of Representatives. https://history.house.gov/Historical-Highlights/1951-2000/The-Highway-Safety-Act-of-1966/
27. U.S. DOT. https://www.transportation.gov
28. Kukkala, V.K., Tunnell, J., Pasricha, S., Bradley, T.: Advanced driver-assistance systems: a path toward autonomous vehicles. IEEE Consum. Electron. Mag. **7**(5), 18–25 (2018)
29. Your AAA Daily. https://magazine.northeast.aaa.com/daily/life/cars-trucks/auto-history/a-seat-belt-history-timeline/
30. Minnesota Medicine. https://www.minnesotamedicine.com/clinical-health-affairs/crash-was-his-name-car-safety-was-his-game/. Accessed 2019

31. National Center for Statistics and Analysis. https://crashstats.nhtsa.dot.gov/Api/Public/ViewPublication/813199. Accessed July 2021
32. California Transportation Commission. https://catc.ca.gov/-/media/ctc-media/documents/tab-4-1-a11y.pdf. Accessed 2018
33. Self Drive Act. https://www.congress.gov/bill/115th-congress/house-bill/3388. Accessed 2017
34. Autonomous Vehicle International. https://www.autonomousvehicleinternational.com/features/adas-3.html. Accessed 2018
35. Daimler Global Media Site. https://media.daimler.com/marsMediaSite/en/instance/ko/DISTRONIC-The-car-learns-to-see.xhtml?oid=41510718
36. Galvani, M.: History and future of driver assistance. IEEE Instrum. Meas. Mag.Instrum. Meas. Mag. **22**(1), 11–16 (2019)
37. Kaur, K., Rampersad, G.: Trust in driverless cars: investigating key factors influencing the adoption of driverless cars. J. Eng. Technol. Manag. **48**, 87–96 (2018)
38. Jin, M., Lu, G., Chen, F., Shi, X.: How driving experience affect trust in automation from level 3 automated vehicles? An experimental analysis. ITSC (2020)
39. Khan, W.Z., Khurram Khan, M., Arshad, Q.U.A., Malik, H., Almuhtadi, J.: Digital labels: influencing consumers trust and raising cybersecurity. IEE (2021)
40. AAA Newsroom. https://newsroom.aaa.com. Accessed 2018
41. Hulse, L.M., Xie, H., Galea, E.R.: Perceptions of Autonomous vehicles: relationships with road users, risk, gender and age. Saf. Sci.. Sci. **102**(1), 1–13 (2018)
42. AAA Foundation for Traffic Safety. https://aaafoundation.org/users-understanding-of-automated-vehicles-and-perception-to-improve-traffic-safety-results-from-a-national-survey/. Accessed 2019
43. Sheng, S., et al.: A case study of trust on autonomous driving, pp. 4368–4373. IEEE (2019)
44. He, Q., Meng, X., Qu, R.: Towards a severity assessment method for potential cyber attacks to connected and autonomous vehicles. J. Adv. Transport. (2020)
45. Garakani, H.G., Moshiri, B., Safavi-Naeini, S.: Cyber security challenges in autonomous vehicle: their impact on RF sensor and wireless technologies. ANTEM (2018)
46. Phantom of the ADAS. https://www.nassiben.com/phantoms. Accessed 2020
47. Chowdhury, A., Karmakar, G., Kamruzzaman, J., Jolfaei, A., Das, R.: Attacks on self-driving cars and their countermeasures: a survey. IEEE Access **8**(1), 207308–207342 (2020)
48. Garrad, P., Gilroy, S.: Developments in connected vehicles and the requirement for increased cybersecurity. Proc. ITRN **7**(1) (2021)
49. Wu, W., et al.: IDH-CAN: a hardware-based ID hopping CAN mechanism with enhanced security for automotive real-time applications. IEEE Access **6**(1), 54607–54623 (2018)
50. CNET. https://www.cnet.com/roadshow/news/2019-automotive-cyber-hack-security-study-upstream/. Accessed 2019
51. Davis, F.D.: Perceived usefulness, perceived ease of use, and user acceptance of information technology. MIS Q. **13**(3), 319–340 (1989)
52. Adams, D.A., Nelson, R.R., Todd, P.A.: Perceived usefulness, ease of use, and usage of information technology: a replication. MIS Q. Manag. Inf. Syst. **16**(2), 227–247 (1992)
53. Davis, F.D.: User acceptance of information technology: system characteristics, user perceptions, and behavioral impacts. Int. J. Man Mach. Stud. **38**(3), 475–487 (1993)
54. Davis, F.D., Venkatesh, V.: A critical assessment of potential measurement biases in the technology acceptance model: three experiments. Int. J. Hum. Comput. Stud.Comput. Stud. **45**(1), 19–45 (1996)
55. Venkatesh, V., Smith, R.: Creation of favorable user perceptions: exploring the role of intrinsic motivation. Theor. Dev. MIS Q. **23**(2), 239–260 (1999)
56. Salkind, N.J.: Exploring Research, 9th edn. Pearson Education Limited, London (2018)

57. Truong, D., Jitbaipoon, T.: How can agile methodologies be used to enhance the success of information technology projects? Int. J. Inf. Technol. Project Manag. **7**(2), 1–16 (2016)
58. Carmines, E.G., Zeller, R.A.: Reliability and Validity Assessment: Sage University Paper Series on Quantitative Applications in the Social Sciences. Planning for Connected and Automated Vehicles (1979)
59. Kimberlin, C.L., Winterstein, A.G.: Validity and reliability of measurement instruments used in research. Am. J. Health-Syst. Pharm. AJHP **65**(23), 2276–2284 (2008)

Check for updates

Advancements in Industry 4.0 Asset Management: Interoperability and Cyber Security Challenges and Opportunities

Usman Ahmad Usmani[1], Ari Happonen[2(✉)], and Junzo Watada[3]

[1] Lebuhraya Ipoh-Lumut Teronoh Perak, Universiti Teknologi Petronas, 32610 Seri Iskandar, Malaysia
usman_19001067@utp.edu.my

[2] LUT University, Yliopistonkatu 34, 53850 Lappeenranta, Finland
ari.happonen@lut.fi

[3] Waseda University, 1-chome-104 Totsukamachi, Shinjuku City 169-8050, Tokyo, Japan

Abstract. The Fourth Industrial Revolution, or Industry 4.0, is transforming the asset management landscape, presenting new challenges and opportunities. The increasing interconnectedness of assets and systems, combined with the growing availability of data, has led to the development of Industry 4.0 platforms for asset management. These platforms provide unprecedented levels of visibility and control, enabling asset managers to optimize performance and reduce costs. However, interoperability and cyber security challenges must be addressed to fully realize the benefits of Industry 4.0 asset management. In this paper, we review the state of the art in Industry 4.0 asset management, and discuss the challenges and opportunities presented by interoperability and cyber security. We propose a framework for addressing these challenges and highlight key research directions for advancing Industry 4.0 asset management. The paper concludes with a discussion of the implications of Industry 4.0 asset management for asset intensive industries, and the role of standards in ensuring interoperability and cyber security.

Keywords: Asset Management · Industry 4.0 · Cyber Security · Interconnected · Digitalization · Optimization · Interconnection · Operations · Artificial Intelligence · Machine Learning · IoT · Big Data · Digital Twin

1 Introduction

The fourth industrial revolution, commonly referred to as Industry 4.0, has brought significant technological advancements into global digitalization [1, 75] and digital design and tool utilization processes [60, 74, 76], business transformation [22] and sustainability development [3, 65], novel technologies [7] and automation with operational transitions [71], better supply chains [8, 48, 54] and especially new movements on manufacturing and asset management [9], both in intercompany [20] and fleet levels [21]. With the emergence of Industry 4.0, a new manufacturing era is being ushered in, characterized by digitalization, automation, and the integration of cyber-physical systems. These

K. Arai (Ed.): FTC 2023, LNNS 816, pp. 468–488, 2023.
https://doi.org/10.1007/978-3-031-47448-4_35

technological advancements have led to a significant increase in the efficiency of the manufacturing process with the use of smart machines and robots, which can communicate with each other and perform tasks autonomously [2, 78]. However, the integration of these technologies with legacy systems, and the increasing complexity of the manufacturing systems, poses significant interoperability and cybersecurity challenges that must be addressed to ensure the integrity and reliability of the systems [61].

One of the primary benefits of Industry 4.0 is the increased efficiency of asset management systems. With the help of Industry 4.0 technologies, asset management systems can be automated, streamlined, and optimized [4]. However, the integration of new technologies with legacy systems can be a complex process, leading to interoperability issues that can impede the efficiency of the system. The integration of multiple systems can lead to the fragmentation of the system, making it difficult to access and share data between systems. This lack of interoperability can result in delays and inconsistencies in data collection, analysis, and decision making. Furthermore, the lack of interoperability can make it challenging to create a unified view of the asset management system, which can affect the accuracy of performance metrics [5, 6].

In addition to interoperability challenges, cybersecurity is also a significant concern in Industry 4.0 asset management systems. As the systems become more connected and reliant on data, they become more vulnerable to cyber threats. The risk of cyberattacks is heightened in Industry 4.0, as it uses interconnected cyber physical systems, which can be exploited by cybercriminals [62, 63]. Any breach in the security of the system can cause significant damage, not only in terms of financial loss but also in terms of reputational damage, and, in some cases, loss of life. This paper focuses on the advancements in Industry 4.0 asset management and the challenges and opportunities presented by interoperability and cybersecurity. The objective of this paper is to identify and analyze the challenges associated with interoperability and cybersecurity in Industry 4.0 asset management systems and to propose solutions to mitigate these challenges. The paper begins with a review of the literature on Industry 4.0 and its impact on asset management systems. The literature review also covers the challenges associated with interoperability and cybersecurity in Industry 4.0 asset management systems. The paper then presents an analysis of the challenges and opportunities presented by interoperability and cybersecurity in Industry 4.0 asset management systems. The analysis focuses on the impact of these challenges on the efficiency and effectiveness of the asset management systems. Finally, the paper proposes solutions to mitigate the challenges associated with interoperability and cybersecurity in Industry 4.0 asset management systems. Mitigating intrusion in middleware software requires a multi-faceted approach that involves the implementation of security best practices throughout the software development lifecycle.

Industry 4.0 has brought about significant technological advancements in asset management systems, improving systems efficiency and effectiveness. However, the new technologies integration with legacy systems and the increasing complexity of the systems have led to interoperability and cybersecurity challenges that must be addressed. Interoperability issues can impede the efficiency of the system, leading to delays and inconsistencies in data collection, analysis, and decision making. Cybersecurity is also a significant concern in Industry 4.0 asset management systems, with the systems becoming more vulnerable to cyber threats as they become more connected and reliant on data.

This paper aims to identify and analyze the challenges associated with interoperability and cybersecurity in Industry 4.0 asset management systems and to propose solutions to mitigate these challenges.

This paper examines the challenges and opportunities presented by interoperability and cybersecurity in Industry 4.0 asset management systems. It identifies the challenges associated with integrating new technologies with legacy systems and the increasing complexity of the systems and highlights the risks of cyber threats in interconnected cyber physical systems. We propose solutions to these challenges to enhance the efficiency and effectiveness of asset management systems, emphasizing the importance of implementing security best practices throughout the software development lifecycle.

2 Industry 4.0 Asset Management: Overview and Advancements

Industry 4.0 has revolutionized the way modern factories operate, allowing them to harness the power of interconnected devices and advanced automation technologies to increase efficiency, reduce costs, and optimize productivity. [10, 11] One of the key areas of focus for Industry 4.0 is asset management, which involves monitoring, maintaining, and optimizing physical assets and equipment to ensure they operate at peak efficiency. As such, this paper provides an overview of Industry 4.0 asset management, highlighting the key advancements that have been made in recent years. We will explore the fundamental concepts that underpin asset management and discuss the role of interoperability and cybersecurity in enabling effective and efficient asset management in an Industry 4.0 context. In addition, we will examine the latest trends and developments in Industry 4.0 asset management and discuss how they are transforming the way businesses operate. By providing a comprehensive overview of the current state of Industry 4.0 asset management, this paper seeks to provide valuable insights and recommendations for businesses looking to optimize their asset management practices and stay ahead of the curve in the Industry 4.0 era.

2.1 Overview of Industry 4.0 Asset Management

The existing solutions to traditional manufacturing systems challenges do not directly apply to Industry 4.0 era. Firstly, Industry 4.0 requires the integration of physical and digital assets, which presents challenges such as cybersecurity risks, compatibility issues between different systems and technologies, and the need for interoperability standards. Secondly, the use of IoT sensors and devices generates a large volume of data that must be collected, processed, and analyzed, requiring sophisticated data management and analytics capabilities. Thirdly, predictive maintenance in Industry 4.0 requires complex algorithms and models that can analyze large volumes of data and make accurate predictions, necessitating a high level of expertise in data analytics and machine learning [19, 23]. Fourthly, Industry 4.0 requires a highly skilled workforce that is trained in the use of new technologies, which may require significant investment in training programs and re-skilling initiatives. Finally, the integration of new technologies in Industry 4.0 can be expensive, and the return on investment may not be immediate, requiring careful consideration of the cost-benefit of implementing new technologies and a long-term strategy for their adoption. Industry 4.0 refers to the current trend of automation and data exchange in manufacturing technologies. Asset management is an essential component of Industry 4.0, as it allows companies to optimize their use of physical and digital assets, thereby improving productivity and reducing downtime. Key components of Industry 4.0 asset management include:

Internet of Things (IoT) Sensors and Devices. IoT sensors and devices can be used to collect data on various aspects of an asset, such as its location, temperature, and performance. This data can then be used to identify potential issues before they become problems, and to optimize the asset's performance [29].

Big Data Analytics. The large volume of data generated by IoT sensors and devices requires sophisticated analysis to extract valuable insights. Big data analytics can be used to identify patterns and trends, and to make predictions about the future performance of assets [13].

Predictive Maintenance. Predictive maintenance uses data analysis to predict when an asset is likely to fail, allowing for maintenance to be performed before the asset breaks down. This can reduce downtime and maintenance costs [14].

Digital Twin Technology. Digital twin technology creates a virtual replica of a physical asset, which can be used to simulate different scenarios and test potential improvements. This allows companies to optimize the performance of assets without risking damage to the physical asset [15].

Artificial Intelligence and Machine Learning. Artificial intelligence (AI) and machine learning (ML) can be used to analyze data from IoT sensors and devices and make predictions about asset performance, helping companies optimize assets use and reduce downtime [12, 16]. Overall, Industry 4.0 asset management relies on the integration of physical and digital technologies, collection and analysis of big data, and the use of advanced analytics and predictive technologies to optimize asset performance.

2.2 Advancements in Industry 4.0 Asset Management

The field of Industry 4.0 Asset Management is constantly evolving with the emergence of new technologies and advancements in existing ones. Some of the recent advancements in Industry 4.0 Asset Management include:

Digital Twin technology. Digital Twin/a virtual model of a physical asset, enables organizations to simulate new scenarios and outcomes without risking damage to the actual asset. With Digital Twin technology, companies can optimize the performance of their assets and improve efficiency [17].

Edge Computing. Edge computing involves data processing at the network edge, closer to the IoT devices that collect it [18]. This can reduce latency, improve data security and privacy, and enable faster decision making by processing data in real time.

Predictive Maintenance. Predictive Maintenance uses AI/ML algorithms to analyze data from IoT sensors and predict when an asset is likely to fail, allowing for maintenance to be performed before the asset breaks down. Predictive Maintenance can reduce downtime and maintenance costs [18].

Blockchain. Blockchain technology can be used to securely store and share asset data and transactions between different stakeholders, such as manufacturers, suppliers, and customers. This can increase transparency, trust, and collaboration in the supply chain [19]. Overall, these advancements in Industry 4.0 Asset Management demonstrate the potential for emerging technologies to improve the efficiency and performance of physical assets, and to transform the way organizations manage and leverage data.

3 Interoperability Challenges in Industry 4.0 Asset Management Systems

Interoperability is a crucial aspect of Industry 4.0 asset management systems, which involve the integration of multiple technologies and systems to monitor and manage assets. However, the use of different communication protocols, data formats, and hardware/software systems can make interoperability a significant challenge in Industry 4.0. Interoperability issues can impede the efficiency of the system, leading to delays and inconsistencies in data collection, analysis, and decision making. In this section, we will discuss the interoperability challenges faced by Industry 4.0 asset management systems in the context of data collection and decision making, as well as possible solutions to mitigate these challenges [66, 77].

3.1 Interoperability Challenges in Data Collection

Interoperability challenges in data collection can arise due to the heterogeneity of technologies, systems, and data sources used in Industry 4.0 asset management. Data sources can include various types of sensors, monitoring systems, databases, and other devices, which can use different communication protocols, data formats, and hardware/software systems. This can lead to inconsistencies in the data collection process, making it difficult to analyze and monitor assets effectively. In addition, the lack of standardization in data collection can make it challenging to integrate and share data between different systems. For example, legacy systems may use proprietary data formats or communication protocols that are incompatible with newer systems that use more modern and open standards [67]. As a result, data integration can be difficult, and the system may become less efficient and effective.

To address these challenges, the use of middleware can provide a solution for facilitating interoperability in data collection. Middleware is software that provides a communication layer between different systems, allowing data to be exchanged and shared between them. Middleware can translate data between different communication protocols, data formats, and hardware/software systems, enabling seamless data exchange between different technologies. Another solution is the use of standardization to ensure that data is collected and transmitted using a common format and protocol. This can make it easier to integrate and share data between different systems, reducing interoperability challenges in data collection [24]. Standardization can also help to ensure that data is collected consistently across different data sources, reducing errors and inconsistencies. Overall, this sub-section highlights the importance of addressing interoperability challenges in data collection for Industry 4.0 asset management systems. By using middleware and standardization, these challenges can be mitigated, improving the efficiency and effectiveness of the system.

3.2 Interoperability Challenges in Decision Making

In Industry 4.0 asset management systems, decision making is a critical process that requires the integration and analysis of data from multiple sources. Interoperability challenges in decision making can arise due to the heterogeneity of data sources and the different analytical tools used for data analysis. This sub-section focuses on the challenges that arise when attempting to integrate different data sources with various analytical tools. The sub-section will explore how these interoperability challenges can affect the accuracy and reliability of the data analysis process and how it can lead to inefficiencies in decision making [25].

One main challenge of interoperability in decision making is the integration of data from different sources with various formats, structures, and granularity. It is difficult to compare and analyze data across different data sources, leading to inaccuracies in data analysis and making it difficult to make informed decisions. Additionally, the use of different analytical tools can further exacerbate interoperability challenges, as different tools may have different data requirements and output formats, making it challenging to integrate data and results between them. Another challenge is the need for real-time data analysis, which is necessary to make quick decisions in rapidly changing situations. However, integrating real-time data from different sources with different communication protocols and formats can be difficult, particularly when the data is transmitted over high speed networks. This can lead to delays and inconsistencies in decision making, making it difficult to respond to changing conditions effectively.

To address these challenges, middleware and standardization can be used to facilitate interoperability in decision making. Middleware can provide a communication layer between different data sources and analytical tools, allowing data to be exchanged and shared between them. Middleware can also provide data transformation services to enable the integration of data from different sources with different formats and structures. Standardization can also help by defining a common data format and communication protocol that can be used by all systems. This can ensure that data is collected and transmitted in a consistent manner, making it easier to integrate and share data between different systems. Standardization can also help to ensure that data analysis is performed consistently across different data sources, reducing errors and inconsistencies. Overall, the sub-section highlights the importance of addressing interoperability challenges in decision making for Industry 4.0 asset management systems. By using middleware and standardization, these challenges can be mitigated, improving the accuracy and reliability of data analysis and decision making [26].

4 Cybersecurity Challenges and Opportunities in Industry 4.0 Asset Management

As Industry 4.0 asset management systems become increasingly connected and reliant on data, the importance of cybersecurity has become more critical than ever before. The integration of new technologies with legacy systems and the complexity of the systems have led to significant cybersecurity challenges to be addressed. Cybersecurity threats such as external attacks, insider threats, and supply chain attacks have significant impact on asset management systems, potentially leading to data breaches, system downtime, and operational disruptions. Here we explore cybersecurity challenges faced by Industry 4.0 asset management systems and the strategies and technologies that can be used to mitigate these risks. Specifically, we will focus on two sub-sections: "Threats to Cybersecurity in Industry 4.0 Asset Management Systems" and "Mitigating Cybersecurity Risks in Industry 4.0 Asset Management Systems."

4.1 Threats to Cybersecurity in Industry 4.0 Asset Management Systems

Industry 4.0 asset management systems are vulnerable to various cybersecurity threats, impacting the operations and data of an organization. One of the most significant threats is external attacks from malicious actors such as hackers, cybercriminals, and state-sponsored attackers. These attackers can exploit vulnerabilities in the system to gain unauthorized access, steal sensitive data, and disrupt operations. Another threat is insider threats from employees or contractors who may intentionally or unintentionally compromise the system's security. These insiders may use their access to the system to steal data, disrupt operations, or carry out other malicious activities. Additionally, supply chain attacks are another emerging threat, where attackers target the suppliers or vendors of the system to gain unauthorized access or introduce malware into the system [27], which demands continuation in supply chain performance evaluation [38] and security model development side. The impact of these threats on Industry 4.0 asset management systems can be severe. Data breaches can result in the theft of sensitive information, including proprietary data, financial information, and personally identifiable information (PII). System downtime can lead to significant disruptions in operations, causing delays and financial losses. As such proper innovation networks for fast solution and gainsharing model based collaboration with real trust models [28] is needed to minimize operational disruptions as those can result to reputational damage, leading to a loss of customer trust and confidence.

To mitigate these cybersecurity threats, organizations must implement a robust cybersecurity strategy that includes access controls, authentication mechanisms, encryption, and intrusion detection and prevention systems. These strategies help to prevent unauthorized access to the system, protect data in transit and at rest, and detect and respond to potential security incidents. Additionally, employee training and awareness programs can help to prevent cybersecurity incidents by educating employees about best practices for maintaining a strong security posture and identifying and reporting potential security incidents.

4.2 Mitigating Cybersecurity Risks in Industry 4.0 Asset Management Systems

As Industry 4.0 asset management systems become increasingly connected and reliant on data, the importance of cybersecurity has become more critical than ever before. These systems are vulnerable to various cybersecurity threats that can have a significant impact on the operations and data of an organization. Organizations must, therefore, implement a robust cybersecurity strategy that includes access controls, authentication mechanisms, encryption, and intrusion detection and prevention systems to mitigate these risks. This section will explore the strategies and technologies that can be used to mitigate cybersecurity risks in Industry 4.0 asset management systems.

Implement Access Controls and Authentication Mechanisms. Access controls and authentication mechanisms are essential components of a strong cybersecurity strategy. Access controls help to ensure that only authorized personnel can access the system, while authentication mechanisms help to verify the identity of users. Implementing access controls and authentication mechanisms can prevent unauthorized access to the system and protect against insider threats [30]. One of the most common authentication mechanisms used in Industry 4.0 asset management systems is two factor authentication (2FA). This mechanism requires users to provide two forms of authentication to access the system, typically a password and a unique code generated by an authentication app or sent via SMS. This method makes it more difficult for unauthorized users to gain access to the system, as they would need to have both the user's password and access to their mobile device.

Use Encryption to Protect Data in Transit and at Rest. In Industry 4.0 asset management systems, encryption is used to protect data in transit and at rest. Data in transit is the data moving between different devices or systems. Data at rest refers to data stored on a device or system. [31] To protect data in transit, organizations use e.g. transport layer security (TLS) or secure sockets layer (SSL) protocols to encrypt data as it is transmitted between devices or systems, preventing unauthorized users from intercepting and reading the data. To protect data at rest, e.g. full disk encryption or file level encryption can be used.

Implement Intrusion Detection and Prevention Systems. Intrusion detection and prevention systems (IDPS) are software and hardware tools monitoring network traffic for signs of potential security breach and anomalies [17]. They detect and prevent unauthorized access, malware, and other types of cybersecurity threats. IDPS can be classified into two types: host based, and network based. Host based IDPS is installed on individual devices and monitors the device's system and application logs for suspicious activity. Network based IDPS is installed on the network and monitors network traffic for suspicious activity. IDPS can generate alerts for suspicious activity, allowing prompt investigation respond to potential security incidents [32].

Monitor and Control Third Party Access. Third party access to Industry 4.0 asset management systems poses additional cybersecurity risks. To mitigate these risks, organizations should monitor and control 3rd party access to the system, e.g. by implementing access controls, auditing, and monitoring tools to track 3rd party access and activities [33]. Additionally, organizations can implement a vendor risk management program to assess the cybersecurity posture of their 3rd party vendors. The program includes assessments of the vendor's security controls, policies, and procedures, as well as audits and evaluations of the vendor's cybersecurity posture.

Provide Employee Training and Awareness Programs. One of the most significant cybersecurity risks in Industry 4.0 asset management systems is the human factor. Employees unaware of cybersecurity best practices can inadvertently compromise the system's security by falling victim to phishing attacks, clicking on malicious links, or using weak passwords. One of the essential strategies for mitigating cybersecurity risks in Industry 4.0 asset management systems is to provide employee training and awareness programs [35]. By providing employee training and awareness programs, organizations can help employees understand the importance of cybersecurity and the role they play in protecting the system. Training and awareness programs should cover topics like best practices for password management, social engineering, and email security.

Employer should automate software and operating systems up dates, and minimize user level software or applications installs. Training programs should be designed to be accessible and engaging, using a variety of teaching methods, including online tutorials, webinars, and in person training sessions. The content of the training needs regular updates to reflect emerging threats and new best practices. Employees needs to be aware of the cyber security threats and understand the potential impact of a security breach. Employee awareness programs can promote a good culture of security within the organization.

In addition to training and awareness programs, organizations should consider implementing cybersecurity policies and procedures. These policies outline organization's approach to cybersecurity and measures in place to protect the system. Policies should cover areas such as password management, network access controls, and incident response procedures. It is also important to regularly review and update these policies to ensure that they remain relevant and effective. Cybersecurity policies should be communicated clearly to all employees, and they should be required to acknowledge that they have read and understood the policies.

Finally, it is important to establish a culture of cybersecurity within the organization. This can be achieved by promoting a sense of shared responsibility for cybersecurity and creating an environment where employees feel comfortable reporting security incidents or concerns. Regular security awareness training and ongoing communication can help foster a culture of security within the organization. In summary, employee training and awareness programs are essential for mitigating cybersecurity risks in Industry 4.0 asset management systems. By providing employees with the knowledge and tools they need to identify and respond to cyber threats, organizations can strengthen their security posture and reduce the risk of a successful cyberattack. Regularly reviewing and updating policies and procedures and establishing a culture of security can further reinforce an organization's cybersecurity efforts [36, 37].

The mitigations of cybersecurity risks in Industry 4.0 are different from those in Industry 3.0 because Industry 4.0 involves a higher level of connectivity, integration, and automation of cyber-physical systems (CPS) compared to Industry 3.0. Industry 4.0 systems are characterized by a large number of connected devices, sensors, and machines that generate and exchange data in real-time. The increased connectivity and data exchange increase the attack surface, making Industry 4.0 systems more vulnerable to cyber-attacks. Moreover, Industry 4.0 systems are built on open standards, making them more accessible to attackers. As a result, the mitigations of cybersecurity risks in Industry 4.0 must be more sophisticated and comprehensive, involving techniques and technologies such as access controls, authentication mechanisms, encryption, and intrusion detection and prevention systems.

5 Integration of New Technologies with Legacy Systems

As technology continues to advance, organizations are faced with the challenge of integrating new technologies with their existing legacy systems. Legacy systems are often deeply embedded within an organization's operations and contain critical business data, making it difficult to replace them entirely. However, with the introduction of new technologies, such as cloud computing, artificial intelligence, and the internet of things, organizations can gain a competitive advantage and improve their operations. Thus, integrating new technologies with legacy systems has become a critical aspect of modernizing an organization's IT infrastructure. In this section, we will explore the challenges and benefits of integrating new technologies with legacy systems, as well as best practices and strategies for a successful integration.

5.1 Challenges of Integrating New Technologies with Legacy Systems

Integrating new technologies with legacy systems can present several challenges for organizations. One of the primary challenges is the compatibility of the new technology with the existing legacy system. Legacy systems are often designed to work with specific hardware, software, and operating systems, and may not be compatible with the latest technology solutions. This can lead to issues such as data loss, system crashes, and reduced performance. Another challenge is the need for specialized skills and expertise to implement and maintain the new technology. This can be a particular issue if the organization's IT team is not familiar with the latest technologies or lacks the resources to learn and implement them. Additionally, there may be a lack of vendor support or technical documentation available for older legacy systems, which can further complicate the integration process [51].

Data security is also a concern when integrating new technologies with legacy systems. Legacy systems may have outdated security protocols or lack the necessary features to protect against modern cyber threats. This can result in data breaches or other security vulnerabilities that can be costly and damaging to the organization. Finally, integrating new technologies with legacy systems can be a time consuming and resource intensive process. Legacy systems often contain critical business data and may be deeply embedded within an organization's operations, making it difficult to replace them entirely. As a result, organizations must carefully plan and execute the integration process to minimize disruption and ensure a successful outcome [39].

5.2 Benefits of Integrating New Technologies with Legacy Systems

Integrating new technologies with legacy systems can offer numerous benefits to organizations, including improved efficiency, increased agility, and better decision-making capabilities. In this section, we will explore some of the key benefits of integrating new technologies with legacy systems. One of the primary benefits of integrating new technologies with legacy systems is improved efficiency. Legacy systems are often outdated and may not be optimized for modern business operations. For example, they may not be able to handle large volumes of data or may require manual processes that are time consuming and prone to error. By integrating new technologies, organizations can automate processes, streamline operations, and increase productivity [40, 41].

Cloud computing is an example of a new technology that can improve efficiency when integrated with legacy systems. Cloud computing allows organizations to store and access data and applications over the internet, eliminating the need for on premise hardware and software. This can reduce the costs and maintenance required for legacy systems, while also providing greater scalability and flexibility. For example, a company could use cloud-based storage to securely store and manage customer data, while also allowing employees to access it from anywhere in the world. Another benefit of integrating new technologies with legacy systems is increased agility. Legacy systems can be inflexible and difficult to modify, making it challenging for organizations to adapt to changing business needs. By integrating new technologies, organizations can enhance their ability to respond to new opportunities and challenges.

Artificial intelligence (AI) is an example of a new technology that can increase agility when integrated with legacy systems. AI enables machines to learn and make decisions based on data, enabling organizations to automate processes and gain insights into their operations. For example, a company could use AI powered software to analyze customer data and provide personalized product recommendations, improving customer satisfaction and sales. Integrating new technologies with legacy systems can also provide better decision-making capabilities. Legacy systems may not have the necessary features or data analysis tools to provide accurate and timely insights into business operations. By integrating new technologies, organizations can gain greater visibility into their operations, allowing them to make more informed decisions [42].

The internet of things (IoT) is an example of a new technology that can improve decision making capabilities when integrated with legacy systems. IoT involves connecting everyday objects to the internet, enabling them to collect and transmit data. By integrating IoT devices with legacy systems, organizations can gain real time insights into their operations and use data to make more informed decisions. For example, a manufacturing company could use IoT devices to monitor equipment performance, identify potential issues, and optimize production processes. Integrating new technologies with legacy systems can also provide a competitive advantage. Legacy systems can limit an organization's ability to innovate and compete with other businesses. By integrating new technologies, organizations can stay ahead of the curve and differentiate themselves from their competitors [43, 44].

Blockchain technology is an example of a new technology that can provide a competitive advantage when integrated with legacy systems. Blockchain enables secure, decentralized data storage and transactions, providing greater transparency and accountability. For example, a financial institution could use blockchain technology to securely and transparently process and store customer transactions, improving customer trust and satisfaction. Finally, integrating new technologies with legacy systems can provide a better customer experience. Legacy systems may not provide the necessary features or flexibility to meet modern customer expectations. By integrating new technologies, organizations can improve the customer experience, leading to greater customer satisfaction and loyalty [45].

Chatbots using AI and natural language processing, are an example of technology that can improve the customer experience when integrated with legacy systems. For example, a retail company could use a chatbot to provide customers with personalized product recommendations, track orders, and answer frequently asked questions. In conclusion, integrating new technologies with legacy systems can provide numerous benefits to organizations. By improving efficiency, increasing agility, providing better decision-making capabilities, providing a competitive advantage, and improving the customer experience, organizations can enhance their operations and meet the demands of the modern business landscape. However, it is important for organizations to carefully consider the potential challenges and risks associated with integrating new technologies with legacy systems. This requires thorough planning, communication, and a commitment to ongoing maintenance and support.

5.3 Strategies and Best Practices for Successful Integration

The suggested strategies and best practices outlined in the previous answer are focused on mitigating the risks and challenges associated with integrating new technologies with legacy systems. The new approaches outlined in this answer are different in that they provide specific tactics and technologies that organizations can use to facilitate the integration process. It is crucial to carefully plan and execute integration projects to mitigate risks and challenges and ensure successful outcomes. The suggested strategies and best practices, such as developing a clear integration strategy, conducting a thorough assessment of the legacy system, and involving stakeholders in the integration process, can help organizations achieve their integration goals. Additionally, implementing appropriate security and compliance measures is also critical to the success of integration projects. Regular monitoring, evaluation, and ongoing maintenance and support are also necessary to ensure the integrated systems continue to operate effectively over the long term. Here are some new approaches that organizations can consider when integrating new technologies with legacy systems:

Microservices Architecture. Instead of integrating new technologies directly into the monolithic legacy system, organizations can consider implementing a microservices architecture. This approach involves breaking down the monolithic system into smaller, independent components (microservices) that can be updated and replaced more easily and are more manageable as components, enabling easier integration. Microservices help to minimize the impact of continued use of legacy systems while enabling organizations to easily adopt new technologies. A clear integration strategy can help ensure that all stakeholders are aligned and focused on achieving the same goals and can help to mitigate the risks and challenges associated with integration and microservice approach utilization [46].

DevOps Practices. Organizations can adopt DevOps practices to streamline the integration process and ensure that all teams are aligned and working collaboratively and reduce the utilized time for error correcting and bug fix problems. DevOps can help automate processes, enable continuous integration and delivery, and improve communication and collaboration between development, operations, and other teams.

Conduct a Thorough Assessment of the Legacy System. Before integrating new technologies, it is important to conduct a comprehensive assessment of the legacy system to identify any potential issues or limitations that may impact the integration. This involves reviewing the existing systems and processes, identifying areas of inefficiency or redundancy, and determining which legacy systems are suitable for integrating new technologies [47].

Low-Code/No-Code Platforms. Low-code/no-code platforms can help organizations rapidly build and deploy new applications and workflows without requiring extensive programming knowledge. These platforms can enable business users to build and integrate new technologies with legacy systems more easily, reducing the time and effort required for integration.

Identify the Right Technologies for Integration. Once the legacy system has been assessed, it is essential to identify the right technologies for integration. This involves evaluating a range of new technologies and selecting those compatible with the legacy system and capable of achieving the desired objectives. Considering the new technologies' long-term impact and compatibility with existing systems and processes is important.

API-First Approach. Involves building APIs (Application Programming Interfaces) as the primary way to access and integrate legacy systems with new technologies, providing a standardized interface for communication and data exchange to facilitate a seamless communication between the two.

Ensure Data Compatibility. New technology and legacy system integration involves multisource data integration. Ensuring data compatibility is critical for successful integration. This involves mapping data between the legacy and new systems, ensuring data quality and accuracy, and resolving data conflicts [49].

Cloud and Hybrid Cloud Solutions. Adopting a hybrid cloud approach is to integrate new technologies with legacy systems, e.g. a combination of public and private cloud infrastructure to enable greater flexibility, compatibility, scalability, and security.

Use Containerization Technology. Containerization technology can help to simplify the integration process by packaging applications and services into portable containers. This approach can help to reduce the complexity of the integration process and minimize the impact on legacy systems.

Involve Stakeholders in the Integration Process. Involving stakeholders to the integration process so as to ensuring that the integration meets the needs of the business.

Includes engagement with key stakeholders, including business leaders, IT teams, and end users, to understand their needs and concerns and ensure the integration aligns with the overall business strategy [50].

Provide Training and Support. Providing training and support to end users is critical to the success of the integration. This involves developing training materials, providing training sessions, and offering ongoing support to ensure that end users are comfortable and proficient with the new technologies.

Monitor and Evaluate the Integration. Monitoring and evaluating involves tracking progress against the objectives and timeline, identifying any issues or areas for improvement, and adjusting the integration strategy as needed [51]. Regular monitoring and evaluation ensure process to remain on track and delivers the expected benefits.

Implement a Phased Approach. A phased approach helps to minimize the risks and challenges associated with integration. This involves breaking the integration project into smaller, more manageable phases, each with its objectives, timeline, and resources. Helps mitigate the risks associated with integrating complex systems and ensure that progress is made toward achieving the overall integration goals.

Leverage Machine Learning and AI. Machine learning and AI technologies can help to streamline the integration process by automating tasks such as data mapping, data transformation, and data validation. These technologies can help to reduce errors and improve the accuracy of the integration process.

Plan for Ongoing Maintenance and Support. Continuing maintenance and support involves developing a maintenance plan that outlines the responsibilities, resources, and timelines for ongoing maintenance and support of the integrated systems. One needs to ensure that the necessary resources and support are in place to ensure the integrated systems continue operating effectively over the long term.

Explore Blockchain Technology. Blockchain technology can offer a secure and decentralized approach to integrating new technologies with legacy systems. This approach can help to ensure data security and privacy while enabling seamless communication between the two systems.

Conduct a Security and Compliance Assessment. Before beginning the integration process, it's essential to conduct a thorough assessment of your organization's security and compliance posture. This will help identify potential vulnerabilities, gaps in compliance, or other areas requiring attention.

Establish Clear Security and Compliance Policies. Having well-defined policies governing security and compliance across your organization is crucial. These policies should outline expectations, responsibilities, and consequences for non-compliance.

Conduct Regular Audits. Regular audits can help identify areas of weakness or non-compliance and enable your organization to take corrective action promptly [53, 55, 79].

Use Security and Compliance Tools. Many security and compliance tools can help automate and streamline successful work flows. These tools can include intrusion detection systems, vulnerability scanners, and compliance management software.

Incorporate User-Centered Design. Incorporating user-centered design principles into the integration process can help to ensure that the new technologies are user friendly and meet the needs of end-users. This approach can help to improve user adoption and acceptance of the new technologies.

Implement Appropriate Access Controls. Access controls should be implemented to ensure that only authorized individuals can access sensitive data or systems [56–59]. This can include two-factor authentication, role-based access controls, and least privilege access.

Use Secure Communication Channels. When sharing sensitive data or communicating about security and compliance, it's crucial to use secure communication channels. This can include encrypted messaging platforms, secure file-sharing tools, and virtual private networks (VPNs).

These approaches can help organizations successfully integrate new technologies with legacy systems while mitigating the risks and challenges associated with integration. It is important to carefully evaluate these approaches and determine which ones are best suited for your organization's needs and objectives.

6 Discussions and Conclusions

The Fourth Industrial Revolution, or Industry 4.0, has been transforming the asset management landscape by introducing new challenges and opportunities. This transformation is due to the increasing interconnectedness of assets and systems, as well as the growing availability of data. Industry 4.0 asset management platform development now provides unprecedented visibility and control. With these platforms, asset managers can optimize asset performance and reduce costs. However, challenges must be addressed to realize the full benefits of Industry 4.0 asset management, particularly interoperability and cybersecurity. Current research mainly focuses on intercompany and trusted partner collaboration in Industry 4.0 and cybersecurity. Compared to other industries, it would seem to be highly beneficial for the industry to look for new opportunities in digital citizen science fields [68], following proper taxonomies [69] and considering why people would be motivated and be like to contribute [70] to the corporate cybersecurity and industrial development needs. However, with all the development, challenges still remain, that must be addressed to realize the full benefits of Industry 4.0 asset management, particularly interoperability and cybersecurity. This paper reviews the state-of-the-art in Industry 4.0 asset management, discussing the challenges and opportunities presented by interoperability and cybersecurity. The authors propose a framework for addressing these challenges and highlight key research directions for advancing Industry 4.0 asset management. The paper concludes with a discussion of the implications of Industry 4.0 asset management for asset intensive industries and the role of standards in ensuring interoperability and cybersecurity. Looking at current trends, constantly changing and developing attack patterns, more complicated industry 4.0 assets and diversifying user base, we would like to suggest more studies on gender diversity, neutrality and technology era [34], how to get females into technology studies [73] to prepare societies for industry 5.0 transformation times [72].

Interoperability is one of the primary challenges in Industry 4.0 asset management. Use of different technologies and standards can result in incompatible systems, which can lead to difficulties in data exchange and limit the effectiveness of asset management. The authors suggest that developing a common data model and using open standards can facilitate interoperability. The use of a common data model can ensure that data is shared and used effectively across systems, while open standards can enable different systems to communicate with each other seamlessly. Another significant challenge in Industry 4.0 asset management is cybersecurity. With the increasing interconnectedness of assets and systems, the risk of cyber-attacks also increases. Cybersecurity threats can result in significant financial losses, reputational damage, and safety risks. The authors suggest that cybersecurity should be integrated into asset management systems from the outset and should be an ongoing concern throughout the asset lifecycle. They propose that a risk-based approach to cybersecurity can be effective, with different risk levels requiring different levels of protection. The authors propose a framework for addressing the challenges of interoperability and cybersecurity in Industry 4.0 asset management. The framework includes several steps, including identifying key use cases, defining data models and standards, selecting appropriate technologies, and designing security controls. The authors suggest that a risk-based approach should be used to prioritize these steps and allocate resources effectively.

We also highlighted key research directions for advancing Industry 4.0 asset management, including developing new algorithms and analytics tools for data analysis, exploring the use of blockchain for secure data exchange, and improving the human machine interface for better decision making. The implications of Industry 4.0 asset management for asset-intensive industries are significant. Industry 4.0 platforms use enables asset managers to optimize asset performance, reduce costs, and improve safety. However, interoperability and cybersecurity challenges must be addressed to realize these benefits fully. The authors suggest that standards will play a critical role in ensuring interoperability and cybersecurity in Industry 4.0 asset management.

In conclusion, Industry 4.0 is transforming the asset management landscape by providing unprecedented levels of visibility and control. We believe that our proposed framework, which includes several steps such as identifying key use cases, defining data models and standards, selecting appropriate technologies, and designing security controls, can guide the development of more effective Industry 4.0 asset management solutions. However, interoperability and cybersecurity challenges need addressing to realize the full benefits of Industry 4.0 asset management. As such, we agree that a risk-based approach to cybersecurity can be an effective means of prioritizing resources and ensuring that different risk levels receive the appropriate level of protection. For future research, we suggest investigation of the use of artificial intelligence and machine learning algorithms [52, 64] to identify and mitigate cybersecurity risks in real-time and exploring the use of advanced analytics tools to better understand asset performance and predict potential failures before they occur. Moreover, we could delve into the development of more secure data exchange protocols, such as the use of blockchain technology, to ensure the integrity and confidentiality of sensitive data. Additionally, we suggest follow up research to consider the development of more user-friendly human-machine interfaces to improve decision-making and enhance collaboration between humans and machines.

References

1. Happonen, A., Ghoreishi, M.: A mapping study of the current literature on digitalization and industry 4.0 technologies utilization for sustainability and circular economy in textile industries. In: Yang, X.-S., Sherratt, S., Dey, N., Joshi, A. (eds.) Proceedings of Sixth International Congress on Information and Communication Technology. LNNS, vol. 217, pp. 697–711. Springer, Singapore (2022). https://doi.org/10.1007/978-981-16-2102-4_63
2. Raja Santhi, A., Muthuswamy, P.: Industry 5.0 or industry 4.0S? Introduction to industry 4.0 and a peek into the prospective industry 5.0 technologies. IJIDeM **17**, 947–979 (2023)
3. Minashkina, D., Happonen, A.: Decarbonizing warehousing activities through digitalization and automatization with WMS integration for sustainability supporting operations. In: E3S Web of Conferences, vol. 158, Article ID: 03002, pp. 1–7 (2020). https://doi.org/10.1051/e3sconf/202015803002
4. Kortelainen, H., Happonen, A.: From data to decisions – the re-distribution of roles in manufacturing ecosystems. VTT Blog, p. 1 (2017). https://doi.org/10.13140/RG.2.2.35922.91843
5. Yang, L., Zou, H., Shang, C., Ye, X., Rani, P.: Adoption of information and digital technologies for sustainable smart manufacturing systems for industry 4.0 in small, medium, and micro enterprises (SMMEs). Technol. Forecast. Soc. Change **188**, 122308 (2023)
6. Santos, B.P.: An investigation upon Industry 4.0 implementation: the case of small and medium enterprises and Lean organizations (2023)
7. Ghoreishi, M., Happonen, A.: Key enablers for deploying artificial intelligence for circular economy embracing sustainable product design: three case studies. In: AIP Conference Proceedings, vol. 2233, no. 1, pp. 1-19 (2020). https://doi.org/10.1063/5.0001339
8. García Alcaraz, J.L., et al.: Effect of green supply chain management practices on environmental performance: case of mexican manufacturing companies. Mathematics **10**(11), Article ID: 1877, pp. 1–19 (2022). https://doi.org/10.3390/math10111877
9. Metso, L., Happonen, A., Rissanen, M., Efvengren, K., Ojanen, V., Kärri, T.: Data openness based data sharing concept for future electric car maintenance services. In: Advances in Asset Management and Condition Monitoring. Smart Innovation, Systems and Technologies, vol 166, Chapter 36, pp. 429–436 (2020). https://doi.org/10.1007/978-3-030-57745-2_36
10. Farahani, B., Monsefi, A.K.: Smart and collaborative industrial IoT: a federated learning and data space approach. Digital Commun. Netw. (2023)
11. Kamali Saraji, M., Streimikiene, D.: Assessment of Industry 4.0 adoption for sustainability in small and medium enterprises: a Fermatean approach. In: Sustainable Manufacturing in Industry 4.0: Pathways and Practices, pp. 187–212 (2023)
12. Usmani, U.A., Happonen, A., Watada, J.: Enhancing artificial intelligence control mechanisms: current practices, real life applications and future views. In: Proceedings of the Future Technologies Conference (FTC), vol. 1, pp. 287–306, October 2022. https://doi.org/10.1007/978-3-031-18461-1_19
13. Attaran, M., Celik, B.G.: Digital twin: benefits, use cases, challenges, and opportunities. Decis. Anal. J. 100165 (2023)
14. Rucki, M.: Recent development of air gauging in industry 4.0 context. Sensors **23**(4), 2122 (2023)
15. Ramanujan, D., Bernstein, W.Z., Diaz Elsayed, N., Haapala, K.R.: The role of Industry 4.0 technologies in manufacturing sustainability assessment. J. Manuf. Sci. Eng. **145**(1), 010801 (2023)

16. Sanchez Londono, D., Barbieri, G., Fumagalli, L.: Smart retrofitting in maintenance: a systematic literature review. J. Intell. Manuf.Intell. Manuf. **34**(1), 1–1 (2023)
17. Usmani, U.A., Happonen, A., Watada, J.: A review of unsupervised machine learning frameworks for anomaly detection in industrial applications. In: 2022 Computing Conference, vol. 2, pp. 158–189, July 2022. https://doi.org/10.1007/978-3-031-10464-0_11
18. Alsudani, M.Q., et al.: Smart logistics with IoT based enterprise management system using global manufacturing. J. Comb. Optim.Optim. **45**(2), 57 (2023)
19. Usmani, U.A., Happonen, A., Watada, J.: Enhanced deep learning framework for fine-grained segmentation of fashion and apparel. In: Intelligent Computing: Proceedings of the 2022 Computing Conference, vol. 2, pp. 29–44, July 2022. https://doi.org/10.1007/978-3-031-10464-0_3
20. Kortelainen, H., Happonen, A., Kinnunen, S.-K.: Fleet service generation—challenges in corporate asset management. In: Koskinen, K.T., et al. (eds.) Proceedings of the 10th World Congress on Engineering Asset Management (WCEAM 2015). LNME, pp. 373–380. Springer, Cham (2016). https://doi.org/10.1007/978-3-319-27064-7_35
21. Kinnunen, S.-K., Happonen, A., Marttonen-Arola, S., Kärri, T.: Traditional and extended fleets in literature and practice: definition and untapped potential. Int. J. Strateg. Eng. Asset Manag. **3**(3), 239–261 (2019). https://doi.org/10.1504/IJSEAM.2019.108467
22. Kortelainen, H., Happonen, A., Hanski, J.: From asset provider to knowledge company—transformation in the digital era. In: Mathew, J., Lim, C.W., Ma, L., Sands, D., Cholette, M.E., Borghesani, P. (eds.) Asset Intelligence through Integration and Interoperability and Contemporary Vibration Engineering Technologies. LNME, pp. 333–341. Springer, Cham (2019). https://doi.org/10.1007/978-3-319-95711-1_33
23. Usmani, U.A., Haron, N.S., Jaafar, J.: A natural language processing approach to mine online reviews using topic modelling. In: COMS2 2021, Rev. Sel. Pap., pp. 82–98, May 2021
24. Kurniawan, T.A., Meidiana, C., Othman, M.H.D., Goh, H.H., Chew, K.W.: Strengthening waste recycling industry in Malang (Indonesia): lessons from waste management in the era of Industry 4.0. J. Clean. Product. **382**, 135296 (2023)
25. Mezgebe, T.T., Gebreslassie, M.G., Sibhato, H., Bahta, S.T.: Intelligent manufacturing ecosystem: a post COVID-19 recovery and growth opportunity for manufacturing industry in Sub-Saharan countries. Sci. Afr. e01547 (2023)
26. Akandere, G., Khajeh, E., Paksoy, T.: The internet of things and cyber physical systems. Smart Sustain. Oper. Supply Chain Manag. Ind. **4**, 277–305 (2023)
27. Suferi, N.S.M., Rahman, M.M., Yusof, I.M.: Industry 4.0 in construction: a conceptual framework. In: AIP Conference Proceedings, vol. 2643, no. 1, p. 030006, January 2023
28. Happonen, A., Siljander, V.: Gainsharing in logistics outsourcing: trust leads to success in the digital era. Int. J. Collab. Enterp. **6**(2), 150–175 (2020). https://doi.org/10.1504/IJCENT.2020.110221
29. Santti, U., Happonen, A., Auvinen, H., Räsänen, T., Eskelinen, T.: Sustainable business model innovation for digital remote monitoring: a follow up study on a water Iot service. In: BIOS Forum 2020, St. Petersburg, Russia, 10/2020, pp. 98–106 (2020). ISBN 978-5-86983-861-2, https://doi.org/10.5281/zenodo.4290135
30. Shrestha, R., Mishra, A., Bajracharya, R., Sinaei, S., Kim, S.: 6G network for connecting CPS and industrial IoT (IIoT). In: Cyber-Physical Systems for Industrial Transformation, pp. 17–38 (2023)
31. Hussain, M.Z., Hanapi, Z.M.: Efficient secure routing mechanisms for the low powered IoT network: a literature review. Electronics **12**(3), 482 (2023)
32. Azeez, N.A., Bada, T.M., Misra, S., Adewumi, A., Van der Vyver, C., Ahuja, R.: Intrusion detection and prevention systems: an updated review. in data management, analytics and innovation. Adv. Intell. Syst. Comput.Intell. Syst. Comput. **1042**, 685–696 (2020). https://doi.org/10.1007/978-981-32-9949-8_48

33. Casalet, M.: Challenges and opportunities of digitalization in mexico. digital and sustainable transformations in a post COVID World: Economic, Social, and Environmental Challenges, pp.451–474 (2023)
34. Kovaleva, Y., Happonen, A., Kindsiko, E.: Designing gender-neutral software engineering program. stereotypes, social pressure, and current attitudes based on recent studies. In: GE@ICSE '22 IEEE/ACM International Conference on Software Engineering, pp. 43–50 (2022). https://doi.org/10.1145/3524501.3527600
35. Martek, I., Hosseini, M.R., Edwards, D.J., Elghaish, F.: 33. Has the internet of things made a real difference to the management of projects? In: Research Handbook on Complex Project Organizing, p. 335 (2023)
36. Cardona, L.A.L., Buritica, L.A.B., Paredes, O.R., Nuñez, C.L.V., Muñoz, F.C., Ocampo, D.F.F.: Industry 4.0 and artificial intelligence and its contribution to modern companies: a systematic study. J. Positive Psychol. Wellbeing 351–365 (2023)
37. Giannelli, C., Picone, M.: Editorial "Industrial IoT as IT and OT convergence: challenges and opportunities." IoT **3**(1), 259–261 (2022)
38. Oubrahim, I., Sefiani, N., Happonen, A.: Supply chain performance evaluation models: a literature review. Acta logistica **9**(2), 207–221 (2022). https://doi.org/10.22306/al.v9i2.298
39. Haleem, A., Javaid, M., Singh, R.P., Rab, S., Suman, R.: Perspectives of cybersecurity for ameliorative Industry 4.0 era: a review based framework. Ind. Robot Int. J. Robot. Res. Appl. (2022)
40. Tsaramirsis, G., et al.: A modern approach towards an industry 4.0 model: from driving technologies to management. J. Sens. (2022)
41. Ahmed, I., Jeon, G., Piccialli, F.: From artificial intelligence to explainable artificial intelligence in industry 4.0: a survey on what, how, and where. IEEE Trans. Ind. Inform. **18**(8), 5031–5042 (2022)
42. Javaid, M., Haleem, A., Singh, R.P., Suman, R., Gonzalez, E.S.: Understanding the adoption of Industry 4.0 technologies in improving environmental sustainability. Sustain. Oper. Comput. (2022)
43. Happonen, A., Tikka, M., Usmani, U.A.: A systematic review for organizing hackathons and code camps in Covid-19 like times: literature in demand to understand online hackathons and event result continuation. In: 2021 International Conference on Data and Software Engineering (ICoDSE), pp. 7–12, November 2021. https://doi.org/10.1109/ICoDSE53690.2021.9648459
44. Shah, K., Patel, N., Thakkar, J., Patel, C.: Exploring applications of blockchain technology for Industry 4.0. Mater. Today Proc. **62**, 7238–7242 (2022)
45. Ahmad, T., Zhu, H., Zhang, D., Tariq, R., Bassam, A., et al.: Energetics systems and artificial intelligence: applications of industry 4.0. Energy Rep. **8**, 334–361 (2022)
46. Shaheen, B.W., Németh, I.: Integration of maintenance management system functions with Industry 4.0 technologies and features—a review. Processes **10**(11), 2173 (2022)
47. Bisht, D., et al.: Imperative role of integrating digitalization in the firms finance: a technological perspective. Electronics **11**(19), 3252 (2022)
48. García-Alcaraz, J.L., Sosa, L.A., Ramírez, C.S., Realyvásquez-Vargas, A., Soto, K.C.A., Happonen, A.: System dynamics in supply chains: a bibliometric review. In: 7th North American International Conference on Industrial Engineering and Operations Management, 12–14 July 2022, Orlando, Florida, USA, pp. 435–446 (2022)
49. Javaid, M., Haleem, A., Singh, R.P., Suman, R.: Enabling flexible manufacturing system (FMS) through the applications of industry 4.0 technologies. Internet Things Cyber Phys. Syst. **2**, 49–62 (2022)
50. Zheng, X., Lu, J., Kiritsis, D.: The emergence of cognitive digital twin: vision, challenges and opportunities. Int. J. Product. Res. **60**(24), 7610–7632 (2022)

51. Mourtzis, D., Angelopoulos, J., Panopoulos, N.: A literature review of the challenges and opportunities of the transition from Industry 4.0 to Society 5.0. Energies **15**, 6276 (2022)
52. Usmani, U.A., Usmani, M.U.: Future market trends and opportunities for wearable sensor technology. Int. J. Eng. Technol. **6**(4), 326 (2014)
53. Valero, M.R., Newman, S.T., Nassehi, A.: Link4Smart: a new framework for smart manufacturing linking Industry 4.0 relevant technologies. Procedia CIRP **107**, 1594–1599 (2022)
54. Oubrahim, I., Sefiani, N., Happonen, A.: The influence of digital transformation and supply chain integration on overall sustainable supply chain performance: an empirical analysis from manufacturing companies in Morocco. Energies **16**(2), 1–25 (2023). https://doi.org/10.3390/en16021004, Article ID: 1004
55. Majid, M., et al.: Applications of wireless sensor networks and internet of things frameworks in the industry revolution 4.0: a systematic literature review. Sensors **22**(6), 2087 (2022)
56. Yang, J.: The impact of Industry 4.0 on the world of work and the call for educational reform. In: Frontier of Education Reform and Development in China: Articles from Educational Research, pp. 285–298 (2023)
57. Gadekar, R., Sarkar, B., Gadekar, A.: Assessment of risks for successful implementation of Industry 4.0. Recent Adv. Comput. Sci. Commun. (Formerly: Recent Patents Comput. Sci.) **15**(1), 111–113 (2022)
58. Ayerbe, E., Berecibar, M., Clark, S., et al.: Digitalization of battery manufacturing: current status, challenges, and opportunities. Adv. Energy Mater. **12**(17), 2102696 (2022)
59. Choi, D., Gao, Z., Jiang, W.: Invited editorial comment: climate change implications for the asset management industry. J. Altern. Invest. **23**(1), 8–11 (2020)
60. Happonen, A., Stepanov, A., Piili, H., Salminen, A.: Innovation study for laser cutting of complex geometries with paper materials. Phys. Procedia **78**, 128–137 (2015). https://doi.org/10.1016/j.phpro.2015.11.025
61. Ryalat, M., ElMoaqet, H., AlFaouri, M.: Design of a smart factory based on cyber physical systems and internet of things towards Industry 4.0. Appl. Sci. **13**(4), 2156 (2023)
62. Gupta, R.: Industry 4.0 adaption in Indian banking sector—a review and agenda for future research. Vision **27**(1), 24–32 (2023)
63. Asadi, S., Ghobakhloo, M., Nilashi, M., Iranmanesh, M., et al.: A hybrid approach based on fuzzy logic and dematel to evaluate Industry 4.0 Adopt. SMEs (2023). SSRN 4331127
64. Usmani, U.A., Jaafar, J.: Machine learning in healthcare: current trends and the future. In: International Conference on Artificial Intelligence for Smart Community: AISC 2020, pp. 659–675, 2022, November
65. Happonen, A., Santti, U., Auvinen, H., Räsänen, T., Eskelinen, T.: Digital age business model innovation for sustainability in University Industry collaboration model. In: E3S Web of Conferences, vol. 211, Article ID: 04005, pp. 1–11 (2020). https://doi.org/10.1051/e3sconf/202021104005
66. Nam, J.H., Lee, S.J., Park, S.P., Lee, I.G.: IP dLedger-decentralized ledger for intellectual property administration. Technol. Forecast. Soc. Chang. **186**, 122105 (2023)
67. Sassanelli, C., Taisch, M., Terzi, S.: History and future of manufacturing. In: Advances in Digital Manufacturing Systems: Technologies, Business Models, and Adoption, pp. 13–36 (2023)
68. Palacin, V., Ginnane, S., Ferrario, M.A., et al.: SENSEI: harnessing community wisdom for local environmental monitoring in Finland. In: CHI Conference on Human Factors in Computing Systems, pp. 1–8 (2019). https://doi.org/10.1145/3290607.3299047
69. Vaddepalli, K., Palacin, V., Porras, J., Happonen, A.: Taxonomy of data quality metrics in digital citizen science. In: Nagar, A.K., Singh Jat, D., Mishra, D.K., Joshi, A. (eds.) Intelligent Sustainable Systems. Lecture Notes in Networks and Systems, vol. 578, pp. 391–410. Springer, Singapore (2023). https://doi.org/10.1007/978-981-19-7660-5_34

70. Palacin, V., Gilbert, S., Orchard, S., Eaton, A., Ferrario, M.A., Happonen, A.: Drivers of participation in digital citizen science: case studies on Järviwiki and safecast. Citizen Sci. Theory Pract. **5**(1), Article: 22, pp. 1–20 (2020). https://doi.org/10.5334/cstp.290
71. Abdelsalam, A., Happonen, A., Kärhä, K., Kapitonov, A., Porras, J.: Toward autonomous vehicles and machinery in mill yards of the forest industry: technologies and proposals for autonomous vehicle operations. IEEE Access **10**, 88234–88250 (2022). https://doi.org/10.1109/ACCESS.2022.3199691
72. Chin, S.T.S.: Influence of emotional intelligence on the workforce for Industry 5.0. Journal of Human Resources Management Research **2021**, 1–7 (2021). https://doi.org/10.5171/2021.882278. Article ID: 882278
73. Kovaleva, Y., Hyrynsalmi, S., Saltan, A., Happonen, A., Kasurinen, J.: Becoming an entrepreneur: a study of factors with women from the tech sector. Inf. Softw. Technol. **155**, 1–12 (2023). https://doi.org/10.1016/j.infsof.2022.107110, Article ID: 107110
74. Piili, H., et al.: Digital design process and additive manufacturing of a configurable product. Adv. Sci. Lett. **19**(3), 926–931 (2013). https://doi.org/10.1166/asl.2013.4827
75. Ghoreishi, M., Happonen, A.: The case of fabric and textile industry: the emerging role of digitalization, internet-of-things and Industry 4.0 for circularity. In: Yang, X.-S., Sherratt, S., Dey, N., Joshi, A. (eds.) Proceedings of Sixth International Congress on Information and Communication Technology. LNNS, vol. 216, pp. 189–200. Springer, Singapore (2022). https://doi.org/10.1007/978-981-16-1781-2_18
76. Widmaier, T., Juhanko, J., Piili, H., Kuosmanen, P., Salminen, A., et al.: Digital design and manufacturing process comparison for new custom made product family - a case study of a bathroom faucet. Est. J. Eng. **19**(1), 76–89 (2013). https://doi.org/10.3176/eng.2013.1.07
77. Tyagi, A.K., Dananjayan, S., Agarwal, D., Thariq Ahmed, H.F.: Blockchain—internet of things applications: opportunities and challenges for Industry 4.0 and Society 5.0. Sensors **23**(2), 947 (2023)
78. Czvetkó, T., Abonyi, J.: Data sharing in Industry 4.0 AutomationML, B2MML and international data spaces based solutions. J. Ind. Inf. Integr. 100438 (2023)
79. Konstantinidis, F.K., Myrillas, N., Mouroutsos, S.G., Koulouriotis, D., Gasteratos, A.: Assessment of industry 4.0 for modern manufacturing ecosystem: a systematic survey of surveys. Machines **10**(9), 746 (2022)

A Scientometric Review of Present and Future Trends of Embedded Systems in the Built Environment

Matthew Ikuabe[1](✉), Clinton Aigbavboa[1], Ayodeji Oke[1], and Douglas Aghimien[2]

[1] SARChI in Sustainable Construction Management and Leadership in the Built Environment, Faculty of Engineering and the Built Environment, University of Johannesburg, Johannesburg, South Africa
ikuabematthew@gmail.com

[2] School of Art, Design and Architecture, Faculty of Arts, Design and Humanities, De Montfort University, Leicester, UK

Abstract. This study presents a scientometric review conducted to define and delineate utilization of embedded systems in construction project delivery. This was done with the view of providing directions for future studies as well as stimulating a wider debate among construction stakeholders on the use of embedded systems in the industry. The study adopted an interpretivist philosophical view using an inductive approach and a grounded theory strategy. The data used were secondary in nature and were gathered from the Scopus database using specific related keywords. Co-occurrence maps were further created based on the bibliographic data gathered using the VOSviewer text mining software. Three clusters of co-occurring keywords were formed from the analysis, and these are labelled as sensor network systems and models, real-time systems and designs, and construction process automation. The study opens a new vista in the deliberations of the studies on technological innovations in the built environment.

Keywords: Built environment · Embedded systems · Real-time systems · Scientometric review · VosViewer

1 Introduction

The importance of the construction industry to the economy and development of any nation cannot be overemphasized. The industry is considered to be key for its influence in the provision of jobs as well as the multiplier of the GDP growth of any country [1]. It serves as a multiplicand to other industries with the provision of infrastructure at the center of other developmental needs of the society. With the glaring importance of the industry, it faces its own fair share of challenges hindering it from harnessing its optimum potentials to yield optimized benefits. One of the major challenges facing the construction industry is its slow pace of adoption of innovations in information technology and digitization [2]. This has placed the industry behind other industries that are early adopters of digital technologies and hence has limited its capacity to attain competitive

K. Arai (Ed.): FTC 2023, LNNS 816, pp. 489–499, 2023.
https://doi.org/10.1007/978-3-031-47448-4_34

gains and optimization of processes. Halim [3] opined that no industry, organization or profession can successfully solve developmental problems except their strategies, visions and aspirations are grounded in the creation of knowledge and systematically propelled by digital technology. In abating the challenges posed to the industry's full realization of its untapped benefits, there must be a strong drive in leveraging emerging information and communication technologies.

Businesses and industries have made frantic efforts to optimize efficiency, and this has increasingly brought about advancement in their methods and operations. The last century experienced the deployment of various operational methods for increased efficiency of processes; however, only a handful attained the heights of dominance. As earlier stated, the construction industry is still inefficient. According to Hampson *et al.*, [4], the construction industry in the last 50 years has only attained approximately half of the efficiency developments when compared with other industries. To a large extent, these results from time and cost overruns, issues pertaining to quality, safety issues, inefficient processes, outdated methods and repetition of processes [5–7]. These shortcomings could have been surmounted if the industry had grown with the advancement of other industries in the last 50 years. The daunting challenges plaguing the industry are usually a derivative of inherent industry features such as building processes and its fragmented nature, models for business based on objectives set out in the short term, ineffective flow of information among economic actors and ineffective procurement routes [8, 9]. Furthermore, Ballard [10] stated that the challenges facing the construction industry emanate from the fact that the engaging process of construction production is seemingly different from monotonous manufacturing process due to the nature of the product. The distinguishing attributes of the construction industry are cogent reasons for the survival strategies as adopted by actors in the construction landscape.

Over the years, advancement in information technology and computing has garnered a significant increase in attention as well as its slow adoption in the construction process. This offers good opportunities for improving and automating the process delivery and performance of the construction industry in comparison with other industries [11, 12]. One of the distinguishing technological advancement adopted for construction is the utilization of embedded systems, which serves a combination of computer processors, devices involving input/output and the computer memory. They are usually a part of a whole device which includes hardware in the form of electronics as well as mechanical parts. According to Jahromi and Kundur [13], the term embedded systems refer to devices that comprise of a synergized cyber (software and processor) and physical (electrical and/or mechanical) constituents for the delivery of a defined task.

Due to the nature of construction activities, embedded systems such as wireless sensors and networks are mostly deployed. According to Jang and Skibniewsky [12], this aids in the methodologies for the acquisition of data. The gathering of field data through wireless means improves the management of construction activities by gaining access to real-time information systems. The application of embedded systems in construction activities, processes and management have been investigated by many studies [14, 15]. A large pool of these studies highlighted the inherent benefits of the adoption of embedded systems in the delivery of construction processes, these spans from monitoring of equipment [16], use of radio frequency identification (RFID) for material tracking [12], tracking progress of work on site [17]. With the clear benefits and advantages brought about with the utilization of embedded systems in construction activities and management, this study reviews the studies on embedded systems in the built environment with a view to understanding what has been accomplished and identifying future research directions.

2 Methodology

Adopting the post-constructivist philosophical thinking, using an inductive approach, this study assesses the trends of published works using the searched keywords "embedded systems" AND "construction industry" OR "built environment" in relation to studies pertaining to construction and its allied fields. Text mining was used to construct meaning and understanding of the research footprint of embedded systems in construction/buildings. Scopus database was chosen for mining researches in this area, as it has been adjudged to have a wide coverage of science-related research [18]. With the adoption of a bibliometric approach, the identification and mining of published intellectual works in the area of embedded systems for construction related fields was carried out. Furthermore, the mapping out of chronological flow and trend pattern of the study area was outlined. The time limitation for the extracted studies was 15 years (2007–2022); and there was a limitation to language type, as all the documents were in English language except for four documents in Spanish and two documents in Chinese which were expunged. There was no filtering of the output with respect to Access type, Subject area, Document type, Publication stage, Source title, Keyword, Affiliation, Funding sponsor, Country/territory and Source type. This is due to the fact that studies in this domain are still in their nascent stage, especially in developing countries, and to provide a better view of the current trends. A total of one hundred and sixty-four (164) documents were extracted after the refinement of the output. The extracted documents were analysed using VOSviewer (version 1.6.17); a software that is easy to use and its approach for visualization is distance-based rather than graph-based [19]. Also, it gives a bibliometric network outlay in nodes that are depicted in a graphic interface that is user friendly.

3 Results and Discussion

3.1 Publication on Yearly Basis

With an extraction of 164 documents which spanned from 2007 to 2022, an analysis was carried out to show the trend of publications within this time frame. As shown in Fig. 1, there is a rise in the number of publications from 1 to 8 within the time frame of 2007 to 2009. This coincides with the introduction of the term "cyber-physical systems" in 2007 by Helen Gill at the National Science Foundation in USA. The next significant rise in the number of publications was in 2012 (6 publications) which connotes the period the concept of Industry 4.0 was initiated in Hannover, Germany. This ushered in a new drive towards innovative applications in system processes and production. In a similar move to that of Industry 4.0 conceptualized in Germany, the introduction of 'Made in China' in 2015 which aimed at attaining similar feats as that of Industry 4.0, spurred on more publications which culminated in the attainment of the number of publications in 2020 (20 publications). The trajectory followed up till 2021 experiencing a total number of 36 publications. The study was done in August 2022, which explains why 13 publications are presented for 2022.

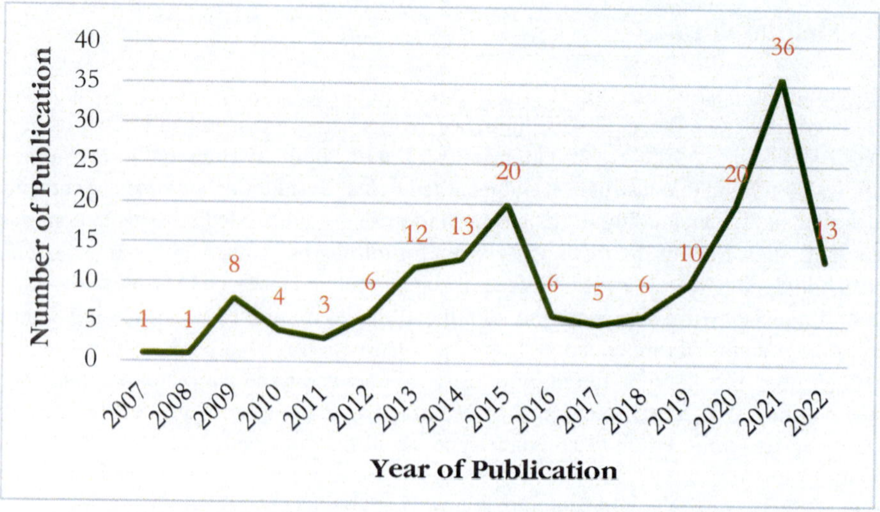

Fig. 1. Publication on yearly basis

3.2 Publication From Countries

Analysis was carried out on the publications emanating from various countries using a threshold of a minimum of 2 publications. However, it was observed that some of the extracted publications had overlap with respect to country of publication (affiliation of authors). Table 1 shows an outline of the various countries having publications including the number of documents and also their citations. The countries with the most published documents are the United States (48 papers, 531 citations), China (41 papers, 209 citations), Germany (14 papers, 154 citations), France (12 papers, 133 citations) and Russia Federation (12 papers, 14 citations). Others are United Kingdom (11 papers, 200 citations), Netherlands (10 papers, 9 citations), Italy (9 papers, 44 citations), Brazil (8 papers, 42 citations), Canada (8 papers, 46 citations), and India (8 papers, 42 citations).

3.3 Citation Rating of Publications

An assessment of the most cited documents was carried out after the extraction of the total documents. With the adoption of the threshold of a minimum of 8 cited documents, a total of 44 documents scaled through the sieving process. Table 2 shows the top 10 extracted documents portraying the author(s) of the documents, the titles, number of citations, and the area of focus. As shown in the table, the most cited document is the

Table 1. Table captions should be placed above the tables

Country	Documents	Citations	Country	Documents	Citations
United States	48	531	Portugal	5	12
China	41	209	Spain	5	54
Germany	14	154	Austria	4	195
France	12	133	Czech Republic	4	34
Russian Federation	12	14	Ireland	4	30
United Kingdom	11	200	Poland	4	43
Netherlands	10	19	Hong Kong	3	7
Italy	9	44	Mexico	3	5
Brazil	8	42	Sweden	3	15
Canada	8	46	Denmark	2	24
India	8	42	Finland	2	7
South Korea	8	38	Hungary	2	10
Taiwan	8	44	Malaysia	2	119
Australia	6	46	Norway	2	2
Switzerland	6	38	Serbia	2	7
Greece	5	23	Turkey	2	66
Japan	5	42	Ukraine	2	3

study of Adam and Markiewiez [20] with 209 citations, this is followed by Motawa and Almarshad [21] having 201 citations. The third and fourth most cited documents are the studies of Zhuang *et al.*, [22] having 200 citations and Göçer *et al.*, [23] having 97 citations.

3.4 Catchment Based on Co-occurring Keywords

Using co-occurring keywords from the extracted documents, the bibliographic data was deployed in the production of a co-occurrence map. A threshold of a minimum of 10 number of occurrence was set, Aghimien *et al.*, [26] used a threshold of a minimum of 4 while the default threshold of VOSviewer is 5. Thus the implication of this is that the choice of the extraction of a particular keyword is dependent on the minimum of 10 occurrences. As shown Fig. 2, findings from the analysis indicated that a total number of 33 keywords were extracted using the adopted threshold of a minimum of 10 occurrences, and 3 clusters were formed. The most frequent keywords are 'embedded systems' occurring 235 times, 'cyber-physical systems' occurring 59 times, 'intelligent buildings' occurring 30 times, 'construction industry' occurring 27 times, architectural design occurring 22 times and 'building information model' occurring 17 times. Figure 2 portrays the map for the co-occurring keywords network visualization of the extracted 33 keywords, and also giving an outlay of the 3 formed clusters.

Table 2. The top 10 extracted documents

Authors	Title	Citations	Area of Focus
Adam and Markiewicz [20]	Energy from earth-coupled structures, foundations, tunnels and sewers	209	Environmental Engineering; Foundations
Motawa and Almarshad [21]	A knowledge-based BIM system for building maintenance	201	BIM, Case-Based Reasoning
Zhuang *et al.*, [22]	Digital twin-based smart production management and control framework for the complex product assembly shop-floor	200	Assembly shop-floor; Big data
Göçer *et al.*, [23]	Completing the missing link in building design process: Enhancing post-occupancy evaluation method for effective feedback for building performance	97	BIM; Building performance
Pereira and Pereira [24]	Embedded image processing systems for automatic recognition of cracks using UAVs	46	Cracks detection; Image processing; Unmanned Aerial Vehicle
Bhattacharya *et al.*, [25]	Automated metadata construction to support portable building applications	41	Portable building applications; Sensor metadata

Cluster 1 (Sensor Network Systems and Models). This cluster has key words such as sensor networks, radio frequency identification (RFID), network security, robotics, mathematical models and embedded software. It has a purple coloration and labelled as *Sensor network systems and models*. The cluster emphasizes the significance of the use of sensor network systems in embedded systems. Shaikh and Pathan [27] noted that sensor network systems are characterized by the capability of effective communication coupled with well-defined networking capabilities. This forms an important aspect of embedded systems. This has led to research in this area being extended to military, commercial and scientific applications which further includes the monitoring of industrial processes, agriculture and biological habitats. Heidemann and Govindan [28] assessed the application of sensor networks in embedded systems and highlighted the important services to include the development of code tools and operating systems, node localization, time synchronization, resource discovery, storage services and databases, remote programming and security. Furthermore, Domdouzis *et al.*, [29] highlighted the prospects

of wireless sensor networking in construction industry, and affirmed the innovativeness brought about with the inculcation of the system in the delivery of construction activities.

Cluster 2 (Real Time Systems and Designs). This cluster has a pictorial of green coloration and having keywords such as real time systems, computer software, energy efficiency, decision making, hardware, design and buildings; and labelled as *Real time systems and designs.* This cluster gives credence to the recent drive for the adoption of innovative systems such as cyber-physical systems and internet of things in construction processes and management. Raj *et al.*, [30] noted that the processing of real time complex events is a core functionality of embedded systems in the analysis of building construction and management.

Cluster 3 (Construction Process Automation). This cluster has a yellow pictorial representation with keywords as big data, structural design, information theory, construction industry, project management, building information modeling, internet of things, automation, and semantics, and labelled as *Construction process automation.* The automation of processes involved in construction activities has been highly encouraged as a result of the immense benefits from such adoption. The overwhelming advantages attained from the automation of construction process spans from pre-construction to post-construction phases.

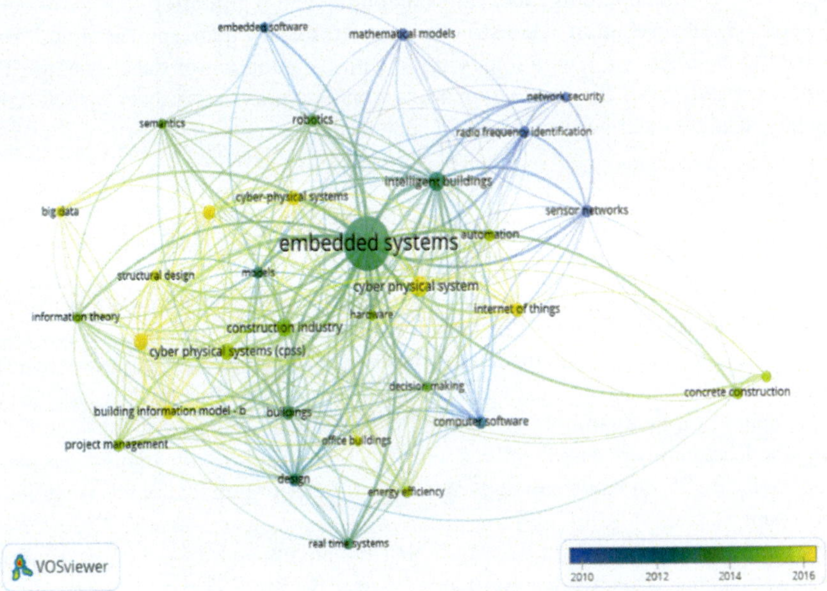

Fig. 2. Overlay visualization of co-occurring keywords

3.5 Past Trends and Future Applications

The visual outlay map of the co-occurring keywords in Fig. 2 also depicts the publication trends of these keywords over the years. As shown in the map, Cluster 1 (Sensor networks systems and models) appeared as the most keywords from 2010 to 2012. The keywords are sensor networks, radio frequency identification (RFID), network security, robotics, mathematical models and embedded software. From the period of 2012 to 2014, Cluster 2 (Real time systems and designs) formed the most occurring keywords within that time frame and includes words such as real time systems, computer software, energy efficiency, decision making, hardware, design and buildings. The third cluster (Construction process automation) makes up the last cluster which spans from 2014 to 2016. The keywords making up this era are big data, structural design, information theory, construction industry, project management, building information modeling, internet of things, automation and semantics.

The construction industry has much to gain from future applications of the different types of embedded systems for construction project execution and management. As mentioned earlier, there is a gradual espousal of systems such as sensor networks, radio frequency identification (RFID), cyber-physical systems and building information modeling. However, future applications should include not just the aforementioned systems but also embedded software technologies, system of systems integration platforms for digitalization, artificial intelligence, edge computing and high-performance computing among others. The execution of construction projects and the management of built facilities tend to experience a revolutionary drive with the adoption of these systems, thus serving as a shift from the conventional and contemporary approaches which would eventually lead to better performance and enhanced delivery.

4 Conclusion

This study has examined the trend of publications pertaining to embedded systems within the built environment. Adopting a bibliometric method and extracting documents published in a span of 15 years, documents were extracted from Scopus database and were subjected to analysis with the aid of VOSviewer. Findings from the study revealed that research publications on embedded systems in the construction industry and the built environment has a fluctuating trend within the time under review with 2007 and 2008 having the least number of publications with 1; while 2021 have the highest number of publication with 36. In between these years, the number of publications is surprisingly inconsistent.

Furthermore, it is shown in this study that the countries with the highest number of publications are USA, China, Germany, France, Russia Federation and the United Kingdom. It is noteworthy that there is no representation of any published work from the African continent. This ultimately showcases a gulf in the research focus between developed countries and Africa, thereby highlighting a potential future direction of research needed in Africa. This is important in bridging the knowledge gap and Africa and developed countries, and enhancing construction processes and the effective management of built facilities. This study revealed three clusters which make up the research direction of the extracted publications; these are sensor network systems and models, real time systems and designs, and construction process automation. The trend shows that research in this domain has evolved from studies on radio frequency identification (RFID), sensor networks and mathematical models, and is currently focusing on cyber-physical systems, building information modeling and big data.

The contribution made by this study to the body of knowledge centres on its unraveling of the research area and focus with respect to embedded systems in the built environment. The evolution of the research focus shows how the various publications have navigated through the different facets making up embedded systems. Moreover, a clear distinction has been made on the strengths of the various countries involved in this area; this includes countries from North America, Europe and Asia. Also, it showcases the dearth in research studies in most developing countries with particular reference to Africa. This reflects the current state of construction project delivery in the continent which is still hugely characterized with outdated systems thereby contributing to the stagnant posture of the construction industry. It is without doubt that as researches of this stature are engaged in, the inherent features of the construction industry peculiar to Africa would be addressed which in turn would aid in abating the perennial challenges in the delivery of construction projects. It is important to state that the study was limited to the use of Scopus database. Future studies can explore the other databases for a comparative analysis.

References

1. Thwala, W.D., Mathonsi, M.: Selection of procurement systems in the South African construction industry: an exploratory study. Acta Commercii **12**(1) (2012). https://doi.org/10.4102/ac.v12i1.127
2. Ikuabe, M.O., Aghimien, D.O., Aigbavboa, C.O., Oke, A.E.: Exploring the adoption of digital technology at the different phases of construction projects in South Africa. In: Proceedings of the International Conference on Industrial Engineering and Operations Management, Dubai, UAE, March, pp. 10–12 (2020)
3. Halim, H.C.: Assessment of the application of information and communication technology in real estate practice (a case study of Lagos Metropolis), MSc. thesis, Department of Estate Management, University of Nigeria, Enugu (2010)
4. Hampson, K., Kraatz, J., Sanchez, A.: R&D Investment and Impact in the Global Construction Industry. English. Routledge, Abingdon (2014)
5. Adafin, J., Rotimi, J.O.B., Wilkinson, S.: An evaluation of risk factors impacting project budget performance in New Zealand. J. Eng. Des. Technol. **19**(1), 41–61 (2021)
6. Gharouni Jafari, K., Ghazi Sharyatpanahi, N.S., Noorzai, E.: BIM-based integrated solution for analysis and management of mismatches during construction. J. Eng. Des. Technol. **19**(1), 81–102 (2021)
7. Ikuabe, M.O., Oke, A.E.: Contractors' opportunism: construction professionals' awareness of influencing factors. J. Eng. Des. Technol. **17**(1), 102–114 (2019). https://doi.org/10.1108/JEDT-03-2018-0054
8. Akintoye, A., Goulding, J., Zawdie, G.: Construction innovation and Process Improvement. English. Wiley-Blackwell, West Sussex (2014)
9. Jonassen, J.O.: Report on integrated practice. Changing business models in BIM-driven integrated practice, English. In: American Institute of Architects (AIA) (2010)

10. Ballard, G.: Should project budgets be based on worth or cost? In: Proceedings, 20th Annual Conference of the International Group for Lean Construction, San Diego, CA, pp. 761–770 (2012)
11. Ikuabe, M.O., Aigbavboa, C., Anumba, C., Oke, A.: Business environment factors influencing the use of cyber-physical systems for facilities management – a confirmatory factor analysis approach. J. Green Build. **18**(1), 169–188 (2023). https://doi.org/10.3992/jgb.18.1.169
12. Jang, W., Skibniewski, M.: Embedded systems for construction asset tracking combining radio and ultrasound signals. J. Comput. Civil Eng. (ASCE) **23**(4), 221–229 (2009). https://doi.org/10.1061/(ASCE)0887-3801(2009)23:4(221)
13. Jahromi, A., Kundur, D.: Fundamentals of cyber-physical systems. In: Anumba, C.J., Roofigari-Esfahan, N. (eds.) Cyber-Physical Systems in the Built Environment (2020). https://doi.org/10.1007/978-3-030-41560-0_1
14. Akanmu, A., Anumba, C.: Cyber-physical systems integration of building information models and the physical construction. Eng. Constr. Archit. Manag. **22**(5), 516–535 (2015). https://doi.org/10.1108/ECAM-07-2014-0097
15. Ranaweera, K., Ruwanpura, J., Fernando, S.: Automated real-time monitoring system to measure shift production of tunnel construction projects. J. Comput. Civ. Eng. **27**(1), 68–77 (2013). https://doi.org/10.1061/(ASCE)CP.1943-5487.0000199
16. Lu, M., Chen, W., Shen, X.S.: Positioning and tracking construction vehicles in highly dense urban areas and building construction sites. Autom. Constr. **16**(5), 647–656 (2007). https://doi.org/10.1016/j.autcon.2006.11.001
17. Bohn, J.S., Teizer, J.: Benefits and barriers of construction project monitoring using high-resolution automated cameras. J. Constr. Eng. Manag. **136**(6), 632–640 (2010). https://doi.org/10.1061/(ASCE)CO.1943-7862.0000164
18. Hosseini, M.R., Maghrebi, M., Akbarnezhad, A., Martek, I., Arashpour, M.: Analysis of citation networks in building information modeling research. J. Constr. Eng. Manag. **144**, 04018064 (2018). https://doi.org/10.1061/(ASCE)CO.1943-7862.0001492
19. Van Eck, N.J., Waltman, L.: Software survey: VOSviewer, a computer program for bibliometric mapping. Scientometrics **84**, 523–538 (2010). https://doi.org/10.1007/s11192-009-0146-3
20. Adam, D., Markiewicz, R.: Energy from earth-coupled structures, foundations, tunnels and sewers. Geotechnique **59**(3), 229–236 (2009). https://doi.org/10.1680/geot.2009.59.3.229
21. Motawa, I., Almarshad, A.: A knowledge-based BIM system for building maintenance. Autom. Constr. **29**, 173–182 (2013). https://doi.org/10.1016/j.autcon.2012.09.008
22. Zhuang, C., Liu, J., Xiong, H.: Digital twin-based smart production management and control framework for the complex product assembly shop-floor. Int. J. Adv. Manuf. Technol. **96**, 1149–1163 (2018). https://doi.org/10.1007/s00170-018-1617-6
23. Göçer, T., Hua, Y., Göçer, K.: Completing the missing link in building design process: Enhancing post-occupancy evaluation method for effective feedback for building performance. Build. Environ. **89**, 14–27 (2015). https://doi.org/10.1016/j.buildenv.2015.02.011
24. Pereira, F.C., Pereira, C.E.: Embedded image processing systems for automatic recognition of cracks using UAVs. IFAC-PapersOnLine **28**(10), 16–21 (2015). https://doi.org/10.1016/j.ifacol.2015.08.101
25. Bhattacharya, A.A., Hong, D., Culler, D., Ortiz, J., Whitehouse, K., Wu, E.: Automated metadata construction to support portable building applications. In: Proceedings of the 2nd ACM International Conference on Embedded Systems for Energy-Efficient Built, pp. 3–12 (2015). https://doi.org/10.1145/2821650.2821667
26. Aghimien, D., Aigbavboa, C., Oke, A., Thwala, W.: Mapping out research focus for robotics and automation research in construction-related studies: a bibliometric approach. J. Eng. Des. Technol. **18**(5), 1063–1079 (2020)

27. Shaikh, A., Pathan, S.: Research on wireless sensor network technology. Int. J. Inf. Educ. Technol. **2**(5), 476–479 (2012). https://doi.org/10.7763/IJIET.2012.V2.183
28. Heidemann, J., Govindan, R.: An overview of embedded sensor networks. In: Hristu-Varsakelis, D., Levine, W.S. (eds.) Handbook of Networked and Embedded Control Systems. Springer, Boston (2004). https://doi.org/10.1007/b137198
29. Domdouzis, K., Anumba, C., Thorpe, A.: Wireless sensor networking in the construction industry - prospects and problems. In: Khosrowshahi, F. (ed.) 20th Annual ARCOM Conference, 1–3 September 2004, Heriot Watt University. Association of Researchers in Construction Management, vol. 2, pp. 1107–1120 (2004)
30. Raj, R., Sahu, R., Chaudhary, B., Prasad, B., Agarwal, S.: Real time complex event processing and analytics for smart building. In: Conference on Information and Communication (IEEE) (2017). https://doi.org/10.1109/INFOCOMTECH.2017.8340593

Author Index

K. Arai (Ed.): FTC 2023, LNNS 816, pp. 501–502, 2023.
https://doi.org/10.1007/978-3-031-47448-4